THREATENED ISLAND NATIONS

Rising seas are endangering the habitability and very existence of several small island nations, mostly in the Pacific and Indian oceans. This is the first book to focus on the myriad legal issues posed by this tragic situation. If a nation is under water, is it still a state? Does it still have a seat at the United Nations? What becomes of its exclusive economic zone, the basis for its fishing rights? What obligations do other nations have to take in the displaced populations, and what are these peoples' rights and legal status once they arrive? Should there be a new international agreement on climate-displaced populations? Do these nations and their citizens have any legal recourse for compensation? Are there any courts that will hear their claims, and based on what theories? Leading legal scholars from around the world address these novel questions and propose answers.

Michael B. Gerrard is Andrew Sabin Professor of Professional Practice at Columbia Law School, where he teaches courses on environmental and energy law and directs the Center for Climate Change Law. He is also Associate Chair of the Faculty of Columbia's Earth Institute. Before joining the Columbia faculty in January 2009, he was partner in charge of the New York office of Arnold & Porter LLP, where he continues as Senior Counsel. He is author or editor of ten books, two of which were named Best Law Book of the Year by the Association of American Publishers: *Environmental Law Practice Guide* (twelve volumes, 1992) and *Brownfields Law and Practice: The Cleanup and Redevelopment of Contaminated Land* (four volumes, 1998). His other recent books are *The Law of Green Buildings* (with Cullen Howe, 2010), *The Law of Clean Energy: Efficiency and Renewables* (2011), and *The Law of Adaptation to Climate Change: U.S. and International Aspects* (with Katrina F. Kuh, 2012).

Gregory E. Wannier is a judicial clerk for the Hon. S. James Otero, United States District Court for the Central District of California. He largely edited this book while serving as Deputy Director of the Center for Climate Change Law at Columbia Law School. His research focuses include EPA regulation of greenhouse gases under the Clean Air Act, legal mechanisms for adapting to the effects of global climate change, implications of electricity grid reliability protections for energy resource development, the viability of market solutions to climate change, and national and international trade protections. He received his JD from Stanford Law School, where he served as president of the Environmental Law Society and editor-in-chief of the *Stanford Journal of Law, Science & Policy*, and he holds a Master's Degree in Environmental Policy from the E-IPER program at Stanford University, School of Earth Sciences.

(2013)

Threatened Island Nations

LEGAL IMPLICATIONS OF RISING SEAS AND A CHANGING CLIMATE

Edited by

Michael B. Gerrard

Columbia Law School

Gregory E. Wannier

U.S. District Court for the Central District of California

CAMBRIDGE
UNIVERSITY PRESS

CAMBRIDGE UNIVERSITY PRESS
Cambridge, New York, Melbourne, Madrid, Cape Town,
Singapore, São Paulo, Delhi, Mexico City

Cambridge University Press
32 Avenue of the Americas, New York, NY 10013-2473, USA

www.cambridge.org
Information on this title: www.cambridge.org/9781107025769

First published 2013

Printed in the United States of America

A catalog record for this publication is available from the British Library.

Library of Congress Cataloging in Publication Data

Gerrard, Michael, 1951–
Threatened island nations : legal implications of rising seas and a changing climate / Michael B. Gerrard, Gregory E. Wannier.
p. cm.
Includes index.
ISBN 978-1-107-02576-9 (hardback)
1. Sea level – Environmental aspects – Islands of the Pacific. 2. Climatic changes – Environmental aspects – Islands of the Pacific. 3. Climatic changes – Political aspects – Islands of the Pacific. 4. Environmental refugees – Legal status, laws, etc. – Islands of the Pacific. 5. Sea level – Environmental aspects – Islands of the Indian Ocean. 6. Climatic changes – Environmental aspects – Islands of the Indian Ocean. 7. Climatic changes – Political aspects – Islands of the Indian Ocean. 8. Environmental refugees – Legal status, laws, etc. – Islands of the Indian Ocean. 9. Liability for climatic change damages. 10. Islands – Law and legislation. I. Wannier, Gregory E. II. Title.
K3585.5.G47 2013
344.14′2046–dc23 2012029782

ISBN 978-1-107-02576-9 Hardback

Contents

Color plates appear after page 300.

Contributors

Judah Ariel is an Associate in the International Trade and Dispute Resolution practice at Sidley Austin LLP. He was previously the trade and foreign policy advisor to a senior member of the U.S. Congress. Ariel has a JD from New York University School of Law.

Leonito (Jun) Bacalando Jr. has been Assistant Attorney General of the Federated States of Micronesia (FSM) for the past four years. He previously served as state attorney and acting attorney general for Pohnpei, FSM. He provides advice to the FSM President on constitutional issues.

Daniel Bartz received his BA summa cum laude from Wright State University in Dayton, Ohio, in 2006, and his JD from the University of Oregon School of Law in 2010.

Dean Bialek is the Director for UN Climate Change and Natural Resources at Independent Diplomat, and negotiator with the delegation of the Republic of the Marshall Islands in international climate talks under the UNFCCC. Dean has a PhD (International Law) from the University of Melbourne and an LLM from University of London.

Maxine A. Burkett is an Associate Professor of Law at the William S. Richardson School of Law. Burkett attended Williams College and Exeter College, Oxford University, and received her law degree from Boalt Hall School of Law at the University of California, Berkeley.

Elizabeth Caldwell joined Baker & McKenzie's Climate Change and Environmental Markets Practice as an Associate in 2010. Prior to joining Baker & McKenzie, Elizabeth was an Australian Youth Ambassador for Development in Port Moresby, Papua New Guinea. She is currently undertaking an MSc in Environmental Economics and Climate Change at the London School of Economics and Political Science.

Mary-Elena Carr, Associate Director, Columbia Climate Center, coordinates education and research initiatives at the boundary of social and physical sciences. She has a BSc and MSc from the University of Barcelona and a PhD from Dalhousie University. Prior to joining Columbia University, she did research at CalTech's Jet Propulsion Laboratory and was Associate Program Director at the National Science Foundation.

Kristin Casper is Legal Counsel at Greenpeace International. She advises on international environmental law and legal aspects of campaigning in the areas of climate change, energy, toxics, and water. Prior to coming to Greenpeace International, she served as a climate and clean energy campaigner in the United States and Australia for ten years.

David Freestone is Executive Director of the Bermuda-led Sargasso Sea Alliance and Adjunct Professor at The George Washington University Law School. He previously worked in the Legal Office of the World Bank, retiring in 2008 as Deputy General Counsel/Senior Adviser. He holds an LLD from the University of Hull and an LLM from the University of London.

Catherine Gascoigne is an Associate with Baker & McKenzie's Environment and Environmental Markets Practice and a Research Assistant to the Law Faculty of the University of New South Wales. Gascoigne graduated from the University of Oxford with a BCL with Distinction and from the University of Sydney with First Class Honours with both a BA and an LLB.

Alice Graff earned her BS in Environmental Studies and Architectural Studies from Tufts University, and her MSc in Environment and Development from The London School of Economics and Political Science. While at the LSE, she won the George and Hilda Ormsby Prize for her research on the framing of climate change appeals.

David Hodgkinson is Special Counsel with Australian law firm Clayton Utz and an Associate Professor in the Law School at the University of Western Australia. Earlier in his career he was Senior Legal Research Officer at the High Court of Australia where he completed most of his work for Justice Sir William Deane. He has also been Director of the Legal Department at IATA (the organisation of the world's airlines) in Montreal.

Michele Klein Solomon is the Permanent Observer of the International Organization for Migration (IOM) to the United Nations. Klein Solomon received her JD and MSFS degrees from the Georgetown University Law Center and Georgetown University School of Foreign Service in 1988. Prior to joining IOM, she served as an Attorney Adviser with the U.S. Department of State, Office of the Legal Adviser, from 1989 to 2000.

Stephen Leonard is President of the Climate Justice Programme and a legal practitioner in Australia.

Siobhán McInerney-Lankford is a human rights lawyer at the World Bank, currently serving as Senior Policy Officer in its Operations Policy and Country Services Vice-Presidency, and formerly as Counsel in its Legal Vice-Presidency (2002–2010). Before joining the World Bank she worked in private practice in Washington, D.C. McInerney-Lankford holds an LLB from Trinity College, Dublin, an LLM from Harvard Law School, and a BCL and DPhil in EU human rights law from Oxford University.

Ilona Millar is a Senior Associate with Baker & McKenzie's Environment and Environmental Markets Practice. Prior to joining Baker & McKenzie in May 2008, she worked for the Foundation for International Environmental Law and Development in London. Millar teaches the international climate law course at the Australian National University.

Nicola Peart received her MA (Cantab) in 2008 and MSc (Imperial College London) in 2009, and is currently studying at University College London.

Ann Powers is a full-time faculty member of Pace Law School's Center for Environmental Legal Studies, where she focuses on international environmental law, oceans and coasts, and environmental diplomacy. Prior to arriving at Pace, she worked for the Chesapeake Bay Foundation, a major regional U.S. environmental group, and for the U.S. Department of Justice.

Rosemary Rayfuse is a Professor of International Law at the University of New South Wales, Australia, and a Professor of International Environmental Law at Lund University, Sweden. She is a member of global research projects on the constitutionalization and fragmentation of international law, arctic governance, and global environmental governance; a member of the IUCN Commission in Environmental Law; and co-chair of its sub-group on high seas governance.

Maketo Robert is Secretary of the Department of Justice and former Attorney General of the Federated States of Micronesia (2008–2011).

Madeleine Rubenstein, formerly the Research Coordinator for the Columbia Climate Center, is a graduate of Barnard College. Her research with the Climate Center focused on energy policy and climate change communication, as well as the intersection of science, society, and ethics. She is currently with the Frankfurt Zoological Society in Ethiopia, developing business plans for community conservation areas.

Kristína Šabová is a lawyer with public interest law organization Environmental Law Service, based in the Czech Republic. She works on CO_2 litigation in the EU countries and is involved in the EU climate and energy policy advocacy work.

Clive Schofield is Professor and Director of Research at the Australian National Centre for Ocean Resources and Security (ANCORS). He holds a PhD in Geography from the University of Durham, UK, and an LLM in international law from the University of British Columbia. He presently holds an Australian Research Council (ARC) Future Fellowship.

Jan Šrytr is a lawyer with public interest law organization Environmental Law Service, based in the Czech Republic. He leads a team of lawyers focused on CO_2 litigation and EU climate and energy policy advocacy. He has five years of experience with major cases of environmental litigation in the Czech Republic and other EU countries.

Leslie A. Stein is a Scholar in Residence at Pace Law School. He was previously a Visiting Scholar at the Center for Climate Change Law at Columbia Law School. He was Chief Counsel for the development of the Metropolitan Strategy for Sydney, was a judge in land use disputes for ten years, and has been a professor at various law schools teaching land use and urban development.

Jenny Grote Stoutenburg, LL.M.; Maître en Droit, is a PhD candidate at the University of Hamburg, Germany, and a former Visiting Scholar at UC Berkeley Boalt Hall School of Law. Her publications include articles relating to small island states, international environmental law, ocean law, and environmentally induced migration. Her dissertation is entitled "Disappearing Island States in International Law."

Christopher Stucko received his JD from Columbia Law School, where he was a Harlan Fiske Stone Scholar, and he holds a Bachelor of Science in Foreign Service from Georgetown University's Edmund A. Walsh School of Foreign Service. He is currently an associate at Freshfields Bruckhaus Deringer in New York.

Jasper Teulings is General Counsel and Advocate in the Legal Unit of Greenpeace International in Amsterdam. He also serves on the Board of EarthRights International. He speaks and publishes on a wide range of topics, such as climate change litigation, free speech and peaceful protest, and Greenpeace's corporate campaigns. Before joining Greenpeace International in 2004 he practiced as a media lawyer and civil litigator in Amsterdam for ten years.

Diego Villarreal, Staff Associate, Columbia Climate Center, has been developing quantitative models to quantify the impact of policies on GHG emissions since 2010. He also works on the communication of scientific research to non-technical audiences. He holds an MA in Climate and Society from Columbia University and a dual degree in Chemical Engineering and Chemistry from Purdue University.

Koko Warner is the Head of the Environmental Migration, Social Vulnerability, and Adaptation Section at UNU-EHS. Warner is a lead author for IPCC's 5th Assessment Report, Working Group 2 on Adaptation. She works extensively in the context of the UNFCCC climate negotiations on adaptation (particularly in risk management and migration), and loss and damage.

Jacob David Werksman directs the Institutions and Governance Program at the World Resources Institute (WRI). He is currently an Adjunct Professor of Law at New York University and at Georgetown University, and an active Member of the State Bar of California. Werksman has previously held posts at the Rockefeller Foundation, United Nations Development Programme, and the Foundation for International Environmental Law and Development (FIELD) in London.

Mary Christina Wood is Philip H. Knight Professor and Faculty Director, Environment and Natural Resources Law Program, at the University of Oregon School of Law.

Katrina M. Wyman is a Professor of Law at New York University School of Law. Wyman has a BA, MA, and LLB from the University of Toronto and an LLM from Yale Law School. Her research interests include environmental issues, such as the consequences of climate change for human settlement and migration.

Lucy Young is a solicitor with Legal Aid Western Australia. She is the coauthor of a number of journal articles on climate change displacement and has a particular research interest in principles of justice and international climate change agreements.

Preface and Acknowledgments

In late 2009, Phillip H. Muller, then the Ambassador and Permanent Representative of the Republic of the Marshall Islands to the United Nations (and now the Minister of Foreign Affairs), approached Columbia Law School's Center for Climate Change Law. He posed a startling scenario: the seas are rising, and some decade – no one knows which – his country of twenty-nine coral atolls and five islands, located midway between Hawaii and Australia, is going to be under water. When that happens, a number of novel legal questions will arise. If a country is under water, is it still a state? Does it still have a seat at the United Nations? What becomes of its exclusive economic zone, and the fishing rights on which it depends for much of its livelihood? What countries will take in its displaced people, and what rights will they have when they arrive? Do they have any legal recourse against those states whose greenhouse gas emissions caused this plight?

We concluded that we could not answer these questions on our own, but that they had so much both practical importance and intellectual interest that they warranted an international conference of legal scholars. So Columbia and the Republic of the Marshall Islands decided to put on such a conference at Columbia. Columbia's Earth Institute provided a seed grant, and ultimately The World Bank, Australia, Israel, and South Korea's Global Green Growth Institute agreed to provide funding so that people from the Marshall Islands and other affected regions of the world could travel to New York and participate in the conference.

The conference took place May 23–25, 2011. A total of 272 people registered from thirty-nine different countries; numerous Pacific nations as well as Asia, Africa, Europe, and the Americas were represented, and an additional 308 people registered for the live Webcast. Serious scholarly discussion occurred, and many connections were made as people from around the world met for the first time and discovered their common goals. A high point came during the formal conference dinner in the ornate rotunda of Columbia's Low Library. The Marshall Islands delegation (twenty-three strong) mounted the stage and sang a series of their nation's songs in Marshallese, led by then-President Jurelang Zedkaia on the ukulele. This highlighted the cultures that stand to be lost if the world does not solve this problem.

We invited the scholars who participated in the conference to contribute formal papers. This volume is the result. It is the first book written about the legal issues faced by island nations that are threatened by sea level rise, and it explores legal theories whose novelty, imagination, and complexity are as great as the problems they seek to tackle. All the chapters were written by lawyers (except for Chapter 2, which was written by scientists who present what is known about the nature and timing of sea level rise as it affects the island nations). Each chapter represents the views of its authors alone; the authors of other chapters, and the editors, may disagree with some of them, but in the spirit of open academic inquiry and the urgency of finding legal approaches to the crises felt by the affected populations, we felt it important to give a full airing to the theories presented.

Many people contributed to the creation of this book. Ambassador Muller and Caleb W. Christopher, Legal Advisor, were our primary liaisons to the Marshall Islands government and also helped organize M.B.G.'s memorable and illuminating trip to the Marshall Islands atolls of Majuro and Eniwetok in December 2010. Multiple members of Columbia Law School and University staff provided invaluable assistance organizing the conference that inspired this book. Julia Errea Ciardullo and Danielle Sugarman, two very talented lawyers, assisted with the preparation of the manuscript. Columbia University's Earth Institute, Environmental Law Clinic, and Climate & Society Program helped the Center for Climate Change Law. Klaus H. Jacob, Terry Liu, and Kristina Stege made special contributions. John Berger and David Jou of Cambridge University Press and Peggy M. Rote of Aptara, Inc., helped the book come into existence. Numerous Columbia University students performed cite checking and research tasks – Aurore Bouyer, Daniel Bregman, Kaitlin Cassel, Julia Christian, Nate Cross, Briana Dema, Angelina Liang, Sarah Steinfeld, Narayan Subramanian, Jessica Wentz, and Rose Winer – as well as Sarah Goldmuntz of Hamilton College. We are most grateful for all of this invaluable assistance.

M.B.G. would like to express his love and gratitude to his wonderful family for their help throughout this process – Barbara, David, and William.

G.E.W. is, as always, grateful to his family for their constant support. He also thanks Michael Wara for providing early-career guidance and advice in the arena of climate change law.

Part I INTRODUCTION

1 Overview

Gregory E. Wannier and Michael B. Gerrard

1. Introduction

In December 2008, tidal surges during the seasonal high-tide period ("King Tide") covered Majuro, an atoll serving as capital of the Republic of the Marshall Islands, which is situated just north of the equator and midway between Hawaii and Australia. These tides washed out roads, low-lying houses, and other coastal installations; severely damaged freshwater-dependent plant life; and caused $1.5 million in damages (about 1% of the national economy). This event was not unique: the Republic of the Marshall Islands is one of the lowest lying countries in the world, sitting only a few meters above static sea level at its highest point, and as a result Majuro has become accustomed to such tidal events every decade or so. However, as carbon emissions continue to increase around the world, sea levels continue to rise and extreme weather events become more numerous and intense, flooding events such as the flooding in Majuro will become more common. The Marshallese people can cope with such flooding events every few years, but perhaps not if they occur every few months. Even more intense flooding than already experienced could imperil the lives and certainly the livelihood of much of the population. If current trends in greenhouse gas emissions continue, it is entirely possible that life on the Marshall Islands could become untenable by the end of the twenty-first century.

The Republic of the Marshall Islands is just one of several nations facing existential threats; other nations, mostly in the Pacific and Indian oceans, also face threats to the habitability of all or most of their territory. These threats raise serious legal questions about the continued viability of these States as well as the provision of protections for individuals who may need to relocate.

This book enlists leading scholars from around the world in considering whether these threatened nations can continue to exist under these circumstances; in what form they would exist; where their citizens can and should move, if at all; how to pay for their adaptation and relocation; and who can be held responsible. In answering these questions, the authors have largely focused their analysis on practical implications of the legal issues discussed. Threatened nations must prepare themselves legally for a future without habitable territory, and there are resulting diplomatic and political steps each nation could pursue to strengthen its legal standing into the future. The chapters in this book highlight current legal authorities and then explain how they can be used. The volume aims to provide a comprehensive summary of the legal issues at play and of legal options for the future, which can then guide constructive legal and political activities.

2. Scientific Summary and Impacts

As the oceanographer Mary-Elena Carr and colleagues show in Chapter 2, by the end of the twenty-first century a one-meter rise in sea levels appears to be inevitable, regardless of efforts to reduce greenhouse gas emissions. A two-meter rise is possible, but a rise of four meters is unlikely. For an island that is mostly or entirely one or two meters above sea level at mean high tide, these are extremely worrisome projections, especially when one considers storm surges that can raise the waters still higher, as well as the possibility of a catastrophic tsunami.

Chapter 2 also demonstrates that sea level rise is only one of the ways that climate change will affect small island nations. Tropical cyclones are projected to have greater intensity. Both droughts and floods may occur more frequently. The droughts lead to shortages of drinking water and impair food supplies; the floods damage infrastructure, contaminate freshwater and agricultural soils with saltwater, and may compel evacuation. Both can damage local ecosystems and reduce biodiversity. The absorption of carbon dioxide by the oceans is leading to ocean acidification, which harms marine organisms with shells and the coral reefs that are critical physical barriers for atoll islands, serving as protection against storm surges and high waves.

Human activities on these islands may exacerbate some of these trends. Coastal construction often relies on dredging and other reef-degradation activities, which may aggravate the impacts of rising sea levels. Within an island nation, degrading conditions in smaller islands may lead to population migration to larger islands; for example, with the flight from the outer islands, the population of Majuro is increasing rapidly, straining its already limited resources.

Chapter 2 explains these and other issues, setting the stage for the inevitable legal questions that then arise.

3. Migration

In the coming decades and beyond, it is likely that the large populations of these islands, as well as those in other climate-sensitive areas of the world, will need to resettle in other areas. Estimating the magnitudes of this migration is very difficult. Sea level rise, increasingly severe weather patterns, and desertification are the environmental influences that are most likely to force people from their homes, but population movement often results from a combination of environmental, social, political, and economic factors. Some communities have more resilience and ability to stay in place than others in the face of external threats.

In one widely cited estimate, Norman Myers predicts that there will be approximately 200 million climate-change migrants by 2050 across the globe.[1] There is no widely cited estimate for migration specifically

1 Norman Myers, *Environmental Refugees: An Emergent Security Issue*, 13th Economic Forum, May 2005, Prague. www.osce.org/documents/eea/2005/05/14488-en.pdf.

from the island nations. However, as Katrina Wyman notes in Chapter 11, the total population of the four most threatened island nations is approximately 500,000, which is about one-third of the total number of migrants accepted annually by the United States, Australia, and New Zealand. Thus, although displacement of the population of threatened island nations will cause great hardship for the affected people and will necessitate legal and political innovation, it should not itself significantly strain the existing world immigration system if countries are willing to take in these people. The greater concern is the cumulative impact of climate-induced migration from much more heavily populated areas such as Bangladesh and portions of Africa.

4. Sovereignty and Territory

One fundamental question that will affect these islands is what happens to the nations themselves if their island territories become uninhabitable. Most of the small island nations have histories as colonies, trust territories, or the like, and have only achieved statehood and sovereignty in recent decades. United Nations membership was attained by the Republic of the Maldives in 1965, the Republic of Vanuatu in 1981, the Republic of the Marshall Islands and the Federated States of Micronesia in 1991, the Republic of Palau in 1994, the Republic of Kiribati and the Republic of Nauru in 1999, and Tuvalu in 2000.[2] Sovereignty in the international system is of course a source of pride for any society, but it provides more practical benefits as well. It allows countries to establish systems of laws and government to govern themselves, to levy taxes, and to incur debts to finance public projects. Membership in the United Nations gives nations the ability to negotiate agreements, secure funding for their people, vote for measures that benefit them in the short and long term, participate in international fora, and otherwise acquire and exert political influence

2 United Nations, Member States of the United Nations, http://www.un.org/en/members/index.shtml.

in the international community. In short, statehood provides access to privileges that many of the most threatened islands rely on to provide or acquire crucial services for their citizens.

Given the importance of sovereignty, the question then becomes how such sovereignty can be preserved; the answer involves both legal and equitable considerations. As Jenny Grote-Stoutenburg argues in Chapter 3, in the international law of statehood there is a tension between the principle of effectiveness – which asks whether a State has a territory, population, government, and independence – and the principle of legality, which holds that the extinction of States must not violate some fundamental norms of international legal order, called *jus cogens* norms. In other words, it is very possible that some of these countries may no longer meet the traditional requirements for statehood (permanent territory and population); however, other nations may well continue to recognize these nations for equitable reasons (and in fact may be legally obligated to do so), meaning that the indices of statehood can likely be preserved.

If these indices are preserved, it then becomes necessary to think about how they should be organized after the population has evacuated. This might most effectively happen via an ex-situ arrangement, as outlined by Maxine Burkett in Chapter 4, whereby country representatives are given full power as national leaders in international law and would manage and distribute national resources to a scattered population, or diaspora. Such a situation would necessitate the establishment of a government system, whose main task would be the administration of national assets for the benefit of its people; that is, a so-called trusteeship system. As Professor Burkett notes, one way to administer this trusteeship could be to establish it under the United Nations Trusteeship system, which has overseen similar arrangements in the past (albeit with mixed success). Key here would be constant and active engagement with the diaspora and its chosen representatives in administering the system, thereby respecting the sovereignty of the nation ex situ.

The extent of the resources available to any nation ex situ depends heavily on its ability to continue to access marine territories, which provide critical fishing and mineral rights. As currently set by the Law of the Sea Convention (LOSC), exclusive economic zones (EEZs) ordinarily extend 200 nautical miles from a nation's low-tide mark. However, the LOSC does not delineate permanent boundaries, and so traditionally EEZs would recede along with the coast if sea levels rose. Of more concern to small island nations, substantial marine territory – as much as 40,000 square nautical miles – could be threatened by the abandonment of a single island, because the LOSC clearly disallows marine territory for uninhabitable rocks. As Ann Powers and Christopher Stucko discuss in Chapter 5, the loss of marine territory could severely affect revenue sources for island nations, for some of which fishery revenues account for up to 42% of their national economies.

In response, Clive Schofield and David Freestone argue in Chapter 6 that precedent elsewhere may support the artificial preservation (or bulwarking) of islands to preserve existing claims. Japan most (in)famously bolstered Okinotorishima Island from a rock to a radar station that serves as a basis for a huge claim of territory to the south. Although other nations have repeatedly challenged this claim, it might be more difficult to mount a challenge to formerly habitable islands, for diplomatic reasons as well as equitable considerations.

In addition to physical responses to preserve existing land, innovations may be possible in the definition of baselines and territories. As Rosemary Rayfuse urges in Chapter 7, nations should define their baselines advantageously to avoid any loss of territory even as sea levels rise.

5. Protections and Solutions

If certain small island nations become uninhabitable, their populations will need to relocate somewhere. However, it remains unclear where they would go. The existing human rights regime and the patchwork of international protections for displaced peoples do not provide much direct guidance on this question, as Siobhan McInerney-Lankford makes clear in Chapter 8. Refugee law as established in the 1951 Convention on Refugees probably would not apply to climate migrants (although subsequent clarifying agreements applying to Africa and the Americas might), and there is no direct international obligation for any particular country to take in such migrants. Similarly, protections in the United States and Europe for victims of environmental disasters are temporary and leave no path to full residency. Human rights law may provide at least an avenue for assertions of rights, however; as McInerney-Lankford details, the patchwork of human rights standards, including the obligation to respect other nations, protect against human rights violations, and fulfill human rights in other countries, exerts at least a sort of legal obligation on large carbon-emitting States. However, human rights obligations are largely unenforceable in practice unless States accept them, taking away some utility (though not all, because of possible political/diplomatic effects) from the possibility of demonstrating human rights violations in the climate context.

Options also exist for international institutions to provide more aid and support to climate-displaced peoples. Traditional institutions that could be integral to this effort include the International Organization on Migration and the United Nations High Commissioner for Refugees. The United Nations Framework Convention on Climate Change (UNFCCC) may also be of potential use in organizing resettlement activities, as described by Michele Klein-Solomon and Koko Warner in Chapter 9. This is particularly true following the outcome of the 2010 Conference of the Parties in Cancun, which recognized the importance of "measures to enhance understanding, coordination and cooperation with regard to climate change induced displacement ... at national, regional and international levels."[3] Ilona Millar and colleagues further suggest in Chapter 14 that the UNFCCC, more specifically its loss

3 United Nations Framework Convention on Climate Change, Decision 1/CP.16, The Cancun Agreements: Outcome of the work of the Ad Hoc Working Group on Long-term Cooperative Action under the Convention (December 2010) ¶14(f).

and damage mechanism, could be used to harness private sector funding and insurance protection for vulnerable parties.

As described earlier, existing frameworks offer some opportunities to arrange funding and resettlement options for the people of small island nations. However, it is almost universally acknowledged that existing institutions do not, and indeed cannot, provide a perfectly tailored solution to what will be a very difficult set of issues to resolve in the coming decades. In response, some scholars argue that a new international convention is required that would be tailored to the problem of climate migration. This group notably includes David Hodgkinson and Lucy Young, who outline their own proposal for a new climate-displacement treaty in Chapter 10 and highlight some of the major efforts made by other academics in the field. These and other proposals vary in their application and administrative structure, but all create refugee-like protections for qualifying environmental victims.

Other scholars, most prominently Jane McAdam, have argued instead that reinterpretation and utilization of existing treaties and institutions are the most effective and viable strategy, especially in view of the many varieties and causes of population migration and the political challenges inherent in reaching a new international agreement.[4] Katrina M. Wyman makes this argument in Chapter 11 and turns the focus to existing agreements and relationships (formed for reasons other than climate change) between threatened nations and potential destination countries that allow migration. Domestic immigration laws in certain countries may also be used, and new bilateral agreements might be negotiated. The main arguments made by this set of scholars against reliance on a new international convention focus on the political difficulty associated with negotiating and obtaining individual countries' approval for a new convention; they argue that even if a new convention were approved it would be watered down to the point of ineffectiveness or simply not be adopted by the relevant countries, and they are concerned that efforts to

4 Jane McAdam, *Climate Change, Forced Migration, and International Law* (2012).

secure a new international convention will detract from more practical efforts to use and improve on existing channels.

Finally, if resettlement becomes unavoidable, that process must be organized. Preparations should be made far in advance of any actual movement and should focus on preserving both physical and financial security and cultural norms. The provision of basic housing and life-supporting infrastructure should be planned. Leslie Stein highlights in Chapter 12 that certain housing obligations and standards exist in both international and domestic law that would apply to any new community. Equally importantly, the political relationships between displaced nationals and host States would need to be resolved, addressing both the displaced communities' relationship with host nations and their involvement in the planning process. The experience of the resettlement of Alaskan villages in Newtok, where community leaders have successfully led the relocation process (without mandates from outside), as contrasted with less successful relocations of island populations in Chagos and elsewhere, suggests that community involvement in the process is critical for the success of any relocation activity. This involvement is important largely because new communities must do more than provide housing. They should be structured to promote livelihoods and preserve critical familial and community bonds, and leaders of the displaced communities are best placed to structure their resettlement process accordingly. These concepts apply primarily in the context of an orderly migration of an entire community; movement that occurs a few families at a time to one place or to many different places, or in urgent response to a sudden crisis, may not allow for this kind of rational planning.

6. Accountability and Responsibility

If island populations are forced to resettle, many have argued that they should be able to recover damages for harms received. However, the authority for such litigation remains unclear. Several possible bases for establishing a substantive violation of international law have been posited. One possible method for imposing liability that has garnered particular attention is breach of treaty claims under the UNFCCC. As Jacob Werksman discusses in Chapter 13, such claims may have merit in large part because of the UNFCCC's focus on climate harms (recognizing harm that is associated with climate emissions) and its linkage to UN dispute-settlement provisions, including its own conciliatory dispute-settlement body under Article 14.6. However, although some sort of declaration of noncompliance can be envisioned, no clear path to obtaining monetary damages or other substantive relief is apparent.

Another interesting possibility, as described by Dean Bialek and Judah Ariel in Chapter 15, would to be to base a claim on ocean acidification, which could kill off tropical coral species, deplete fish reserves, and potentially further undermine the physical stability of coral atolls. This allows a line of litigation under the LOSC, which could be advantageous because the LOSC offers a compulsory dispute-settlement mechanism, and there is little dispute about the chain of causation from greenhouse gas emissions to ocean acidification to harm to marine species.

A third set of ideas is presented by Mary Christina Wood and colleagues in Chapter 16. The two major possibilities these authors lay out for establishing substantive liability are the duty under the World Heritage Convention to preserve natural and human heritage and theories under the public trust doctrine.

Finally, advocacy strategy that has gained some traction is to challenge environmental impact assessments for failing to consider climate impacts internationally. Maketo Robert, Jasper Teulings, Kristin Casper, and others describe one detailed example in Chapter 17 that garnered significant attention. The Federated States of Micronesia (FSM), with Greenpeace as counsel, challenged a proposed coal-fired power plant as inadequately accounting for its transboundary climate impacts. Their action was in the form of submissions to the Czech Ministry of Environment rather than a lawsuit. These submissions captured the attention of the ministry and other units of the Czech government, receiving worldwide media attention, but did not yield a favorable substantive decision.

With all of these theories, finding a court to hear such claims and persuading it to grant substantive relief are significant challenges that no party has yet surmounted. As Werksman and McInerney-Lankford discuss, the International Court of Justice is empowered by Chapter XIV of the UN Charter as the principal judicial organ of the UN, but it has limited powers, including the ability to issue advisory opinions when requested by the UN General Assembly; as noted earlier, certain treaties, including the UNFCCC in Article 14.6, offer similarly advisory commissions that could perhaps hear such cases. Any resulting decisions could be of considerable symbolic significance and would no doubt figure prominently in international negotiations, but they would not themselves force a reduction in greenhouse gas emissions or put money in the coffers of the threatened island nations to help them with their adaptation or migration efforts.

7. Wrap-Up

As described in this overview, this book examines a number of legal questions that island nation officials, representatives, and residents must face over the next several decades. Knowledge of these nations' legal status and rights will be critical for assessing options and taking the right preparatory steps. However, to prepare for this changing world, small island governments will also need to take several actions domestically, including updating existing institutions to prepare administratively for sea level rise and possible relocation. Such actions are beyond the scope of this book but may include projects to move, fortify, or adjust buildings and infrastructure to cope with anticipated climatic changes and to educate their populations about what is happening and what is coming. Changes in property and insurance systems may also be warranted.

All this is easy enough to say. However, many small developing States barely have the capacity to deal with day-to-day administration. Coping with an unprecedented threat to national existence is more than can reasonably be expected of these countries with their current resources.

Because this is fundamentally a global problem, the burden to resolve these issues falls squarely on the countries with the world's largest greenhouse gas emissions. Through no fault of their own, entire nations could soon be lost to the ocean. They have attempted, and will continue to attempt, to ease the pain of any transition through legal innovations and active planning – and in those activities this book will, we hope, serve as a useful guide – but they will still need considerable help. It is our moral duty as a society to prevent anthropogenic climate change to the extent that we can and to help these threatened nations cope with the climate change that will occur despite our best efforts.

2 Sea Level Rise in a Changing Climate

What Do We Know?

Mary-Elena Carr, Madeleine Rubenstein, Alice Graff, and Diego Villarreal

1. Introduction

Rising sea level is a commonly discussed consequence of global climate change. It is of particular concern for nations with low elevation above mean sea level, such as small island States or deltaic regions (e.g., Bangladesh). In the case of small island States, this concern is exacerbated by high population density, concentration of infrastructure and settlements at the coast, geographic isolation, and limited economic and financial resources. This chapter aims to provide a context for the other chapters in this volume on what is known about sea level, how it is projected to change as a result of anthropogenic climate change, and how these changes will affect small island States, with a particular focus on those in the tropical Pacific.

2. How Is Sea Level Defined, What Affects It, and How Is It Measured?

Sea level is defined in absolute terms relative to the Earth's center of mass,[1] but it is experienced relative to land. Changes in sea level occur when there is a change in either the mass or the volume of water in the ocean. Total mass changes as water is added to or removed from the

1 Glenn A. Milne, W. Roland Gehrels, Chris W. Hughes, & Mark E. Tamisiea, *Identifying the Causes of Sea Level Change*, 2 Nature Geoscience 471, 472 (2009).

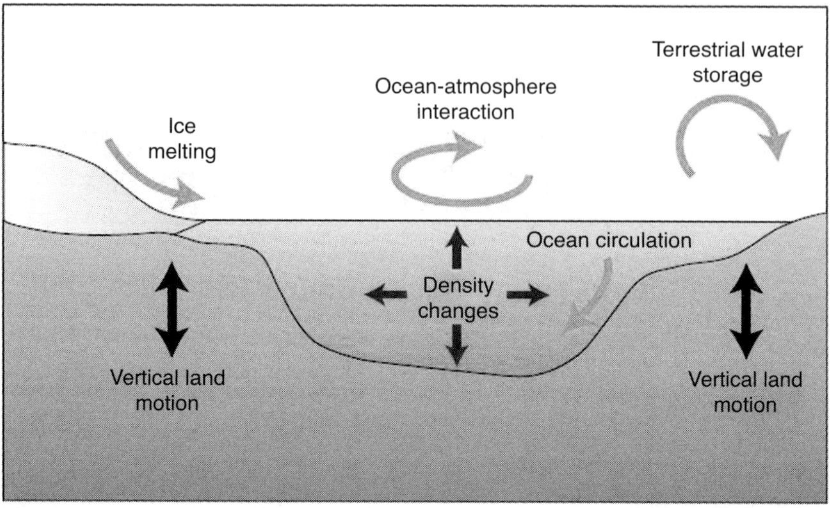

Figure 1. The factors that contribute to changes in sea level. Reproduced from Glenn A. Milne, W. Roland Gehrels, Chris W. Hughes, & Mark E. Tamisiea, *Identifying the Causes of Sea Level Change*, 2 Nature GeoScience 471, 472 (2009). Reproduced with permission from Nature Publishing Group.

ocean, such as in a net imbalance between evaporation and precipitation, or when water flows from land to sea, either as rivers or ice. Mass can be redistributed by currents or winds. The same mass of seawater changes in volume, expanding when it warms or freshens (Figure 1). Sea level also changes with vertical land motion (Figure 1). Such motions can occur over very long timescales, as in the case of the adjustment following the redistribution of the mass of ice and water when the massive ice sheets covering land at the end of the last glacial maximum melted to form lakes or flowed out to sea, where the water moved gradually through the ocean basins. This slow movement of both continents and ocean basins is known as glacial isostatic adjustment (GIA).[2] Land also undergoes vertical motion over short timescales caused by groundwater extraction or tectonic activity. Although these processes have occurred throughout Earth's history, humans affect sea level rise directly by manipulating

2 *Id.*

the flow of ground and surface water and indirectly through greenhouse gas emissions that raise the average global temperature.[3] This warming affects both the mass and volume of seawater through multiple processes: the primary effects are increased melting of land ice and higher ocean temperatures, both of which contribute to a global rise in sea level.

Sea level is traditionally measured using tide gauges, which measure water height relative to land and thus must be corrected for land motion. In the absence of additional geologic data, it is hard to estimate land movement beyond the glacial adjustment that can be modeled. To avoid these errors, tide gauges must be sited carefully to avoid land motion.[4] Tide gauges, in use since the late nineteenth century, provide the longest running instrumental record of sea level rise. The Global Sea Level Observing System compiles observations from high-quality tide gauges (i.e., with a minimum accuracy of one centimeter under all wave conditions). Although tide gauge coverage was sparse in 1900, the number of tide gauges has more than tripled since their introduction in the late nineteenth century, improving our ability to quantify global and regional changes in sea level.[5]

Satellite altimeters, first launched in 1993, enable almost global coverage, allowing unprecedented understanding of the spatial patterns of sea level. These satellite sensors emit a microwave pulse that reflects off the ocean surface and is registered by the altimeter and provides a direct measure of absolute sea level. The distance between the satellite and ocean surface is calculated by the time it takes the microwave to be sent and received.[6] The position of the satellite is determined by averaging orbit calculations from multiple observing stations. The ocean height

3 Nathaniel L. Bindoff et al., *Observations: Oceanic Climate Change and Sea Level*, *in* Climate Change 2007: The Physical Science Basis. Contribution of Working Group I to the Fourth Assessment Report of the Intergovernmental Panel on Climate Change 385, 408 (S. Solomon, D. Qin, M. Manning, Z. Chen, M. Marquis, K. B. Averyt, M. Tignor & H. L. Miller, eds., 2007).

4 *Id.*

5 *Id.* at 431–32.

6 *Id.*

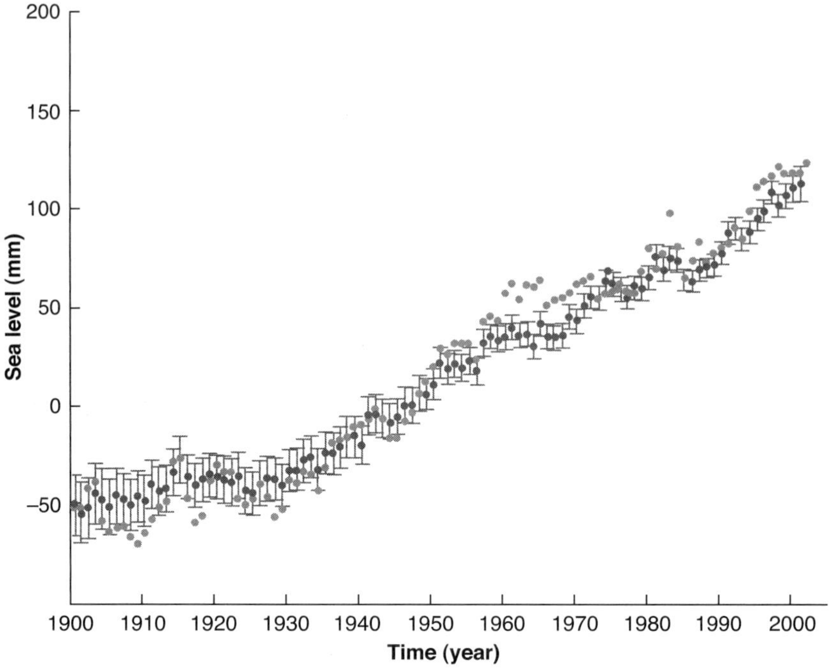

200
150
100
50
0
−50
Sea level (mm)
1900
1910
1920
1930
1940
1950
1960
1970
1980
1990
2000
Time (year)

(a)

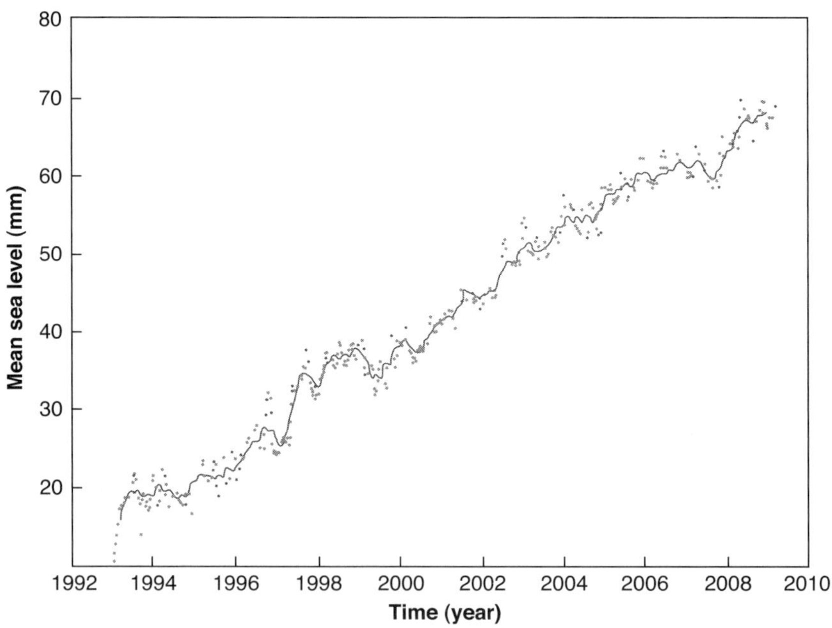

80
70
60
50
40
30
20
Mean sea level (mm)
1992
1994
1996
1998
2000
2002
2004
2006
2008
2010
Time (year)

(b)

is estimated relative to a reference ellipsoid surface that approximates Earth's shape. After accounting for factors that affect the observations, such as tides or atmospheric properties, and averaging the observed ocean height over time and space, the uncertainty is about 0.4 mm/y.[7]

3. Changing Sea Level

The global average sea level measured by tide gauges (Figure 2a) and altimeters (Figure 2b) was relatively constant between 1900 and 1930. Since that time sea level has not only risen but also the rate of this rise has increased over time: tide gauges estimated sea level rise at about 1.8 (±0.3) mm/y between 1930 and 2000, whereas the altimeters measured the rise at 3.1 (±0.4) mm/y between 1993 and 2009.[8] Globally averaged estimates of sea level rise experience variability on interannual and decadal timescales. For example, the altimeter registered a sea level rise and fall of almost 10 mm in 1997–98 associated with ocean warming due to El Niño (Figure 2b).

Measurements of the processes that affect sea level (thermal expansion and melting land ice) can be compared with independent observations of sea level from tide gauges or satellites to obtain a budget of sea level change (a comparison of the observed change in sea level versus the changes in the elements that determine sea level). The Fourth Assessment Report of the Intergovernmental Panel on Climate Change

7 Milne et al., *supra* note 1, at 473.

8 Anny Cazenave & William Llovel, *Contemporary Sea Level Rise*, 2 Ann. Rev. Marine Sci. 145, 147–49 (2010).

Figure 2. Observational record of sea level (a) from tide gauges since 1900 (p. 147); (b) from satellite altimeter since 1993 (p. 149). Reproduced from Anny Cazenave and William Llovel, *Contemporary Sea Level Rise*, 2 Annual Review of Marine Science 145, 147–49 (2010). Annual Review of Marine Science, Copyright 2010 by Annual Reviews, Inc. Reproduced with permission of Annual Reviews, Inc. in the format Other book via Copyright Clearance Center.

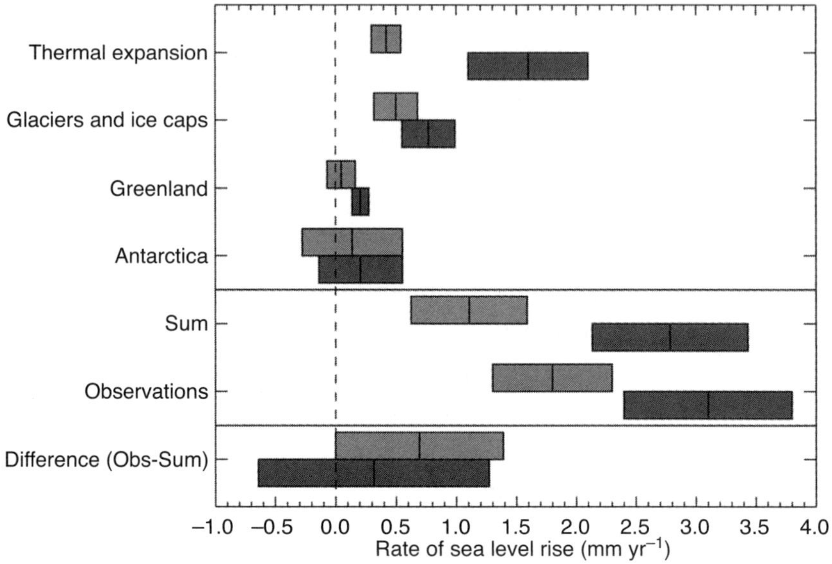

Figure 3. The budget of changing sea level. The bars represent the 90% error range, with blue entries covering 1961–2003 and brown entries covering 1993–2003. The upper four entries are estimates of the contributions to the sea-level-change budget, the middle two sum these contributions with the observed rate of rise, and the lowest bar is the observed rate less the sum of contributions. Reproduced from Nathaniel L. Bindoff et al., *Observations: Oceanic Climate Change and Sea Level*, *in* CLIMATE CHANGE 2007: THE PHYSICAL SCIENCE BASIS. CONTRIBUTION OF WORKING GROUP I TO THE FOURTH ASSESSMENT REPORT OF THE INTERGOVERNMENTAL PANEL ON CLIMATE CHANGE 385, 419 (Solomon, S., D. Qin, M. Manning, Z. Chen, M. Marquis, K. B. Averyt, M. Tignor and H. L. Miller, eds., 2007).

(IPCC AR4) quantified the components of the sea level budget for 1961–93 and 1993–2009[9] (Figure 3). The processes that affect sea level are not only challenging to measure on a global scale but also undergo significant variability; both variability and measurement error are included in the range of uncertainty, which is often comparable to the average value.

9 Bindoff et al., *supra* note 3, at 419.

The IPCC AR4 analysis reveals that between 1961 and 1993, melting mountain glaciers and thermal expansion of the ocean contributed similar amounts to sea level rise (0.4–0.5 mm/y). The sum of thermal expansion, melting of mountain glaciers, and melting of the ice sheets of Antarctica and Greenland was approximately 1.1 mm/y, which is consistent with the observed sea level shift of around 1.7 mm/y. Thermal expansion made the largest contribution (in excess of 1.5 mm/y) between 1993 and 2003, whereas mountain glaciers represented 0.8 mm/y. Both Antarctic and Greenland ice sheet melt increased slightly, but only in the case of Greenland did the increase exceed the uncertainty of the earlier measurement. The sum of the components was about 2.8 mm/y, which is very close to the independently observed sea level rise from altimeters of 3.1 mm/y for this time period. Cazenave and Llovel updated the IPCC AR4 budget for the period 2003–8.[10] They found that thermal expansion of the ocean has slowed, with an average contribution of approximately 0.3 mm/y (which may be due to the La Niña of 2007 in which average Pacific temperatures were cooler; see the following for an explanation of La Niña),[11] whereas the contributions of the mountain glaciers and of Antarctic and Greenland ice sheets together have increased to about 1 mm/y. This is approximately double the total of approximately 0.5 mm/y sea level rise caused by ice sheets between 1993 and 2003.

Although thermal expansion and mountain glaciers dominated the budget of sea level change from 1960–93, polar ice sheets have been an increasing source of sea level rise since 2003, indicating the potential shift in the sea level change budget. The bulk of ice on land (>99%) is in found at the poles; this polar land ice holds the potential for an increase in sea level of around 63 m (Table 1).[12] Ice sheets are formed from the accumulation of many seasons of snowfall; the weight of icc serves to spread the massive sheets toward the coast where they melt

10 Cazenave & Llovel, *supra* note 8, at 81.

11 *Id.* at 86.

12 Peter Lemke et al., *Observations: Changes in Snow, Ice, and Frozen Ground, in* Climate Change 2007, *supra* note 3, at 337, 341.

Table 1. *Percentage of Land Ice in Each Component and Potential to Contribute to Global Sea Level Rise*[13]

Land Ice	Volume [%]	Sea Level Rise [m]
Snow and permafrost	0.1	0.7
Mountain glaciers	0.35	0.26
Polar ice sheets	99.5	63.9
Greenland	11.4	7.3
Antarctica (WAIS)	88.1	56.6 (5)

and blocks of ice break off, in a process called calving, to form icebergs (Figure 4).[14] The slow movement of ice sheets is reasonably well understood, as are surface melting and runoff into the sea. However, there are also fast-moving ice streams or outlet glaciers that flow more rapidly into the ocean; ice loss through ice streams and outlet glaciers has recently been observed to increase rapidly.[15] The budget for the mass of ice can be estimated from changes in surface elevation (Figure 5), by comparing observations of snow accumulation versus ice flow and surface runoff, or by gravity measurements over the ice sheets that reveal changes in mass.[16]

The net mass balance is negative (i.e., greater melting) for the Greenland ice sheet and the West Antarctic Ice Sheet (WAIS). East Antarctica has high average altitude and registers the coldest global temperatures; its mass balance is close to zero[17] and is not expected to undergo significant melt in the next few hundred years. (The "mass balance" is the net sum between accumulation and loss of the ice, quantified by mass. If accumulation exceeds melting, then the mass balance is positive; if accumulation is less than ice loss, then the mass balance is negative.) Melting of the ice sheets in Greenland and WAIS could jointly lead to a 11- to 12-m sea level rise.

13 *Id.* at 342, Figure 4.1.

14 *Id.* at 342.

14 *Id.* at 361.

16 *Id.*

17 *Id.* at 365.

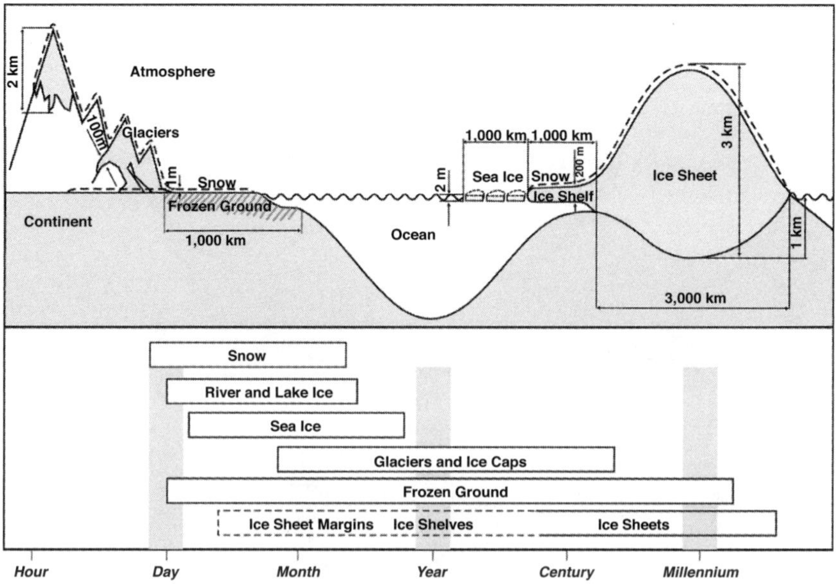

Figure 4. Schematic representation of the cryosphere: snow, river and lake ice, sea ice, mountain glaciers, ice shelves, ice sheets, and frozen ground. Only ice on land can affect sea level. Reproduced from Peter Lemke et al., *Observations: Changes in Snow, Ice, and Frozen Ground*, *in* Climate Change 2007: The Physical Science Basis. Contribution of Working Group I to the Fourth Assessment Report of the Intergovernmental Panel on Climate Change 337, 363–65 (Solomon, S., D. Qin, M. Manning, Z. Chen, M. Marquis, K. B. Averyt, M. Tignor and H. L. Miller, eds., 2007).

Rignot et al. estimate the change in mass of the Greenland and Antarctic ice sheets between 1992 and 2010 by quantifying the mass balance through ice thickness and discharge rates and by measuring changes in gravitational fields.[18] Both approaches suggest that the ice sheets are consistently losing mass despite significant seasonal and interannual variability. The total contribution to sea level rise from the ice sheets is comparable in magnitude to the loss from mountain glaciers, approximately 1.3 mm/y. Both mountain glaciers and polar ice sheets were found to be

18 E. Rignot et al., *Acceleration of the Contribution of the Greenland and Antarctic Ice Sheets to Sea Level Rise*, 38 Geophysical Res. Letters 1, 1 (2011).

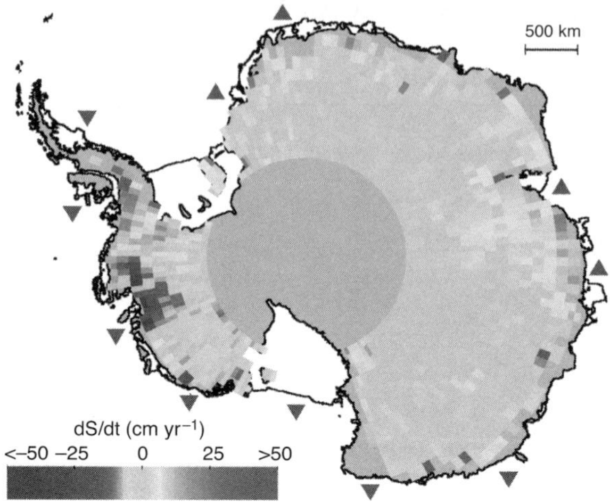
500 km
dS/dt (cm yr^{-1})
<−50 −25 0 25 >50

(a)

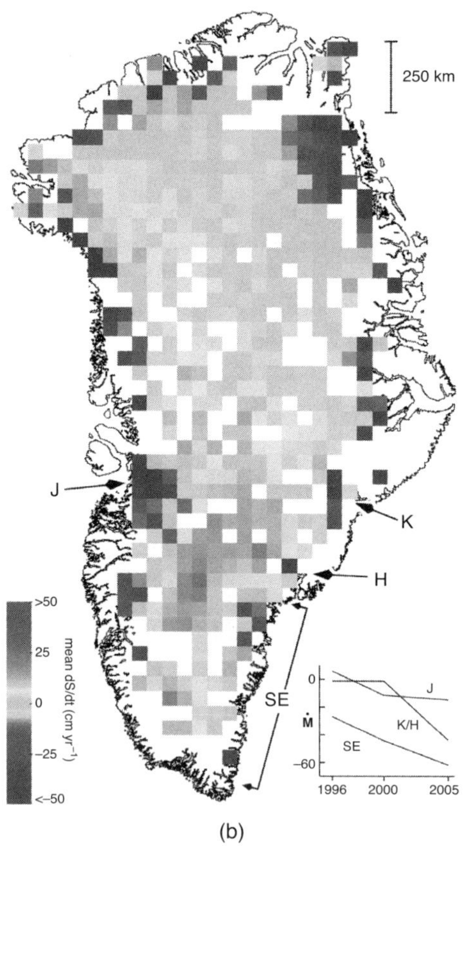
250 km
J
K
H
SE
>50
25
0
−25
<−50
mean dS/dt (cm yr^{-1})
0
−60
$\dot{M}$
J
K/H
SE
1996
2000
2005

(b)

losing mass at increasing rates, but the acceleration in melting of polar ice sheets was three times larger than that of glaciers.[19] This suggests that melting of polar ice sheets will increasingly contribute to sea level rise.

4. Regional Variability in Sea Level

Satellite altimeter measurements, which enable a spatial snapshot of sea level change over a specific observation period, show strong regional variability in sea level rise from 1993–2007, caused primarily by a combination of thermal expansion and ocean currents (Figure 6).[20] For this period sea level rise was greatest in the western Pacific, the central Indian Oceans, south of Greenland, and south of ~40°S in the Indian and Atlantic basins. By contrast, sea level increased less than the global average in the eastern Pacific and the western North Atlantic oceans.

"Virtual" tide gauges, which combine information from multiple physical tide gauges to represent regional average sea level from 1860–2008, also show spatially non-uniform changes (Figure 7).[21] Although sea level increases in most regions started in the first half of the twentieth century, the patterns of change are not consistent between regions. There is, for example, essentially no long-term increase in Baltic sea level and very little in the southwest Atlantic. As was noted from the snapshot of a shorter time period observed by the altimeter (Figure 6), sea level

19 *Id.* at 4.
20 Milne et al., *supra* note 1, at 473–74.
21 *Id.* at 474.

Figure 5. Rate of change in surface elevation in (a) Antarctica, from 1992 and 2003. Red triangles indicate locations of thickening of more than 30 cm per year, whereas purple triangles show thinning of the same magnitude, p. 365; and (b) Greenland, from 1996–2005, p. 363. Reproduced from Peter Lemke et al., *Observations: Changes in Snow, Ice, and Frozen Ground*, *in* CLIMATE CHANGE 2007: THE PHYSICAL SCIENCE BASIS. CONTRIBUTION OF WORKING GROUP I TO THE FOURTH ASSESSMENT REPORT OF THE INTERGOVERNMENTAL PANEL ON CLIMATE CHANGE 337, 363–65 (Solomon, S., D. Qin, M. Manning, Z. Chen, M. Marquis, K. B. Averyt, M. Tignor and H. L. Miller, eds., 2007).

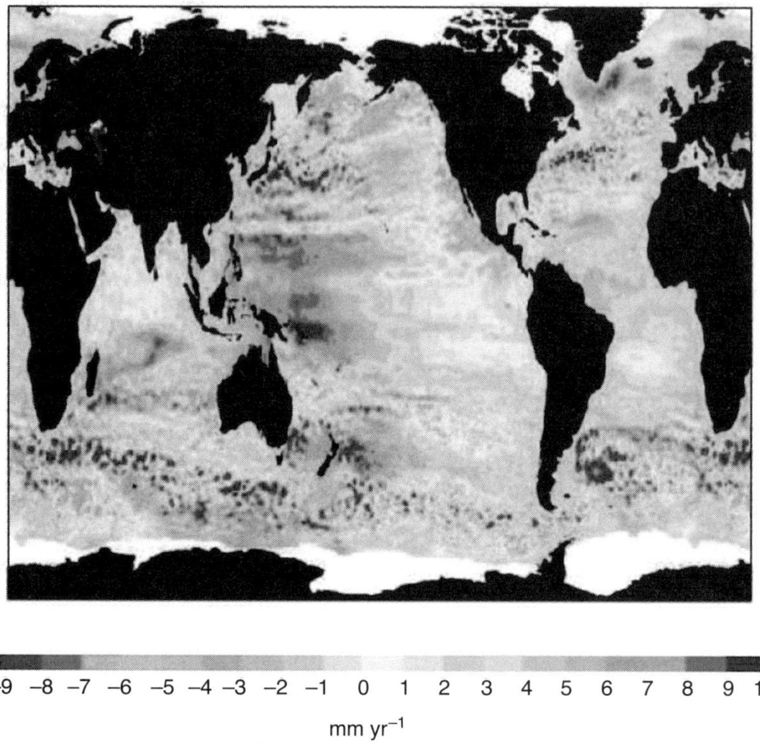

Figure 6. The spatial pattern of sea level rise between 1993 and 2007, with the global average of 3.1 mm/y removed. Reproduced from Glenn A. Milne, W. Roland Gehrels, Chris W. Hughes, & Mark E. Tamisiea, *Identifying The Causes of Sea Level Change*, 2 NATURE GEOSCIENCE 471, 472 (2009). Reproduced with permission from Nature Publishing Group.

rise is lower in the eastern and central Pacific and greater in the western Pacific basin. Seasonal and interannual variabilities, which in most cases surpass the long-term trend observed to date, are prominent in tide gauge measurements.

Sea level variability in the Pacific Ocean is dominated by the El Niño Southern Oscillation (ENSO; Figure 8), under which weather varies annually between normal conditions, El Niño conditions, and La Niña conditions. Under normal conditions, the prevailing easterly winds push water along the equator toward the western Pacific. As the surface waters

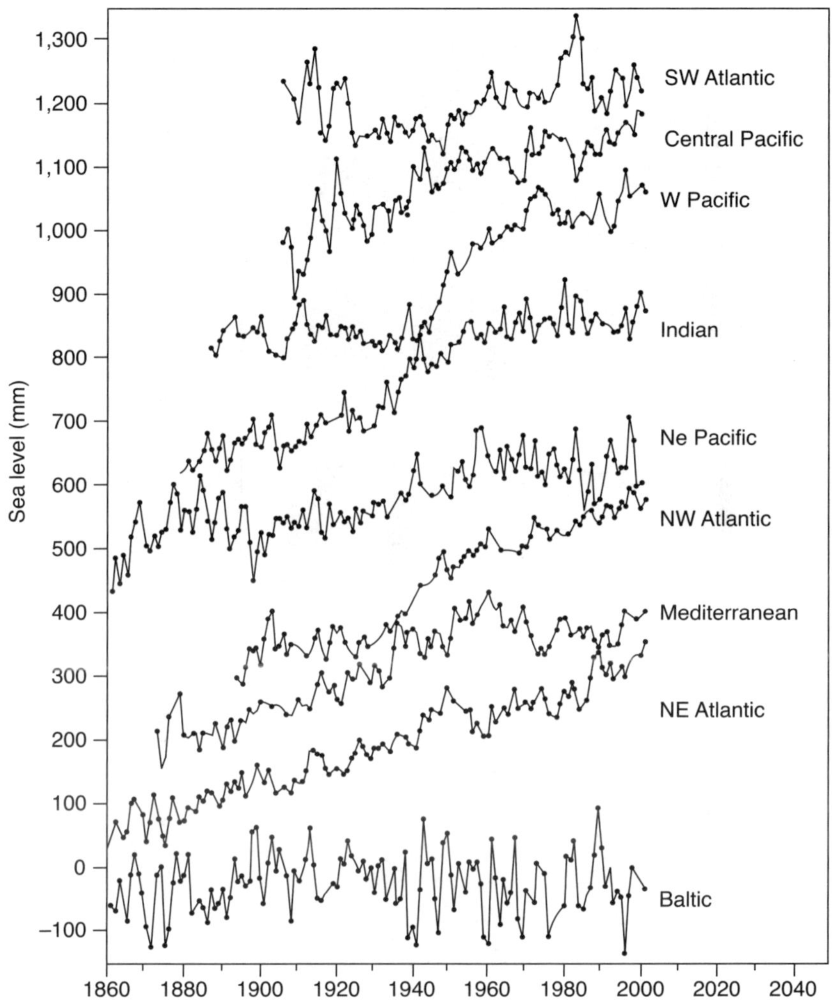

Figure 7. Time series of sea level for "virtual" tide gauges from 1860 to 2007. Reproduced from Glenn A. Milne, W. Roland Gehrels, Chris W. Hughes, & Mark E. Tamisiea, *Identifying the Causes of Sea Level Change*, 2 NATURE GEOSCIENCE 471, 472 (2009). Reproduced with permission from Nature Publishing Group.

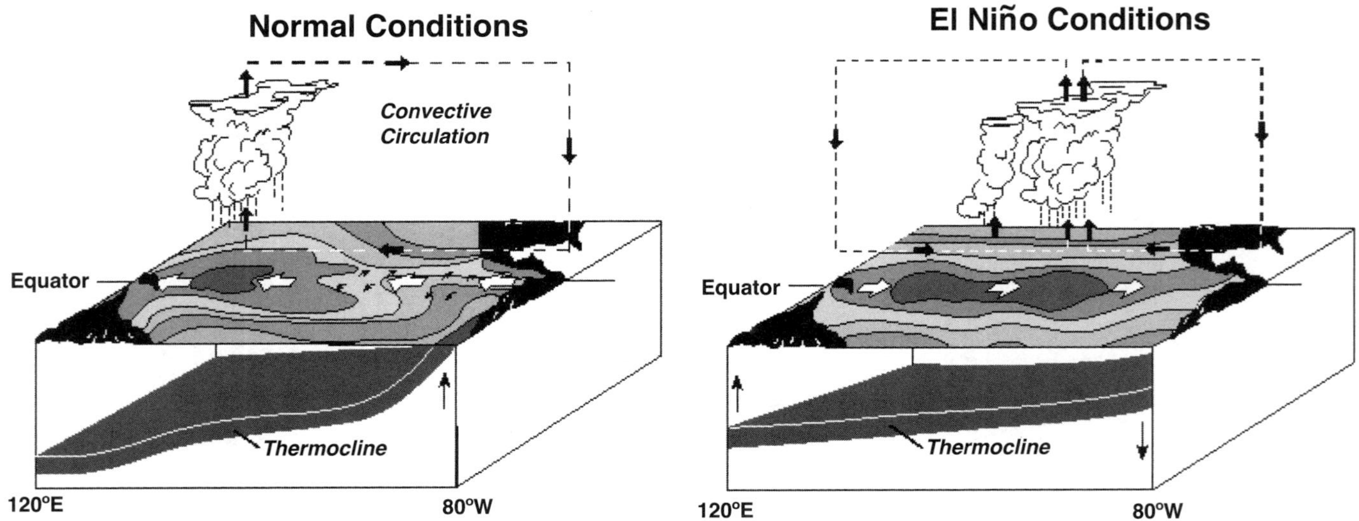

Figure 8. Schematic of normal and El Niño conditions. Reproduced from the website of Pacific Marine Environmental Laboratory (PMEL) of the National Oceanic and Atmospheric Administration (NOAA), http://www.pmel.noaa.gov/tao/elnino/nino_normal.html (last visited June 2, 2011).

travel west they are subject to intense solar heating, such that sea surface temperature (sea level) increases as waters move from east to west. The warm waters of the western tropical Pacific undergo intense evaporation, leading to the formation of clouds and accompanying atmospheric convective processes. Under El Niño conditions, the easterly winds lessen or cease, as does the surface westward current, which can even reverse. The warmest water is then found in the central and eastern Pacific, displacing the convective circulation eastward, with falling sea level in the western Pacific. Under La Niña conditions, the easterly winds are stronger and the warmest waters, highest sea levels, and atmospheric convection are found farther west than during normal conditions. ENSO occurs at irregular intervals 3 to 7 years apart.[22] Given the large exchange of heat between the ocean and atmosphere, ENSO events affect global mean temperatures.[23] The shifts in atmospheric patterns can affect precipitation and temperatures in regions as distant as Africa.

ENSO affects sea level globally, regionally, and locally. The 1997–98 ENSO event, the strongest in the instrumental record, changed global sea level by almost 10 mm (Figure 2b). The regional impact for the tropical Pacific and Indian oceans was about 20 mm (Figure 9).[24] On a local scale (Figure 10), the impact of ENSO is even greater: the records for Kwajalein (8.4N, 167.4E) in the Republic of the Marshall Islands (RMI) show ENSO-related fluctuations between 10 cm (for 1997–98) and 20 cm[25] (for 1982–83).[26]

22 Kevin E. Trenberth et al., *Observations: Surface and Atmospheric Climate Change, in* Climate Change 2007, *supra* note 3, at 235, 287.

23 *Id.* at 288.

24 John A. Church et al., *Sea Level Rise at Tropical Pacific and Indian Ocean Islands*, 53 Global & Planetary Change, 155, 160 (2006).

25 Large-scale processes, such as ENSO, have varying regional manifestations. Kwajelein experienced lower sea level during the 1982 El Niño than during the 1997 event.

26 Bindoff et al., *supra* note 3, at 413–14.

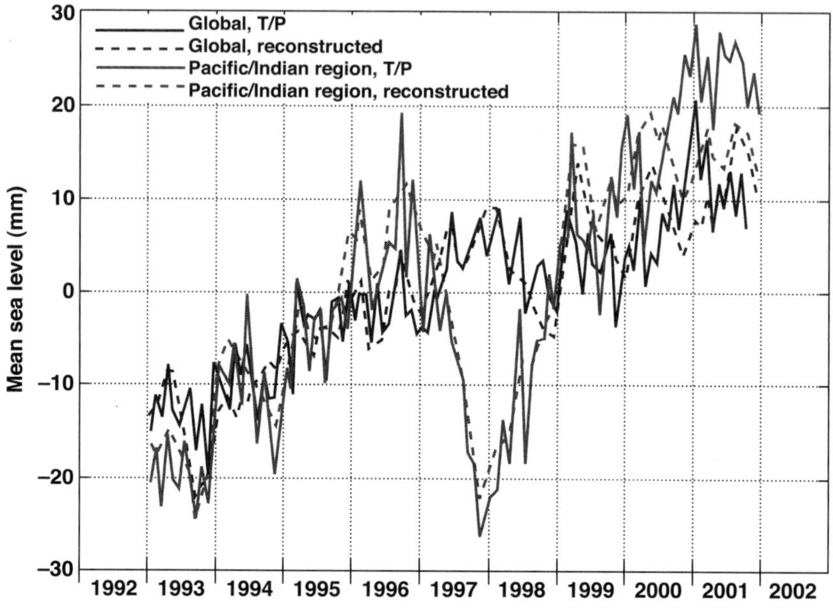

Figure 9. Tide gauge and altimeter records for the western Pacific and Indian tropical oceans from 100E to 120W and from 40S to 40N. This region is the same one depicted in Figure 11. Reproduced from John A. Church et al., *Sea Level Rise at Tropical Pacific and Indian Ocean Islands*, 53 Global and Planetary Change 155, 160 (2006), with permission from Elsevier.

Sea level variability in the tropical Pacific measured by tide gauges (see Figure 10) has doubled since the 1970s, consistent with the stronger and more prolonged El Niños indicated by independent indices of ENSO.[27] Although there are few long-term tide gauge records for Pacific islands (six locations with records over 50 years, and twenty-three with records over 24 years) and even fewer in the Indian Ocean (only seven open-ocean island sites),[28] the observational record is consistent both in

27 Church et al., *supra* note 24, at 166; Bindoff et al., *supra* note 3, at 413.

28 Church et al., *supra* note 24, at 162.

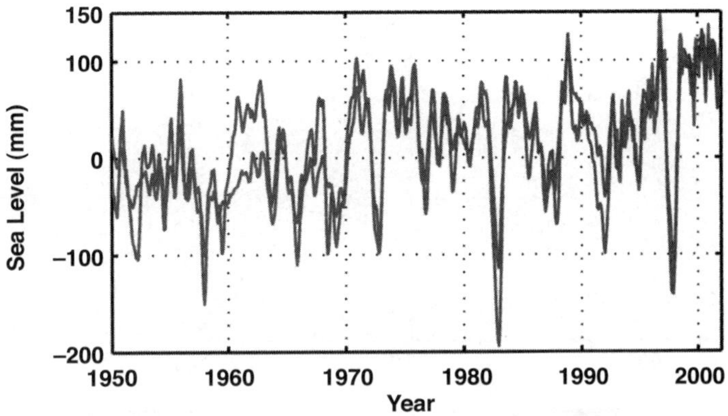

Figure 10. Tide gauge (blue) and altimeter (green) sea level record at Kwajalein (8.4N 167.4E) from 1950–2000, with annual and semi-annual signals removed. The reconstructed sea level indicates ENSO sea level variability exceeding 10 cm. Reproduced from Nathaniel L. Bindoff et al., *Observations: Oceanic Climate Change and Sea Level, in* CLIMATE CHANGE 2007: THE PHYSICAL SCIENCE BASIS. CONTRIBUTION OF WORKING GROUP I TO THE FOURTH ASSESSMENT REPORT OF THE INTERGOVERNMENTAL PANEL ON CLIMATE CHANGE 385, 414 (Solomon, S., D. Qin, M. Manning, Z. Chen, M. Marquis, K. B. Averyt, M. Tignor and H. L. Miller, eds., 2007).

magnitude and spatial pattern with that measured by the altimeter for 1993 to 2001 (Figure 11).[29]

5. Projections of Future Sea level Rise

The IPCC AR4 projected sea level rise in 2100 as the sum of thermal expansion, melting of mountain glaciers, and surface melting only from polar ice sheets; the projections assumed a constant and small rate of change of polar ice sheet mass.[30] Consequently the projected sea level change comes primarily from thermal expansion and melting mountain

29 *Id.* at 159.

30 Gerald A. Meehl et al., *Global Climate Projections, in* CLIMATE CHANGE 2007, *supra* note 3, at 747, 820–21.

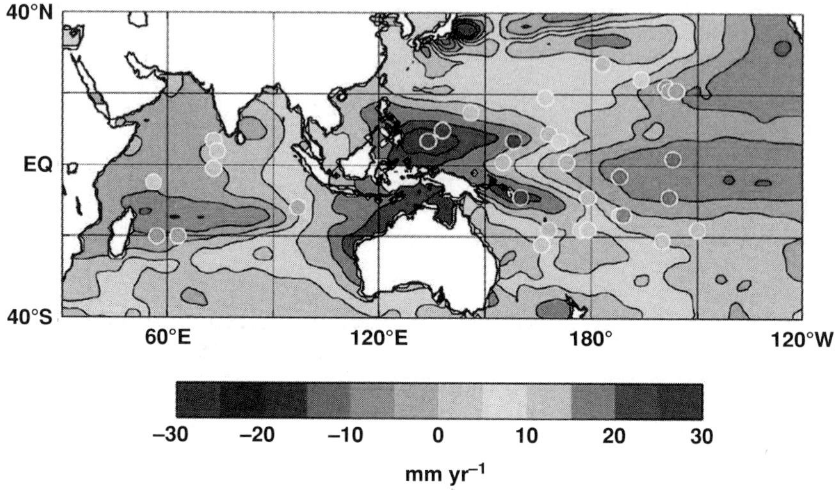

Figure 11. Tide gauge and altimeter data from 1993 to 2001 for the tropical Pacific and Indian Oceans. Reprinted from John A. Church et al., *Sea Level Rise at Tropical Pacific and Indian Ocean Islands*, 53 GLOBAL AND PLANETARY CHANGE 155, 159 (2006), with permission from Elsevier.

glaciers (Figure 12).[31] The IPCC AR4 projection for rise in sea level in 2100 ranges between 18 and 60 cm for the different models and emission scenarios.[32] Projections of sea level rise are regionally variable just as the different contributions to sea level vary regionally in the present. The projected sea level rise for 2100 from the IPPC AR4 is highest in the Arctic, consistent with projections of greater polar warming (Figure 13).[33] An additional complication will ensue over long timescales (of hundreds to a thousand years) from complete melting of the massive ice sheets at the poles, which would modify Earth's distribution of mass. If Greenland were to melt completely, sea level rise would be greater in the southern hemisphere, and conversely, the melting of the WAIS would lead

31 *Id.*
32 *Id.* at 821.
33 *Id.* at 813.

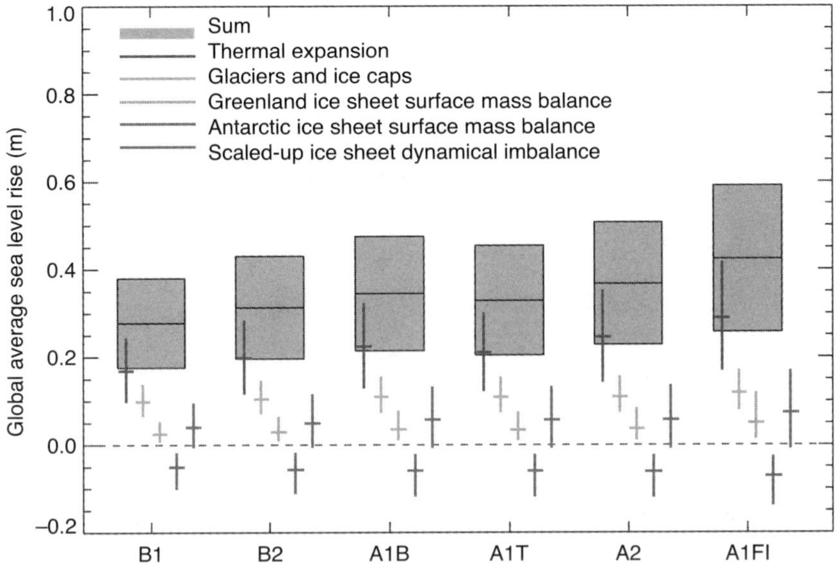

Figure 12. Future projections of sea level rise. B1, B2, A1B, A1T, A2, and A1F1 are different emission scenarios; the range of values corresponds to output from different models. The crosses indicate the range of values corresponding to thermal expansion, melting glaciers, and polar ice sheets. The contribution of polar ice sheets was held constant in time and for each scenario. Reproduced from Gerald A. Meehl et al., *Global Climate Projections*, *in* Climate Change 2007: The Physical Science Basis. Contribution of Working Group I to the Fourth Assessment Report of the Intergovernmental Panel on Climate Change 747, 821 (Solomon, S., D. Qin, M. Manning, Z. Chen, M. Marquis, K. B. Averyt, M. Tignor and H. L. Miller, eds., 2007).

to greater sea level rise in the northern hemisphere, especially in Pacific mid-latitudes (Figure 13).[34]

Any shifts in the rates of ice sheet loss, including accelerated discharge processes such as the calving of icebergs, were not included in the IPCC AR4 projection because they were absent from the climate models due to an inadequate understanding of ice sheet dynamics. Many strongly criticized the IPCC projection for omitting the potential for the

34 Milne et al., *supra* note 1, at 472.

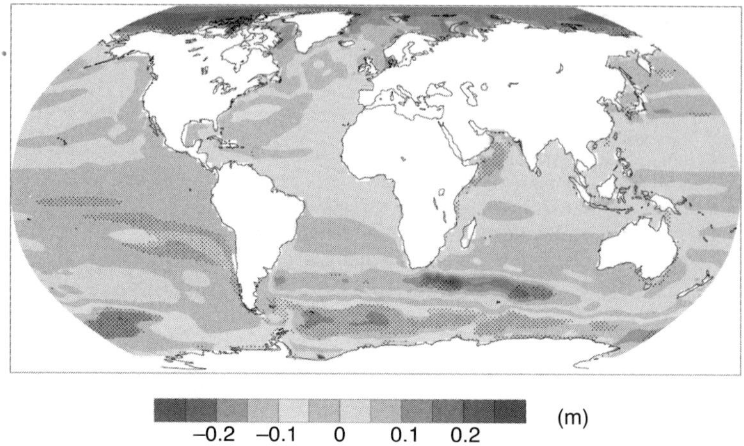

(a)

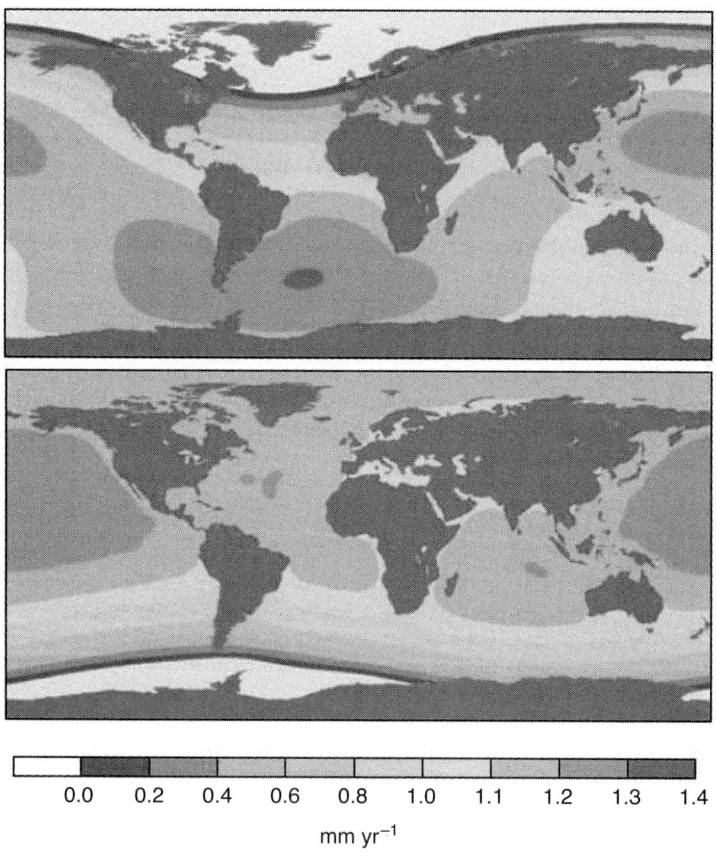

(b)

rapid loss of ice. Since the publication of the AR4, the understanding of the underlying physics of ice sheets and the ability to model their dynamics have significantly improved.

Another approach to projecting changes in sea level relies on statistical relationships between an observable climate variable, such as global mean temperature, and global sea level. Projections of this observable variation can then be used to quantify future sea level rise by assuming that the same quantitative relationship will hold in the future. In one such study, Vermeer and Rahmstorf applied the relationship between global average temperature and sea level to temperature projections for three emission scenarios and nineteen models. They estimate global sea level rise between 75 and 190 cm in 2100 (Figure 14).[35] Because this estimate is based on a relationship between observed changes, it includes all components of the sea level budget, to the degree that they contributed during the period in which the relationship is derived. The uncertainty in this projection depends in part on the likelihood of accelerated melting of polar ice sheets. It is important, thus, to quantify the bounds on the projected contribution from Greenland and Antarctica.

35 Martin Vermeer & Stefan Rahmstorf, *Global Sea Level Linked to Global Temperature*, 106 Proceedings of the Nat'l Acad. Sci. 21527, 21530 (2009).

Figure 13. Spatial pattern of projected sea level rise: (a) projected in 2100 relative to a global average rise, calculated as the difference between the global averages for 1980–99 and 2080–99; and (b) expected from the melting of the ice sheets in Greenland (top) and West Antarctic (bottom). (a) reproduced from Gerald A. Meehl et al., *Global Climate Projections*, *in* Climate Change 2007: The Physical Science Basis. Contribution of Working Group I to the Fourth Assessment Report of the Intergovernmental Panel on Climate Change 747, 813 (Solomon, S., D. Qin, M. Manning, Z. Chen, M. Marquis, K. B. Averyt, M. Tignor and H. L. Miller, eds., 2007). (b) reproduced from Glenn A. Milne, W. Roland Gehrels, Chris W. Hughes, and Mark E. Tamisiea, *Identifying the causes of sea level change*, 2 Nature GeoScience 471, 472 (2009). Reproduced with permission from Nature Publishing Group.

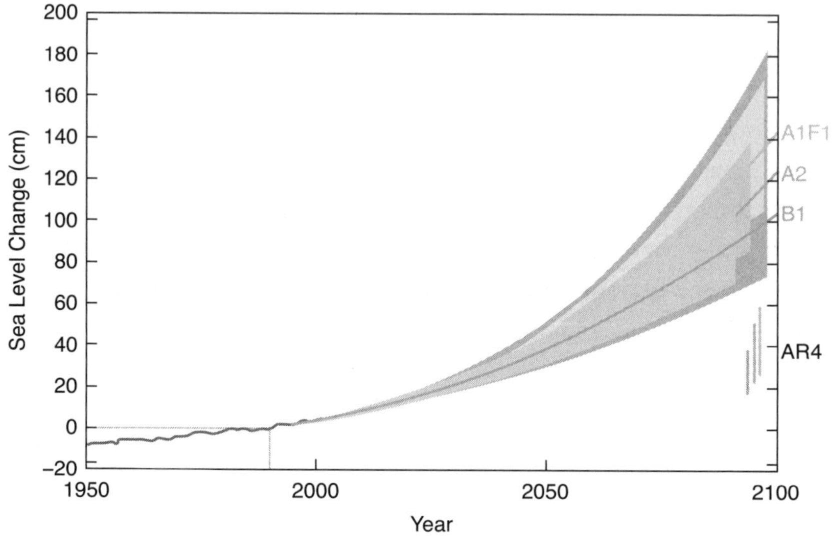

Figure 14. Projected sea level rise from 1990 to 2100, estimated from a statistical relationship between temperature and sea level, applied to nineteen models for each of three different emission scenarios: B1, A2, and A1F1. The red line shows sea level change for the observed temperature record. The green (B1), blue (A2), and yellow (A1F1) lines refer to the standard carbon cycle setting under each scenario, while the corresponding uncertainty range encompasses the high and low settings for the carbon cycle. The gray shading is an additional uncertainty associated with the statistical relationship between temperature and change in sea level rise. The projected sea level change for the three scenarios from AR4 is shown as vertical lines. Reproduced from Martin Vermeer and Stefan Rahmstorf, *Global Sea Level Linked to Global Temperature*, 106, Proceedings of the Nat'l Acad. of Sci. 21527, 21531 (2009).

Two studies provide constraints on polar ice sheet ice loss. If the rate of satellite-based acceleration of mass change of Greenland and Antarctica measured since 1993 continues without change, Rignot et al. estimate that polar ice sheets alone will contribute 56 cm to global sea level rise by 2100.[36] In another estimate, Pfeffer et al. note that rapid ice loss through calving of icebergs is restricted to glaciers where the

36 Rignot et al., *supra* note 18, at 4.

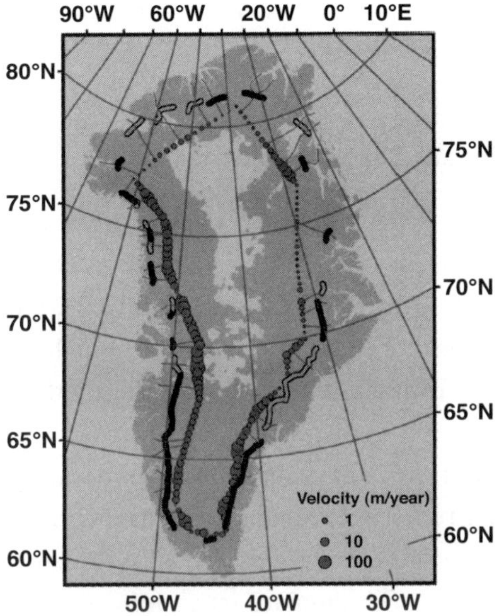

Figure 15. Map showing outlet glacier gates in Greenland: gates that are below sea level are shown as dark green and those above sea level as black. Reproduced from W. T. Pfeffer, J. T. Harper, and S. O'Neel, *Kinematic Constraints on Glacier Contributions to 21st-Century Sea Level Rise*, 321 SCIENCE 1340, 1341 (2008). Reprinted with permission from the American Association for the Advancement of Science.

underlying land is below sea level,[37] and they project 30–120 cm of sea level rise from Greenland and WAIS by 2100.[38]

By constraining the estimate of Vermeer and Rahmstorf for these two estimates of polar ice sheet loss, we conclude that 60 to 80 cm is an appropriate estimate for the contribution of the melting of polar ice sheets to sea level rise by 2100. Given the anticipated rise caused by thermal expansion, a lower bound of 1 m for sea level rise seems unavoidable. Conversely, although there may be an increase of 2 m, it is unlikely that it would be 4 m or higher, because Pfeffer et al. found that the

37 W. T. Pfeffer, J. T. Harper & S. O'Neel, *Kinematic Constraints on Glacier Contributions to 21st-Century Sea Level Rise*, 321 SCIENCE 1340, 1341 (2008).

38 *Id.* at 1342.

spatial restriction on rapid ice discharge provides an upper bound on ice sheet loss.

6. Additional Climate Change Projections

The impacts of anthropogenic climate change are not limited to changes in sea level. Average air temperature warmed between 1901 and 2004, increasing 0.6–1.0°C over the Pacific between 20N and 30S, accompanied by a greater number of hot days and warm nights.[39] Changes in climate are modeled through general circulation models (GCMs), which use equations that represent physical processes in the ocean, atmosphere, ice, and land surfaces to reproduce temperature, humidity, rainfall, wind, and currents over time for a three-dimensional grid over the globe. Although these equations are very similar to those used for weather prediction, GCMs simulate much longer time periods to study climate.

However, GCMs are limited, because some physical processes occur at scales smaller than that of the grid. In addition, even small uncertainties, such as in cloudiness, can have a large impact over the long periods simulated. Generally GCMs reproduce precipitation with less accuracy than they simulate temperatures.[40] Most models project that in a warmer world, wet areas will become wetter, whereas dry areas will become drier.[41] Using coarse resolution GCMs to assess impacts on regional scales is particularly challenging for small islands because they are frequently smaller than the model grid size; the model assumes that

39 N. L. Mimura et al., *Small Islands*, *in* Climate Change 2007: Impacts, Adaptation and Vulnerability. Contribution of Working Group II to the Fourth Assessment Report of the Intergovernmental Panel on Climate Change 687, 691 (M. L. Parry, O. F. Canziani, J. P. Palutikof, P. J. van der Linden & C. E. Hanson, eds., 2007).

40 *Id.* at 694.

41 John E. Hay et al., Climate Variability and Change and Sea level Rise in the Pacific Islands Region: A Resource Book for Policy and Decision Makers, educators and Other Stakeholders, South Pacific Regional Environment Programme vi (2003), *available at* http://www.sprep.org/climate/doc/documents/Prelims.pdf.

the surface is ocean, instead of land.[42,43] The ENSO signal is a dominant mode of variability for the Pacific; projections of ENSO frequency and intensity vary between models.

Despite the uncertainties of applying GCMs for small island nations, the IPCC AR4 projects increased air temperature for all seasons and increased frequency of the most extreme temperatures.[44] These increases will most likely be accompanied by more variable precipitation patterns.[45] Models from the Third Assessment Report of the IPCC predict fewer annual rainy days and increased daily precipitation.[46] Jones et al. found that the 1 in 10-year precipitation event (100 mm/day) would occur every 5 years by 2060 for the tropical western Pacific.[47] The combination of these changes in the frequency and intensity of precipitation suggests that both droughts and floods could become more frequent in small island States.[48]

The effects of anthropogenic climate change on extreme weather events, such as hurricanes and cyclones, are less certain. Although models disagree on the number and tracks of storms, they generally project increased severity of tropical cyclones in a warmer climate.[49,50] Recent results show a consensus on the storms' greater intensity after 2100; the level of agreement is particularly good in the western Pacific, where intensity is projected to increase by ~20%.[51] This trend seems to have

42 R. N. Jones, *An Analysis of the Effects of Kyoto Protocol on Pacific Island Countries*, South Pacific Regional Environment Programme 17 (1999).

43 J. H. Christensen, et al., *Regional Climate Projections, in* CLIMATE CHANGE 2007, *supra* note 3, at 848, 916.

44 Mimura et al., *supra* note 39, at 694.

45 *Id.*

46 L. A. Nurse, et al., Small Island States, *in* CONTRIBUTION OF THE SECOND WORKING GROUP TO THE THIRD ASSESSMENT REPORT OF THE INTERGOVERNMENTAL PANEL ON CLIMATE CHANGE 845, 852 (R. Payet & S. Saeed, eds., 2001).

47 Jones, *supra* note 42, at 17.

48 Mimura et al., *supra* note 39, at 694.

49 *Id.* at 695.

50 Hay et al., *supra* note 41, at vi.

51 Thomas R. Knutson et al., *Tropical Cyclones and Climate Change*, 3 NATURE GEOSCIENCE 157, 161 (2010).

already started: Webster et al. found that the percentage of Category 4 or 5 hurricanes in the western Pacific increased from 25% in 1975 to 1989 to 41% from 1990 to 2004.[52]

The ocean is warming as a consequence of climate change. In the Pacific, the strong increase in sea surface temperature in the equatorial Pacific has lessened in recent years: the western Pacific is cooler and the eastern Pacific warmer. This change is consistent with more prolonged and stronger El Niños.[53] In addition to warming, the ocean is absorbing atmospheric carbon dioxide and acidifying. Ocean acidification (discussed in greater detail in Chapter 15) affects the chemistry of carbonate and negatively affects the rates of calcification of marine organisms with shells or corals.[54] Healthy coral reefs are critical for atoll islands in the context of sea level rise because reefs are physical barriers to currents, waves, and storm surges and are sources of sediment for island accretion.

7. Vulnerability and Resilience of Small Island States

The IPCC has concluded that low-lying small island States are among the countries in the world most vulnerable to the effects of climate change. This vulnerability is a function of both their high exposure to climate change and their limited capacity to adapt.[55] Those islands with minimal elevation above sea level, especially atolls, are particularly vulnerable to the effects of storms and flooding. Small island States share physical, geographic, and socioeconomic characteristics that render them particularly vulnerable to sea level rise and other impacts of climate change. The high ratio of coastline to land area puts large parts of the population at risk to a single event or condition, such as a storm or eroded shoreline.[56]

52 P. J. Webster et al., *Changes in Tropical Cyclone Number, Duration, and Intensity in a Warming Environment*, 16 Science 1844, 1846 (2005).

53 Lemke et al., *supra* note 12, at 420.

54 Woodshole Oceanographic Institute (WHOI), *Frequently Asked Questions about Ocean Acidification*, http://www.whoi.edu/OCB-OA/FAQs/ (last visited June 2, 2011).

55 Nurse et al., *supra* note 46, at 865.

56 Patrick D. Nunn, *Responding to the Challenges of Climate Change in the Pacific Islands: Management and Technological Imperatives*, 40 Climate Res. 211, 211 (2009).

This risk is exacerbated by high population density and a high concentration of population and physical infrastructure at the coasts.[57] Those islands with minimal elevation above sea level, especially atolls, are particularly vulnerable to the effects of storms and flooding. These States often already face limitations on freshwater resources, which are especially troubling given potential shifts in precipitation patterns in a warming world.[58] Heavy reliance on rainwater and a freshwater lens means that the supply of freshwater is vulnerable to variations in those conditions.

Finally, socioeconomic characteristics common to many small island States can limit their ability to adapt to climate change or other environmental hazards, such as cyclones, storm surges, or tsunamis.[59] Small island States are often relatively isolated geographically, tend to rely on a small resource base, frequently are subject to rapid population growth, and can have limited financial, technical, and institutional capacity.[60,61]

8. Impacts of Climate Change on Small Island States

The many anticipated effects of climate change and sea level rise on small islands are expected to negatively influence the physical, economic, and social systems of small island States.

8.1. Temperature

Increases in atmospheric temperature are expected to harm several aspects of terrestrial ecosystems and human health; these impacts include

57 Nurse et al., *supra* note 46, at 867; Mimura et al., *supra* note 39, at 695.

58 Mimura et al., *supra* note 39, at 695.

59 Nurse et al., *supra* note 46, at 867.

60 United Nations Framework Convention on Climate Change (UNFCCC), *Climate Change: Small Island Developing States* 26–27 (2005), http://unfccc.int/resource/docs/publications/cc_sids.pdf.

61 John Barnett & W. Neil Adger, *Climate Dangers and Atoll Countries*, 61 Climatic Change, 321, 322 (2003); Nurse et al., *supra* note 46, at 866.

decreased agricultural yields[62] and increased probability of extreme thermal stress during summers.[63] In addition, warming is expected to lower biodiversity as the ranges of certain species change, leading to an increase in the incidence of non-indigenous invasive species, especially on mid- and high-latitude islands.[64] As heat waves increase in frequency and intensity, and as vector-borne diseases spread, human health will suffer.[65]

8.2. Sea Level Rise

A rise in average sea level will increase the likelihood of erosion, saline intrusions into groundwater, and risk of flooding. Most human settlements and much economic activity are concentrated at or near the coast; erosion of these areas will therefore have broad socioeconomic implications.[66] In atoll nations, most or all of the land is on or near the shore of either the ocean or the lagoon; the width of the islands (i.e., the distance between the ocean and the lagoon) is often in the tens of feet. Shoreline erosion puts many aspects of small island States at risk, including permanent dwellings, infrastructure, beaches, and agricultural lands.

It should be noted, however, that the effects of sea level rise on shorelines are complex: in particular, reef islands are dynamic structures and can experience accretion as well as erosion. In general, isolated events (such as storms) and locally determined processes of sediment transport are the most important factors in whether an island experiences accretion or erosion;[67] the gradual effects of average sea level rise are less influential.[68] That said, it is clear that these islands will undergo geomorphic change as a result of sea level rise and that erosion is a general

62 Nunn, *supra* note 56, at 214.
63 Nurse et al., *supra* note 46, at 852.
64 Mimura et al., *supra* note 39, at 689.
65 UNFCCC, *supra* note 60, at 15.
66 Nunn, *supra* note 56, at 212–13.
67 Colin D. Woodroffe, *Reef-Island Topography and the Vulnerability of Atolls to Sea Level Rise*, 62 Global & Planetary Change 77, 90 (2008).
68 *Id.*

concern in the region, although the degree and magnitude of change will vary from island to island.[69] In this light, erosion should be considered in conjunction with a range of other factors that result in geomorphic change, including land use.[70]

In addition to contributing to erosion, rising sea levels are expected to cause saltwater to penetrate the groundwater, which is usually limited to a freshwater lens. Freshwater lenses are formed on atolls as freshwater floats atop denser salt water;[71] they are often narrow and found just below the surface. On the Marshall Islands, for example, the lenses are accessible only 25–200 cm below the surface.[72] A rise in sea level is expected to turn the water brackish and further compromise an already strained supply of freshwater.[73] Population migration from the outer islands to Majuro, the Marshall Islands' capital island, is further straining freshwater supply there, and captured rainwater runoff from the airport runway has become a major source of potable water.

Sea level rise is also expected to contribute to more frequent and severe flooding of low-lying and coastal areas.[74,75] This flooding will be particularly harmful for the infrastructure, settlements, and agricultural lands that lie at or near the coast of most islands and atolls.[76]

Finally, damage to coastal ecosystems as a result of sea level rise, especially mangrove forests and coral reefs, is of serious concern to small islands. Although a healthy reef or mangrove may not be threatened by sea level rise per se (because healthy coral, for example, can grow up to

69 Arthur P. Webb & Paul S. Kench, *The Dynamic Response of Reef Islands to Sea Level Rise: Evidence from Multi-Decadal Analysis of Island Change in the Central Pacific*, Global and Planetary Change 12 (2010).

70 *Id.* at 11–12.

71 Barnett & Adger, *supra* note 61, at 323; Republic of the Marshall Islands (RMI) Environmental Protection Authority, *Initial Communication under the United Nations Framework Convention on Climate Change* 49 (2000), http://unfccc.int/resource/docs/natc/marnc1.pdf.

72 RMI, *supra* note 71, at 46.

73 *Id.*

74 Woodroffe, *supra* note 67, at 90.

75 Mimura et al., *supra* note 39, at 689.

76 Nurse et al., *supra* note 46, at 863.

5–8 mm/y, in keeping with an estimated 4–8 mm/y of sea level rise),[77] a combination of sea level rise with other climatic changes (such as rising sea surface temperature and ocean acidification) as well as nonclimate stressors will collectively reduce the resilience of these ecosystems and put them at risk of serious degradation.[78]

8.3. Precipitation Variability

Longer and more severe periods of drought are of particular concern for small island States, because such periods will reduce both the amount of rainwater collected directly and the recharge of the freshwater lens. Decreased precipitation will jeopardize the supply of freshwater, with negative impacts on human health, agricultural yields, and socioeconomic development. Increased intensity of precipitation is expected to result in decreased soil fertility and general soil degradation, again suggesting negative impacts for agriculture and food security.[79]

8.4. Sea Surface Temperature and Ocean Acidification

Coral reefs, in addition to being key marine ecosystems, play a critical role as physical barriers and as producers of sediment for atoll island accretion. Corals are at siege from large-scale stressors, primarily rising ocean temperatures[80] and the ensuing risk of coral bleaching, as well as ocean acidification,[81] which affects rates of calcification. Coral bleaching is a process by which the algal symbiont of the calcareous coral organism dies or is expelled with an accompanying whitening and loss of color; repeated or sustained bleaching events may lead to death of the coral

77 RMI, *supra* note 70, at 49.

78 Mimura et al., *supra* note 39, at 690.

79 *Id.* at 698.

80 Alasdair J. Edwards, *Impact of Climate Change on Coral Reefs, Mangroves, and Tropical Seagrass Ecosystems*, *in* Climate Change: Impact on Coastal Habitation 214 (Doeke Eisma ed., 1995).

81 Barnett & Adger, *supra* note 61, at 325.

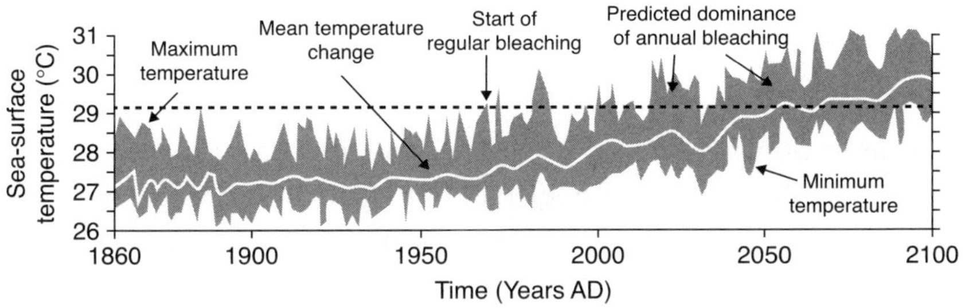

Figure 16. Historical and projected risk of bleaching events for Tahiti, shown through sea surface temperatures from 1860 to 2100. By 2050, projected temperatures suggest that annual bleaching episodes will occur. Reproduced with permission from Inter-Research from Patrick D. Nunn, *Responding to the Challenges of Climate Change in the Pacific Islands: Management and Technological Imperatives*, 40 CLIMATE RES. 211, 213 (2009).

itself. A significant trigger for bleaching is increased temperature. There was a worldwide bleaching event during the 1998 El Niño. The risk of coral bleaching for Tahiti from episodic events has increased since 1970 and is projected to occur annually after 2050 (Figure 16).[82]

In addition to damaging coral reefs, a rise in sea surface temperature is expected to decrease the fish catch[83] as changes in the temperature, salinity, and turbidity of ocean water lead to changes in the distribution and abundance of fish[84] as well as to the fish stock's migratory patterns and depth.[85]

82 Nunn, *supra* note 56, at 213.

83 Murari Lal, Hideo Harasawa, & Kiyoshi Takahashi, *Future Climate Change and its Impacts over Small Island States*, 19 CLIMATE RES. 179, 188 (2002).

84 RMI, *supra* note 71, at 56; IPCC AR4 Ch. 16, p. 689 ("Increasing sea surface temperature and rising sea level, increased turbidity, nutrient loading and chemical pollution, damage from tropical cyclones, and decreases in growth rates due to the effects of higher carbon dioxide concentrations on ocean chemistry are very likely to affect the health of coral reefs and other marine ecosystems which sustain island fisheries. Such impacts will exacerbate nonclimate-change stresses on coastal systems.").

85 UNITED NATIONS FRAMEWORK CONVENTION ON CLIMATE CHANGE (UNFCCC), *Vulnerability and Adaptation to Climate Change in Small Island Developing States* 14 (2007), http://unfccc.int/files/adaptation/adverse_effects_and_response_measures_art_48/application/pdf/200702_sids_adaptation_bg.pdf.

8.5. Extreme Weather Events

An increase in the incidence and severity of extreme weather events, such as cyclones, poses a threat to small islands. In addition, any increases in the intensity or frequency of El Niño events would have significant consequences for small island States, many of whose climates are largely influenced by such large-scale ocean-atmosphere events.[86] Low-lying areas and areas that have suffered erosion and lost the protection of reefs and mangroves will be especially vulnerable to storm damage. In addition to causing damage to infrastructure, buildings, and agricultural lands, storms cause erosion and migration of island land mass, leading to further damage.[87] Extreme storms and the consequent flooding and high winds can cause considerable loss of life.

9. Interactions between Climate Change and Local Human Activities

The impacts of climate change and sea level rise do not exist in isolation from human activities. The threat posed by climate change is especially high for small island States precisely because of the many human activities and systems that will be negatively affected by a changing climate, including water supplies, agriculture, and infrastructure. In addition, many aspects of social and economic development are expected to exacerbate the impacts of climate change. These additional stressors include rapid population growth and increasing population density, damage caused by coastal infrastructure and construction, water overuse and contamination, degradation of coastal ecosystems, and limits imposed by financial, technical, institutional, and socioeconomic capacity.

86 *Id.* at 4.
87 Webb & Kench, *supra* note 69, at 11.

9.1. Population Growth

Two of the primary human factors of vulnerability to climate change on small island States are population growth and increasing population densities. In the RMI, for example, both population growth and increasing population density in the urban center of Majuro (caused in large part by migration from the outer islands) affect environmental health: the estimated population growth rate in 2011 in the RMI stood at 1.954%[88] (exceeding the world average of 1.09%[89]), and around 72% of the population resides in urban areas (compared with 50.5% globally). Population growth and increasing urbanization strain water resources due to overpumping of groundwater, excessive damming in the case of larger island States and delta nations, and increasing pollution.[90] In addition, population growth is expected to threaten food security, utilities, waste disposal facilities, and infrastructure, as well as place a strain on coastal settlements (the destination for much migration).[91]

9.2. Infrastructure and Construction

Most of the critical physical infrastructure of small island States, including international airports and road networks, is found at or near the coasts, rendering them more vulnerable to the average sea level rise of 3.5 mm/y, to storm surges,[92] and to damages from inundation, flooding, and coastal land loss.[93] The risks to transportation infrastructure,

88 Central Intelligence Agency (CIA), *The World Factbook: Marshall Islands*, https://www.cia.gov/library/publications/the-world-factbook/geos/rm.html (last visited June 2, 2010).

89 Central Intelligence Agency (CIA), *The World Factbook: Population Growth Rate*, https://www.cia.gov/library/publications/the-world-factbook/fields/2002.html (last visited June 2, 2011).

90 Mimura et al., *supra* note 39, at 697.

91 *Id.* at 701.

92 *Id.*

93 *Id.* at 703.

including roads, airport, bridges, and port facilities,[94] will affect multiple sectors, such as the delivery of health services, agriculture, food security, and tourism.

Recent construction in some of the Pacific islands has moved away from traditional building methods, which tend to be resilient in the face of storm damage; much new construction is less resilient and is more vulnerable to thermal stress.[95] In addition to the general increase in coastal infrastructure, structures such as seawalls, groins, and causeways have been built along the shorelines of many small island States, leading to a "hardening" of the coastline.[96] Although these developments are intended to prevent erosion and control island structure, they in fact often result in net erosion by interfering with the natural patterns of sediment flows.[97] These structures may therefore actually impede any adaptation that would otherwise naturally occur.

9.3. Water Use and Contamination

Freshwater resources on small island States are generally limited to rain collection and a freshwater lens. This lens is easily contaminated by ocean water and pollution and can be quickly depleted by a lack of rain.[98] Climate change, with concomitant changes in precipitation patterns and sea level rise, therefore represents a serious threat to water resources on small island States. Even slight changes in precipitation or sea levels have a disproportionately large impact on the freshwater lens: a 25% reduction in rainfall and a 50-cm rise in sea level in Tarawa Atoll, Kiribati, for example, is expected to reduce the size of the freshwater lens by 65%.[99,100]

94 *Id.* at 702.
95 *Id.* at 701.
96 RMI, *supra* note 71, at 48.
97 *Id.*
98 Barnett & Adger, *supra* note 61, at 322.
99 Mimura et al., *supra* note 39, at 697.
100 RMI, *supra* note 71, at 46.

Water resources are already stressed by growing demand on many islands, primarily as a result of population and economic growth (especially growth in tourism). In addition to limited quantities of freshwater, waste, due to population growth and urbanization, and runoff from agriculture have the potential to further compromise water quality.[101] Poor water quality, a contributor to water-borne illnesses, is a major factor in the overall human health of small islands.[102]

9.4. Coastal Ecosystems

Mangroves and coral reefs play an important role in the overall health of tropical Pacific islands by protecting against flooding and erosion and forming habitats for the fish and shellfish that provide sustenance and revenue to the islanders.[103,104] These kinds of coastal ecosystems will be less able to successfully respond to climatic changes when they are already stressed by human activities. Local stressors of coral reefs include coastal development, marine pollution, runoff of sediment and nutrients, overfishing, mining, and dynamiting.[105,106]

Because of the important role that coral reefs play in the well-being of these islands, coral reef degradation will have broad socioeconomic impacts, including compromised food security and reduced revenue from tourism.[107] A decrease in fish stocks as a result of lost coral reef habitat could compromise the health of island populations and their economic viability and could lead to additional loss of revenue from tourism.[108,109] In addition, coral reef degradation and ensuing coastal

101 *Id.*
102 Mimura et al., *supra* note 39, at 693.
103 Nunn, *supra* note 56, at 212.
104 Nurse et al., *supra* note 46, at 859, 863.
105 Mimura et al., *supra* note 39, at 699.
106 Nunn, *supra* note 56, at 212.
107 *Id.* at 213.
108 RMI, *supra* note 71, at 48–49.
109 Mimura et al., *supra* note 39, at 698.

erosion will lead to amplified effects from flooding events, net erosion of land (because corals reefs serve as barriers to storm surges and as sediment sources), and the contamination of freshwater aquifers by saline intrusions.[110]

Damage to mangroves, a widespread ecosystem in tropical coastlines, has similarly important ramifications, including impacts on human health, food security, and general economy activity. Although mangrove forests are thought to adapt naturally to changes in climate, this ability is compromised when the forests are under stress due to human activities.[111] Globally, mangrove forests are severely affected by urbanization, agriculture, aquaculture, cutting for timber, and pollution.[112] When the adaptive capacity of forests is compromised, the likelihood of serious degradation as a result of climate change is increased.

9.5. Agriculture and Food Security

Anthropogenic climate change will affect agriculture on small islands in several ways. Changes in precipitation patterns and ensuing variations in the water table are expected to negatively affect agricultural yields.[113] In addition, some islands' staple food crops (including breadfruit, coconuts, pandanus, and taros) are sensitive to salt concentrations; saline intrusion into the groundwater and soil salinization could therefore lead to declines in coastal agriculture production and compromise food security.[114] Finally, the agricultural lands of many small

110 Barnett & Adger, *supra* note 61, at 325.

111 Mimura et al., *supra* note 39, at 706.

112 K. Kathiresan, *Threats to Mangroves: Degradation and Destruction of Mangroves*, *in* United Nations University Syllabus for "Training Course on Mangroves and Biodiversity" 479–81 (2010), *available at* http://ocw.unu.edu/international-network-on-water-environment-and-health/unu-inweh-course-1-mangroves/Degradation-and-destruction-of-mangroves.pdf.

113 RMI, *supra* note 71, at 49.

114 Peter Roy & John Connell, *Climate Change and the Future of Atoll States*, 7 J. COASTAL RES. 1057, 1064 (1991); Mimura et al., *supra* note 39, at 689.

islands, as well as countries like Bangladesh, are concentrated in low-lying areas, along coasts, and in delta plains. These lands and their agricultural yields will therefore be further threatened by rising sea levels and inundations.[115]

For many island inhabitants, fish represents the most significant source of protein intake. Fishery stocks are threatened in multiple ways by climate change (including through changes in sea surface temperature) and ocean acidification, as well as through the loss of habitat as coral reefs and mangrove forests are degraded.[116] Loss of fish stocks threatens economic revenue and tourism as well as nutritional health.

9.6. Financial, Economic, and Institutional Capacity

The United Nations Framework Convention on Climate Change (UNFCCC) estimates that insufficient financial, technical, and institutional capacities hinder the adaptive capacity of small island developing States.[117] The cost of many adaptation measures, especially "hard" engineering measures, is frequently beyond the means of national governments of many small island States; even when such projects can be financed internationally, doing so can risk diverting development aid from other priorities.[118]

The economic vulnerability of small island States to climate change is amplified by their relative isolation from other major markets, which decreases the competitiveness of their exports, and by their heavy reliance on exploitation of natural resources and systems (including forests, fisheries, and beaches) to spur economic growth.[119] This reliance on natural resources, often accompanied by relatively high

115 Roy & Connell, *supra* note 114, at 1057.
116 UNFCCC, *supra* note 60, at 19.
117 *Id.* at 14.
118 *Id.* at 27.
119 Mimura et al., *supra* note 39, at 693.

unemployment levels and income inequality, renders their economies especially vulnerable to natural disasters. In addition, the islands' significant dependence on trade results in greater sensitivity to external pressures and economic shocks[120] such as financial and currency crises.[121]

In addition, many small island States are heavily reliant for national economic growth on tourism, a sector that adds stress to environmental systems and infrastructure.[122] Growing consumption coupled with population growth is anticipated to increase per capita waste production and resource demand, as well as place greater stress on food and water supplies, water quality, and waste disposal systems.[123] Reliance on tourism for economic growth increases economic vulnerability because of the potential climate-related loss of beaches, coral reefs, and infrastructure loss as a result of storms.[124] Additionally, milder climates in North America and northern Europe could decrease the demand for vacations in these primarily tropical islands.[125]

Moreover, small island States are often hindered in their adaptation efforts by limited institutional capacity. Because of an absence of reliable data and a lack of information on the projected changes in climate at the necessary scale, climate change vulnerability assessments have not always been carried out.[126] The wide-ranging impacts of climate change necessitate a comprehensive, integrated approach to adaptation measures, but small island nations possess inadequate resources for effective planning for climate change as they face multiple socioeconomic challenges.[127]

120 UNFCCC, *supra* note 60, at 14.
121 Mimura et al., *supra* note 39, at 693.
122 UNFCCC, *supra* note 60, at 23.
123 RMI, *supra* note 71, at 38.
124 UNFCCC, *supra* note 60, at 23.
125 *Id.*
126 *Id.* at 26.
127 *Id.*

9.7. Health

Finally, the health impacts of climate change provide a clear instance of the interaction between the climate and human systems. Many small island States will experience negative health effects as a result of climate change because most lie in tropical or subtropical zones, and an increase in average temperature is expected to increase the spread of diseases, such as malaria, dengue, filariasis, schistosomiasis, and cholera, via vectors, food, and water.[128] Warmer sea surface temperatures are linked to Ciguatera outbreaks, and increased air temperatures are expected to increase the incidence of heat stress, strokes, and ensuing mortality and morbidity.[129]

This deterioration in public health may be exacerbated by existing weaknesses in the national health systems, in which the rates of many climate-sensitive diseases are in fact already increasing.[130] The lack of resources prevents the health systems of most Pacific Island countries from dealing effectively with existing health problems, and these islands are thus unlikely to be able to respond to the additional stress of climate change.[131]

10. Conclusions

Climate change affects sea level primarily through the warming of the ocean and the melting of mountain glaciers and polar ice sheets. These forms of melting have the potential to significantly raise global sea levels, although regional variations will be significant. Independent estimates of the contribution of polar ice sheets to rising sea levels range from

128 Mimura et al., *supra* note 39, at 701.

129 Jon Barnett, *Dangerous Climate Change in the Pacific Islands: Food Production and Food Security*, *in* Regional Environmental Change S235 (2011).

130 Mimura et al., *supra* note 39, at 701.

131 Barnett, *supra* note 129, at S235.

60–80 cm; combined with thermal expansion from IPCC models, at least 1 m of average global sea level rise by 2100 appears unavoidable. Sea level rise in 2100 in excess of 4 m is unlikely due to constraints in the rates and patterns for ice loss from polar ice sheets. However, actual flooding experienced at any given location is the sum of long-term changes and short-term variability (such as storms and ENSO). By combining tidal ranges and likely interannual variability, most Pacific islands will be subject to more intense and frequent flooding by 2100. In addition to a rise in average global sea level, anthropogenic climate change is expected to raise atmospheric and ocean temperature, contribute to increased precipitation variability with the possibility of more frequent and more severe drought, increase the intensity of extreme weather events such as tropical cyclones, and contribute to greater ocean acidification.

The implications for small island States of rising sea level and other changes in the climate system will be particularly acute because of the physical and socioeconomic vulnerabilities of these nations. In particular, sea level rise is expected to stress the supply of freshwater; increase flooding, inundation, and erosion and/or island migration; decrease agricultural yields and fish stocks; and negatively affect human health. These impacts, in combination with other climatic changes like precipitation variability, extreme weather events, and ocean acidification, will interact with human stressors such as ecosystem degradation and pollution to collectively reduce these islands' capacity to adapt to and thrive in a changing climate.

Part II SOVEREIGNTY AND TERRITORIAL CONCERNS

3 When Do States Disappear?

Thresholds of Effective Statehood and the Continued Recognition of "Deterritorialized" Island States

Jenny Grote Stoutenburg

1. Introduction

The failure of the post-Copenhagen era to produce new meaningful greenhouse gas emission reduction obligations for States renders it more likely that unmitigated climate change and its potentially most destructive consequence, sea level rise, will threaten the sovereign existence of low-lying island States such as the Marshall Islands. This chapter addresses the question of when exactly these sovereign States would cease to exist. Customary international law stipulates four objective criteria of statehood: a State must possess a defined territory, a permanent population, a government, and a certain measure of independence.[1] Small island States whose territory became completely inundated and whose entire population had to relocate to other countries would lose these traditional insignia of statehood. But just like sea level rise itself, inundation and emigration will be a gradual process. The chapter defines the thresholds at which the loss of personal and territorial effectiveness would presumably occur.

1 The 1933 Montevideo Convention on the Rights and Duties of States names as the fourth criterion the "capacity to enter into relations with the other States." Due inter alia to its circularity, this requirement has not become customary international law. *See* Montevideo Convention on the Rights and Duties of States, Dec. 26, 1933, *entered into force* Dec. 26, 1934, 165 L.N.T.S. 19.

Although sea level rise might cause low-lying island States to lose their effective statehood, this does not necessarily mean that they would cease to exist as international legal subjects. In a second line of reasoning, the chapter argues that other States have a moral and perhaps even a legal duty to continue recognizing island States that have lost their effective statehood. This argument draws support from a tension that characterizes the international law of statehood: that between the two competing principles of effectiveness and legality. The principle of effectiveness, on which the first part of the chapter is based, focuses on the factual situation: what matters is whether a State has a territory, population, government, and some degree of independence. The principle of legality in contrast stipulates that the extinction of States must not violate certain fundamental norms of the international legal order, so-called *jus cogens* norms.

The principle of effectiveness evolved as a revolutionary turn away from notions of monarchic legitimacy, pursuant to which inheritance and historic rights rather than actual authority determined a territory's status. In combination with the constitutive theory of recognition that prevailed in the positivist nineteenth century, which gave existing States the power to decide who would join their ranks, the law of statehood served to exclude large parts of the world from the "international community." The principle of effectiveness in contrast was markedly egalitarian: every entity possessing the insignia mentioned earlier was deemed to be a State, irrespective of its recognition by other States, which as a result was accorded only declaratory effect. Yet the increasing consolidation of modern international law in the second half of the twentieth century promoted the realization that effectiveness could clash with fundamental principles of international law. This was true for the creation of territorial entities; for example, the South African Bantustans or "homeland States" that were established for the sole purpose of furthering the policy of apartheid. But more importantly in the present context, it also applied to the extinction of States, as evidenced by the example of States rendered ineffective through belligerent occupation or illegal annexation in violation of the prohibition of the use of force, such as Kuwait in 1990.

When serious violations of fundamental international norms are involved in the creation or extinction of States, the principle of legality supersedes the principle of effectiveness, resulting in the denial of the status of statehood to effectively existing entities and the continued granting of the status of statehood to entities that have effectively disappeared. Thus, if a State is *created* in violation of fundamental norms, other States have a legal duty not to recognize it as a State. This is what happened with the Bantustans, but also for example with the Turkish Republic of Northern Cyprus, which was created through the unlawful use of force. If a State is, however, *extinguished* in violation of fundamental norms, other States have a legal duty not to recognize the new effective situation, but to continue recognizing the legal personality of the State that has factually disappeared. This was practiced with the Baltic States, which reemerged after more than 50 years of Russian annexation, or with Kuwait, which survived the annexation by Iraq.

The extinction of sovereign island States due to anthropogenic climate change also affects fundamental norms of international law, such as the right to self-determination and permanent sovereignty over natural resources. But although these *jus cogens* norms will be *affected* by the disappearance of small island States, meaning that their exercise will be factually impaired, it is difficult to establish that States actually *breach* these norms in the technical sense of the law of State responsibility. The multicausal and protracted nature of climate change, as well as the weak obligations contained in the climate regime, make it difficult to assign legal responsibility for the inundation of an island State to one or even several States. Yet even if a legal duty of continued recognition cannot be established, considerations of international justice and solidarity provide if not a legal, then at least a moral imperative for the continued recognition of "deterritorialized" island States.

2. Thresholds of Effective Statehood

This section identifies the thresholds at which an island State would be deemed to have lost the traditional statehood criteria of defined territory, permanent population, government, and independence.

2.1. Defined Territory

Although possession of a defined territory is generally identified as a fundamental prerequisite of statehood, international law does not prescribe a minimum area necessary for a State to exist on. Sea level rise could render the already minute territory of low-lying atoll States such as the Marshall Islands, the Maldives, Tuvalu, or Kiribati uninhabitable, reduce it to the dimensions of a rock in the sense of Article 121(3) of the UN Law of the Sea Convention (LOSC),[2] turn it into a low-tide elevation in the sense of LOSC Article 13(1), or even submerge it completely. In the first two scenarios, the territorial effectiveness criterion would per se remain fulfilled, because uninhabitable islands and even rocks qualify as land. The case would be different if only a low-tide elevation remained of the original island. In the *Qatar/Bahrain Case* of 2001, the International Court of Justice (ICJ) refuted Bahrain's contention "that low-tide elevations by their very nature are territory," finding instead that the existing rules "do not justify a general assumption that low-tide elevations are territory in the same sense as islands."[3] Arguably, this holds true even

2 United Nations Convention on the Law of the Sea, Dec. 10, 1982, *entered into force* Nov. 16, 1994, 1833 U.N.T.S. 397 [hereinafter LOSC].

3 Case concerning Maritime Delimitation and Territorial Questions between Qatar and Bahrain (Qatar v. Bahrain) (Merits), Judgment, I.C.J. Rep. 2001, 40, at 100 et seqq. (Mar. 16). The ICJ confirmed this approach in the Malaysia/Singapore Case of 2008, in which it held that sovereignty over the disputed low-tide elevation of South Ledge belonged to the State in whose territorial sea it was located. *See* Case concerning Sovereignty over Pedra Branca/Pulau Batu Puteh, Middle Rocks and South Ledge (Malaysia v. Singapore) (Merits), Judgment, I.C.J. Rep. 2008, 12, at 99 et seqq. (May 23).

when the low-tide elevation is built up by a lighthouse or similar artificial installation.[4]

Yet even in the instances where some land persists above water, it is not certain that this would suffice to fulfill the defined territory criterion. Most likely, the island State would become uninhabitable long before it became completely submerged. A functional or teleological interpretation of the statehood definition, however, suggests that uninhabitable land does not qualify as State territory. After all, the postulate of a defined territory was not introduced in the statehood definition as an end in itself, but as a means to an end: it is the physical basis that ensures that people can live together as organized communities. Only land that hosts a permanent population can serve its function to delimit societies and their national legal spheres from each other and from the global commons. According to a functional interpretation, only land that is inhabited therefore qualifies as State territory.

Evidently, the situation is different where an existing State owns a remote uninhabited island. Barren pieces of land (and also barren islands) that form part of a larger territorial unit share the legal status of the whole and hence count as territory. What is at issue here is the distinction between the acquisition of territory by existing States and the territory necessary for a State to come into being or, in the present context, to survive. Thus, if the island State maintained territorial existence through the acquisition of substitute territory from another State or through the merger or federation with another State, it could continue to exercise sovereignty over its remaining island territory just as other States do with regard to remote and uninhabited islands, although in the last two scenarios this right would inure to the State with which the island State merged or to the new State created by the fusion. In the absence of such survival strategies, however, the island State would cease

4 *See* Robert Lavalle, *Not Quite a Sure Thing: The Maritime Areas of Rocks and Low-Tide Elevations under the UN Law of the Sea Convention*, 19 Int'l J. Mar. & Coastal L. 43, 53 (2004).

to satisfy the territorial prerequisite for statehood once an uninhabitable piece of land was the only *terra firma* left.

To retain a State territory, it is thus of importance for island States to prevent at least some part of their island from becoming uninhabitable. This aim could be achieved by protecting the land through the building of seawalls or similar "hard" defense structures. "Soft" protection measures, such as beach nourishment or the artificial elevation of the island with material procured from the ocean floor or other (abandoned) islands, are another option (for more information on options for protecting islands against sea level rise, see Chapter 2). Although there might exist some controversy in extreme cases, where the island would have disappeared were it not for the artificial protection, it is generally agreed that the artificial protection of naturally formed island territory does not turn an island into an artificial island in the sense of LOSC Article 60(8),[5] given that the island State would not try to *generate* maritime entitlements through artificial means, but only aim to *preserve* its already recognized rights. But even if the island State decided to build a completely new artificial island, as the Maldives have done with the island of Hulhumalé, this artificial island would still constitute territory. A look at the genesis of the artificial islands regime, from the 1930 Hague Codification Conference to the International Law Commission's draft for the first Law of the Sea Conference in 1958, reveals that historically, artificial islands were always assimilated to natural islands when it came to their quality as territory, although their capacity to generate maritime zones was eventually abrogated at UNCLOS I in 1958 (for more information

5 *See*, *e.g.*, A. H. A. Soons, *The Effects of a Rising Sea Level on Maritime Limits and Boundaries*, 37 Neth. Int'l L. Rev. 207, 222 8 (1990); Clive R. Symmons, *Some Problems Relating to the Definition of "Insular Formations" in International Law: Islands and Low-Tide Elevations*, IBRU Maritime Briefing Vol. 1 No. 5, Durham, 3 (1995); David Freestone, *International Law and Sea Level Rise*, *in* International Law and Global Climate Change, 109, 113 (Robin Churchill & David Freestone eds., London et al., 1991). The distinction is important because artificial islands do not generate any maritime zones but are only entitled to a safety zone of generally not more than 500 meters around them (*see* LOSC, *supra* note 2, at Article 60).

on maritime zones, see Chapters 5–7). Although its status under the law of the sea might differ from that of a natural island, even an artificial island would therefore count as "defined State territory."

The situation would be more complicated if the small island State tried to replace its territory with an artificial installation. President Anote Tong of Kiribati, for example, has considered buying so-called floating islands from Japan on which to relocate the I-Kiribati.[6] These floating islands would be qualified as artificial installations rather than artificial islands, because they would not be heaped on the seafloor, but would only be connected to the seafloor through stilts, moorings, or similar devices. Such artificial installations were historically never assimilated to islands, and they do not qualify as territory. However, it seems likely that if an inundated island State claimed a floating island as its new territory, other States would acquiesce in treating the artificial substitute as territory on grounds of fairness and equity, as long as the island State did not claim any additional maritime zones.

2.2. Population

If land only counts as territory for the purposes of the statehood definition when it harbors a permanent population, it becomes essential to define the threshold at which a group of people can no longer be qualified as such.

An inductive approximation of a minimum quantitative standard can be derived by comparison with the smallest non–self-governing territory that has been offered independence by the UN. The Pacific island territory of Pitcairn harbors a permanent population of about fifty people. In numerous UN General Assembly resolutions, the UN has acknowledged the right of the Pitcairn inhabitants to become a sovereign State in

6 Anote Tong, President of Kiribati, Speech at the Fifth Global Conference on Oceans, Coasts, and Islands, hosted by UNESCO and the government of France in Paris, France (May 3, 2010).

exercise of their right to self-determination.[7] Although the analogy is slightly lopsided, referring as it does to a case of State creation instead of continuation, it is relevant that a group of fifty people has been internationally recognized as forming a – potential – State population. In a similar vein, Jon van Dyke argues that a stable community of around fifty people is required to distinguish islands in the sense of LOSC Article 121(1) from uninhabitable rocks in the sense of LOSC Article 121(3).[8]

However, in addition to this quantitative standard, there are also qualitative criteria that characterize a permanent population. This has been confirmed by a German administrative court that dealt with the statehood of the "Principality of Sealand," a former World War II sea fort located in international waters off the coast of Great Britain on which a number of individuals claimed to have erected a new sovereign State.[9] The court rejected the submission by a German national that he had lost his citizenship by acquiring that of the Principality, holding that the Principality could not have validly bestowed nationality on the claimant because it did not qualify as a State for want of a defined territory and a permanent population. The court acknowledged that the 106 persons claiming to be "Sealand nationals" could in principle constitute a population, given that international law does not prescribe a minimum number of inhabitants. It nonetheless held that the purported State lacked this statehood criterion because its alleged subjects did not lead a communal life. According to the court, a permanent population must not only have a basic infrastructure at its disposal providing for the vital necessities of human life, but it must also have the will to live together as a community that jointly masters all aspects of communal existence.

7 *See, e.g.*, U.N.G.A. Res. 2869 (XXVI) (Dec. 20, 1971).

8 *See* Jon M. Van Dyke & Robert A. Brooks, *Uninhabited Islands: Their Impact on the Ownership of the Oceans' Resources*, 12 Ocean Dev. & Int'l L. J. 265, 286 (1983); Jon M. Van Dyke, Joseph R. Morgan, & Jonathan Gurish, *The Exclusive Economic Zone of the Northwestern Hawaiian Islands: When Do Uninhabited Islands Generate an EEZ?*, 25 San Diego L. Rev. 425, 437 (1988).

9 Verwaltungsgericht Köln [VK] [Administrative Court of Cologne], May 3, 1978, Case No. 9 K 2565/77, Deutsches Verwaltungsblatt 510, at 511 et seq., 1978 (Ger.).

Given that only some people were permanently on the fort in order to guard its facilities while the other "nationals" restricted their presence to sporadic visits, that the sea fort offered casinos and entertainment facilities but little else, and that the purported Sealand nationals had acquired their new nationality mainly to pursue certain economic and fiscal interests, the court found this qualitative element to be missing.

Applied to the case of "disappearing" island States, this means that as long as enough of the original island infrastructure subsisted to enable an – even small – number of people to continue living on the remaining island territory as a community (if necessary with outside help, because the examples of present-day small island States and territories also show that communities do not have to be self-sufficient in all respects), this community could form a "population nucleus" capable of upholding the personal element of statehood. This population nucleus would serve as legal anchor to the probably much larger diaspora of island people, enabling the relocated citizens to visit the remnants of their home country and providing them protection from statelessness.

However, if an island State can maintain a population in the true sense of the term by enabling a population nucleus to remain on some part of the island, this also implies that if some islanders stayed on the residual island territory not because they regarded it as the territorial basis of their daily lives, but to carry out more narrowly circumscribed tasks such as the maintenance of certain facilities (for instance a lighthouse) or the pursuit of certain economic projects (such as the exploitation of natural resources), they would objectively no longer count as a population. Although the presence of some caretakers might suffice to prevent the island from being downgraded to a "rock" in the sense of LOSC Article 121(3),[10] the personal component of effective statehood would be lost.

10 Note, however, that many authors are of the view that to qualify as sustaining human habitation, an island must harbor permanent (civilian) inhabitants, not just research, military, or government personnel.

On the other hand, it is unlikely that other States would declare the island State nonexistent as long as it still had a government and some caretakers remained on the island. In fact, with the State of the Vatican City, there already exists an example of a State with a pure "caretaker population." The inhabitants of the Vatican City live there only because and as long as they hold office with the Holy See. Yet although the Vatican does not possess "a human society stably united in its territory,"[11] it is recognized by other States as a State. The example of the Vatican therefore proves that a territory that does not fulfill the population requirement can nevertheless qualify as a State if recognized as such by other States. Whereas in the case of the Vatican City, the reasons for this recognition are to be found in its special relationship with the Holy See,[12] States would likely be prompted to accord continued recognition to small island States that have lost their population due to sea level rise for reasons of international justice and solidarity. It can hence be concluded that even if the island State no longer fulfilled the population requirement, it would probably continue to exist as long as it kept the semblance of a population by ensuring that some people stayed on the remaining territory. The recognition by other States would in this case be constitutive instead of only declaratory, meaning that it would not acknowledge the fulfillment of the objective statehood criteria but would instead preserve the statehood of the island State in the sense of a legal fiction.

2.3. Government

A State as an effective legal entity needs to possess a government capable of prescribing, implementing, and enforcing governmental authority vis-à-vis the State's subjects (internal dimension) and of executing its obligations under international law (external dimension). Although

11 *See* J. Duursma, *Fragmentation and the International Relations of Micro-States – Self-Determination and Statehood*, Cambridge, at 411 et seq. (1996). This is a book and should be formatted accordingly.

12 *Ibid.*, at 416 et seq.

historically the introduction of the requirement of governmental effectiveness had marked the departure from notions of monarchic dynastic legitimacy independent of actual territorial control, in the second half of the twentieth century the principle of effectiveness itself underwent important modifications caused by the process of decolonization and the right to self-determination. The requirement of effective governmental authority was qualified to account for new realities in the newly emerging States, and colonial entities possessing the right to self-determination were recognized as States before they had reached traditional levels of governmental effectiveness. This loosening of the effectiveness standard for the newly emerged States was due to the importance attached in international law to an entity's right to exercise authority over a given territory. Where a claim to independence and autonomous governmental control is considered legitimate, as was the case for the former colonial entities, the right to exercise authority takes precedence over the actual ability to exercise this authority.[13]

The same considerations would likely play a role with regard to island States threatened by sea level rise. These island States, having themselves emerged from colonization, possess the uncontested right to exert control over their territories, the actual exercise of which is however in danger of being frustrated through external circumstances. Although what is at issue here is a case of State extinction and not creation, the fact that small island States have a legitimate right to govern their territory would likely influence the standard of effectiveness applied to their continued existence. Consequently, as long as a remaining "population nucleus" maintained some form of political organization through a governing organ that possessed the authority to impose and enforce obligations on the population and that could function as a contact point for international relations, a small island State threatened by sea level rise

13 *See* J. Crawford, The Creation of States in International Law (2d ed., Oxford, 2006), at 57. This again attests to the influence of considerations of legality on the definition of statehood.

would still fulfill the requirement of governmental effectiveness necessary for statehood.

One can also imagine a scenario where the stressors of climate-change-induced displacement, especially if such displacement was precipitated by a sudden-onset disaster like a major flood, would lead to the breakdown of governmental order in the island State. In this case, it could be argued that the State, even if it still featured some territory and a remaining population, would cease to exist for want of what some authors consider the single most important criterion of effective statehood: governmental effectiveness. But in view of the fact that even so-called failed States – in which internal conflicts and power struggles have undermined State institutions to the point that the internal functions of government are suspended and external relations rendered impossible – continue to exist as States, the same would likely apply to a disordered island State, especially because its governmental collapse would not (only) be caused by internal factors but be precipitated by externally imposed climate change. Unlike failed States however, a threatened island State could not ensure continued international visibility through the mere presence of its (ever diminishing) territory. To compensate for the loss of territorial effectiveness, it would therefore need to maintain a government that could defend its interests in the international arena.

Should the impacts of climate change render the maintenance of organized governmental institutions on the island itself impossible, the government of the island State could attempt to establish itself in a host State as a government in exile. With the concept of governments in exile, international law accepts governments that are detached from the requirement of territory, or at least locality. Traditionally, governments have been driven into exile either by external causes such as belligerent occupation or illegal annexation, or because of internal circumstances such as civil war or revolutionary changes. In the first case of unlawful external intervention, the continued international recognition of the occupied or annexed State and its government is mandated by considerations of legality due to the *jus cogens* nature of the prohibition of the use of force. This international legal protection of continued agency is missing in the second case of internal upheavals, because States are free to choose their own form of government and the recognizing States risk interfering in the internal affairs of the – effective – in situ authority that has taken the ousted government's place. In both cases however, a government in exile that has been de jure recognized as continuing to represent its State maintains far-reaching competences, such as the power to conclude treaties and perform unilateral acts on behalf of its State, entertain formal diplomatic relations, represent its State in international fora, dispose of State property abroad, exercise personal jurisdiction over its nationals abroad, and protect them inter alia through the provision of consular representation, the lodging of protests, and the provision of passports – in short, the government in exile maintains all the powers that the de jure government of a State can exercise extraterritorially. The reason for this is that governments in exile do not constitute some sort of *sui generis* subject of international law with a distinct legal personality, but remain an organ of the State they represent. Yet this also means that governments in exile depend for their existence on the continued existence of their State.

The recognition of an island government as a government in exile would pose no legal problems as long as a "population nucleus" remained on the island (i.e., the criteria of population and territory were still effectively met). Although the other States would have to disregard the fact that generally, governments in exile are seen as a temporary institution, whereas in the case of the island State the momentum would not be toward an eventual return home, but toward permanent diaspora, the recognition would not create a new *sui generis* subject of international law. Although the rationale for accepting the government as a government in exile would be new, the legal structure would be the same as in the traditional cases, because the government would continue to represent an existing State.

The situation would become more complicated once the impacts of climate change forced the last population nucleus to leave the remaining island territory. The recognition of a government in exile without a State to which it belongs would bestow international legal personality on an organ of a legal subject that no longer exists. A limited precedent for this kind of action dating from World War I can be found in the recognition of certain émigré organizations, such as the Czechoslovak National Council or the Polish National Committee, as "authorities in exile" before the State they purported to represent even existed as such. Yet in contrast to actual governments in exile, which enjoy a fixed set of competences and maintain their sovereignty even while abroad, authorities in exile can exercise only delegated authority and enjoy only those competences that the host State is willing to grant them. It is therefore questionable whether the island government could still be considered to be independent under such a scenario.

However, this dilemma would be avoided if the other States continued to recognize the "deterritorialized" island State *as a State* even after the departure of the last population nucleus. In this case, the island State's government would continue to represent its State (maintained alive through constitutive recognition) and could keep exercising all those competences that "real" governments in exile enjoy (for more information on possible mechanisms for a nation ex situ, see Chapter 4).

2.4. Independence

The 1933 Montevideo Convention on the Rights and Duties of States names as the last statehood criterion the "capacity to enter into relations with the other States." Because, however, *legal* capacity to this effect depends precisely on recognition of an entity as a State, rendering the criterion circular, whereas the *factual* capacity is already covered by the external dimension of governmental effectiveness, most authors and international jurisprudence rely not on the formulation employed by the Montevideo Convention, but require an entity's independence as proof of its statehood. Just as does the capacity criterion, the notion of independence has a factual and a legal component. A State is factually independent if it is essentially self-sufficient; this self-sufficiency is in turn contingent on its size and the resources at its disposal. Factual independence is therefore inherently contingent on the existence of a territorial basis. Legal independence in contrast means that a State is not subject to the legal authority of another State. Reduced to its legal component, the notion of independence loses its naturalistic overtones and becomes a conventional notion that can be awarded regardless of an entity's territorial status. It is disputed whether both the factual and the legal component must be present for a State to be truly independent.

A number of authors require States to possess both factual and legal independence. This approach has led some commentators to question the independence of island "micro-states" without a large territorial basis. The problem would come to a head if sea level rise reduced the island State to a residual territory whose remaining population nucleus was dependent on outside help to ensure its viability. Even though the other statehood criteria might still be effectively fulfilled under such a scenario, according to this opinion, the island State would lose its independence once its survival became reliant on foreign aid.

Most authors, however, hold the view that only legal independence is a criterion for statehood. Under this view, the island State would not lose its independence even if its diminished territory prevented it from upholding its previous level of self-reliance, as long as it did not subject its legal order to that of another State. This would even be the case if its government established itself in exile, because governments in exile representing an existing State continue to enjoy legal independence, even if the factual exercise of this independence might be restricted in some circumstances by the territorial sovereignty of the host State. Yet even under this more generous interpretation of the last statehood criterion, independence would become precarious once the evacuation of the last population nucleus from the remaining island territory would cause the other effective statehood indicia to lapse. As explained earlier, if the island government and the former island population continued to exist merely as some sort of émigré organization, this organization would no longer be endowed with a circumscribed set of legal rights. Instead, its existence and scope of authority would depend on the goodwill of the host State or States. In this case, it would have lost not only its factual but also its legal independence. Again, this consequence could be avoided if the island State continued to be recognized *as a State* despite being "deterritorialized." This constitutive recognition, as a voluntary act by other States, would bestow statehood, and thus legal independence, to the State that has factually ceased to exist. Legal independence would then indeed become a consequence of, rather than a precondition for, recognition. As a result, the fixed set of guaranteed legal competences that are ascribed to governments in exile would continue to inure to the island State's government. Legal independence of "deterritorialized" island States is therefore conditioned by the continued – constitutive – recognition of the island State as a State.

3. A Duty of Continued Recognition Based on Considerations of Legality?

This section turns to the question of whether other States have a legal or at least moral duty to continue recognizing island States that have lost their effective statehood.

3.1. Statehood and Legality

As explained in the introduction, the loss of the effective indicia of statehood does not terminate a State's legal existence if that loss is brought about in serious violation of a fundamental norm of the international legal order. The concept of fundamental international norms received concrete shape and legal form through the notion of *jus cogens*, or peremptory norms of international law. The notion of *jus cogens* was originally confined to the sphere of treaty law. Article 53 of the Vienna

Convention on the Law of Treaties (VCLT)[14] provides that a treaty is void if, at the time of its conclusion, it conflicts with a peremptory norm of general international law. The same article defines peremptory norms as "a norm accepted and recognized by the international community of States as a whole as a norm from which no derogation is permitted and which can be modified only by a subsequent norm of general international law having the same character." However, it is now generally accepted in international legal doctrine and practice that the scope of *jus cogens* extends beyond the law of treaties. Most notably, the concept was included in the International Law Commission's Draft Articles on the Responsibility of States for Internationally Wrongful Acts (DASR) of 2001.[15] Articles 40 and 41 DASR establish particular consequences for a serious breach of an obligation arising under a peremptory norm of general international law. A breach of *jus cogens* is defined as serious in Article 40(2) DASR "if it involves a gross or systematic failure by the responsible State to fulfill the obligation." If such a serious breach occurs, States are inter alia prohibited from recognizing the resulting situation as lawful (Article 41(2) DASR). This duty of non-recognition is central to the application of the concept of *jus cogens* to questions of territorial status: it has resulted in the denial of the status of statehood to effective territorial entities created through fundamentally illegal circumstances and in the continued attribution of statehood to States whose effectiveness has been extinguished in violation of *jus cogens*.

State practice has denied the status of statehood to effective entities created by aggression or the use of force, in violation of the right to self-determination, or through the application of a system of racial

14 Vienna Convention on the Law of Treaties, May 23, 1969, *entered into force* Jan. 27, 1980, 1155 U.N.T.S. 331.

15 U.N. GAOR, International Law Commission, *Draft Articles on Responsibility of States for Internationally Wrongful Acts, with Commentaries*, 56th Sess. Supp. No. 10, 43 (2001).

discrimination. The earliest instance of the first situation concerned the putative State of Manchukuo, forcibly established by Japan on Chinese territory in 1931, which the League of Nations refused to recognize. The Turkish Republic of Northern Cyprus, which came into being in 1983 in the wake of Turkish military intervention and which Turkey is alone in recognizing as a State, is a more recent case in point. The instances of States created in violation of the right to self-determination and through racial segregation are exemplified by the white minority regime established in Rhodesia in 1965 in disregard of the will of the majority of the population, and by the already mentioned South African Bantustans. In all of these cases, the resulting duty of nonrecognition was implemented collectively through resolutions by the political organs of the League of Nations, and subsequently the UN. The ICJ has confirmed the existence of a duty of nonrecognition in two advisory opinions: the *Namibia Advisory Opinion* of 1971[16] and the *Wall Advisory Opinion* of 2004.[17] In a third case, the *East Timor Case* of 1995, the duty had been invoked by Portugal, but the court was prevented from pronouncing on the matter for lack of jurisdiction.[18]

State practice with regard to the continued existence of States whose effectiveness has been extinguished in violation of peremptory norms of *jus cogens* remains confined to cases of States illegally occupied or annexed through the unlawful use of force, simply because so far, the only cases of involuntary State extinction due to foreign influence have resulted from the forcible intervention of other countries. Examples include the States annexed in the period between 1936 and 1940 – among them Austria, Poland, Czechoslovakia, and the Baltic States – which

16 Legal Consequences for States of the Continued Presence of South Africa in Namibia (South West Africa) notwithstanding Security Council Resolution 276, Advisory Opinion, I.C.J. Rep. 1971, 16, at 54 et seqq. (Jun. 21).

17 Legal Consequences of the Construction of a Wall in the Occupied Palestinian Territory, Advisory Opinion, I.C.J. Rep. 2004, 136, at 196 and 199 et seq. (Jul. 9).

18 Case Concerning East Timor (Portugal v. Australia), Judgment, I.C.J. Rep. 1995, 90 (Jun. 30).

nevertheless continued to enjoy recognition and international legal personality. The Baltic States in particular illustrate the principle that loss of effectiveness through illegal annexation does not extinguish statehood, having "reemerged" as the same legal entities after more than fifty years without any territorial control. Likewise, the occupation and annexation of Kuwait by Iraq in 1990 were declared "null and void" by the UN Security Council, which called "for the restoration of Kuwait's sovereignty, independence and territorial integrity and of the authority of its legitimate government."[19] Concomitant with the duty of nonrecognition of the unlawfully established new regime, there therefore also exists a duty to continue recognizing the international legal personality of the occupied or annexed State.

It can be argued that this protection of a State's personality against extermination also extends to the extinction of States in violation of other peremptory norms than the unlawful use of force. A general customary rule to this effect cannot be established due to the lack of State practice. Nevertheless, Article 41(2) DASR extends the duty of nonrecognition to *all* serious violations of *jus cogens*, without distinction as to the particular peremptory norm that has been breached. It seems therefore that this provision expresses a general principle of modern international law according to which there is a duty not to recognize the extinction of a State caused by a serious violation of *jus cogens*, and thus to continue recognizing the international legal personality of the affected State.

The crucial question is then whether the extinction of low-lying island States due to anthropogenic climate change constitutes a serious violation of peremptory norms of general international law. There are indeed several fundamental international norms that would be affected by the disappearance of an island State. The nontechnical term "affected" is used intentionally, because there is a difference between saying that a right has been affected (i.e., that its exercise has been factually impaired)

19 U.N. SC Res. 662 (Aug. 9, 1990) and U.N. SC Res. 674 (Oct. 29, 1990).

and that it has been violated through a breach attributable to another State. After a brief overview of the fundamental norms involved, the final part of this chapter turns toward the question of whether a serious breach of these norms by another State in the technical sense of the law of State responsibility can be established.

3.2. Fundamental Norms Implicated in the Disappearance of Small Island States

The loss of their territorial basis of statehood would first impair the islanders' right to self-determination. Originally perceived as a mere political dogma, the right to self-determination has evolved into an established norm of international law. Not only that, but it has acquired the highest possible status in the rudimentary hierarchy of international law. In the *East Timor Case*, the ICJ held that "Portugal's assertion that the right of peoples to self-determination, as it evolved from the Charter and from United Nations practice, has an *erga omnes* character, is irreproachable,"[20] it being generally understood that this *erga omnes* effect of the right to self-determination stems from its qualification as a peremptory norm of international law.[21] Common Article 1(1) of the two International Human Rights Covenants of 1966[22] stipulates that by virtue of their right to self-determination, all peoples "freely determine their political status and freely pursue their economic, social and cultural development." Climate-change-induced sea level rise will damage island State economies, tear up the social fabric, and uproot island cultures. By destroying the physical basis for the exercise of any kind of activity, the inundation of a small island State's territory would impair the ability of

20 *East Timor Case*, *supra* note 18, at 102.

21 However, there exists some doubt whether the right to self-determination also represents *jus cogens* with regard to the people of sovereign States, as opposed to dependent peoples that have not yet attained statehood.

22 International Covenant on Civil and Political Rights, Dec. 19, 1966, *entered into force* Mar. 23, 1976, 999 U.N.T.S. 171; International Covenant on Economic, Social and Cultural Rights, Dec. 16, 1966, *entered into force* Jan. 3, 1976, 993 U.N.T.S. 3.

the islanders to determine their political, economic, social, and cultural future.

Equally affected will be another fundamental right that is closely related to the right to self-determination, namely the right to permanent sovereignty over natural resources. In addition to being contained in a long line of UN resolutions, this right was codified in the second paragraph of common Article 1 of the Human Rights Covenants, which provides that all peoples may freely dispose of their natural wealth and resources, adding that in no case may a people be deprived of its own means of subsistence. The reference to permanent sovereignty over natural resources in many other multilateral treaties and instruments has led the ICJ to recognize its importance and to declare it a principle of customary international law.[23] In view of the fact that it has frequently been labeled an "inalienable right"[24] and given the respective *opinio juris* of – particularly developing – States, there is good reason to support the qualification of the right to permanent sovereignty over natural resources as possessing *jus cogens* character. In a dispute with Australia before the ICJ, which did not reach the merits stage, Nauru relied on the right as a cause of action to achieve compensation for excessive phosphate mining during Australia's mandate and trusteeship over the island.[25]

This right could also be applied to climate-linked events: if the right to permanent resource sovereignty and ownership of one's means of subsistence was affected by the overexploitation of one particular resource on which the inhabitants of Nauru depended, it could be argued, this right would *a fortiori* be impaired by the destruction of the entire resource base of a country that climate change will bring about. This

23 Armed Activities on the Territory of the Congo Case (Democratic Republic of the Congo v. Uganda), Judgment, I.C.J. Rep. 2005, 168, at 251 (Dec. 19).

24 *See, e.g.*, U.N.G.A. Res. 1803 (XVII), *Declaration on Permanent Sovereignty over Natural Resources* (Dec. 14, 1962).

25 Case Concerning Certain Phosphate Lands in Nauru (Nauru v. Australia) (Preliminary Objections), Judgment, I.C.J. Rep. 1992, 240, at 243 (Jun. 26). *See also* Antony Anghie, *"The Heart of My Home": Colonialism, Environmental Damage, and the Nauru Case*, 34 (2) HARV. INT'L L.J. 445, at 472 et seqq. (1993).

is especially true because the scope of the right extends not only to land but also to marine resources. These marine resources will be lost with the shrinking or disappearance of an island State's maritime zones caused by sea level rise, a consequence that could be prevented only by amending the law of the sea to provide for stable maritime zones.[26]

Third, the enjoyment of other, more individual, human rights will be impaired by the impacts that climate change will have on island populations. In a study on the relationship between climate change and human rights, the Office of the UN High Commissioner for Human Rights (OHCHR) concluded that, in addition to the right to self-determination and permanent sovereignty over natural resources, the rights to life, adequate food, water, health, and adequate housing will be affected.[27] For the populations of island States faced with the loss of their homeland, the right to a nationality as proclaimed in Article 15(1) of the Universal Declaration on Human Rights of 1948[28] might equally be compromised. In contrast, the "freestanding" human right to a healthy environment has not yet entered general customary international law. Of the human rights just named, only the right to life has a recognized *jus cogens* status. It could, however, be argued that the human rights to food, water, health, and housing, whose enjoyment is vital for the realization of the *jus cogens* right to life and which protect the minimum material and social preconditions for human existence, should equally be accorded peremptory force.[29]

26 *See* Jenny Grote Stoutenburg, *Implementing a New Regime of Stable Maritime Zones to Ensure the (Economic) Survival of Small Island States Threatened by Sea-Level Rise*, 26 Int'l J. Mar. & Coastal L. 263 (2011). *See also* Chapter 7 for other proposals on how to preserve maritime zones.

27 Report of the U.N. High Commissioner for Human Rights on the Relationship between Climate Change and Human Rights, U.N. A/HRC/10/61, at paras. 21–38 (Jan. 15, 2009) [hereinafter OHCHR].

28 U.N.G.A. Res. 217 (III) A, (Dec. 10, 1948).

29 Admittedly, this interpretation is not easily compatible with the derogable status of the rights to health, food, water, and housing under Art. 4 ICESC, which allows States to subject them to certain limitations.

Lastly, one other doctrine of international law unrelated to human rights plays a role in the disappearance of small island States. In its *Nuclear Weapons Advisory Opinion* of 1996, the ICJ referred to the "fundamental right of every State to survival."[30] Although it is not entirely clear how this fundamental right relates to the concept of *jus cogens* or even what its exact implications are, it is obvious that it would be affected by the physical disappearance of an island State due to anthropogenic climate change.

3.3. Serious Breach of a Peremptory Norm of General International Law?

The fact that the exercise of these rights will be impaired does not necessarily indicate that they will have been breached in the technical sense of the term. The distinction is highlighted by the already mentioned OHCHR Report, which confirms that climate change will have implications for the enjoyment of human rights, but falls short of concluding that the effects of climate change can be qualified as human rights violations in a strict legal sense (i.e., that a breach of these human rights by another State can be established).[31] Although this reticence might partly be explained by the fact that a different position would have raised the thorny issue of the extraterritorial scope of human rights obligations, it cannot be ignored that efforts to assign State responsibility for the detrimental effects of climate change face many legal challenges. This is particularly so in the present case, where not "only" State responsibility for damages caused by climate change, but a serious breach of the norms of peremptory international law discussed earlier needs to be proved, because only a serious breach of *jus cogens* would entail the duty of continued recognition of "deterritorialized" small island States.

30 Legality of the Threat or Use of Nuclear Weapons, Advisory Opinion, I.C.J. Rep. 1996, 226, at 263 (Jul. 8).

31 *See* OHCHR, *supra* note 27, at para. 70.

An internationally wrongful act in the sense of Article 2 DASR requires two elements: first, the attribution of a certain conduct (which can be an act or an omission) to a State under international law, which must secondly constitute a breach of an international obligation of the State. Regarding the first element of attribution, it could be said that the attributable conduct consists in the emission of greenhouse gases (GHGs) that contribute to climate change. This would render attribution problematic, because these emissions are mainly caused by individuals and privately owned industries and not State organs, and the conduct of private persons is only attributable to a State under narrowly defined circumstances that are not given in this case. However, what is at issue is not the emission of GHGs per se, but the failure of States to regulate it. It is the *omission* of ensuring that GHG emissions within its territory do not exceed the allowed limits that is attributable to a State.

To entail a duty of continued recognition of "deterritorialized" island States, the failure to regulate GHG emissions must constitute a serious breach of the *jus cogens* norms discussed earlier. A breach of these norms by other States could be established if it could be shown that their failure to regulate GHG emissions is directly causal for the disappearance of low-lying island States.

Because only those emissions that States fail to regulate despite an obligation to do so can be taken into account, it is not enough to say that all historic and present emissions of GHGs by States in general cause the inundation of an island State. International law foresees strict State responsibility (i.e., liability for damage caused by lawful behavior) only for certain conventional regimes dealing with inherently dangerous activities where damage is caused not through the normal conduct of activities but by way of accidents, such as the operation of space objects, nuclear energy generation, or oil pollution.[32] In contrast, the emission of GHGs – which in itself is an activity not prohibited by international law and which

32 Only the liability regime for damage caused by space objects provides for strict *State* liability; other liability regimes are generally civil liability regimes that sometimes foresee some form of residual State liability.

is generally seen as not inherently dangerous (although this assessment might change in the future) – only becomes unlawful conduct if it violates certain legal standards of permissible behavior.[33] These standards are provided by the climate change regime – the UN Framework Convention on Climate Change (UNFCCC)[34] with its Kyoto Protocol[35] – as well as by the customary "no-harm rule" (for more information on the responsibilities of States for GHG emissions, see Chapters 8 and 13).

The first step toward a collective response to global climate change, the UNFCCC of 1992 sets the objective of achieving "stabilization of greenhouse gas concentrations in the atmosphere at a level that would prevent dangerous anthropogenic interference with the climate system." In line with the concept of common but differentiated responsibilities, it obliges the developed and newly industrialized countries listed in its Annex I to adopt national policies to mitigate climate change by limiting GHG emissions and protecting and enhancing GHG sinks and reservoirs. Although its "aim" of returning the GHG emission levels of Annex I countries to 1990 levels by the year 2000 is generally not seen as an enforceable legal commitment, the case can well be made that the UNFCCC obliges Annex I countries to reverse their long-term emission trends. Consequently, Annex I countries that have continued to increase their emission levels since their ratification of the UNFCCC can be seen to be in breach of their obligations under the UNFCCC.

Quantified emission limitation and reduction commitments for industrialized States were achieved with the Kyoto Protocol of 1997, which obliges Annex I countries to reduce their overall emissions by on

33 This is already apparent from the definition of attributable conduct given earlier: the failure to regulate GHG emissions can only be defined by reference to a certain standard of allowed emissions. As long as this standard is not zero emissions, which would result in strict State liability for any GHG emissions, conventional or customary emission standards need to be identified.

34 U.N. Framework Convention on Climate Change, May 9, 1992, *entered into force* Mar. 21, 1994, 1771 U.N.T.S. 107.

35 Kyoto Protocol to the U.N. Framework Convention on Climate Change, Dec. 10, 1997, *entered into force* Feb. 16, 2005, (1998) 37 I.L.M. 32.

average 5.2% below 1990 levels in the commitment period covering the years 2008 to 2012. States failing to adhere to their reduction commitments during that time span will be in breach of the Kyoto Protocol. The Copenhagen Accord of 2009[36] in contrast only invites pledges for emission targets from developed countries and for nationally appropriate mitigation actions from developing countries to reach its stated objective of holding global temperature increases below 2 degrees Celsius. Because of their voluntary nature, the Copenhagen pledges do not constitute binding emission standards.

A customary emission standard can be found in the prohibition against causing transboundary environmental harm, also called the "no-harm rule," according to which States have the responsibility to ensure that activities within their jurisdiction or control do not cause damage to the environment of other States or of areas beyond the limits of national jurisdiction.[37] The no-harm rule is not a strict liability standard, but requires States to act with due diligence. States fail to act with due diligence only when they fail to prevent harm that is foreseeable. Most commentators agree that 1990, the year of the first Assessment Report of the Intergovernmental Panel on Climate Change, should be adopted as the date from which the dangers of climate change damage were clearly foreseeable for States. Thus, it could be argued that States that, since 1990, have not undertaken proportionate measures to protect other States and the global commons against the consequences of climate change are in breach of their obligations under the no-harm rule. However, with the UNFCCC and the Kyoto Protocol, States have agreed on a multilateral conventional regime designed specifically to address the problem of global climate change. Although the standards contained in the UNFCCC most likely are too vague to supersede the no-harm rule, the reduction commitments of the Kyoto Protocol could be regarded as

36 Copenhagen Accord, Mar. 30, 2010, Decision 2/CP.15, FCCC/CP/2009/11/Add.1.
37 *Cf.* Nuclear Weapons Advisory Opinion, *supra* note 30, at 242.

lex specialis[38] to the customary obligation. Those States that comply with their Kyoto targets would then be protected from any further reaching claims under the no-harm rule.[39] Moreover, given the fact that the multilateral climate regime exempts developing countries from emission reduction obligations in line with the principle of common but differentiated responsibilities, it seems unlikely that this value judgment could be undermined by holding them responsible for failing to curb emissions under the no-harm rule.[40]

Thus, although it is possible to establish breaches by States of their conventional and customary duties to regulate GHG emissions, the preceding remarks also show that only a small fraction of all GHGs released worldwide has been and is being emitted in violation of legal standards. Annex I States can be held responsible under the UNFCCC for that part of their emissions that exceeds their obligation to reverse their long-term emission trends. Annex I parties that have taken on quantified emission reduction obligations under the Kyoto Protocol can be held responsible for those emissions that exceed the Kyoto targets. The industrialized States that have not taken on obligations under the Kyoto Protocol, essentially the United States, can be held responsible for those emissions

38 The *lex specialis derogat legi generali* ("*lex specialis*") rule solves conflicts between concurrent legal obligations by giving preference to the special obligation over the more general one.

39 On the other hand, the governments of Fiji, Kiribati, Nauru, and Papua New Guinea made the following declaration on signing the UNFCCC: "The Government… declares its understanding that signature of the Convention shall, in no way, constitute a renunciation of any rights under international law concerning state responsibility for the adverse effects of climate change, and that no provisions in the Convention can be interpreted as derogating from the principles of general international law." Similar statements were made by the Cook Islands, Kiribati, Nauru, and Niue on signing the Kyoto Protocol. Still, it is difficult to imagine that the ICJ, for instance, would hold a State that actually complied with its Kyoto targets responsible for violation of the no-harm rule.

40 This might change if, in the wake of the Copenhagen Accord, developing countries start to take on binding reduction commitments, but then the question would arise again whether these commitments concretize the no-harm rule.

exceeding their obligation under the no-harm rule to take proportionate measures to protect others from the negative impacts of climate change. In addition, this responsibility is limited in time: the law of State responsibility cannot be applied retroactively. An act of a State therefore only constitutes a breach of an international obligation if the State is bound by the obligation in question when the act occurs (Article 13 DASR). The obligation of conduct to reverse emission trends under the UNFCCC only applies from the point in time where the convention has become effective for a certain State – at the earliest therefore from 1994 on, the date of its entry into force. The Kyoto Protocol requires Parties to show progress toward meeting its targets from 2005 on; its actual commitment period only extends from 2008 to 2012. And the no-harm rule is limited in its temporal applicability by the requirement of foreseeability, which is generally deemed not to have been fulfilled before 1990. Thus, the large bulk of historic emissions released before the 1990s do not violate any international legal emission standards, because no such standards were in place at the time the emissions occurred.

Even leaving aside the problem of cumulative causation – the fact that not one State alone, but all States together (mainly of course the largest emitters) contribute to climate change – the chances that only those GHG emissions that States have emitted in contravention of their legal duties outlined earlier would cause the inundation of an island State are slight. Yet only the "unlawful emissions" can be taken into account in determining whether a State (or perhaps a group of States[41]) has breached the peremptory norms that are affected by the disappearance of low-lying island States. Moreover, only *serious* breaches of peremptory norms entail the duty of nonrecognition of the resulting situation according to Article 40(2) DASR, with a serious breach being defined as a gross or systematic failure by the responsible State to fulfill the

41 Although some authors have explored the idea of holding States jointly and severally liable for climate change damage, it seems more likely that this concept is not applicable where the establishment of the actual breach of an obligation is concerned, even if this breach, as in the present case, is characterized by the infliction of damage.

obligation. The refusal by many States to commit to scientifically necessary emission reductions can certainly be seen as a systematic denial of the need to address climate change in a meaningful manner. Unfortunately, this strategy is successful, because the nonexistence of stricter international mandatory emission reduction obligations makes it impossible to hold States legally accountable for the consequences of their actions.[42] Because of the laxity of the existing standards, the gross and systematic failure to address the climate change problem does not translate into a gross and systematic violation of international legal obligations.

Yet even if a legal duty of continued recognition of "deterritorialized" island States should founder on these difficulties, States could accept moral responsibility for their disproportionate use of the world's atmosphere and continue recognizing inundated island States out of a sense of international justice and solidarity.

The island State, although no longer possessing the effective insignia of statehood, would then continue to exist *as a State*, meaning that it would continue to benefit from the presumption of sovereign equality of States. This would distinguish it from other nonterritorial so-called special sovereign entities such as the Sovereign Order of Malta, which after the loss of their territory have been reduced to subjects with only relative international legal personality and functionally circumscribed rights. In contrast, the government of the island State would continue to represent its State in international fora on an equal footing with other States and would keep exercising all those international rights and competences

42 *See* Bruce Pardy, *The Kyoto Protocol: Bad News for the Global Environment*, 14 J. Envtl. L. & Pract. 27 (2004), who criticizes the Kyoto Protocol for stymieing meaningful action on climate change, concluding that "[e]nvironmentally, legally, and politically, the world would be better off with no international greenhouse gas agreement at all." For a possible path of action *see* Alexander Gillespie, *Small Island States in the Face of Climatic Change: The End of the Line in International Environmental Responsibility*, 22 UCLA J. Envtl. L. & Pol'y 107, at 129 (2003/04), who suggests requesting an advisory opinion from the ICJ on whether the climate negotiations are being conducted in bad faith.

of States that do not require a territorial basis. Most importantly, it could continue to protect its nationals, who would not be stateless (at least de jure), and it could perhaps also continue to administer the island State's maritime zones, if those were maintained as stable despite sea level rise.

4. Concluding Remarks

If the scientific predictions are correct, at some point in the future low-lying island States might lose their effective statehood. As has been explained in this chapter, this would likely be the case once the last "population nucleus" would have to leave the remaining island territory. Island States seeking to maintain a claim to effective statehood should therefore try to ensure that such a scenario does not occur, if necessary through the artificial fortification or elevation of a part of their homeland that could continue to host a group of people that live together as a community. However, if this should prove to be impossible, claims to continued statehood by threatened island nations based on lesser displays of effectiveness would probably also be accommodated by other States, as long as the semblance of territorial and personal effectiveness was maintained for instance with the help of an artificial installation or the deployment of some caretakers on some residual territory. This would be the case at least as long as the island State's government, if necessary from its new location in exile, claimed continued existence, because it is unlikely that other States would pronounce an island State nonexistent against its will. In contrast, a legal duty to continue recognizing island States that have lost their effective statehood founders on the problem of establishing a serious violation of the affected *jus cogens* norms by other States, because the nature of climate change itself and of the legal regime devised to address it renders the fruitful application of the law of State responsibility exceptionally difficult. However, moral restraints might prevent the other members of the "international community" from sanctioning the death of one of their own that has disappeared as a result of their collective action.

The ascription of full status and participation rights usually reserved for "traditional" States to genuinely nonterritorial entities would require the greatest structural transformation international law has seen since the establishment of the so-called Westphalian order. However, if the international community does not act quickly enough to prevent small island States from losing their effective statehood, the acknowledgment of their entitlement to survive as a legal community is the least that is owed to them.

4 The Nation Ex-Situ

Maxine A. Burkett

1. Introduction

It is plausible that by the close of this century the impacts of climate change will render certain States in the Pacific and Indian Oceans uninhabitable. The geopolitical consequences – not to mention the devastation to the nations and peoples at issue – will be great and unprecedented. These consequences will be significant despite the relatively small number of individuals affected. This loss of land may create a conundrum for international law, which is only equipped to deal with neater circumstances of stateless individuals or expiring States. Optimally, there would be sufficient political will and commitment at the international level to reduce emissions aggressively and work toward the continued habitability of all, although as Professor Carr addresses in Chapter 2, sea levels will rise regardless. Absent any indication of such swift action and given scientific eventualities, international law should now find a suitable resolution for the nation-states that will lack a territory. In response to this phenomenon, I argue that international law

I thank Michael Gerrard, Robert Perkinson, Joshua Stanbro, Gregory Wannier, the participants of the Center for Climate Change Law's Spring 2011 conference, *Threatened Island Nations: Legal Implications of Rising Seas and a Changing Climate*, and the Climate Law editors and referees, for their invaluable feedback. For research assistance, I thank Amy Brinker, Trevor Tamashiro, and Kylie Wager. This chapter is adapted from Maxine Burkett, *The Nation Ex-Situ: On Climate Change, Deterritorialized Nationhood, and the Post-Climate Era*, 2 Climate Law 1 (Fall 2011).

can accommodate a new category of international actors: the Nation Ex-Situ.[1]

Ex-situ nationhood would be a status that allows for the continued existence of a sovereign State, which would be afforded all the rights and benefits of sovereignty among the family of nation-states, in perpetuity. It would protect the peoples forced from their original place of being by serving as a political entity that remains constant even as its citizens establish residence in other States. It is a means of preserving the existing State and holding the resources and well-being of its citizens – in new and disparate locations – in the care of an entity acting in the best interest of its people.[2]

In practice, this new category would require the creation of a government framework that could exercise authority over a diffuse people. Other scholars have suggested a resurrection of the political trusteeship system as a means of administering the duties of a deterritorialized government. Here, I elaborate on those proposals and recommend a possible hybrid structure that provides a permanent space for long-distance, and perhaps collaborative, governance of Nations Ex-Situ.

In this chapter, therefore, I seek to complete two quite different but intimately related tasks. First, I demonstrate that the recognition of a Nation Ex-Situ is a novel but justified interpretation of the State, one that international law could tolerate and incorporate into its existing definitions. To the extent that critics deem this an impossible departure from existing norms, I argue that the exigencies of the soon-to-be-landless

1 In this chapter, I use "nation" or "nationhood" to describe the unified peoples described earlier. I use "State" or "statehood" to describe the entity that is the constitutive element of the international community. The nation-state describes the sovereign State comprised of, or inhabited by, a unified people.

2 Here, I borrow conceptually from the practice of ex-situ conservation, which literally means "off-site conservation," and describes the practice of protecting an endangered population outside of its natural habitat. The analogy is not drawn between species and peoples but between species and this creature of international law – the nation-state – which is an elemental organism of the international law taxonomy.

favor new and creative ways for States to exist in the international community. Second, I demonstrate that a modified political trusteeship system is a viable model for effective participation in the international community, given the scenarios that those displaced by climate change face. A key element of this arrangement would be the continued self-governance of those nations by trustees who are elected by their peers and who act for the benefit of the ex-situ nation. (This role would not be a reprise of their – in some cases – prior trustee relationships.) I then detail integral elements of this arrangement as applied to the peculiar circumstances of endangered States.

The chapter proceeds as follows: after providing a brief background on climate-induced migration and the urgent plight of small islanders and threatened nation-states, Part 2 argues for a reenvisioning of the State, in which a deterritorialized State, or Nation Ex-Situ, is an alternative form of statehood recognized as consistent with, or a natural extension of, existing international law norms. In supporting this assertion, I do three things: I demonstrate that alternative forms of the State are not novel and briefly point to historical examples of entities that provide a precedent. I then identify elements of cosmopolitan theory, as well as the evolving conditions of various diasporas, that show that a landless – or "virtual" – State is appropriate here and is, at the same time, an inevitable evolution of contemporary citizenship and statehood. Finally, I argue that if none of these arguments avails, the extraordinary circumstances of threatened States alone could justify the proposed unique departures from the norm.

Part 3 provides a framework for how the Nation Ex-Situ might operate. I first describe what a trusteeship arrangement would look like for deterritorialized States in transition and the establishment of permanent ex-situ status. I then identify the advantages that would inure to both the new State, as a political entity, and to its diasporic citizens. I also briefly discuss how this kind of arrangement may affect the States that host the deterritorialized citizens of the Nation Ex-Situ. I conclude with an exercise in which I apply the new arrangement to a supposed scenario for the

Republic of the Marshall Islands – one of the atoll nation-states that may ultimately be landless. Playing out the implications of "The Republic of the Marshall Islands *ES*"[3] makes thorny issues of history and sovereignty become clearer.

To conclude, I discuss endangered nations as a quintessential example of the fundamental shifts that may continue to confront the law because of climate change. The legal questions raised regarding the meaning of the "State," as defined by international law, present novel and complex challenges to contemporary jurisprudence. International law operates within the narrow band of principles for statehood and sovereign relationships that marks the contemporary Westphalian organization of the global community. Although there is room for elements of creativity and novel interpretation, there is an expectation of consistency and coherence that climate change fundamentally calls into question.

To give theoretical structure and integrity to this fundamental shift in understanding that the law might undergo, I argue that we are embarking on a new legal and political epoch: the postclimate era.[4] This shift is not unique to international law. International law is simply among the legal fields first in the queue for analysis, deconstruction, and evolution.

3 The Republic of the Marshall Islands *ES* would be the short form used for the ex-situ nation, in which "*ES*" stands for ex-situ.

4 This is parallel to, perhaps, the Anthropocene. Humanity's impact on the Earth's climate and ecosystems has inspired some to suggest that the most recent geological epoch, the Holocene, was abruptly ended by the launch of the Anthropocene during the industrial revolution. *See*, e.g., Elizabeth Kolbert, *Enter the Anthropocene – Age of Man*, Nat'l Geographic, March 2011, *available at* http://ngm.nationalgeographic.com/2011/03/age-of-man/kolbert-text. This shift in nomenclature suggests, then, that humans have had an impact on the global environment comparable to a new ice age. "Postclimate" does not describe a theory useful once climate change impacts have ceased. Indeed, for all intents and purpose on timescales relevant to human beings, climate change is irreversible. See Susan Solomon et. al., *Irreversible Climate Change due to Carbon Dioxide Emissions*, 106 Proc. Nat'l Acad. Sci. 1704 (2008).

2. Reenvisioning the State

2.1. The Challenges of Sovereignty and Statelessness

For some small island nations climate change will yield environmental change so "dramatic and all-encompassing of livelihoods that ... most or all inhabitants will be forced to migrate."[5] This will not occur instantaneously, but over time. The relatively near-term dissolution of a country's habitable territory due to climate change, which leads to so-called endangered States,[6] raises novel questions that may challenge the very foundation of Westphalian, or nation-state, sovereignty.[7] The international personality of States, particularly their creation and disappearance, is admittedly one of the most complex areas of law. Although there exist international law regimes that govern issues such as the deprivation of nationality following the succession of a State, none governs circumstances in which no successor State exists and the predecessor State has been rendered uninhabitable or has physically disappeared.[8]

5 Roger Zetter, *The Role of Legal and Normative Frameworks for the Protection of Environmentally Displaced People*, *in* MIGRATION, ENVIRONMENT AND CLIMATE CHANGE: ASSESSING THE EVIDENCE 396 (Frank Laczko & Christine Aghazam eds., 2009). Four completely atoll nations in the Pacific are particularly vulnerable. They are Kiribati, Tuvalu, and the Republic of the Marshall Islands.

6 Endangered States include those identified as "sinking island states" by the United Nations High Commissioner for Refugees. See U.N. High Commissioner for Refugees, CLIMATE CHANGE AND STATELESSNESS: AN OVERVIEW (May 15, 2009), http://www.unhcr.org/refworld/docid/4a2d189d3.html.

7 Less dramatically, though significant, with sea level rise, coastlines may shift or submerge with potential implications for the rights of coastal States over their maritime zones. See Charles Di Leva & Sachiko Morita, *Maritime Rights of Coastal States and Climate Change: Should States Adapt to Submerged Boundaries*, (World Bank Law & Dev. Working Paper Series, Paper No. 5, 2008), *available at* http://preventionweb.net/go/9194. Crucial questions raised as a result will be whether maritime boundaries can shift and whether a country's right to explore offshore resources will be affected, among others. *Id.*

8 See UNHCR, CLIMATE CHANGE, *supra* note 6 (explaining that there is no precedent for the loss of an entire territory or the exile of an entire population); Jane McAdam, *Environmental Migration Governance*, (U.N.S.W. Faculty of L. Res. Series, Working Paper

There is evidence, however, that although statehood is a legal concept with a determinate content, it is flexible.[9] Notably, substantial changes in territory, population, or government or even a combination of all three do not necessarily extinguish a State.[10] In fact, in the period since the signing of the UN Charter, there have been very few cases of State extinction and almost no cases of involuntary extinction.[11] Further, each of the elements that are central to the concept of a State "may present significant problems in unusual situations,"[12] suggesting that statehood as a flexible legal concept is appropriate. That possibility for flexibility coupled with the strong presumption that favors the continuity and disfavors the extinction of an established State[13] suggests that acceptance of creative interpretations of law to recognize the continued existence

No. 1, 2009) [hereinafter *Environmental Migration Governance*] (explaining that it is unprecedented for a State to disappear altogether, as opposed to being reconstituted as another through extinction or merger, for which laws of State succession apply).

9 James R. Crawford, *The Creation of States in International Law* 718 (2006). For a strong critique of the current understanding of the "state," see Guido Acquaviva, *Subjects of International Law: A Power-Based Analysis*, 38 VAND. J. TRANSNAT'L L. 345 (2005) (challenging the definition of the term "state" that is commonly accepted in legal scholarship as the basis for assessing whether an entity is a subject of international law and concluding that the only essential element of a subject of international law is its sovereignty).

10 Crawford, *supra* note 9, at 700; *see also* Jane McAdam, *"Disappearing States," Statelessness and the Boundaries of International Law, in* CLIMATE CHANGE AND DISPLACEMENT: MULTIDISCIPLINARY PERSPECTIVES (J. McAdam ed., 2010) [hereinafter *Disappearing States*]. McAdam explores the creation of States in the climate change context to determine at what point the absence of certain criteria of statehood might lead other States to deny a State's continued existence. *Id.*, at 110. Indeed, an effective government may be the most important criterion for statehood, since all others depend on it. *Id.*, at 113.

11 Crawford, *supra* note 9, at 715. Relevant to this article, Crawford notes, "It is significant that almost all the cases of extinction listed above involved either entities that were ephemeral or whose independence was not clearly established or were instances of voluntary extinction, when a people ... or their representatives ... decided to put an end to their State and to opt for a different future." *Id.*

12 Restatement (Third) of Foreign Relations 201 (1987); *see also* Acquaviva, *supra* note 9, at 349 ("Uncommon situations test the veracity and reliability of the definition itself, at least if the definition is to serve any practical purpose.").

13 UNHCR, *Climate Change*, *supra* note 6 (stating that to the extent that statelessness is foreseeable, efforts should focus on preventing it from arising, consistent with the

of a State – particularly in this "unusual situation" – is plausible (for a thorough discussion of theories of statehood, see Chapter 3).[14]

Climate change will require meaningful interpretations of the law and may sometimes justify significant departures from precedent or the legal status quo. Accordingly, new positive rules that recognize a novel category of "State" as a consequence of the changing landscape may be essential.[15] This continued recognition is important as a practical matter to help such nations favorably resolve existing maritime entitlements, for example,[16] as well as to confront the deeper dilemma of threats to sovereignty and culture.[17] I argue that a deterritorialized statehood is an appropriate means for continued existence.[18]

principle of prevention of statelessness recognized in international law as a corollary to right to a nationality); McAdam, *Disappearing States, supra* note 10. McAdam argues that the lack of all four indicia of statehood (i.e., a defined territory, a permanent population, an effective government, and the capacity to enter into relations with other States) may not mean the end of a State.

14 Crawford, *supra* note 9, at 701. *But* see UNHCR, *Climate Change, supra* note 6 (stating that although there is no internationally agreed definition of what constitutes a State, there is agreement in the existing doctrine that there must be territory inhabited by a permanent population under control of an effective government). For further discussion of the propriety of deterritorialized statehood, see discussion *infra*.

15 Although the "what" and "how" remain unresolved, it is agreed that new rules of international law will be necessary to provide "stability, certainty, and a future" to endangered nation-states. Rosemary Rayfuse, *International Law and Disappearing States: Utilising Maritime Entitlements to Overcome the Statehood Dilemma, in* University of New South Wales Faculty of Law Research Series, at 1 (Univ. of New South Wales, Paper No. 52, 2010) [hereinafter *International Law*].

16 Several scholars have written on the possible impacts to existing maritime entitlements. See, e.g., Rayfuse, *International Law, supra* note 15; Di Leva & Morita, *supra* note 7. Recognition of States "is the act through which a subject of international law indicates its willingness to enter into inter-state relations with another subject of international law and thus is evidence – but not proof – that the latter has acquired international personality." Acquaviva, *supra* note 9, at 350. Consistent with his overall critique, Acquaviva goes on to explain why recognition may not logically establish the international personality of States and is, in fact, "merely a political choice, with no effective bearing on the qualification of an entity as a subject of international law." *Id.* at 364. The key, again, is sovereignty.

17 UNHCR, *Climate Change, supra* note 6, at 2; McAdam, *Disappearing States, supra* note 10, at 113.

18 There are, of course, other possibilities, including acquiring new territory from a distant State by treaty of cession. *See* UNHCR, *Climate Change, supra* note 6, at 2;

Deterritorialized statehood as an alternative future for endangered States is not a new concept. Others have contemplated such a successor entity.[19] In particular, Rosemary Rayfuse has briefly introduced the possibility, arguing that "[u]ltimately, a more equitable solution may lie in recognition of a new category of deterritorialised State."[20] No one, however, has elaborated on how to further define and justify this kind of statehood, or a "new category of international personality . . . in the context of existing States."[21]

In this section I identify alternative forms of the State and entities that are tantamount to State actors. I then identify elements of cosmopolitan theory, as well as the evolving conditions of various diasporas, that show that a landless State is appropriate here and is, at the same time, an inevitable evolution of contemporary citizenship and statehood under the circumstances. Finally, I argue that if none of these justifications is persuasive, the extraordinary circumstances of threatened nation-states alone can justify such unique departures from the norm.

2.2. Alternative Forms of the State as Precedent

Noteworthy historical instances of deterritorialized governance demonstrate that the deterritorialized State is neither new nor inconceivable

Rayfuse, *International Law, supra* note 15, at 7. However, Rayfuse rightly points out the practical impediments to such a solution – namely that it is difficult to imagine any State agreeing to cede any portion of its land. "The political, social and economic ramifications of ceding valued and/or inhabited territory may simply exceed the capacities – and courage – of existing governments." Rayfuse, *International Law, supra* note 15, at 8.

19 Rayfuse, *International Law, supra* note 15 (introducing the concept of "deterritorialized state" and arguing for its application as the basis for continuing recognition of sovereignty of disappeared States over their preexisting maritime zones and resources therein).

20 *Id.* Rayfuse acknowledges that a full analysis of the issue is beyond the scope of the article. *Id.*, at 9.

21 Rayfuse explicitly states that nothing proposed in her article is "intended to suggest a new category of int'l personality available to peoples, however defined, raising current or future self-determination claims in context of existing states." *Id.* at 10.

under current international law. Three existing categories of international actors provide clear precedent for deterritorialized States: (1) countries that have governed without a territory, (2) "failed states"[22] and governments in exile, and (3) economic entities that serve quasi-governmental roles. Other scholars have elaborated on the relevance of these types of governance to the phenomenon of endangered States.[23] I briefly summarize each here.

Turning first to governments without territory, Rayfuse identifies and elaborates on the Sovereign Military Order of Malta and the Holy See as governance entities without a territory that participate in international relations on a par with landholding States.[24] International law recognizes the former as a sovereign subject, although it is headquartered in Rome and lacks traditional markers of statehood – namely, a permanent population and territory.[25] The Holy See (over certain periods of history) is similar.[26] It has consistently enjoyed full legal personality under international law, maintains diplomatic relations with most States,

22 *But see* Crawford, *supra* note 9, at 721 (discussing the thorny issues surrounding the term "failed State," its differing perception among developed and developing States, and the ways in which the designation has been used to justify "questionable exercises in international intervention"). Crawford clarifies the conceptual confusion behind the use of "failed States," explaining that the situations described "are, evidently, crises of government . . . or governance." *Id.* at 721–22.

23 *See, e.g.*, Rayfuse, *International Law, supra* note 15; Rosemary Rayfuse, *(W)hither Tuvalu? International Law and Disappearing States, in* University of New South Wales Faculty of Law Research Series (Univ. of New South Wales, Paper No. 9, 2009) (on file with the author) [hereinafter *(W)hither Tuvalu?*]; McAdam, *Disappearing States, supra* note 10.

24 Rayfuse, *(W)hither Tuvalu?, supra* note 23, at 10.

25 *See* Sovereign Order of Malta, http://www.orderofmalta.org/?lang=en (last visited June 16, 2011). "The Sovereign Order of Malta is a sovereign subject of international law. The Order – which is based in Rome, in via Condotti – has its own Government, an independent magistracy, bilateral diplomatic relations with 104 countries and is granted the status of Permanent Observer in many international organisations, such as the United Nations." *Id.*

26 *See* Acquaviva, *supra* note 9, at 353–57 (explaining that for a certain period after 1808, and then again between 1870 and 1929, the Pontiff had no jurisdiction over any territory at all before ruling Vatican City).

and participates in intergovernmental and international agreements.[27] Importantly, historical changes in territorial control, as well as population, have not affected the international personality of the Holy See.[28]

Governments in exile are examples of functional, yet nonterritorial entities that international law also recognizes. Palestinians and indigenous nations, such as the Maori or Tibetans, are examples of these communities that have been internally dislocated or formally deterritorialized by the processes of invasion or colonization.[29] Indeed, history is "replete with examples" of governments operating in the territory of other States.[30] Assuming no interference in the exiled government's functions, its independence is preserved despite the inability to govern within its own land. The international community has also continued to recognize "failed states" even during the period in which they are effectively failing.[31]

Finally, economic entities, such as the European Union (EU) and Taiwan, enjoy recognition of their sovereignty at the international level. Importantly, EU citizenship confers tangible, additional benefits to its citizens.[32] Similarly, international law recognizes Taiwan's exercise of functional sovereignty on the international level, particularly in international economic affairs, although it is not recognized as a State.[33]

27 *See Holy See*, US Department of State, http://www.state.gov/r/pa/ei/bgn/3819.htm (last visited June 16, 2011).

28 *See* Acquaviva, *supra* note 9, at 354–55.

29 Rayfuse, *(W)hither Tuvalu?, supra* .note 23 Indeed, Tibet's recent election of its newest Prime Minister provides an example of how the mechanics of exiled governance can operate. See Jim Yardley, *Tibetan Exiles Elect Scholar as New Prime Minister*, N.Y. TIMES, April 27, 2011. Admittedly, governments in exile have been contemplated as temporary and exceptional. McAdam, *Disappearing States, supra* note 10, at 112. I contemplate a deterritorialized State that exists in perpetuity subject to renewal by its diasporic citizens. *See* discussion *infra*. Further, the circumstances of the endangered certainly justify exceptional responses. *See* discussion *infra*.

30 McAdam, *Disappearing States, supra* note 10, at 116.

31 *Id*. at 111.

32 Rayfuse, *(W)hither Tuvalu?, supra* note 23, at 11.

33 For a relevant and comprehensive discussion of Taiwan's international personality, *see* Acquaviva, *supra* note 9, at 369–75.

2.3. Diaspora, Cosmopolitanism, and the Shifting State

Both cosmopolitan theory and the experience of the diaspora serve as a further counterweight to neat understandings of statehood, land, and citizenship.[34] The Westphalian system fails to capture the experience of contemporary deterritorialized groups and individuals. The conventional conception of citizenship is one in which belonging to a nation-state affords an individual the ability to exercise civil, political, and social rights. According to Michelle McKinley, however, contemporary global movement forces a rethinking beyond the State that is traditionally marked by borders.[35] Indeed, the experience of members of diverse and multiple diasporas makes the cosmopolitan framework uniquely relevant for the twenty-first-century experience of land and State for groups and individuals. Under this rubric, it becomes clear that the concept of deterritorialized statehood is not far-fetched; in fact a rough blueprint for its theoretical and normative infrastructure may already exist in the diverse scholarship of cosmopolitan theory.[36]

2.3.1. Diaspora and Cosmopolitanism

Cosmopolitanism espouses the belief that all human beings, irrespective of State affiliation or political boundaries, belong to a single global community that should be acknowledged and cultivated. For the

34 *See* generally Michelle A. McKinley, *Conviviality, Cosmopolitan Citizenship, and Hospitality*, 5 UNBOUND 55 (2009) (examining the relationship between refugeehood, statelessness, and cosmopolitan citizenship and using the prism of statelessness to explore the commonality between cosmopolitanism and nationalism with emphasis on deterritorialized modes of citizenship). McKinley argues that various non-Westphalian conceptions of the nation depart significantly from the territorially bounded framework of citizenship. *Id.* at 56.

35 *Id.* at 67. Indeed, McKinley's project vis-à-vis the cosmopolitan framework is to "advance transboundary solidarities that at the same time account for the compelling nature of locally produced and negotiated identities. All cosmopolitans have their personal histories that seem to justify their self-identified positions as global citizens." *Id.*

36 Political institutions, moral norms, or relationships, as well as shared markets or forms of cultural expression, are among the diverse foci of cosmopolitan theory. *See* generally *Cosmopolitanism*, Stanford Encyclopedia of Philosophy, http://plato.stanford.edu/entries/cosmopolitanism/ (last visited March 27, 2011).

cosmopolitan, therefore, the notion that "blood and soil" must be inextricably intertwined is anathema.[37] Further, this interlinkage is not only outmoded because it fails to describe contemporary phenomena but it also has very real consequences, particularly for climate migrants.

Refugees in particular "have always embodied the cosmopolitan ideal; their appeal to humanitarian principles resonates with broader calls for global citizenship."[38] This is particularly true if applied to the unusual circumstances of climate-displaced persons. These people enjoy no special legal status – including refugee status – to assist in their plight.[39] As McKinley recounts, critics argue that the narrow focus on refugee status – consistent with the more narrow understanding of the State, rights, and affiliation – by organs like the UN Refugee Agency has detracted from the international community's ability to advocate for people in "genuine conditions of distress" because they are not strictly political refugees.[40] In fact, McKinley argues that cosmopolitan citizenship provides the normative foundation for an "alternative refugee regime" with an "expanded menu of rights," which the Westphalian model currently truncates.[41] This expansion may well benefit climate-displaced people and provide viable responses to the legal knot that the concept of "environmental refugees" generally introduces.

Even beyond the relief that cosmopolitan theory might provide for the protection of climate-displaced persons throughout the globe, cosmopolitanism provides a framework for the postcolonial, diasporic experience that already marks the identity of many today, including

37 *Id.* at 72 ("this invocation of blood and soil in tandem with citizenship demarcates the dividing line between nationalists [whether academic or popular] and cosmopolitans. Indeed, for the cosmopolitan, the combination of blood, soil, and citizenship seems to converge somewhere between the parochial and the eerily totalitarian.").

38 *Id.* at 63. Even though the appellation is unavailable to climate migrants for reasons discussed in Burkett, *supra* note a, the similar experience of being displaced offers the same opportunity for reflecting on the contemporary relevance of the cosmopolitan framework.

39 Burkett, *supra* note a.

40 *Id.* at 75.

41 *Id.* at 76.

those from island nations. Postcolonial migration has produced a particular kind of citizenship. Although citizenship is generally understood to be territorially bound, postcolonial citizenships are deterritorialized and maintained through, among other things, travel, pilgrimage, labor migration, and "complex residential patterns that traverse boundaries."[42] Further, territorial boundaries coupled with an effective national identity poorly reflect the contemporary experience in which global migration has produced "citizens" that exercise substantive citizenship in at least two States[43] – as well as citizens who enjoy long-distance bonds with nations from which they may have long emigrated.

Indeed, the postcolonial diaspora describes a contemporary experience of a deterritorialized nation, at least in part. As McKinley explains, the diaspora is another process of imaginary nation-building born of displacement, shares many characteristics of deterritorialized or long-distance nationalism,[44] and is a "politically transformative space where legal boundaries matter less than transcendent solidarities."[45] The expatriate experience of Samoans and Tongans in Australia and New Zealand may well be instructive here (for more information on diasporic communities and their experience, see Chapter 11).[46]

42 *Id.* at 56. *See also* Francois Gemenne & and Shawn Shen, EACH-FOR, Environmental Change and Forced Migration Scenarios, Case Study Report of *Tuvalu and New Zealand: Case Study Report* (2009) *available at* http://www.each-for.eu/documents/CSR_Tuvalu_090215.pdf (describing the triple objective of Tuvalu's pursuit of migration opportunities in Australia, Fiji, and New Zealand: alleviating overpopulation in the capital, increasing remittances, and responding to fears of climate change).

43 *Id.* Further, McKinley explains, the ethnocultural nationalist similarly recognizes that people belong to other social groups beyond the State and often view membership in those groups as important. *Id.* at 72.

44 *Id.* at 81.

45 *Id.* at 68.

46 *See* generally Ilana Gershon, *Viewing Diasporas from the Pacific: What Pacific Ethnographies Offer Pacific Diaspora Studies*, 19 Contemp. Pac. 474 (2007); Helen Morton Lee, Tongans Overseas: Between Two Shores (2003); L.F. Va'a, Saili Matagi, Samoan Migrants in Australia (2001); Deborah C. Gough, *Mobility, Tradition and Adaptation: Samoa's Competitive Advantage in the Global Marketplace*, 4 Grad. J. Asia-Pac. Stud. 31 (2006).

The critical lesson is that diasporic communities do not necessarily relinquish their substantive citizenship in their birth countries or other transboundary allegiances to people and culture. The case of diaspora is useful for imagining a viable nation-state that is both dispersed and deterritorialized. The analogy, admittedly, is not complete. For most members of the diaspora, the existence of a tangible home remains, even if one never makes a visit home – and for some Pacific islanders, land and territory do matter greatly, which I discuss in more detail later.[47] However, for the purposes of the international law discourse, the trend of dispersed residence and transboundary loyalties has already disrupted the "neat correlation between citizenship and residence."[48] To the extent that it was ever a strict condition, international law can abandon territory as a requirement of statehood and meaningful citizenship.[49]

2.3.2. Roots, Routes, and the Social Network

The deterritorialized State must still pass muster for its prospective citizens; for although a virtual State bound by a social network may be clever and legally plausible, it must be satisfactory to its members. I suggest tangible benefits of maintaining a virtual State in the form of a Nation Ex-Situ in Part 3, with a particular focus on the shared desire to retain sovereignty, culture, and local traditions.[50] Here, I test the consistency

47 *See* discussion *infra*. McKinley also admits that it would be "irresponsible, if not cavalier, to proclaim effacement of borders. Borders do matter to those in situations of economic desperation and personal persecution." McKinley, *supra* note 34, at 68.

48 *Id.* at 83.

49 Before concerns of heading down a rabbit hole in which the State itself becomes meaningless arise, McKinley provides important analysis to allay those fears. Arguing that transmigrants "destabilize the cozy picture of the people-state-territory trilogy," McKinley follows with, "[b]ut only so much. Transnationalism still depends on an extended nationalist imaginative project, stopping short of civic cosmopolitanism. The Transmigrant engages in nationalism, albeit long distance nationalism." *Id.* at 81.

50 *See* Rayfuse, *supra* note 15, at 11 (describing the desires of Tuvalu and the Maldives to retain their sovereignty, culture, and traditions, including sovereignty over their maritime zones).

and viability of this arrangement against the experiences to date of the citizens of endangered States.

The "roots and wings"[51] of the contemporary cosmopolitan is a particularly poignant metaphor for the Pacific islander and provides support for the deterritorialized State. Although the Pacific island region is notably heterogeneous, for the Pacific islander, broadly speaking, land is vitally important, and for many cultures it is inseparable from those who "belong" to it. The islanders' "roots" in the land are literal in meaning for the cultures in which the word for "land" describes both the soil and the placenta.[52] For the Marshallese, for example, the land possesses a "sacred or spiritual quality" and is the "very root of their worldly existence."[53] Climate change will result in considerable spiritual and cultural loss, and any international response to the loss of land and the concomitant rootedness and spirituality would do well to be mindful of that interrelationship.

There is a duality to the Pacific experience that is also relevant. Pacific island communities are marked by both rootedness and migration. In his discussion of the meaning and importance of land in the Pacific, John Campbell cautions that it is important not to represent Pacific peoples and communities as being physically bounded by their location.[54] In fact, in many Pacific communities "mobility and stability" coexist, and the "paradoxical notion of both roots and routes" is an important aspect of Pacific island lives. This hybridity exists, while people's sense of belonging and connection to a place remains intact, even if they are at quite a physical distance from home.[55]

51 McKinley, *supra* note 34, at 79 (citing Jeremy Waldron, *What is Cosmopolitan?*, 8 J. Pol. Phil. 227, at 229–30 (2000)).

52 *See* generally John Campbell, *Climate-Induced Community Relocation in the Pacific: The Meaning and Importance of Land, in* Climate Change and Displacement: Multidisciplinary Perspectives, *supra* note 10 [hereinafter *Climate-Induced Community*].

53 *Id.* at 60.

54 *Id.* at 63.

55 *Id.* Citing M. Jolly, Campbell explains, "But venturing on routes does not suggest the loss of roots, the perhaps less tangible, but nonetheless visceral, bonds that exist and

The Pacific region today is marked by migration, and Polynesia in particular has historically experienced significant movement.[56] The number of islanders who live in other Pacific Rim countries is already quite significant.[57] Despite this distance from home, an extraterritorial sense of nationhood is maintained through remittances as well as family reunions.[58] This model of a vibrant social web across borders and regions may be useful for the leadership of small island nations as they contemplate a landless future.

Kiribati provides one example of where such a model could be useful. Indeed, "merits"-based migration to Australia and New Zealand is the long-term strategy of the government of Kiribati.[59] This movement would replicate existing routes that the people of Kiribati ("i-Kiribati") have traveled. Pockets of i-Kiribati communities abroad will become larger and more numerous and could maintain traditions and culture and provide a smoother transition for newly arriving climate migrants.[60]

Of course, the actual loss of the land to return to is an unprecedented and potentially devastating exercise of migration and hybridity.[61] To the

connect people with their place. Many migrants, even long-term ones, still consider themselves to belong to their land, even if they are physically dislocated from it." *Id.*

56 *See* generally Gershon, *supra* note 46; Jean Louis Rallu, *Recent Trends in International Migration and Economic Development in the South Pacific*, 11 ASIA-PAC. POP. J. 23 (1996). See also Gemenne & Shen, *supra* note 42 (stating that migration is a significant pattern of lifestyle, and even a social routine at times). Of migrants interviewed for research, many had migrated several times during their lifetimes, some up to eight times, which confirmed the unique and common "migratory" nature of Tuvaluans. This, the authors contend, is common among Pacific islanders. *Id.* at 9, 20.

57 Campbell, *Climate-Induced Community, supra* note 52, at 64.

58 *Id.* at 65.

59 Jane McAdam, *Swimming against the Tide: Why a Climate Change Displacement Treaty Is Not the Answer*, 23 INT'L J. REFUGEE L. 2, at 20 (2011). This kind of migration would model current labor migration. I-Kiribati would undergo job training in sectors needing labor assistance in potential host States.

60 *Id.*

61 *See* generally John Campbell, CLIMATE CHANGE AND POPULATION DISPLACEMENT, at 2 (November 2009) (manuscript on file with author) [hereinafter CLIMATE CHANGE]. Campbell states that nearly all Pacific island communities that have a significant number of emigrants sustain a home population. "This notes of keeping the home fires

extent that seas will continue to encroach on islands as a result of climate change, the key concern becomes what is the best way for the international community to facilitate the transition. A virtual nation-state, held together by a social network, may help ease the rootlessness that an uninhabitable territory will engender.[62] This would not be an unreasonable departure from current international law. In fact, during previous times of resource scarcity, Pacific islanders could move freely to other islands for relief; however, "the legal ... barriers to entry imposed by States today restrict freedom of movement."[63] Because of these barriers, the movement that has previously helped these cultures adapt and thrive is now foreclosed. Removing the legal impediments that limit free movement will avoid compounding the already difficult transition that the people of endangered States might have to make.

2.4. Appropriate Interpretation of International Law and Extraordinary Circumstances Militate in Favor of the Deterritorialized State

If the deterritorialized experience of other international governance entities and the diasporic modifications of the current Westphalian system are not themselves convincing, a reasonable extension of existing law applied to the extraordinary circumstances that climate change

burning is [sic] important in the context of communities that are very mobile ... these roots enable many Pacific people to embark on journeys along routes. There is always somewhere to return to. But we have few examples, if any, of what happens when the roots no longer have land on which to become anchored." *Id.* at 5–6.

62 *See* discussion *infra*. Further, Campbell suggests that climate change might not devastate some Pacific cultures. He explains, "Culture is ever-changing, and the fluidity of cultures and social structures in the Pacific (as elsewhere) cannot be denied. It is possible after numerous generations that relocated people may find peace, no longer yearn for the land which was part of their ancestors, and build successful and prosperous lives in other environs. But, given the strength, tenacity and indivisibility of the person-community-land bond, relocation is bound to be disruptive for a considerable length of time over many generations."

63 McAdam, *Disappearing States, supra* note 10, at 109.

introduces might suffice. Quite simply, the arguments underlying the critical presumption in international law of the continuity of existing States are equally applicable to the current endangered nations scenario.[64] Similarly, appropriate interpretation of existing laws and the spirit in which they were drafted and ratified supports expansive interpretation for the benefit of the peoples of endangered States. For example, if the objective of the UN Convention on the Law of the Sea is to create and maintain stability, certainty, and fairness in the governance of oceans, then a freezing of the baselines or the outer limits of maritime zones would be a consistent – and the most just – means to preserve endangered States' rights to their marine resources.[65] Similarly, the purpose of the Peace of Westphalia was ostensibly to provide security and stability, by way of recognizing borders and nation-states. That Westphalian system undergirds international organization today. Allowing the deterritorialized State – bound by culture and experience – to maintain a place in the international community would meet the objective purpose of the Westphalian system at a time when climate change poses an extraordinary threat to territorial integrity, as well as to global peace and security.

Interpretation of international law without attention to context, object, and purpose can stymie the potential for the law to reach the optimal – and perhaps most transformative – result.[66] For endangered States, legal interpretation and analysis that do not have the principles of fairness and stability at root would be unduly cramped and dangerously unresponsive. Indeed, if the deterritorialized-state solution is not deemed viable or within the reasonable parameters of the current international law regime, then international law's relevance and legitimacy are questionable and could render current law a dangerous

64 *Id.* at 115–16.

65 Rayfuse, *International Law, supra* note 15, at 6.

66 *See* Aurelie Lopez, *The Protection of Environmentally-Displaced Persons in International Law*, 37 Envtl. L. 365, at 391 (2007) (quoting James C. Hathaway, The Rights of Refugees under International Law 49 (2005)).

obstacle, thereby worsening the already dim climate forecast for endangered States.[67]

3. The Nation Ex-Situ: Political Trusteeship and the "Postclimate" State

The Nation Ex-Situ would be a new entity in the international arena, but the framework for its execution is one of fair precedent. The political trusteeship system provides a helpful model for how to govern a nation without a habitable territory. The government of ex-situ nations would sit in a permanent location and manage the affairs of the State at a distance. Other scholars have introduced similar arrangements. In particular, Rosemary Rayfuse has suggested that an "authority" could continue to manage maritime zones to the benefit of the displaced population, through resource rents that could fund relocation and livelihood in the new host State.[68] Although the role of the government ex situ would have a similar mandate, greater emphasis would be placed on preserving all other elements of the nation-state that should endure extraterritorially – key among them including the persistence of culture, connections among its people, and the security and well-being of its citizens. As the citizens of the ex-situ nation will almost certainly be scattered across the globe, its government can also serve as a vital political and cultural nucleus.

67 One can understand these kinds of exacerbations as legal feedback mechanisms, or non-natural aggravations of climate change. Drawing on the positive and negative feedback loops that climate scientists have postulated and observed, Professor David Caron provides a trenchant analysis of laws that are not themselves natural, but are human-made elements of the existing legal order that, without amendment, "aggravate the suffering that will accompany climate change." Caron explains, "The greater the change, the greater the aggravation." David D. Caron, *When Law Makes Climate Change Worse: Rethinking the Law of Baselines in Light of a Rising Sea Level*, 17 Ecology L. Q. 621, at 623 (1990). There are myriad examples of these changes that are extant and uniquely relevant to climate change and the experience of the most vulnerable. I elaborate on these in a forthcoming article, *Climate Change and the Law: On Legal Feedback Mechanisms and Just Compensation*.

68 Rayfuse, *(W)hither Tuvalu?*, *supra* note 23, at 11.

In this section, I discuss the UN international trusteeship system, on which I base this proposal. I then describe the mechanics of the Nation Ex-Situ. I conclude with a discussion of the benefits this structure provides to both the governments of the ex-situ nations and their citizens.

3.1. The Trusteeship Arrangement for Deterritorialized States

Trusteeship arrangements have been a consistent feature of international law. Consistent with common law trust rules, political trustees administer the territories they hold in trust for the benefit of the inhabitants of that territory.[69] The League of Nations, the United Nations, and other ad hoc international trusteeships have demonstrated the viability of these arrangements under international law.[70] A critical difference in the contemporary application of the trusteeship system is that the purpose of the arrangement would be to maintain self-governance and self-determination, and elected citizens of the Nation Ex-Situ themselves would serve as trustees.

To ensure peace and security in transitions to self-government and self-determination, the United Nations established the international trusteeship system for certain designated trust territories.[71] Because independence or self-governance was the purpose of each trust, the trustee was to develop the territory for self-government "while respecting the right of the peoples of the territory to permanent sovereignty over indigenous natural wealth and resources."[72] Although the objects of focus in 1945 were quite specific, and territory was often of central concern, the basic objectives of the trusteeship are applicable to the current climate scenario. Most relevant to the endangered States are the goals of

69 Henry H. Perritt, Jr., *Structures and Standards for Political Trusteeship*, 8 UCLA J. Int'l L. & Foreign. Aff. 385, at 389 (2003).

70 *Id.* at 387–89.

71 The UN trusteeship system focused primarily on postcolonial transitions. *See id.* at 396. The endangered States scenario is quite different in provocation, but could benefit from the structure and lessons learned from the UN system.

72 *Id.*

furthering international peace and security, promoting the advancement of the peoples affected consistent with their freely expressed wishes, and ensuring the equal treatment in social, economic, and commercial matters for all Members of the United Nations.[73]

The UN Trusteeship System expressly excluded the territories that had become Members of the United Nations. This exclusion was consistent with the principle of sovereign equality, which the UN Trusteeship System was to respect.[74] In the current climate change context, however, the introduction of a new kind of trusteeship system for current member States is not only appropriate but it is also necessary to recognize the sovereign equality of the endangered State. In other words, for these States to persist as international personalities, they must contemplate a revised trust arrangement. The UN Trusteeship System serves as a precedent to demonstrate the viability of this kind of removed governance and provides a structure that the international community can most easily replicate given earlier experience. A few departures from the original UN Trusteeship System will produce a more appropriate structure for endangered States.

The key departure will be the extent to which the United Nations, particularly the Security Council or the General Assembly, directs the terms and functioning of the trusteeship arrangement and the subsequent Nation Ex-Situ.[75] Under the traditional arrangement, another State, a group of States, or the United Nations itself has held the territory under supervision in trust.[76] In this new arrangement, the UN's absence in the internal structuring of the arrangement would be essential. To respect the continued sovereign equality of the endangered State, the United Nations and member States would act *only* to support the

73 UN Charter, ch. XII, art. 76.

74 UN Charter, ch. XII, art. 78.

75 This kind of oversight was an element of the UN Trusteeship System. *See, e.g.*, UN Charter ch. XIII, arts. 86–87. For a specific recommendation on exact legal mechanism for the creation of these authorities, see discussion *infra*.

76 Perritt, *supra* note 69, at 390.

transition to, and establishment of, ex-situ nationhood. At the same time, if it is anchored in a UN structure, the establishment of the ex-situ nation can benefit from having the international community play a significant supportive role in administering and financing the ministerial affairs of the deterritorialized State. In practice, the United Nations might establish an office with the express purpose of facilitating the transition and long-term governance of endangered States, all of which are marked by their limited resources. A secretariat would serve as the locus of the UN administrative support and might also serve as a hub for all Nations Ex-Situ to work collaboratively and synergistically for the benefit of current and future generations.[77]

With these modifications, the structure of the historical trusteeship system remains instructive. In the case of endangered States, appointed members of these States supported by the UN would serve as the political trustee. The trustee could first serve as an interim body that governs alongside or as an office of the existing in-situ government of the endangered State, to facilitate an orderly transition for the peoples and governance mechanisms of that State.[78] Drawing on relevant international interventions,[79] the United Nations could help establish, and other States

77 The Trusteeship System contemplated "ambiguous" beneficiaries, including future generations. *Id.* at 390. What this meant in practice is that the beneficiary in Kosovo, for example, could be a State that subsequently came into existence, such as an independent Kosovo. The obvious parallel here would be the beneficiaries, or future citizens, of the ex-situ nation. See discussion of the United Nations Interim Mission in Kosovo *infra*.

78 This kind of intermediate body or office has precedent. *See id.* at 389. Perritt describes intermediate sovereignty and phased recognition as a kind of "soft trusteeship" and provides historical examples of its application. *Id.*

79 *See* generally *id.* Here a relevant parallel is the UN Interim Mission in Kosovo (UNMIK) that served as trustee, holding legal title over the *res* – that is, the property comprising Kosovo. For the UNMIK, the equitable title vested in the people of the trust territory as beneficiaries. *Id.* at 390. Perritt identifies this as the best example of a contemporary political trusteeship. *Id.* at 402. Perritt identifies four additional international interventions after the formal termination of the UN Trusteeship System that, he argues, exhibit the continued viability of the political trusteeship concept. *Id.* at 398.

recognize, the Interim Mission of a given endangered State. For example, the Interim Mission for the Republic of the Marshall Islands *ES* would have the authority to administer all provisions related to the transition of the endangered State from its current status to a purely deterritorialized entity.[80] Again, of course, the existing leadership of an endangered State may opt to serve as the interim authority. The decisions that the interim body would make might include the following: (1) determining appropriate modifications to the current in-situ political and economic institutions; (2) enacting legislation for continued citizenship as well as distribution of monies from resource rents, adaptation funding, or compensation from other international entities,[81] among other things; (3) resolving property disputes and appropriate regimes for compensation; and (4) developing a mechanism for determining what is in the best interest of the diaspora and representing the people in international affairs accordingly.[82] The most significant element of this authority would be to ensure diplomatic protection of scattered citizens living in various host States.[83] This would fall under the general ability to enter into negotiation and arrangements with other States and intergovernmental organizations as a sovereign entity.

80 *See* further discussion *infra*.

81 *See* generally Maxine Burkett, *Climate Reparations*, 10 MELBOURNE J. INT'L L. 509 (2009). The Norwegian Refugee Council similarly contemplates compensation for property and land lost, as well as assistance in resettling and reestablishing livelihoods and residence elsewhere of those affected. Vikram Kolmannskog, *Climate of Displacement, Climate for Protection?*, DIIS Brief, Migration Seminars 12 (2008), *available at* http://www.diis.dk/sw69538.asp.

82 *See* Perritt, *supra* note 69, at 421 (describing the plenary authority exercised by the trustee). Rayfuse envisions an "authority" that would continue to represent the State at the international level as well as the rights and interests of its citizens vis-à-vis their new host State or States. Rayfuse, *(W)hither Tuvalu?, supra* note 23, at 11. These rights could include the right of dislocated citizens to maintain their original personal, property, cultural, linguistic, and nationality rights for themselves and descendants, while simultaneously being granted full citizenship rights in the new "host"' State or States. *Id.*

83 *See* McAdam, *Disappearing States, supra* note 10, at 117.

Once the moment of complete territorial dislocation occurs for the relevant population, the interim body would combine with the dislocated governing body to form a single locus of power.[84] This would be the new central government of the Nation Ex Situ. Although this new body would be distinct in name, the manner in which it operates – namely for the benefit of its people, even if outside of a territory and entirely diasporic – will be almost identical to that of the in-situ government. The Nation Ex-Situ will coordinate all social, economic, and commercial matters related to its former territory for the benefit of its dispersed citizens.

The endangered States electing to participate, and their citizens, will determine the specific terms of the ex-situ governance through iterative processes of evaluation and amendment.[85] The Ex-Situ Agreement, similar to trustee agreements if established, would detail how the endangered State will be governed from the time of transition and throughout the period of uninhabitability. The latter period could in theory exist indefinitely. It is important to note, however, that a key element of the Agreement is that, although the nation would be designed to exist in perpetuity, it would be subject to dissolution at any time if the disembodied so choose via whatever voting procedures that nation chose

84 It is possible that certain nations will be in a liminal space for quite some time. There may not be a distinct moment of clear deterritorialization. In fact, the current uncertainty of climate forecasts could not necessarily rule out the possibility that a nation would be substantially compromised by rising sea levels, yet be able to host a population of citizens, even if quite small, for generations. The question then becomes, how do you administer an interim mission for, say, fifty years? Is this even desirable? In response, I would suggest that each State and its citizens would have the latitude to make the decision for themselves whether to dissolve or exist in parallel through the instruments available to them in the Interim Mission. I thank Robert Perkinson for raising this important logistical challenge.

85 This and similar recommendations discussed later draw from the Trusteeship System detailed in the UN Charter. This particular recommendation mirrors Article 79 of the UN Charter. See also Perritt, *supra* note 69, at 432. Perritt explains that the political trusteeship must accurately ascertain the will of the local population and whether the people want to form their own State. If the population desires to form a separate State, the trusteeship must then determine whether statehood is feasible and if so, prepare the population and territory for legitimate statehood under international law. *Id.*

to use.[86] Thus, the trusteeship would establish an administering authority for the interim, as well as a permanent ex-situ entity that would be responsive to the necessarily evolving needs of its diaspora.[87] Again, the United Nations would serve a supporting administrative role for as long as the Interim Mission and the Nation Ex-Situ choose to exist.

Of course, the legitimacy of the Nation Ex-Situ would be subject to the acquiescence of the international community, including the presumably multiple host States to which islanders will migrate. In his discussion of the structures and standards of political trusteeship, Henry Perritt explains that international legitimacy is necessary "so that a critical mass of the international community will recognize and protect the trustee and its successors. Without international legitimacy, the trustee will likely lack necessary resources, and may face State-sponsored resistance."[88] The unique reasons for deterritorialized statehood and methods by which the Nation Ex-Situ status would emerge favor international acceptance. Further, the character and functional purpose of the Nation Ex-Situ buttress the legitimacy of this new kind of statehood. In addition to delivering tangible benefits and services to its citizens, the ex-situ nation will help maintain the personal and group pride, ideology, and

86 Like other nations, the Nation Ex-Situ should not have a prescribed expiration date at its inception. This is critical to my proposal. The Nation Ex-Situ should be functionally indistinct from its territorialized State counterparts in that a decision to dissolve is integral to the exercise of its sovereignty. Therefore, the State and its nationals are the only ones that can legitimately make this decision. Importantly, Rayfuse imagines something quite different. Indeed, she argues that it is useful to view the trustee-type arrangement as transitional, lasting one generation or one human lifetime. Rayfuse, *(W)hither Tuvalu?, supra* note 23, at 13. In this 30- to 100-year period it is likely that much else in the international legal regime will have to be reconsidered and reconfigured, she maintains. *Id.* Although I agree with her assessment regarding the inevitable and significant changes to the law that will necessarily emerge over the next few decades, I argue that guarantees of sovereignty and self-determination require that ex-situ nationhood remain constant.

87 *See, e.g.*, UN Charter, ch. XII, art. 81.

88 Perritt, *supra* note 69, at 424. Similarly, internal legitimacy is necessary "so that the population will not, to borrow Gandhi's term, 'withdraw its support' from the government." *Id.* Perritt identifies factors to determine internal legitimacy, including charismatic leadership, group pride, and the end to national stage conflict. *Id.* at 442.

traditional custom of the territorialized entity it succeeds.[89] Ultimately, however, the extent to which the international community believes that this new status is consistent with international law will likely be determinative. The presumption in favor of the continued existence of the State and prevention of statelessness for individuals supports acceptance. Further, other scholars maintain that State practice suggests that the international community might well recognize ongoing statehood "even when facts no longer seem to support the State's existence."[90] Nation ex-situ status would be a definitive response to the dilemma facing an endangered State; it also ensures self-determination, governmental effectiveness, economic stability, and the protection of rights.[91]

In light of this need for recognized international legitimacy and the value of this new status, it will be important for the international community to institutionalize international acceptance through a legal instrument – such as a UN resolution or treaty – that unambiguously vests sovereignty in the Nation Ex-Situ.93[92]

3.2. The Legitimate Postclimate State and Citizenship and Nationality for the Postclimate Diaspora

As discussed earlier, a new political and legal environment brought about by climate change will force the creation of the Nation Ex-Situ. If properly planned and supported, there could be clear benefits for the new State and, in particular, its citizens. In addition to maintaining an effective participation of endangered States in the international community, the Nation Ex-Situ can guard against the slow failure of

89 *Id.* at 425 (identifying modern criteria for both international and internal legitimacy).

90 McAdam, *Disappearing States, supra* note 10, at 117.

91 *See* Perritt *supra* note 69, at 428, 430 (identifying factors that increase international legitimacy).

92 *See e.g., id.*, at 420. The international community can do this categorically. In other words, the "community of states ... can lay down rules in treaties or in customary international law that automatically confer statehood when certain requirements are met" for endangered states. *See id.*

a State that is overcome by increasing outward migration. The loss of human capital that accompanies large and sustained emigration can render a State unable to provide basic social services and engage in productive activities.[93] Walter Kalin, the Representative of the UN Secretary-General on the Human Rights of Internally Displaced Persons, has stated, "People from islands and territories will start to migrate, legally or with an irregular situation, and overall the society will slowly disintegrate. For a certain time there will be a government, but it will be a fiction. It will be a slow process of whole nations dying in the social sense in addition to the geographical sense."[94] As climate change compromises the vital coastal, water, and agricultural resources of a State, coupled with the negative effect of outward migration,[95] State failure is a very real possibility. Premature large-scale resettlement may also weaken a country's claims to compensation for the socioeconomic impacts introduced by climate change.[96]

A clear deterritorialized successor entity can ensure a robust nation-state while facilitating the necessary resettlement of its people. During the resettlement process, the interim office would oversee its efficacy and ensure that human rights are respected and that vulnerability is not increased because of the move. Once resettled, the diaspora would maintain citizenship in an ex-situ State that has the well-being of its dispersed nationals as a singular concern.

93 *See* UN Secretary-General, *Climate Change and Its Possible Security Implications*, at 17, UN Doc. A/64/350 (September 11, 2009).

94 Rachel Morris, *What Happens When Your Country Drowns?*, Mother Jones, Nov./Dec. 2009, *available at* http://motherjones.com/environment/2009/11/tuvalu-climate-refugees.

95 McAdam makes a similar suggestion, positing, "Paradoxically, planned and staggered migration over time – the solution favored by Pacific islanders if in situ adaptation to climate change is not possible – may ultimately start to erode longer-term claims to continued sovereignty and statehood, since the state's 'disappearance' may begin once the bulk of the permanent population has moved abroad and obtained a legal status in a new country (either through naturalization or by being born a citizen there)." McAdam, *Disappearing States, supra* note 10, at 129.

96 *See* Gemenne & Shen, *supra* note 42. See also Kolmannskog, *supra* note 81, at 12.

As a practical matter, the designation of nationality allows the Nation Ex-Situ to exercise jurisdiction over its citizens.[97] Absent that recognized affiliation, other States could seek to exercise jurisdiction over a stateless individual. In the case of the climate migrants from endangered States, they could enjoy citizenship in their host State, depending on the laws of the host State involved, as well as in their Nation Ex-Situ.[98] The benefit of this kind of dual citizenship mirrors the general advantage of citizenship – that is, State protection from harassment.[99] By the recognition of host-State citizenship, the climate migrants will enjoy some semblance of equal membership in their new host State of residence. At the same time, the ex-situ nation will also advocate on behalf of its citizens, particularly in circumstances in which they are less well off because of resettlement.[100]

The benefits of citizenship are coupled with the psychosocial benefit of the perpetuation of the culture and traditional lifeways indigenous to the endangered State. There is also some hope for continued community, which can occur in at least three ways. First, an identity based on formal membership in a group can be passed down through generations. Even without a physical location, the sense of bonded community will continue

97 *See* Jeffrey L. Blackman, *State Successions and Statelessness: The Emerging Right to An Effective Nationality under International Law*, 19 MICH. J. INT'L L. 1141, 1149 (1998).

98 This claim to nationality in the host State, I argue, would be a reasonable right that would accompany resettlement. The full details of this arrangement, and other resettlement rights, are beyond the scope of this paper. This granting of nationality would be consistent, however, with the individual right to a nationality set out in Article 15 of the Universal Declaration of Human Rights. *See* generally Blackman, *supra* note 97, at 1172, 1175 (discussing the right to nationality and dual citizenship). Further, to prevent temporary statelessness, dual nationality may need to be permitted at least for a transitional period. *See* UNHCR, *Climate Change*, *supra* note 6, at 3.

99 *See* Blackman, *supra* note 97, at 1150.

100 This scenario is not far-fetched. *See* Kolmannskog, *supra* note 81, at 12 (explaining that involuntary relocation and resettlement rarely lead to improvements in the quality of life of those moved; therefore, moving communities in anticipation of climate change may precipitate vulnerability more than it avoids it). This decline in livelihood due to climate-related relocation and resettlement may have precedent. *See* Tangata Vainerere, *Climate Change Refugees Return to Bougainville Atolls*, PACNEWS, July 22, 2009, *available at* http://www.pina.com.fj.

to have multigenerational relevance and strength. Second, the nationals of the ex-situ nation can continue to enjoy a shared purpose. Indeed, this appears to be the most powerful binding force for the Sovereign Order of Malta, whose members remain bonded by history, spirituality, and service.[101] Finally, there is the possibility of reunion in other locations as a method of continuing cultural identity. Of course, the means by which the diaspora maintains its lifeways through ex-situ nationhood would be decided by its members. In the ways described, however, community can persist. Though tenuously, cultural identity can survive territory even at a significant distance. Indeed, to underestimate the benefit of some form of cultural survival could further injure the people of these endangered States. Although political identity is second to "home," deterritorialized statehood may well ensure some semblance of community and culture that is itself integral.[102]

3.3. The Republic of the Marshall Islands Ex-Situ (RMIes)

There is some indication that endangered nations are interested in retaining State identity in order to maintain control over resources and other assets. The President of Kiribati has proposed the establishment of a small government outpost on the State's only high ground for this exact purpose. This arrangement may have differing advantages and disadvantages, depending on the particular history of an endangered State and its climate forecast. Here, I look at the prospects of a Republic of Marshall Islands Ex-Situ (RMI*es*) and identify issues to consider.

The Republic of the Marshall Islands (RMI) is one of four atoll-only nation-states or territories in the Pacific. It is, as a result, particularly vulnerable to climate-related loss of territory.[103] That prognosis alone makes the proposed ex-situ status particularly relevant and the

101 *See* Sovereign Order of Malta, *supra* note 25.

102 Indeed, the Jewish diaspora may well be the most poignant example of enduring community and cultural bonds, despite a long history with deterritorialized existence.

103 Zetter, *supra* note 5, at 396.

emergence of the RMI*es* a distinct possibility. The palatability of that potential future should be judged against the RMI's historical experience with trusteeship systems. Indeed, the RMI, along with a handful of other Pacific island States, operated under the trusteeship agreement establishing the Trust Territory of the Pacific Islands (TTPI).[104] The TTPI was a strategic trust, under which the United States was responsible only to the UN Security Council for its administration and was given preferential treatment in economic matters, among other things.[105] This relationship with the United States succeeded both German colonization and Japanese administration of the islands, and it preceded the Compact of Free Association with the United States that established Marshallese independence in 1986.[106] The Compact provided, among other things, economic support by the United States and gave the United States full authority over the security and defense of the RMI. Indeed, the RMI is quite dependent on the United States, despite its sovereign status.[107] In essence, the RMI has had extensive experience with external administration and a very short history of independent statehood,[108] which itself has been colored by the unique nature of the Compact of Free Association.[109]

Others may argue that the RMI's complex history with its own sovereignty may support a simple return to the familiar trusteeship arrangements with – most likely – the United States. Alternatively, a further amendment of the Compact that contemplates self-governance in

104 *See* generally Peter D. Trooboff, *Foreign State – Republic Of The Marshall Islands – Treaties – Sovereignty – Trust Territory of the Pacific Islands*, 83 Am. J. Int'l L. 583 (July 1989).

105 *Id.* at 584. *See also* Larry Wentworth, *The International Status and Personality of Micronesian Political Entities*, 16 ILSA J. Int'l L. 1 (Spring 1993).

106 The Amended Compact of Free Association came into force in 2004.

107 *See CIA – The World Factbook, Australia-Oceania: Marshall Islands, available at* https://www.cia.gov/library/publications/the-world-factbook/geos/rm.html (providing statistics that demonstrate the degree of U.S. government assistance).

108 The RMI only joined the United Nations in September 1991.

109 *See* Wentworth, *supra* note 105 (describing the basic principles of the Compact).

free association with the United States, with specific negotiation around the eventual loss of territory and resettlement needs, may appear appropriate.[110] However, even if the United States were to absorb the population of the RMI and the United States further chose to represent vigorously the interests of the Marshallese, it seems that the existence of the RMI*es* would remain necessary as a practical matter. First, the United States would need to maintain a relationship with some specific entity for the compact or trust arrangement to have any meaning. In this case, that would most sensibly be the RMI*es*. Further, the RMI*es* may ease the devastation that will almost certainly accompany the inevitable rupture of the characteristic land-person-community bond in the Pacific discussed earlier.

Consideration of the non-institutional relationship the Marshallese have with their current territory may indeed be the strongest argument for establishing the RMI*es*. As discussed in Part 2, the land is imbued with significant spiritual meaning for the Marshallese. For any community or culture, even those of the diaspora, the existence of the home country is pivotal to identity. Yet, although the devastation that will come with the loss of these lands seems an unavoidable consequence, the experience of relocation and resettlement might be eased by a concerted effort to maintain "community lore and law"[111] – which continued sovereignty under RMI*es* can guarantee.[112] The RMI*es* would quite literally be the tie that binds.

Although this arrangement is wholly suboptimal to remaining at home without devastating climate change, I suggest that ex-situ nationhood is the most appropriate response in the Pacific. In fact, many Pacific island communities are marked by their history of remarkable migration

110 *See, e.g.*, McAdam, *Disappearing States, supra* note 10, at 126.

111 Campbell, *Climate Change*, *supra* note 61, at 25.

112 Or at least some semblance of community lore and law. In his discussion of climate change and the importance of land in the Pacific, Campbell contemplates a more favorable transition similar to what I describe here. *See id.*

and relocation over great distances. This would clearly be the next and most challenging frontier.[113]

4. Conclusion: Navigating the Postclimate Era

Climate change has introduced the potential for a seismic shift in the way we have organized human systems on the planet. Our thorough reworking of our environment will necessarily affect the legal, political, and economic systems in which they were conceived, developed, and globalized. At this moment in geological history, when the impacts of climate change are increasingly palpable and understood, I suggest that we are embarking on a postclimate era in law and human society.

In this period, the human systems that developed under the relatively stable climate will confront challenges to their baseline assumptions and rules of participation. Initially individual, disparate, and uncoordinated critiques of prior assumptions are provoked – for example, recent queries as to the inherent sensibility of the law of baselines. Eventually, I contend, a fundamental rethinking of our decision-making frameworks will be required. I suggest that an appropriately unified and thorough postclimate theory will emerge. The details of that theory are beyond the scope of this chapter; however, the instigating environment for this inquiry becomes clear in this explication of ex-situ nationhood. In the postclimate era, our legal systems will not be immune to dramatic change. Indeed, those early signs of disparate critique are emerging first – perhaps unsurprisingly – in the environmental and international law arenas.

Evidence increasingly suggests that effective management of climate-induced migration and disappearing States will benefit both regional stability, through the minimization of humanitarian crises and avoided

113 *See* Maxine Burkett, *In Search of Refuge: Pacific Islands, Climate-Induced Refuge, and the New Legal Frontier*, 98 AsiaPacific Issues 1 (2011) (discussing the legal frontier that climate change introduces).

conflicts and through preserved socioeconomic development.[114] Avoiding the poverty and suffering that may arise from continued inaction compels immediate implementation of multilevel policy interventions and concordant governance systems. Challenges to statehood and the need for substantial, if not total, relocation over a relatively short period of time make small island States distinct enough targets to warrant a swiftly executed initiative tailored to their plight. This will not provide the kind of certainty required of the law.[115] It will, however, set the framework for sequenced and measured responses appropriate for particular scenarios and geographical regions.

Exhaustive elaboration on the theory and implications of a new governance model in an era of climate change is beyond the scope of this chapter. It appears, however, that we may be embarking on a new legal and political epoch, in which uncertainty, inconsistency, and unprecedented phenomena may indeed become the norm. It is not clear what exact changes it will yield, but it will be key to identify the principles we use to engage them.

114 Asian Development Bank, *Addressing Climate Change and Migration in Asia and the Pacific* (2012), *available at* http://www.adb.org/sites/default/files/pub/2012/addressing-climate-change-migration.pdf.

115 McAdam, *Environmental Migration Governance, supra* note 8, at 8.

5 Introducing the Law of the Sea and the Legal Implications of Rising Sea Levels

Ann Powers and Christopher Stucko

1. Introduction

For at least the past twenty years, scholars have been arguing that low-lying island States are existentially threatened by global warming.[1] As global temperatures rise, so the argument goes, melting ice caps and warming waters will cause sea levels to rise, causing low-lying islands to lose significant amounts of territory and, perhaps eventually, be completely submerged.[2] Although such claims may have seemed far-fetched when first articulated, the loss of dry territory to the seas is a real concern. Uninhabited islands began disappearing in 1998,[3] and the sea reclaimed its first inhabited island only eight years later.[4] Today, residents of other

1 Alfred Soons, *The Effects of a Rising Sea Level on Maritime Limits and Boundaries*, 37 NETH. INT'L L.R. 207 (1990).

2 Charles Di Leva & Sachiko Morita, *Maritime Rights of Coastal States and Climate Change: Should States Adapt to Submerged Boundaries?* 2–11 (World Bank Law & Development, Working Paper Series No. 5, 2009).

3 The lowest lying islands of Kiribati were submerged in 1998. Geoffrey Lean, *Disappearing World: Global Warming Claims Tropical Island*, INDEPENDENT, Dec. 24, 2006, *available at* http://www.independent.co.uk/environment/climate-change/disappearing-world-global-warming-claims-tropical-island-429764.html.

4 In 2006, India's Lohachara Island, home to ten thousand people, became the first inhabited island to be wiped from the map. *Id.*

The authors thank research assistants Nicholas Curtiss-Rowlands and Sarah Mielke for their assistance.

inhabited islands are abandoning their homes as rising tides continue to render more land uninhabitable.[5]

As shown in Chapter 2, there is widespread agreement that seas have been rising and will continue to rise through the twenty-first century. If even relatively conservative predictions within these ranges eventuate, low-lying island States face significant or complete territorial loss.[6]

This chapter introduces some of the legal structures and practical issues to help understand the available tools to protect States and to appreciate what is at stake for many nations. First, the chapter provides an overview of the development of the law of the sea, which has culminated in the Law of the Sea Convention (LOSC). Next, it explains what might be lost for threatened States. Finally, it discusses the enforcement of States' maritime rights. Chapters 6 and 7 then discuss options for preserving nationality under existing rules.

2. The Law of the Sea

The historical starting point of the international law of the sea is Grotius's famous 1609 work, *Mare Liberum*, in which he asserted the freedom of the seas doctrine in the context of increasingly important

5 Residents of Papua New Guinea's Carteret Islands, which is expected to be completely submerged by 2015, started abandoning their homes in 2007. John Stewart, *Rising Seas Force Carteret Islanders out of Home*, LATELINE, May 2, 2007, *available at* http://www.abc.net.au/lateline/content/2006/s1840956.htm. *See also* Neil Tweedie, *Carteret Islands: 'The Sea Is Killing Our Island Paradise,'* TELEGRAPH, Dec. 9, 2009, *available at* http://www.telegraph.co.uk/earth/carteret-islands/6771651/The-sea-is-killing-our-island-paradise.html.

6 One report predicts a 1-m rise by 2060. Shaina Stahl, *Unprotected Ground: The Plight of Vanishing Island Nations*, 23 N.Y. INT'L L. REV. 1, 3 (2010). That would be enough to cover 80% of the most populated atoll of the Marshall Islands. Di Leva & Morita, *supra* note 2, at 8. A 2.4-m rise would be enough to submerge completely the Maldives. *See* CIA World Factbook, https://www.cia.gov/library/publications/the-world-factbook/geos/mv.html (listing 2.4 m as the highest elevation for the country).

and competitive maritime trading empires.[7] This doctrine, based largely on preexisting Roman law, declares that "the sea is common to all" and that "no part of the sea can be considered as the territory of any people whatsoever."[8]

As the dynamics of maritime trade among imperial powers changed, John Selden's *Mare Clausum* offered a countervailing theory to that of Grotius, positing that coastal States had a long history of exercising sovereignty over coastal waters.[9] The debate between Grotius and Selden eventually resulted in a widespread preference for open seas, but with a security exception for sovereign control over limited coastal waters.[10] In 1703, this provided legal endorsement for the preexisting "cannon ball rule," whereby States exercised control over waters as far as a cannon ball fired from the shore could reach.[11] In 1782, an Italian jurist proposed that this rule be regularized and quantified as a three-mile territorial sea, a view echoed by President George Washington at the outbreak of war between France and England in 1793.[12]

No formal attempt was made to codify the law of the sea until 1930, when the League of Nations unsuccessfully undertook the task.[13] In 1958, another attempt was made by the United Nations[14] that resulted in four conventions generally endorsing open seas.[15] In 1960, the UN tried and failed to limit the principle of open seas by delimiting zones of maritime jurisdiction where coastal States could exercise varying degrees of sovereignty.[16] These goals were eventually realized in the 1982 Law of the Sea Convention. The LOSC has been in force since November 16,

7 Ram Prakash Anand, *Origin and Development of the Law of the Sea: History of International Law Revisited* 82–89 (1983).

8 *Id.* at 79, 86.

9 *Id.* at 105.

10 *Id.* at 137.

11 *Id.* at 138.

12 *Id.* at 139.

13 *Id.* at 140.

14 *Id.* at 176.

15 *Id.* at 184.

16 *Id.* at 185–90.

1994, and has been ratified by 161 States to date.[17] Despite the recognition by the United States that many LOSC provisions reflect binding customary international law,[18] the United States is one of the few States that has not yet ratified the treaty.[19]

The LOSC divides the seas into the following general zones: internal waters, territorial waters, the contiguous zone, the exclusive economic zone (EEZ),[20] and the high seas.[21] All of these zones are determined by reference to the coastal State's baseline, defined in the treaty as the low-water mark along the shoreline.[22] States have varying degrees of sovereign control over coastal waters depending on the zone. Internal waters comprise all bodies of water, such as lakes, bays, and rivers, that are landward of the baseline.[23] These are considered part of the State's domestic territory, over which the State has full sovereignty.[24] Next is the territorial sea, which extends twelve nautical miles out from the baseline[25] and in which the coastal State retains sovereignty.[26] Certain limitations exist to this sovereignty, such as the right of innocent passage by ships of other States.[27] The contiguous zone extends from twelve

17 *See Status of United Nations Convention on the Law of the Sea*, UNITED NATIONS TREATY COLLECTION, *available at* http://treaties.un.org/Pages/ViewDetailsIII.aspx?&src=TREATY&mtdsg˙no=XXI~6&chapter=21&Temp=mtdsg3&lang=en (last visited April 8, 2011).

18 Stephanie Holmes, COMMENT: *Breaking the Ice: Emerging Legal Issues in Arctic Sovereignty*, 9 CHI. J. INT'L L. 323, 332 (2008).

19 *See id.*

20 These four zones are referred to as maritime zones.

21 There are special rules for maritime jurisdictional zones of archipelagic States. *See* United Nations Convention on the Law of the Sea arts. 46–54, Dec. 10, 1982, 21 I.LM. 1245 [hereinafter LOSC].

22 *Id.* at art. 5. Other methods of delimitation are permissible where this approach is not feasible. *See, e.g., id.* at art. 7 (allowing States to draw "straight baselines" linking "appropriate points" when the coastline is "deeply indented" or formed by a "fringe of islands").

23 *Id.* at art. 8.

24 *Id.* at arts. 2, 8.

25 *Id.* at art. 3.

26 *Id.* at art. 2.

27 *Id.* at arts. 17–26.

to twenty-four nautical miles from the baseline and serves as a zone in which the coastal State may exercise the control necessary to prevent or punish certain legal violations that occurred or may occur in its territory or territorial sea.[28]

The next and largest maritime zone is the EEZ, which extends 188 miles beyond the territorial sea.[29] The defining characteristic of the EEZ is that the coastal State has sovereign rights over living and nonliving natural resources in the water, the seabed, and its subsoil.[30] In addition to control over resources within the EEZ, the coastal State also has the jurisdiction to regulate marine scientific research, preserve the marine environment, and establish artificial islands.[31]

Waters beyond the EEZ are the high seas, in which all States, whether coastal or land-locked, have the freedom of navigation and overflight, as well as the freedom to lay submarine cables, construct artificial islands, conduct research, and fish.[32] The only restriction on these freedoms is that States must exercise "due regard" to the interests of other States.[33]

Although rights to seabed and subsoil resources are granted as a part of the EEZ regime, they are specifically governed by a somewhat duplicative continental shelf jurisdiction regime. Every coastal State is entitled to claim at least 200 nautical miles (nm) of continental shelf jurisdiction[34] regardless of whether a coastal State has any geological

28 Such categories of laws include customs, fiscal, immigration, or sanitary laws and regulations. *Id.* at art. 33.

29 *Id.* at art. 57.

30 *Id.* at art. 56.

31 *Id.* at art. 56. *Id.* at art. 60 para. 8 (establishing an artificial island does not itself create a new EEZ or extend an existing one).

32 *Id.* at art. 87(1).

33 *Id.* at art. 87(2).

34 *See id.* ("The continental shelf of a coastal State comprises the seabed and subsoil of the submarine areas that extend beyond its territorial sea throughout the natural prolongation of its land territory to the outer edge of the continental margin, *or to a distance of 200 nautical miles from the baselines from which the breadth of the territorial sea is measured where the outer edge of the continental margin does not extend up to that distance.*") (emphasis added).

continental shelf.[35] This is the same extent granted under the EEZ regime.[36] The difference between continental shelf and EEZ jurisdiction is that a State's jurisdiction under the continental shelf provisions extends only to the seabed and subsoil resources but not to resources in the water itself.[37] Further, States can extend jurisdiction under the continental shelf regime beyond 200 nm when the natural prolongation of the geological continental shelf extends farther than that limit.[38] Any seabed or subsoil beyond the farthest limit of continental shelf jurisdiction is called the Area.[39]

3. What Happens to a State's Continental Shelf and EEZ When Tides Rise under Current Law?

Under the terms of LOSC Article 76, the continental shelf jurisdiction of a coastal State becomes "permanent" once its delimitation is deposited with the Secretary-General of the United Nations, regardless of whether the delimitation is in accord with the natural geographic features or the 200-nm measurement.[40] This implies that, regardless of the extent to which a baseline recedes, States that have complied with LOSC Article 76 are entitled to maintain their claims at the date of deposit in perpetuity or until a new map is deposited.[41] Article 76, however, does not explicitly establish what happens to the continental shelf jurisdiction of coastal States that have not deposited their claimed delimitation with the Secretary-General.[42] Presumably the maritime zones would be measured

35 Defined as "the natural prolongation of its land territory to the outer edge of the continental margin." *Id.* at art. 76(1).
36 *See supra* note 26–27 and accompanying text.
37 *Id.* at arts. 76 para. 1, 77.
38 *Id.*
39 *Id.* at art. 1 para. 1.
40 *Id.* at art. 76, para. 9; Soons, *supra* note 1, at 216–17.
41 Soons, *supra* note 1, at 216–17.
42 *Id.* at 217, n. 51. As of November 2010, only 53 of the 161 States Parties to LOSC have made the requisite deposits, and notably absent are some of the most threatened island States, including the Maldives and Marshall Islands. According to the UN, only the following States have submitted the required forms: Argentina, Australia,

from the baseline at the time of submission, without regard to the pre-recession geographic features. This nonsubmission may or may not have a significant impact on a State.

On the one hand, nonsubmission would have no legal repercussions if the natural continental shelf of the noncompliant State extends beyond 200 nm and the baseline continues to exist. Maritime zone measurements will be according to geographic features of the seabed and so will not be responsive to sea level rise. In other words, regardless of where the baseline moves, the outer edge of the continental shelf jurisdiction will continue to be determined by the natural edge of the geological shelf itself.

On the other hand, States whose natural continental shelves do not extend beyond 200 nm are at risk of losing jurisdictional continental shelf by not delimiting it in accordance with Article 76. The potential outer limit of a noncompliant coastal State's continental shelf jurisdiction will recede along with the baseline for as long as the delimitation is not deposited with the Secretary-General, and areas that the State could have claimed jurisdiction over will revert to international seabed.[43] Moreover, a claim to any continental shelf jurisdiction at all would be lost if an island State, and thus its baseline, is completely submerged before compliance.[44]

In contrast to the general rule for continental shelves, EEZs do not become permanent at any point.[45] There is no reference at all to

Bahamas, Belgium, Brazil, Chile, China, Costa Rica, Croatia, Cuba, Cyprus, Equatorial Guinea, Fiji, Finland, France, Gabon, Germany, Grenada, Honduras, India, Indonesia, Ireland, Italy, Jamaica, Japan, Kenya, Latvia, Lebanon, Lithuania, Madagascar, Mauritius, Mexico, Myanmar, Nauru, Netherlands, New Zealand, Norway, Pakistan, Palau, Papua New Guinea, Philippines, Romania, Sao Tome and Principe, Saudi Arabia, Seychelles, Spain, Trinidad and Tobago, Tunisia, United Arab Emirates, United Kingdom, Uruguay, Vanuatu, and Vietnam. List of Charts Deposited with the United Nations (last updated Nov. 10, 2010), http://www.un.org/Depts/los/LEGISLATIONANDTREATIES/depositpublicity.htm.

43 Soons, *supra* note 1, at 217, n. 51.

44 *See* David Caron, *When Law Makes Climate Change Worse: Rethinking the Law of Baselines in Light of a Rising Sea Level*, 17 Ecology L.Q. 621, 634 (1990) (noting that maritime boundaries under the LOSC are contingent on the existence of a baseline).

45 *Id.* at 634; Soons, *supra* note 1, at 216.

"permanent" delimitations in the LOSC provisions relevant to the EEZ, despite such references to both continental shelf and deltaic delimitations.[46] This implies that the rules for continental shelves and deltas were intended to be deviations from the generally understood rule that maritime boundaries measured from the baseline will move with the baseline.[47] This argument is bolstered by the recognition that the LOSC's predecessor treaty declared that baselines are "ambulatory."[48] Consequently, although it is not explicitly stated, it appears that LOSC contemplates ambulatory baselines, and therefore EEZ claims will be lost if baselines are submerged.

The jurisprudence of some domestic courts is in accord with this interpretation.[49] Although the ICJ has not directly dealt with the issue, it has issued guidelines for establishing baselines and other maritime boundaries that may leave room for flexibility.[50] International courts have also considered factual circumstances concerning the boundary at issue to be relevant, including the geography of a particular coastline, whether a claim to natural resources such as oil or natural gas has been made, and the security interests of the coastal State.[51] Consequently, despite the consensus that the terms of the LOSC call for ambulatory

46 Caron, *supra* note 44, at 634.

47 *Id.* at 634 (discussing negative implication derived from LOSC language); Soons, *supra* note 1, at 220.

48 Jeffrey Peters, *United States v. Alaska: Section 10 Permits, the Territorial Sea, and Federalism*, 1 Ocean & Coastal L.J. 59, 78 (1994) (noting that the prior convention called for ambulatory territorial seas and established that boundaries may be modified by natural changes and artificial accretion to the coastline).

49 The United States, although not a party to the LOSC, is an example of such a State. In *U.S. v. Alaska* the Supreme Court noted that "the shifts in a low-water line along the shore … could lead to a shift in the baseline for measuring a maritime zone" and that the "State's entitlement to submerged lands beneath the territorial sea would change" accordingly as the baseline shifts. United States v. Alaska, 521 U.S. 1, 31 (1997).

50 The equidistance principle, for example, entails drawing an equidistant line and considering whether there are factors calling for the adjustment or shifting of that line in order to achieve an "equitable result." Land and Maritime Boundary between Cameroon and Nigeria (Cameroon v. Nigeria: Equatorial Guinea Intervening), Merits, Judgment, *ICJ Reports* 2002, 303, para. 288.

51 Ian Brownlie, Principles of Public International Law 217 (2003).

baselines, international courts may be flexible in their application of those terms in the interest of equity, softening the impacts on States that have not complied with LOSC Article 76.

4. Why Does it Matter?

Continental shelf and EEZ jurisdiction are significant because of the natural resources they cover and the resulting economic value they confer. The UN estimates that 87% of the world's undersea hydrocarbon reserves are located in the seabed and subsea soil within coastal State EEZs and continental shelves.[52] More specifically, the U.S. Geological Survey maintains that there are deposits of minerals such as cobalt, titanium, zirconium, nickel, and platinum within the continental shelf/EEZ jurisdiction of at least some Pacific island States.[53] Further, delimitating a State's continental shelf is often an extremely resource-intensive process, which might be cost prohibitive for many developing and least developed States.[54]

Although mineral resources have significant revenue-producing potential, fishery jurisdiction has been far more important in practice for many small island States. The most abundant fishing grounds lie in coastal waters for ecological reasons, so the ability to fish in EEZs is desirable.[55] Under the LOSC's EEZ jurisdiction scheme, coastal States

52 *The United Nations Law of the Sea (A historical perspective)*, UN Division for Ocean Affairs and the Law of the Sea, http://www.un.org/Depts/los/convention_agreements/convention_historical_perspective.htm#Exclusive%20Economic%20Zone (last visited August 26, 2011).

53 *Pacific EEZ Minerals*, Pacific Coastal & Marine Science Center of the U.S. Geological Survey (June 25, 2010), http://walrus.wr.usgs.gov/research/projects/pac_eez_minerals.html (Marshall Islands, Federated States of Micronesia).

54 *See* I Made Andi Arsana, *The Delineation of Indonesia's Outer Limits of its Extended Continental Shelf and Preparation for Its Submission: Status and Problems*, UN Division for Ocean Affairs and the Law of the Sea, Office of Legal Affairs 76, 76 (2007), *available at* http://www.un.org/depts/los/nippon/unnff_programme_home/fellows_pages/fellows_papers/arsana_0708_indonesia.pdf.

55 *The United Nations Law of the Sea, supra* note 52.

can regulate who may fish in their waters, under what conditions, and at what price. Fishing license fees charged by coastal States can thus yield significant revenue,[56] and revenue from fisheries can be a substantial portion of small island States' GDPs. For example, Kiribati collects between $20 million and $35 million annually from fishing licenses, accounting for 13% to 23% of its GDP.[57] In 2007, fishing license revenues accounted for 42% of all government revenues for the country.[58] Tuvalu also derives significant income from its maritime rights and the associated fishing industry.[59] In 2007, for example, the sale of foreign tuna fishing licenses yielded $4.53 million,[60] or 11% of all government revenue.[61] In 2008, these revenues nearly doubled to $8.86 million.[62] The amount of revenue can also be quite significant when compared to the population size. In Nauru, fishing licenses generated $518 per person in 2007.[63] The Marshall Islands, Palau, and the Seychelles are other examples of island States with generally small economies that rely heavily on money derived from fishing licenses and the fishing industry.[64]

An ambulatory baseline could also have a significant impact on conservation and fishery management efforts. If the baseline were to shift

56 *Id.*

57 *Background Note: Kiribati*, US Department of State (February 3, 2011), http://www.state.gov/r/pa/ei/bgn/1836.htm.

58 Robert Gillett, *Marine Fishery Resources of the Pacific Islands*, FAO Fisheries and Aquaculture Technical paper, 44 (2010), http://www.fao.org/docrep/012/i1452e/i1452e00.pdf.

59 *Tuvalu*, CIA World Factbook, https://www.cia.gov/library/publications/the-world-factbook/geos/tv.html.

60 Tuvalu National Biodiversity Strategy and Action Plan, Fourth National Report to the Convention on Biological Diversity (A. J. Tilling & E. Fihaki eds.), http://www.sprep.org/att/IRC/eCOPIES/Countries/Tuvalu/55.pdf.

61 Gillett, *supra* note 58, at 44.

62 Tuvalu National Biodiversity Strategy and Action Plan, *supra* note 60, at 9.

63 Gillett, *supra* note 58, at 44.

64 *Id.*; Ian Kawaley, *Implications of the Exclusive Economic Zone and EEZ Management for Seychelles, a Small Midocean Commonwealth Territory*, 29 Ocean Development & International Law 225, 235 (1998).

substantially landward, as it might for certain low-lying coastal countries or island States, fisheries and ecosystems formerly controlled by States would become international "commons" governed pursuant to the high seas regime. The LOSC provides various policies for conserving resources within the high seas and requires cooperation among States, but all too often valuable common pool resources are poorly protected and the State that formerly had jurisdiction would no longer have any greater claim or priority than other States.

That States consider their maritime jurisdiction and its concomitant natural resources to be important is evidenced by the prevalence of litigation regarding the issue. For example, Bangladesh initiated two LOSC arbitration proceedings regarding the Bay of Bengal against Myanmar and India in October 2009.[65] Bangladesh alleged that "foreign oil companies holding exploration licenses from Bangladesh ... have been intimidated by warships from Myanmar within waters Bangladesh claims for it's [sic] own."[66] In addition, Bangladesh alleged that Myanmar and India prevented Bangladesh "from exploiting potentially huge deposits of oil and natural gas located off its coast."[67] This type of contentious dispute has the potential not only to prevent developing countries such as Bangladesh from economically benefiting from their resources but may also cause friction between nations, and perhaps even armed conflict.

To summarize, if maritime jurisdiction is lost to rising tides as a result of ambulatory baselines, claims to maritime resources will be lost and the potential for international disputes would rise. For many nations, the potential lost revenue could exacerbate other issues caused by sea level rise, such as the significant loss of dry land, mass displacement of their people, and a need to make efforts to retain statehood in the international arena. Addressing these issues has the potential to be expensive, and the loss of maritime resources will make efforts more difficult.

65 Michael A. Becker & Ernesto J. Sanchez, *International Law of the Sea*, 44 Int'l Law. 519, 524–25 (Spring 2010).

66 *Id.*

67 *Id.*

Therefore, States should take steps to ensure that their marine-based revenue sources are secured as soon as possible both to minimize the risk of international disputes and to help secure resources.

5. Enforcement Measures

Effective enforcement of maritime jurisdiction is essential if States are to reap the benefits of their rights. Enforcement, however, has long been a challenge, especially with respect to the EEZ, for two primary reasons. First, a State's EEZ often covers an incredibly vast amount of sea territory, both in absolute terms and relative to the amount of dry land a State possesses.[68] Second, island States with large EEZs often lack the resources necessary to effectively enforce their rights over them.[69] The inability to prevent the overfishing of Alaskan pollock in the Bering Sea and Chilean sea bass in Antarctic waters, despite international regulation, demonstrates the difficulty of enforcing maritime legal regimes.[70]

The traditional methods of EEZ enforcement are sea and air surveillance, sometimes through regional cooperation.[71] The South Pacific Forum Fisheries Agency (FFA), for example, coordinates enforcement

68 B. Martin Tsamenyi, *The Exercise of Coastal State Jurisdiction over EEZ Fisheries Resources: The South Pacific Practice*, 17 Ambio 255, 256 (table 2).

69 *Id.* at 256. The Solomon Islands ranks 172 in global per-capita GDP out of 230 countries, whereas the Marshall Islands ranks 180 and Tuvalu ranks 200. *Marshall Islands*, CIA World Factbook, https://www.cia.gov/library/publications/the-world-factbook/rankorder/2004rank.html?countryName=Marshall%20Islands&countryCode=rm&regionCode=au&rank=180#rm.

70 Regarding Alaskan pollock, *see* Allison Winter, *U.S. Bans Commercial Fishing in Warming Arctic*, N.Y. Times, August 21, 2009, *available at* http://www.nytimes.com/gwire/2009/08/21/21greenwire-us-bans-commercial-fishing-in-warming-arctic-33236.html; FAO, *Review of the State of the World Fishery Resources: Marine Fisheries* 1997, *available at* http://www.fao.org/docrep/003/w4248e/w4248e22.htm. Regarding Chilean sea bass, *see* Commission for the Conservation of the Article Marine Living Resources ("CCAMLR"), General Information, *available at* http://www.ccamlr.org/pu/e/gen-intro.htm; U.S. Dept of Commerce, U.S. Department of State, *Chilean Sea Bass Frequently Asked Questions*, *available at* http://www.nmfs.noaa.gov/trade/chile.pdf.

71 Tsamenyi, *supra* note 68, at 256–57.

among island States in the South Pacific.[72] Even with the pooling of resources, the FFA is only able to mount one surveillance operation per year and only has a handful of ships and aircraft to cover more than ten million square kilometers.[73] In 2009, the operation lasted only 10 days, and the aircraft were able to cover only 800,000 square miles, or 8% of the EEZ for which the FFA was responsible.[74]

Many individual State efforts are underfunded and ineffective. Tuvalu, for example, only has one fishing patrol boat and a monthly air patrol to police its roughly one million square kilometer EEZ.[75] The lack of effective enforcement has effectively allowed widespread illegal and unreported fishing to continue with impunity,[76] reducing the State's ability to collect revenues and also undermining Tuvalu's maritime conservation program.[77] The Seychelles has a "Greenline," which allows citizens to report illegal activity to the government.[78] This system, which is used for reporting violations such as illegal felling of protected trees, is plagued by the slow mobilization of enforcement resources and is ill suited for maritime enforcement.[79]

Although constricted by resource limits, Article 73 of the LOSC allows States to use any enforcement measures "necessary to ensure

72 Monitoring, Control, Surveillance: Introduction, Pacific Islands Forum Fisheries Agency, http://www.ffa.int/monitoring_surveillance.

73 Operation Kurukuru, Pacific Islands Forum Fisheries Agency, http://www.ffa.int/operation_kurukuru.

74 *Id.*

75 U.N. Food and Agriculture Organization [FAO], Tuvalu (April 2002), *available at* http://www.fao.org/fi/oldsite/FCP/en/TUV/profile.htm; Tuvalu National Biodiversity Strategy and Action Plan, *supra* note 60, at 8; Joeli Vetayaki, Pio Manoa, & Alan Resture, Addressing Climate Change and Sea Level Rise in the Pacific Islands, 47 *Kagoshima University Research Center for the Pacific Islands Occasional Papers* 1, 16 (2007).

76 *Id. at* 10.

77 Tuvalu National Biodiversity Strategy and Action Plan, *supra* note 60, at 10, 15.

78 Republic of Seychelles, Third National Report to the Convention on Biodiversity, 97 (January 2010) *available at* http://www.cbd.int/doc/world/sc/sc-nr-03-en.pdf http://www.cbd.int/doc/world/sc/sc-nr-03-en.pdf.

79 *See id.*

compliance," with only a few enumerated legal limitations.[80] Consequently, there is room for States to be creative in enforcement. Some possible solutions include enforcement in international ports, enforcement by other fishing vessels or NGOs, vessel-monitoring systems (VMSs), and satellite surveillance.

Historically, vessels are generally only subject to the laws of the State whose flag they carry.[81] The LOSC modifies this doctrine by allowing enforcement actions to be taken by the coastal State, even when the ship in question is not flying the coastal State's flag. Indeed, coastal State government officials currently use national registration processes to help keep track of vessels fishing within their jurisdiction.[82] This practice, however, only assists in enforcing EEZ jurisdiction against vessels that are already registered while doing nothing to address the more difficult issue of unregistered ships that are fishing illegally.[83] A better way to combat these unregistered ships is to station government officials at international ports to check the paperwork of incoming and outgoing ships. Ships could be subjected to inspection of their licenses and documentation of their catches. To share the costs of this approach, regional consortia like the FFA could be used to staff ports abroad. This approach would require the cooperation of port States, and incentives would therefore need to be aligned accordingly. One way to do this would be to allow the port State to share in the fines collected as a result of granting foreign inspectors access to the port.

80 *See* LOSC, *supra* note 21, at art. 73, para. 2 (stating that arrested vessels and crews shall be "promptly released" on payment of bond) and art. 73, para. 3 (stating that penalties cannot include imprisonment without the consent of the State involved).

81 Moritaka Hayashi, *Enforcement by Non-Flag States on the High Seas Under the 1995 Agreement on Straddling and Highly Migratory Fish Stocks*, 9 Geo Int'l Envtl. L. Rev 1, 3 (1996).

82 FAO, Flag State Responsibilities, http://www.fao.org/DOCREP/005/Y3536E/y3536e07.htm.

83 *Id. See, e.g.*, FFA Observer Program, Pacific Islands Forum Fisheries Agency, http://www.ffa.int/observers.

Another approach would be for coastal States to delegate their enforcement powers to either private vessels or NGOs. Australia has acquiesced to enforcement by NGOs with respect to its antiwhaling laws.[84] Although this practice has been challenged by Japan as eco-terrorism and piracy, those claims do not hold up under international law.[85] Because the NGO actions merely enforce domestic Australian laws in a way tacitly approved by the government itself, the eco-terrorism claim is faulty.[86] Enforcement of domestic laws by NGOs is also not piracy. Under the LOSC, piracy is defined as "illegal acts of violence or detention, or any act of depredation, committed for private ends by the crew or the passengers of a private ship or a private aircraft, and directed (i) on the high seas, against another ship or aircraft, or against persons or property on board such ship or aircraft; (ii) against a ship, aircraft, persons or property in a place outside the jurisdiction of any State."[87] In the Australian context, the NGO is not acting for private ends, so this definition does not apply. Moreover, because the enforcement action would by definition take place only within the EEZ, where the coastal State has "sovereign right" over living and nonliving resources, the definition of piracy is incongruent with this policy.[88] Also, if the vessel in breach leaves the EEZ while being pursued, the LOSC provides for a right of hot pursuit.[89]

Another strategy, vessel monitoring systems (VMSs), involves the use of technology to enforce EEZ jurisdiction. Under this tactic, States require fishing vessels to install and operate a VMS that allows the State

84 *See* Joseph Roeschke, comment, *Eco-Terrorism and Piracy on the High Seas: Japanese Whaling and the Rights of Private Groups to Enforce International Conservation Law in Neutral Waters*, 20 VILL. ENVTL. L.J. 99 (2009).

85 *See id.* at 124–30.

86 *See id.* at 125–28.

87 LOSC, *supra* note 21, at art. 101(1)(a)-(b).

88 *Id.* at art. 56(1)(a).

89 *Id.* at art. 111(1)-(2).

to monitor the ship's location.[90] This system, combined with required reporting of catches, assists the coastal State's enforcement of fishing quotas and licensing rules in its EEZ.[91] The Maldives, for example, has recently ordered a new VMS system to monitor its waters.[92] Such a system, however, requires that the technology be properly installed, functioning, and switched on for monitoring to take place.[93] Thus, ships that have not registered or have not switched on their VMS equipment are invisible to the coastal State, which is a major obstacle to effective enforcement.[94]

Satellite imagery would be a useful supplement to a VMS. Satellites have long been used to detect fish and squid populations[95] and environmental changes,[96] and given the rapid advances in the quality of satellite imagery, the technology would be useful in detecting fishing vessels themselves.[97] Indeed, even publicly available satellite images are of

90 Monitoring Programs, THE AUSTRALIAN FISHERIES MANAGEMENT AUTHORITY, http://www.afma.gov.au/managing-our-fisheries/compliance-activities/monitoring-programs/.

91 *Id. See also* Jay Ellis, *Fisheries Conservation in an Anarchical System: A Comparison of Rational Choice and Constructivist Perspectives*, 3 J. INT'L L. & INT'L REL. 1, n. 88 (2007).

92 FAO, National Fishery Sector Overview-Maldives 2, ftp://ftp.fao.org/FI/DOCUMENT/fcp/en/FI_CP_MV.pdf; "GeoEye Wins Contract to Build Vessel Monitoring System for Republic of Maldives", PR News Wire (Jan. 6, 2011), http://www.prnewswire.com/news-releases/geoeye-wins-contract-to-build-vessel-monitoring-system-for-republic-of-maldives-113023184.html.

93 Monitoring Programs, THE AUSTRALIAN FISHERIES MANAGEMENT AUTHORITY, http://www.afma.gov.au/managing-our-fisheries/compliance-activities/monitoring-programs/.

94 *See, e.g., Annual Domestic Compliance and Enforcement Program*, AUSTRALIAN FISHERIES MANAGEMENT AUTHORITY 3, *available at* http://www.afma.gov.au/managing-our-fisheries/compliance-activities/domestic-compliance-and-enforcement-program-2010–11/.

95 *See* Adrienne Oppenheim, note, *The Plight of the Patagonian Toothfish: Lessons from the Volga Case*, 30 BROOK. J. INT'L L. 293, n. 183 (2004).

96 Samuel Menefee, *The Impact of Sea Level Rise on International Law and Policy*, 9 UCLA J. ENVTL. L. & POL'Y 175, 189–90 (1991); *see generally* Charles Davies et al., *Moving Pictures: How Satellites, the Internet, and International Environmental Law Can Help Promote Sustainable Development*, 28 STETSON L. REV. 1091 (1999).

97 Davies, *supra* note 96, at 1093.

high enough quality to locate fishing vessels.[98] The effectiveness of this approach would depend on the availability of up-to-date satellite images and would benefit from technology that could compare the images to the locations of VMS-registered ships within the EEZ being monitored.[99] If a fishing ship appeared on the image but was not emitting a VMS signal, the coastal State could dispatch a targeted enforcement vessel to investigate under its LOSC Article 73 authority. This would alleviate the need for regular patrol ships and allow a much more efficient enforcement mechanism while providing a means of verifying that VMSs are installed and turned on.

6. Conclusion

The Law of the Sea, as articulated in the LOSC, provides for an extensive and complicated regime of partitioning off areas of maritime jurisdiction for coastal States. As sea levels rise, all coastal States face shifts in their maritime jurisdictional zones. Low-lying island States face an even greater problem: the potential disappearance of all maritime jurisdictional claims. These maritime rights are a potential source of revenue for all coastal States, but are especially important for the comparatively small economies of island States and will become even more important for island States if they are faced with other adaptation costs as a result of climate change.

An inherent inequity of the current situation is that, although climate change will have a severe economic impact on low-lying island States,[100]

98 *See, e.g.*, Flash Earth, http://www.flashearth.com/?lat=24.534682&lon=-81.776514&z=5&r=0&src=msa (click "+" to zoom in fully and observe a small fishing vessel).

99 This would alleviate the otherwise time-consuming need to compare the satellite images to VMS data manually.

100 The World Bank estimates that between 2010 and 2050, increased global temperatures will cost developing countries between US$75 and $100 billion to adapt to climate change. World Bank, *The Cost to Developing Countries of Adapting to Climate Change: New Methods and Estimates*, The Global Report of the Economics of Adaption to Climate Change Study (June 2010), http://siteresources.worldbank.org/EXTCC/Resources/EACC-june2010.pdf.

they are often among the smallest contributors to the problem of climate change and their losses will benefit some of the greatest contributors to the problem. Coastal States will lose their legal authority to collect fishing licensing fees, which means that ships from major fishing countries such as the United States and Japan will no longer need to pay to fish in the former EEZs.[101] In addition, oil and gas companies may be able to argue that they no longer need coastal State permission to build rigs or be required to comply with coastal State regulations in maritime zones that were formerly under a State's jurisdiction.[102] The remaining chapters of this section address ways to prevent these problems from materializing.

101 FAO Tuvalu, *supra* note 75.

102 Some scholars argue that a coastal State may avoid this issue by depositing a delineation of its continental shelf with the Secretary-General of the United Nations, because this process will fix a State's continental shelf. *See* Chapter 7. Others disagree; *see* Chapter 3.

6 Options to Protect Coastlines and Secure Maritime Jurisdictional Claims in the Face of Global Sea Level Rise

Clive Schofield and David Freestone

1. Introduction

It is now widely accepted that significant sea level rise is taking place and that this phenomenon is likely to accelerate in the future. This poses potentially disastrous implications for many coastal States, especially those with large and heavily populated low-lying coastal areas, as well as small low-lying island States. In addition to the essentially terrestrial, inward-looking threat posed to low-lying coastal areas and their associated populations from inundation by rising seas, threats also exist looking outward from the land to the ocean spaces adjacent to such threatened territories. In particular, sea level rise has the potential to significantly affect national claims to maritime jurisdiction all the way to the outward extent of maritime zones.

Generally, the coastal baseline from which States' maritime zones are measured is the "normal" low-water line. From this baseline States may measure their territorial sea, contiguous zone, exclusive economic zone, and continental shelf. Consequently, if the low-water mark recedes as a result of inundation and sea level rise, this may affect the measurement of all of a country's maritime zones. Because coastal States' baselines are predominantly based on the "normal" low-water line, if those "normal" baselines recede as a consequence of sea level rise, so too will the maritime zones measured from such baselines, leading to the erosion of the coastal State's maritime claims. Further, sea level rise has the potential to inundate small islands and other geographical features that may also be used as critical basepoints in defining the limits of claims to maritime jurisdiction. This too may have major impacts on the capacity of an insular feature to generate maritime jurisdictional claims.

This chapter begins by outlining some of the issues and uncertainties associated with the phenomenon of global sea level rise with an emphasis on its uneven impacts both temporally and spatially. Next, the chapter explores potential impacts stemming from sea level rise. It then assesses several responses open to coastal States in the face of rising sea levels. These responses include both the construction of physical defenses, notably in the form of sea defenses, and emerging "soft" engineering strategies and more adaptive planning measures. The final options discussed to preserve maritime jurisdictional claims are legal; among these are options to preserve maritime jurisdictional claims despite the inundation of coasts, including the fixing of normal baselines resulting in fixed maritime jurisdictional limits, or simply setting fixed maritime jurisdictional limits independently. The chapter critically appraises some of these options and proposals open to coastal States seeking to address and adapt to the challenges posed to their maritime jurisdictional rights by sea level rise, drawing on the relevant scholarly literature that has developed on this topic over the past two decades.

2. Sea Level Rise: Complexities and Uncertainties

Significant evidence exists to support the contention that climate change is already having multiple, sustained, extremely serious, and real impacts on the oceans. Indeed, there now exists broad agreement in the scientific community as to the reality of rising sea levels on a global scale. Chapter 2 discusses in detail the current state of scientific knowledge about the pace and causes of sea level rise and the accompanying uncertainties. As it shows, sea level rise is a phenomenon that exhibits marked spatial and temporal variability. Sea level varies diurnally, under the influence of the tides, but also seasonally, regionally and interannually.[1] Moreover, the impacts on a particular coastline are substantially dependent on its particular characteristics, such as the morphology of the seabed immediately offshore.

Uncertainties also arise in determining the extent of actual sea level rise against the backdrop of ongoing long-term cyclical changes. However, even relatively modest sea level rise has the potential to have severe consequences and pose major challenges for coastal States.

3. Threats Posed by Sea Level Rise

3.1. Threats to Land

The most immediate concern for coastal States is the threat that sea level rise could lead to the inundation of substantial, valuable, and populated parts of their land territory. Such a scenario could in all probability result in the large-scale displacements of coastal-dwelling populations and the loss of associated properties and infrastructure. Even without total or periodic inundation, sea level rise would be likely to make such low-lying areas more vulnerable to extreme weather events such as storm surges, thus rendering them less and less habitable. A further consequence could be increasing saltwater intrusion and the loss of valuable habitats such as coastal wetlands, some of which are protected by international treaties.[2] Alarm bells over these threats were rung by a number of scholars at least two decades ago.[3]

1 *See Climate Change 2007: Working Group I: The Physical Science Basis*, Intergovernmental Panel on Climate Change (IPCC), *available at* http://www.ipcc.ch/publications_and_data/ar4/wg1/en/ch5s5-es.html.

2 David Freestone, *International Law and Sea Level Rise*, *in* INTERNATIONAL LAW AND GLOBAL CLIMATE CHANGE, 109, 119–22 (Robin R. Churchill & David Freestone eds., London/Dordrecht: Graham and Trotman/Martinus Nijhoff, 1991); David Freestone & John Pethick, *Sea Level Rise and Maritime Boundaries: international implications of impacts and responses*, in INTERNATIONAL BOUNDARIES; FRESH PERSPECTIVES, Vol. 5, 73–90 (Gerald Blake ed., Routledge, 1994).

3 Freestone, *supra* note 2; F. Bird & J.R.V. Prescott, *Rising Global Sea Levels and National Maritime Claims*, MARINE POLICY REPORTS 177–96 (1989); A.H.A Soons, *The*

These threats pose an especially problematic prospect in light of the fact that there is a general migration trend from rural to urban areas and from interior/highland to coastal areas. Indeed, coastal areas represent the most populated parts of the global landmass. One study estimates that sea level rise of one meter would inundate territory presently occupied by around 60 million people.[4] Significant sea level rise would therefore necessarily pose daunting socioeconomic, environmental, and human security challenges on a global scale.

3.2. Threats to Maritime Jurisdiction

In addition to threats to land territory, sea level rise could potentially result in significant reductions in the spatial extent of national claims to maritime jurisdiction. Changes to coastlines and therefore to baselines leading to the potential submergence of key basepoints from which claims to maritime jurisdiction are measured could have huge and incremental impacts on national claims to maritime jurisdiction. Reductions in the size of zones to which certain coastal States may be entitled are likely to have profound economic and jurisdictional consequences, because rights over these maritime spaces dictate exploitation rights over valuable resources that might also therefore be lost. Low-lying island States, such as Kiribati, the Maldives, the Marshall Islands, and Tuvalu, which have geographically restricted territory,[5] appear to be especially vulnerable to these threats. This type of threat is likely to be exacerbated where island groups are particularly dispersed and critical basepoints, on which broad maritime jurisdictional claims depend, come under threat.

Effects of Sea Level Rise on Maritime Limits and Boundaries, Netherlands Int'l L. Rev., 37, 207–32 (1990); David D. Caron, *When Law makes Climate Change Worse: Rethinking the Law of Baselines in Light of Rising Sea Level*, Ecol. L.Q. 17, 621 (1990).

4 Anil Ananthaswamy, *Going, Going…*, New Scientist, July 4, 2009, at 26, 30.

5 *See The World Factbook, the Maldives, 2010*, Central Intelligence Agency (CIA), *available at* https://www.cia.gov/library/publications/the-world-factbook/geos/mv.html (last visited July 17, 2011). The land areas of these States are as follows: Kiribati (811 km^2), the Maldives (298 km^2), the Marshall Islands (181 km^2), and Tuvalu (26 km^2).

However, just as sea level rise is understood to be a spatially and temporally uneven phenomenon, so too are the impacts of sea level rise in light of the variability and complexity of coastlines. A vertical sea level change applied to a relatively steep coastline will have an insignificant impact in terms of the horizontal position of the normal baseline. Where the gradient of the coastline is shallow, however, even a relatively slight vertical change can result in substantial shifts in the position of the baseline horizontally. However, it is by no means the entire normal baseline that contributes to the outer limits of a maritime claim. Thus if the critical controlling basepoints are not affected, then the relevant maritime limits will also not move.

3.3. Implications for Islands

Sea level rise has significant potential consequences for islands and their inhabitants. Most obviously, it could lead to the inundation of parts or even the entirety of the land territory of certain islands, forcing the displacement of significant numbers, or even all, of their inhabitants. Although the total inundation of inhabited islands as a direct result of sea level rise has apparently yet to occur, in certain cases, such as that of the Carteret Islands of Papua New Guinea, the relocation of threatened populations is already planned and ongoing (see later discussion).

Even if the total inundation of island territory does not take place (at least in the short term), climate-change-related impacts could make it increasingly untenable for island communities to be sustained on certain islands, forcing their populations to leave.[6] Indeed, even a relatively

6 *See* Ilan Kelman, *Island Evacuation*, 31 Forced Migration Rev. 20 (October 2008). (Kelman identifies five, frequently interconnected, climate-change-related factors affecting island habitability: sea level rise, increased severity and frequency of storms, changes to marine resources, increasing acidity of the oceans, and challenges to freshwater resources). *See also* Rosemary Rayfuse, *W(h)ither Tuvalu? International Law and Disappearing States*, *in* Proceedings of the International Symposium on Islands and Oceans, 91–93 (Hiroshi Terashima ed., 2009). (Rayfuse has been, correctly, critical of the use of the term "environmental refugees" in this context: "far from protecting

slight rise in sea level, of a few tens of centimeters, could render certain islands uninhabitable by, for instance, increasing the incidence of severe flooding or curtailing the availability of potable water.[7] For example, although Fongafale Island, the largest island in Funafuti Atoll, Tuvalu, reaches an elevation of 1.2 m, a sea level rise of 0.6 m would submerge the island's central depression, which lies between its elevated storm and lagoonal ridges. Many of its population of around 5,000 live there.[8] Similarly, only one island, Kanton Island, in Kiribati's Phoenix Islands group, is presently inhabited; several other islands in the group supported populations in the past, but no longer can because of a lack of potable water.[9]

Sea level rise also has the potential to threaten the insular status of certain features. For example, an island that is presently always above water level may, as a consequence of sea level rise, disappear during high tide, thus being reduced to the status of a low-tide elevation (features that are exposed at low tide but are submerged at high tide). Under Article 13 of the LOSC, low-tide elevations may act as basepoints only if they fall wholly or partially within the breadth of the territorial sea measured from the baseline of a State's mainland or island coasts.[10] Consequently, such features have been termed "parasitic basepoints."[11] As a result, if

the rights of persons displaced due to sea level rise, the use of the pejorative, essentially negative concepts of refugee law serves only to conclusively disempower the persons being displaced").

7 *See* Joeli Veitayaki, Pio Manoa, & Alan Resture, *Pacifc Islands and the Problems of Sea Level Rise Due to Climate Change*, *in* Terashima, *supra* note 6, at 55.

8 *See* Hajime Kayanne, *Eco-Technological Management of Atoll Islands against Sea Level Rise*, *in* Terashima, *supra* note 6, at 12, 14–17.

9 Veitayaki et al., *supra* note 7, at 55. Note also that the Government of Kiribas has designated the Phoenix Islands a large marine protected area. *See* http://www.phoenixislands.org/index.php (last visited August 2010).

10 Clive Schofield & Chris Carleton, *Developments in the Technical Determination of Maritime Space: Charts, Datums, Baselines, Maritime Zones and Limits*, 3 I.B.R.U. Maritime Briefing No. 3 at 38 (2001); J.R. Victor Prescott and Clive Schofield, The Maritime Political Boundaries of the World 107–08 (The Hague, Martinus Nijhoff, 2005).

11 Clive R. Symmons, *Some Problems Relating to the Definition of Insular Formations in International Law: Islands and Low-Tide Elevations*, 1 Maritime Briefing No. 5, 7 (1995).

sea level rise causes certain features to be reduced to the status of a low-tide elevation or to be submerged entirely, the coastal State concerned could lose significant maritime rights.[12]

Alternatively, sea level rise could lead to the reclassification of a feature from being a 'full' island, from which claims to the full range of maritime zones may be made, to one of the categories of insular formation from which only restricted maritime claims can be made. In particular, Article 121(3) of the United Nations Convention on the Law of the Sea (LOSC) of 1982 states that "[r]ocks which cannot sustain human habitation or an economic life of their own" are incapable of generating exclusive economic zone (EEZ) or continental shelf rights.[13] The significance of such a change of insular status is illustrated by the United Kingdom's reclassification of Rockall from full "island" to "rock" status, resulting in the loss to the United Kingdom of around 60,000 square nautical miles of maritime space previously claimed as part of the United Kingdom's fishery zone.[14]

Whether the threats to the ability of islands to support habitation and thus, ultimately, to the viability of small Island States are solely (or even largely) the consequence of climate change and sea level rise is open to debate. Coral cays, for example, may naturally be subject to evolution and decay over time.[15] Conversely, it may be that some of these features are capable of natural adaption to sea level rise over time; evidence suggests that coral islands are remarkably robust and enduring features, with the geological record indicating that atolls have survived

12 Prescott & Schofield, *supra* note 10, at 406, 409.

13 United Nations Convention on the Law of the Sea, December 10, 1982, 1833 U.N.T.S. 397 [hereinafter LOSC].

14 *See* Clive Symmons, *Ireland and the Rockall dispute: an analysis of recent developments*, 6(1) IBRU Boundary and Security Bulletin, 78 (Spring 1998). *See also* Clive H. Schofield, *Shifting Limits? Sea Level Rise and Options to Secure Maritime Jurisdictional Claims*, 4 Carbon & Climate L. Rev., 405, 406, 409 (2009).

15 Patrick D. Nunn, Oceanic Islands 243–49 (1994); David Hancox & J.R. Victor Prescott, *A Geographical Description of the Spratly Islands and an Account of Hydrographic Surveys amongst Those Islands*, 1 Maritime Briefing No. 6 (1995); Prescott & Schofield, *supra* note 10, at 57.

sea levels a half-meter higher than present levels.[16] Kench indicates that recent evidence from the Maldives suggests that the sea level rose by a half-meter around 2,000–4,000 years ago, after the islands in question were formed, yet the islands developed and remained above sea level.[17] Similarly, Webb and Kench, in an analysis of twenty-seven coral atoll islands in the central Pacific Ocean, found that 86% of these features had either remained stable or increased in area over a 20- to 60-year timescale despite a reported sea level rise in the central Pacific region.[18] This phenomenon is discussed further in Chapter 2, but at the least runs counter to widespread perceptions that sea level rise will inevitably lead to the erosion of island shorelines.

A critical issue in this context is the health of coral island ecosystems, as is also highlighted in Chapters 2 and 15. Rising populations, land use changes, pollution, and building work along the coast, including the construction of breakwaters, seawalls, and other sea defenses, can adversely affect the health of coral reefs. In particular such interventions can reduce calcification rates, leading to greater erosion as opposed to coral build-up. Further, the interruption of natural sediment flow regimes can compromise the ability of coral islands to adapt to changing sea levels, undermining the island's natural ability to adapt to sea rise.[19]

3.4. Total Inundation of States

The complete or significant inundation of low-lying States, particularly island States, poses a number of novel legal problems. If the sea level

16 Paul Kench, *Understanding Small Island Dynamics: A Basis to Underpin Island Management*, in Terashima, *supra* note 6, at 24–28.

17 *Id.*

18 Of the islands assessed, 43% remained stable and 43% increased in area. *See* Arthur Webb & Paul Kench, *The Dynamic Response of Reef Islands to Sea-Level Rise: Evidence from Multi-Decadal Analysis of Island Change in the Central Pacific*, 72 GLOBAL AND PLANETARY CHANGE 3, 234–246 (2010).

19 *See id.*, at 33, 35; Kench, *supra* note 16, at 15–19. *See also* Richard Kenchington, *Maintaining Coastal and Lagoonal Ecosystems and Productivity*, in Terashima, *supra* note 6, at 1–11.

continues to rise steadily, as many scientists predict, this could cause an existential crisis for a number of States. Although it does not seem imminent, a number of States have been identified that could be rendered uninhabitable by a one-meter rise in sea level and a subsequent increase in storms, flooding, and the loss of the freshwater lens; that is, the layer of freshwater accumulated from rainfall floating on top of denser saltwater. The question is whether such a State, such as the Maldives, could continue to exist if its land area is completely submerged by the sea or so much of it is under water that it cannot sustain sufficient population to have "an economic life of its own," as required by Article 121(3) of the 1982 LOSC. That said, it is important to remember the lack of specificity in the language of Article 121(3), coupled with the flexible interpretation of these provisions in State practice.

Such a scenario would raise fundamental questions of continued sovereignty and the ability of such States to continue to function in the international sphere. These issues are discussed by Jenny Grote-Stoutenburg in Chapter 3, and Chapters 8–11 examine the legal status of persons displaced by climate change and the various options for accommodating them in new locations.

It should, however, be borne in mind that the rise in sea level will be a gradual and an episodic process. As long ago as 1989 a Commonwealth Secretariat Expert Group suggested a "variety of ways in which ... it would be possible for many inhabited low lying islands to adapt, albeit with external assistance."[20] Webb and Kench's recent study, mentioned earlier, also suggests that sea level rise may lead to unanticipated outcomes, especially in coral island contexts.[21] Although we should not ignore the problems of complete inundation discussed earlier, planning in the short and medium term is important to facilitate effective cooperation needed to combat the less dire short-term effects of sea level rise. Customary international law, as Soons suggests, does provide

20 Martin Holdgate, *Climate Change, Meeting the Challenge*, Report of the Commonwealth Group of Experts, Commonwealth Secretariat, London (1989).

21 Webb & Kench, *supra* note 18, at 234–46.

mechanisms to meet many of the novel situations presented by the effects of sea level rise.[22]

4. Response Options

4.1. Planned Retreat and Relocation

The easiest option, in one sense at least, is to do nothing. Rather than attempting to protect the coast and stabilize its present location, one possibility would be to simply manage the impacts of rising sea levels. Such an approach would allow coastlines to find their own "natural equilibrium," thus avoiding the need to construct costly sea defenses.[23] The disadvantages of such an essentially passive response, however, are all too obvious. Such an approach would necessarily lead to the inundation of coastal territory, the displacement of populations, the loss of built-up areas and associated infrastructure, and the sacrifice of important coastal habitats such as wetlands.[24] It is therefore doubtful that such a "do-nothing" policy would prove to be politically acceptable.[25] Nonetheless, such approaches, which recognize increased dynamism in the coastal zone, could provide for coastal development that is responsive rather than resistant to change. In this context a "planned retreat" would include removing/relocating coastal developments after defined "setbacks" are encroached on through coastal erosion.[26]

For small Island States, the option of simply allowing the landward encroachment of low water lines is even less attractive than for mainland continental States, because several small Island States have severely restricted territorial extents. There is little scope for the coastline to

22 Soons, *supra* note 3, at 231.

23 Freestone, *supra* note 2, at 119.

24 *Id.*, at 119–22.

25 *Id.* at 119.

26 Timothy F. Smith, Stephen Myers, Dana C. Thomsen, & Johanna Rosier, *Integrated Coastal Zone Management and Planning*, *in* MARINE RESOURCES MANAGEMENT (Warwick Gullett, Clive Schofield, & Joanna Vince eds., 2011) at 115.

retreat before islands may have to be abandoned in their entirety. Indeed, in some States this process is already occurring. In 2005 a decision was made to relocate the 2,600 inhabitants of the Carteret Islands of Papua New Guinea in response to rising sea levels.[27] Similarly, in 2006, the Indian island of Lohachara, located in the Sundarbans region where the Ganges and Brahmaputra rivers empty into the Bay of Bengal and once the home to 10,000 people, was reportedly evacuated because of the impacts of sea level rise, albeit "as a precaution."[28]

4.2. Protecting the Coast

A traditional means of protecting, or at least attempting to protect, a threatened coastline from erosion is implementation of a "bulkhead" policy.[29] Hard engineering options, such as the construction of sea walls, groins, and wave reduction structures – notably revetments, offshore breakwaters, rock armor, and gabions – are intended to stabilize the position of the coast and protect key infrastructure located in the coastal zones.[30] It is generally accepted that coastal States can, by implementing such measures, stabilize portions of their baselines and thereby preserve their associated maritime zone entitlements.[31] While

27 *See* International Organisation of Migration (IOM), *Migration, Climate Change and the Environment*, IOM Policy Brief – May 2009, *available at* http://iom.int/jahia/webdav/shared/shared/mainsite/activities/env_degradation/iom_policybrief_may09_en.pdf (last visited August 27, 2010). *See also* John Stewart, *Rising Seas Force Cartaret Islanders out of Home*, Lateline (ABC television broadcast Feb. 5, 2007), *available at* http://www.abc.net.au/lateline/content/2006/s1840956.htm. *See also Cartaret Islanders to be Resettled to Bougainville*, Solomon Times, November 5, 2008, *available at* http://www.solomontimes.com/news.aspx?nwID=2971. (It was reported that although the first group of islanders were scheduled for relocation in March 2009, many residents had refused to consider leaving the island).

28 Geoffrey Lean, *Disappearing World: Global Warming Claims Tropical Island*, The Independent, December 24, 2006, *available at* http://www.independent.co.uk/environment/climate-change/disappearing-world-global-warming-claims-tropical-island-429764.html.

29 Freestone, *supra* note 2, at 117–19.

30 Schofield, *supra* note 14, at 411.

31 Soons, *supra* note 3, at 222.

tried and tested, this approach has also repeatedly demonstrated that such construction can result in significant and problematic consequential impacts, including interrupting natural sediments flows and causing unexpected erosion and/or deposition to other parts of the coast.[32] Indeed, the impacts of such hard defenses are not confined to the coastal zone they were designed to protect. Instead, the interruption of natural erosional processes in one area of a coast can affect adjoining coastal areas, depriving them of their sediment supply. In areas where there is already coastal erosion, the bulkhead policy may lead to an acceleration of the natural erosion rate. In one well-documented case on the east coast of England, the erosion rate along 15 km of unprotected coast adjoining a section of seawall increased fourfold over a 50-year period.[33] Even in areas of coastal deposition, such as estuaries, coastal wetlands, and sand dunes, the deprivation of sediment caused by a bulkhead policy may reduce or prevent accretion that may, in turn, result in increased flood damage to low-lying coastal hinterlands as sea level rise continues.[34]

These types of impacts are especially problematic in the context of coral islands, which depend on uninterrupted sediment flows to sustain the island-building processes that maintain their integrity: "[t]he physical dynamics of sediment supply and transport are critical factors in the context of management of rocky and sedimentary oceanic islands."[35] Consequently, often such approaches have been criticized as large scale, costly, disruptive, and resulting in "unfortunate and sometimes very serious unintended consequences."[36] Instead, an

32 Freestone & Pethick, *supra* note 2, at 73–90.

33 J. Pethick, *Waves of Change: Coastal Response to Sea Level Rise*, Geographical Analysis, 19, 1–4 (1989).

34 *Id.*

35 Kenchington, *supra* note 19, at 4. *See also* Kench, *supra* note 16, at 22–24. *See also* Schofield, *supra* note 14, at 411. ("The physical dynamics of sediment supply and transport are critical factors in the context of management of rocky and sedimentary oceanic islands").

36 Kenchington, *supra* note 19, at 4.

ecosystem-based and sustainable management approach is especially relevant in the management of coral cays and other islands associated with coral reefs.[37]

Unilateral action by one State may have considerable effects on its neighbors. It can be argued that, by analogy with situations that have arisen along shared or transboundary rivers, action by one national authority in building a bulkhead that deprives another area in another State of sediment could give rise to liability under international law.[38] For instance, sediment transport studies within the North Sea suggest that fine-grained sediment, derived from the cliff erosion on the east coast of England, composes an important source of suspended material in the North Sea that is vital for the accretion of coastal mudflats and marshes bordering Wadden Sea States.[39] In such cases, a bulkhead policy by one State that seeks designed to preserve the current coastal morphology may seriously exacerbate the problems faced by its neighbors. In West Africa, Shannon demonstrated that a combination of seawall construction, harbor building, and damming of inland rivers, by depriving the coast of sediment, has led to accelerated erosion in neighboring States.[40]

Boyle points out that the *Trail Smelter Arbitration*[41] and the *Corfu Channel* cases[42] "have long supported the proposition that no State may cause or permit its territory to be used to inflict serious harm on other

37 *Id.*

38 Freestone & Pethick, *supra* note 2, at 73–90.

39 *Id.*; I. N. McCave, *Mud in the North Sea*, *in* NORTH SEA SCIENCE 75–100 (E.D. Goldberg ed., 1973); D. Eisma, *Supply and Deposition of Suspended Matter in the North Sea*, 5 SPECIAL PUBLICATIONS INTERNATIONAL ASSOCIATION OF SEDIMENTOLOGISTS 415–28 (1981); J. Dronkers, J.S.I.J Van Alphen, & J.C. Borst, *Suspended Sediment Transport Processes in the Southern North Sea*, *in* RESIDUAL CURRENTS AND LONGSHORE TRANSPORT (R.T. Cheng ed., 1990).

40 E. H. Shannon, *Coastal Erosion and Management along the Coast of Liberia*, *in* CHANGING CLIMATE AND THE COAST, 25–48 (IPCC Working Group III, J. Titus ed., 1990).

41 The Trail Smelter (U.S. v .Can.), 3 R.I.A.A. 1905 (Tribunal 1938/1941).

42 The Corfu Channel Case (U.K. v. Alb.), 1949 I.C.J. 4 (April 9).

States."[43] It seems clear, therefore, given the current state of awareness of the global nature of the world ecosystem, that a State does not have the right to behave, even in its own territory, in a way that damages its neighbors. Most international environmental lawyers now agree that the behavior of one State that adversely affects shared areas or even damages its own territory in a manner that affects the global ecosystem will give rise to responsibility by that State to others. Of course, a chain of causation must be proved to establish responsibility, and indirect damage may be virtually impossible to prove.

Faced with increasing criticism of hard engineering structures on the coast, engineers and coastal managers have examined alternatives that use the natural coastline to absorb wave energy and prevent erosion. For example, in 1992 a UK government report stated,

> Increasing realization of the effectiveness [of natural coastal systems] has led engineers to question the value of traditional coastal defences or protective structures. Instead they are examining solutions to coastal problems which seek to control or emulate natural systems rather than to replace them. This has become known as soft engineering – the solution of coastal management problems through environmentally sensitive schemes which involve substantial or total utilization of natural systems as an integral part of the solution. These are more consistent with ecological principles and other conservation interests as well as having greater social acceptability.[44]

Nevertheless, despite the known and notable drawbacks of the bulkhead policy, this hard engineering approach may remain appropriate to protect especially valuable parts of the coast. Such high-value locations may include key coastal locations such as cities (salient, though not

43 A. Boyle, *International Law and the Protection of the Global Atmosphere: Concepts, Categories and Principles*, *in* International Law and Global Climate Change, (R.R. Churchill & D. Freestone eds., 1991).

44 Minister of Agriculture, Fisheries and Food (1992) cited Freestone & Pethick, *supra* note 2, at 84–85.

necessarily ideal examples include the system of levees at New Orleans and the seawall surrounding Malé in the Maldives).[45]

It is also conceivable that States may seek to preserve critical basepoints in this way; Japan's efforts to preserve its southernmost territory, Okinotorishima perhaps represents the most extreme example in this context. Okinotorishima (also known as Douglas Reef) comprises two groups of very small rocks resting on top of a relatively broad coral reef platform. These features are only marginally above the high-water mark and are vulnerable to erosion.[46] Japan takes the view that these features are islands capable of generating EEZ and continental shelf rights. To protect the above-water rocks making up Okinotorishima, in the late 1980s Japan constructed sea defenses, vertically higher than the islets themselves, which form a 360-degree ring around each of the threatened features. It should be noted that Japan's extensive maritime claims from Okinotorishima have been protested by other States. In particular, both China and the Republic of Korea have expressed the view that these features are no more than "rocks" within the meaning of LOSC Article 121(3).[47]

This type of approach, however, is clearly not suited to all coastlines. For example, it has been suggested that the cost of Japan's sea defenses

45 *See* Schofield, *supra* note 14, at 412.

46 Song notes that "highest tide" in these two tide features are only 16 and 6 cm above the surface of the water, respectively. *See* Yann-huei Song, *Okinotorishima: A "Rock" or an "Island"? Recent Maritime Boundary Controversy between Japan and Taiwan/China*, *in* MARITIME BOUNDARY DISPUTES, SETTLEMENT PROCESSES, AND THE LAW OF THE SEA, 145, 148 (Seoung-Yong Hong & Jon M. Van Dyke eds, The Hague, Martinus Nijhoff, 2009). *See also* Joan Brown, Angela Colling, Dave Park, John Phillips, Dave Rothery, & John Wright, *Case Studies in Oceanography and Marine Affairs* 84–85 (Oxford, Pergamon Press, 1991); Prescott & Schofield, *supra* note 10, at 84–85.

47 *Permanent Mission of the People's Republic of China to the United Nations*, Note verbale to H.E. Mr. Ban Ki-Moon, Secretary-General of the United Nations, February 6, 2009, CML/2/2009 (translation), *available at* http://www.un.org/Depts/los/clcs_new/submissions_files/jpn08/chn_6feb09_e.pdf; *Republic of Korea*, Permanent Mission to the United Nations, Note verbale to H.E. Mr. Ban Ki-Moon, Secretary-General of the United Nations, February 27, 2009, MUN/046/09, *available at* http://www.un.org/Depts/los/clcs_new/submissions_files/submission_jpn.htm.

around Okinotorishima exceeded US$200 million.[48] Moreover, Lewis has pointed out the problems of building traditional sea defenses for Fanafuti, the main island of Tuvalu on which 2,700 of the country's total population of 8,500 live:

> The land form of Fanafuti is so narrow and attenuated that in order to protect its 2.5 square kilometers, 54 kilometers of sea defences would have to be constructed. Moreover, so narrow is the land form for much of its length that sea defences on one side would be protecting the back of sea defences on the other with nothing in between.[49]

In addition, the land behind the defense would have to be raised to prevent flooding from within, resulting from the porous character of the coral rock that predominantly constitutes these features. The situation is similar in many other coral islands.[50]

An analogous approach, and one long practiced by low-lying coastal States such as the Netherlands, has been to undertake reclamation projects as a means to build up or extend coasts. A related response by threatened island States could be the construction of artificial or even floating islands. Indeed, the Maldives is actively pursuing precisely this type of strategy. Its objective is reportedly to construct multiple large artificial islands, elevated to approximately 3 m above sea level and reinforced with concrete, to provide safe refuges for the country's

48 Brown et al., writing in 1991, put the estimated cost of the project at 135 million. Silverstein, quoting Japanese sources, states that the three-year project, starting in 1988, was set to cost US$240 million. *See* J Brown, A. Colling, D. Park, J. Phillips, D. Rotehery, & J. Wright, CASE STUDIES IN OCEANOGRAPHY AND MARINE AFFAIRS 85 (Oxford: Pergamon Press, 1991). *See* Andrew L. Silverstein, *Okinotorishima: Artificial Preservation of a Speck of Sovereignty*, 16 BROOK. J. INT'L L. 409, 410 (1990). Tisdall put the figure at US$250 million. *See* Simon Tisdall, *China Aims for US to Hit Rock Bottom in Asia*, THE GUARDIAN, April 8, 2005, *available at* http://www.taipeitimes.com/News/editorials/archives/2005/04/08/2003249626.

49 J. Lewis, *Sea Level Rise: Some Implications for Tuvalu*, AMBIO 18, 58 (1990).

50 R. L. McClean, *Kiribati and Sea Level Rise*, Report on a Field Visit for Commonwealth Expert Group on Climatic Change and Sea Level Rise, Commonwealth Secretariat (1989); A.J. Edwards, *The Implications of Sea Level Rise for the Republic of the Maldives*, Report to the Commonwealth Secretariat (1989).

population.[51] Although it is acknowledged that, in accordance with the terms of Article 60(8) of the LOSC, artificial islands, together with artificial installations and structures "do not possess the status of islands," "have no territorial sea of their own," and are excluded from affecting maritime boundary delimitation, there appears to be nothing to prevent the extension and, arguably, preservation of a naturally formed feature through reclamation works.

4.3. Using Basepoints

At the core of the legal questions relating to potential impacts of sea level rise on national claims to maritime jurisdiction is Article 5 of the LOSC. This article, which is generally taken to represent customary international law, provides that "the normal base line for measuring the breadth of the territorial sea is the low water line along the coast as marked on large scale charts officially recognized by the coastal State."[52] Such normal baselines represent the predominant type of baselines worldwide. As discussed earlier and in Chapter 5, it has long been recognized that parts of the coast can be dynamic and can change location and configuration in relatively short periods of time. Consequently, as normal baselines change or "ambulate," so too will the maritime jurisdictional limits measured from them.[53] Although the 1982 Convention does not touch on this issue, it is possible to reach such a view through a "negative implication" of the LOSC (laid out in depth by Rayfuse in Chapter 7). This represents a critical concern in the context of sea level rise.

As Rayfuse further describes, Article 7 of the LOSC permits a coastal State to use straight baselines "in localities where the coastline is deeply

51 Chris Morris, *Maldives Rises to Climate Challenge*, BBC NEWS ONLINE, March 17, 2009, *available at* http://news.bbc.co.uk/1/hi/world/south_asia/7946072.stm (last visited October 21, 2010).

52 LOSC, *supra* note 13, at art. 5.

53 Michael W. Reed, *Shore and Sea Boundaries: The Development of International Maritime Boundary Principles through United States Practice* 185 (2000); Prescott & Schofield, *supra* note 10, at 100–101.

indented and cut into, or if there is a fringe of islands along the coast in the immediate vicinity."[54] States that use this system, derived from the judgment of the International Court of Justice in the *Anglo-Norwegian Fisheries* case,[55] must comply with other technical conditions[56]: for example, straight baselines may only be drawn from "appropriate points" – islands and other features that are above water at high tide – and low-tide elevations may only be used as basepoints if "lighthouses or similar installations which are permanently above sea level have been built on them."

However, such straight-line baselines are also potentially threatened by sea level rise, because the basepoints, which "anchor" or tie the defined baselines, may also be inundated and may therefore lose their legal effect.[57] In such a case, if a low-water line may be maintained by extensive and, of course, expensive artificial construction, then using similar artificial means to bolster basepoints may also be a legitimate way by which a country can maintain straight baseline points. The artificial maintenance of islets, rocks, and other similar features may prevent them from becoming low-tide elevations. In addition, new artificial installations may be built on low-tide elevations to bring them within the terms of Article 7(4). Indeed, coastal States are likely to use the proviso of Article 7 – that the drawing of baselines to and from low-tide elevations is permissible if it "has received general international recognition" – to argue that if features listed on charts subsequently submerge, they can maintain legal status because they have already received the required recognition.[58]

In circumstances specified by Article 7 of the LOSC, coastal States are allowed to use straight baselines joining headlands, islands, rocks,

54 LOSC, *supra* note 13, at art. 7.

55 Ango-Norwegian Fisheries (U.K. v. Nor.), Advisory Opinion, 1951 I.C.J. 116 (December 18).

56 LOSC, *supra* note 13, generally.

57 *See* Schofield, *supra* note 14, at 413.

58 Bird & Prescott, *supra* note 3, at 177–96.

and other features. The disappearance of key geographical features that are used as basepoints, and on which the limits of such claims depend, as a result of sea level rise could have significant impacts on the States. As noted earlier, this threat to the extent of national maritime jurisdictional claims is especially significant for coastal States, such as Bangladesh, that have large stretches of low-lying coasts. Indeed, where a broad coastline regresses rapidly, as in the case of areas of Bangladesh where erosion rates are up to 140 meters a year,[59] the cumulative effect can be quite substantial. In fact, it was in recognition of Bangladesh's situation that Article 7(2) of the LOSC was drafted. It permits, in restricted circumstances, straight baselines to be maintained notwithstanding the movement of the actual coast:

> Where because of the presence of a delta and other natural conditions the coastline is highly unstable, the appropriate points (i.e. for straight baselines) may be selected along the furthest seaward extent of the low-water line and, notwithstanding the subsequent regression [*sic*] of the low-water line, the straight baseline shall remain effective until changed by the coastal State in accordance with this Convention.

Although this provision was drafted for specific circumstances, it has long been recognized that it could be used in response to sea level rise. Most notably, more than twenty years ago some already argued that there was a risk that in the context of sea level rise this limited exception would be used more widely than was contemplated by, or many might regard as legitimate under, the LOSC regime.[60] The reality of changing baselines in the twenty-first century may, however, prompt attempts to use a much broader application of these provisions.

59 D. R. Stoddard & J. S. Pethick, *Environmental Hazard and Coastal Reclamation: Problems and Prospects in Bangladesh*, *in* Understanding the Green Revolution (T. Bayliss-Smith and E. U. Wanmali, eds., 1984).

60 Bird & Prescott, *supra* note 3, at 177–96.

4.4. Islands

Analogous issues arise with respect to the use of islands as basepoints for the generation of maritime claims. Under Article 121(1), an "island" is "a naturally formed area of land surrounded by water" and is entitled to claim the full suite of maritime zones.[61] This claim is, however, subject to the important qualification in Article 121(3) that "[r]ocks which cannot sustain human habitation or economic life of their own shall have no exclusive economic zone or continental shelf."[62] Although considerable controversy surrounds Article 121(3),[63] there is at least prima-facie evidence that it has been recognized as customary international law.[64] Whatever the merits of that debate, sea level rise might well cause low-lying islands to disappear entirely or to lose so much of their area or their freshwater lens as to become uninhabitable and thus be transformed into "rocks," thereby causing them to lose their entitlement to previously established maritime zones.

Soons' pioneering paper of 1990 explored several options open to a coastal State and islands facing such a situation.[65] The weight of doctrinal opinion suggests that the artificial enhancement of an existing rock, so as to make it habitable or capable of sustaining an economic life of its own, does not transform it into an island capable of generating EEZ and continental shelf rights. However, the artificial maintenance of an island to prevent it from becoming a "rock" for the purposes of Article 121(3) might, Soons thinks, be permissible.[66] Certainly, the downgrading of a feature's classification from a small island, or "islet" under the

61 LOSC, *supra* note 13, at art. 121(1).

62 *Id.*, at art. 121(3).

63 B. Kwiatkowska & A. H. A. Soons, *The Entitlement to Maritime Areas of Rocks which Cannot Sustain Human Habitation or Economic Life of Their Own*, 21 Netherlands Yearbook of International Law 140–82 (1990).

64 Freestone, *supra* note 2, at 112.

65 Soons, *supra* note 3, at 215–29.

66 *Id.*, at 223.

terms of Article 121(1), to a "rock" under Article 121(3) would pose considerable difficulties in determining the status of the maritime zones that it had previously generated, particularly because these maritime zones may have been recognized by other States as being under the sovereignty or jurisdiction of the State to which it belongs. This situation is exacerbated when the maritime claims of States in possession (or, indeed, entirely composed) of low-elevation islands are also under threat from sea level rise – especially where small, remote, and low-lying islands give rise to significant maritime jurisdictional entitlements. In such cases, Soons suggests that we may see the emergence of a new rule of customary international law that "[c]oastal States are entitled, in the case of landward shifting of the baseline as a result of sea level rise, to maintain the outer limits of the territorial sea and of the EEZ where they were located at a certain moment in accordance with the general rules in force at that time."[67]

Moreover, baseline changes could have a bearing on the delimitation of maritime boundaries that are not yet settled. It has long been accepted that, once international maritime boundaries are established by treaty, they do not change except through agreement among the parties concerned. Once fixed, therefore, maritime boundaries generally do not shift position if, for example, a basepoint contributing to the construction of the boundary line in question disappears. This is because boundary treaties have a privileged status in the international law of treaties and are not subject to change even in the case of "subsequent fundamental change of circumstances,"[68] such as the disappearance of an important basepoint.[69]

67 *Id.*, at 225.

68 Vienna Convention on the Law of Treaties, art. 62(2)(a), May 23, 1969, 1155 U.N.T.S. 331 8 1. L. M. 679. *See also* Soons, *supra* note 3, at 222.

69 It remains to be seen, however, whether a maritime boundary would survive should baselines recede or disappear to the extent that the maritime zone of one of the coastal States involved no longer extended as far as the agreed maritime boundary line.

4.5. Fixing Baselines or Declaring Maritime Limits Directly

An alternative to physical intervention is for coastal States to try to preserve their baselines and maritime rights through legal means. This strategy would not resolve the issue of inundation of vulnerable low-lying areas. The legal options, however, might well preserve the State's rights over the valuable marine resources that could offer economic support to populations displaced as a result of sea level rise.

Among these options (laid out in detail in Chapter 7), the most promising methods under existing law would be to unilaterally fix maritime jurisdictional claims, decoupling them from ambulatory normal baselines. The justification here comes from Article 5 of the LOSC, which provides that "the normal baseline for measuring the breadth of the territorial sea is the low water line along the coast as marked on large scale charts officially recognized by the coastal state."[70] This language suggests that the key requirement is that the chart be *recognized* by the coastal States. Consequently, if States do not update their charts to reflect the loss of land territory or basepoints, then the LOSC suggests that the coastal State can be the arbiter of its own baselines. Although this could benefit the coastal State, a policy of not updating charts would pose potential dangers to seafarers as official charts become more and more inaccurate over time. A dual charts system of official charts for maritime jurisdictional purposes and navigational charts, however, could resolve this problem.

The more ambitious option would be to negotiate a formal treaty. There is no clear model for creating such a treaty, although other options exist, some of which are laid out in Chapter 7. The UN Framework Convention on Climate Change (UNFCCC) is one option. It has provisions for creating new protocols but has become almost moribund over the last five years, and the 194 parties are currently struggling to negotiate a regime that will come into effect after the Kyoto Protocol commitment

70 LOSC, *supra* note 13, at art. 5.

period expires in 2012. In 1994, Freestone argued for a sea level rise protocol to be part of the UNFCCC, but such a protocol is unlikely to be a high priority in the current round of negotiations.[71] Another option, although similarly unlikely, is to amend the 1982 Law of the Sea Convention. In a detailed study, Freestone and Oude Elferink have explored the reasons why States have avoided the complex amendment procedures of the LOSC for both pragmatic and political reasons.[72] There is also growing momentum, discussed in Chapter 9, for a treaty or other agreement to deal with climate-displaced persons in a way that could also create a solution to some aspects of the maritime zone dilemma.

5. Conclusions

Sea level rise, as demonstrated in Chapter 2, appears destined to continue and may accelerate, even if the emissions of greenhouse gases are brought under control and their levels ultimately decline. As the global climate system works to bring itself into equilibrium, historical records of times when the global atmosphere had similar levels of greenhouse gases (albeit not from anthropogenic causes) suggest substantially higher sea levels in the decades to come.

The implications of higher sea levels for the territorial integrity of some States are substantial. These implications involve both the inward-looking landward intrusions of water on low-lying areas and also the

71 David Freestone, *From Copenhagen to Cancun: Train Wreck or Paradigm Shift?* 12 ENVTL. L. REV, 87–93 (2010).

72 David Freestone & Alex G. Oude Elferink, *Flexibility and Innovation in the Law of the Sea: Will the LOS Convention Amendment Procedures Ever Be Used?, in* STABILITY AND CHANGE IN THE LAW OF THE SEA: THE ROLE OF THE LOS CONVENTION 163–216 (Alex G. Oude Elferink ed. 2005). ("States have systematically avoided the formal Amendment procedure of the Convention despite the time and effort that went into crafting its systems of check and balances. The effort was deemed worthwhile in order to maintain the carefully balanced 'package deal' in which the demands of some groups of states were reflected in the agreement in return for the inclusion of other provisions favoured by other groups. These political compromises could be undermined by selective amendments").

outward-looking impacts on maritime zones. These threats were well identified by scholars, cited earlier, writing more than two decades ago when the phenomenon of sea level rise was first mooted. The passing of the years has clarified both the reality of this threat and the severity of the long-term trends. It is important, however, to appreciate that, whatever the inevitable very long-term impacts may be in the coming centuries, in the next several decades it will most likely be a gradual and an episodic process. It is also highly likely to be variable both in scale and place – for already it is clear that globally it is not a uniform process – and that both spatial and temporal variability will add considerably to the uncertainties. Although sea level rise clearly has the potential to be cataclysmic, in its early impacts there is some hope that it may be more manageable in certain locations. For example, Webb and Kench's recent study showed that 86% of Pacific Islands studied are stable or have increased in size over the last few decades despite the higher than average reported sea level rise in the central Pacific region. Natural movements of material such as sand and silt that can maintain these processes are, however, impeded by the presence of human structures, and they may not be able to keep up with accelerating sea level rise and with the loss of coral reefs from ocean acidification. Hoping that natural accretion alone will protect the islands and not exploring alternative responses would therefore seem to be foolhardy.

The main response options were also outlined years ago. These include construction of physical defenses and the application of legal options. Some of the options discussed then are repeated here. Physical defenses remain popular and may become more so, yet they are also problematic, not only in terms of cost but also because of their uncertain impacts; more sophisticated engineering solutions are emerging that may possibly mitigate some of the negative externalities of these projects.

From a legal point of view, many have pointed out the problems that sea level rise poses to the complicated maritime zone regime developed by the 1982 Law of the Sea Convention. Because the maritime zones are measured from coastal baselines and basepoints declared or recognized by the coastal State, changes in coastline configuration and land inundation could have serious effects on these zones. Some maritime boundaries, such as those agreed by treaty, are fixed, whereas others may not be. However, it does seem certain that, in the short to medium term at least, States will be reluctant to reduce or abandon any of their maritime claims. Does law therefore stand as an obstacle to reaching an accommodation with sea level rise? We think not: the international legal system has a lot more flexibility than it is often credited with. Although formal amendments to the LOSC are highly unlikely, because of the complexity of the procedures that are involved, as Soons pointed out in 1990, international customary law inevitably evolves, and so a solution or a series of solutions seem possible.

The precipitous effects of sea level rise will force States to begin to use some of the opportunities that legal commentators have already highlighted to retain their existing maritime zones, even if physical land is lost to inundation, or at least to try to minimize their losses. As others have indicated, if the issue becomes pressing, the political organs of the United Nations are likely to become involved. All in all, we remain positive that the gradual and evolutionary nature of sea level rise will facilitate an orderly evolution of the legal regime.

7 Sea Level Rise and Maritime Zones

Preserving the Maritime Entitlements of "Disappearing" States

Rosemary Rayfuse

1. Introduction

Sea level rise is not a new phenomenon. Oceans have waxed and waned across geological time, at times revealing land bridges between continents, while at other times blanketing the earth in oceans up to seven meters deeper than at present. What is new, however, is the presence of nearly seven billion human inhabitants scattered across the globe, one-third of whom live in coastal areas. Significantly, more than 643 million people currently live in low-lying coastal areas that are at a great risk from the ocean-related impacts of climate change, including storm surges, increasingly frequent coastal inundation, and permanent sea level rise. What is also new is the division of the world into geopolitical units called "States," and the division of the ocean and the seabed into juridical units known as "maritime zones." Sea level rise, whether a result of natural processes or caused by anthropogenically induced or

This chapter draws on and updates my earlier work. Rosemary Rayfuse, *W(h)ither Tuvalu? International Law and Disappearing States*, *in* PROCEEDINGS OF THE INTERNATIONAL SYMPOSIUM ON ISLANDS AND OCEANS 91–93 (Terashima ed., 2009); Rosemary Rayfuse, *International Law and Disappearing States: Utilising Maritime Entitlements to Overcome the Statehood Dilemma*, University of New South Wales Faculty of Law Research Series, Working Paper No 52, (2010), *available at* http://ssrn.com/abstract=1704835; Rosemary Rayfuse, *International Law and Disappearing States: Maritime Zones and the Criteria for Statehood*, 41(6) TL. POL'Y & L. 281–87 (2011).

exacerbated climate change, will affect the physical nature of these maritime zones, altering marine habitats and possibly the distribution of marine resources. Sea level rise will also affect the legal nature and extent of these maritime zones by radically altering the coastlines from which these maritime zones are measured, thereby potentially depriving States of sovereignty over ocean areas to which they were previously entitled. Even more dramatically, small low-lying island States, such as Tuvalu, Kiribati, the Marshall Islands, and the Maldives, are at risk of disappearing physically and geopolitically. This potential outcome is referred to as the "disappearing State" phenomenon.[1]

Since the advent of international concern about climate change in the late 1980s, much has been written about sea level rise and its implications for the physical and legal status of maritime zones and of small low-lying island States. Practical and policy approaches to resolving these issues have also been suggested. This chapter reviews the approaches that disappearing States can take to help preserve their maritime claims and provides some new approaches. It should be noted that much of what is discussed here is equally relevant to all coastal States.

It is important to acknowledge at the outset that the term "disappearing State" may be problematic. Scientific literature suggests that atoll islands may be more resilient to sea level rise than originally thought. Analysis of island change in the Central Pacific shows that over the past twenty to sixty years, atoll islands have either remained stable or have expanded in area despite a measurable sea level rise.[2] However, both the rate and the amount of sea level rise measured over the time period of the study were considerably lower than the estimates of future sea level rise, which, according to the Intergovernmental Panel on Climate Change (IPCC), are conservatively said to

1 *See* Rayfuse, *W(h)ither Tuvalu*, *supra* note a.

2 Arthur P. Webb & Paul S. Kench, *The Dynamic Response of Reef Islands to Sea-Level Rise: Evidence from Multi-Decadal Analysis of Island Change in the Central Pacific*, 72 Global and Planetary Change 234–46 (2010).

range between 0.38 and 0.59 meters by 2100.[3] It is unclear whether the processes of natural geomorphic change will be able to keep pace with increasing sea level rise. In addition, expansion does not guarantee habitability, because the geomorphic adjustment in island shape, location, and form may not compensate for the loss of habitable land. Thus, although an island may continue to exist as a physical fact, the island's population may still be forced to evacuate.[4] As discussed infra, uninhabited islands do not generate the full range of maritime zones. Moreover, according to traditional international law doctrine, a permanent population is a prerequisite of statehood. It is the very real possibility of a disappearing population, forced to leave due to sea level rise, that raises the intriguing juridical possibility of the disappearing State.

It must also be acknowledged that, although the issue of disappearing States is an intriguing juridical dilemma, discussion of the issue may be premature, an unhelpful diversion, or both. However, a world that is incapable of envisioning the future is unlikely to have one. Elucidation and consideration of the juridical dilemmas posed by sea level rise are necessary preparations that will assist coastal States and, in particular, potentially disappearing States to establish the legal conditions necessary to maintain their maritime jurisdictional interests and, potentially, their existence as States.

Part 2 of this chapter reviews the international rules relating to the establishment of maritime zones and their application in the case of sea level rise, referred to here as the "baseline dilemma." Part 3 discusses the relevance of the baseline dilemma in the context of the possible inundation of island States by sea level encroachment. The final section discusses a range of options for resolving the dilemma.

3 Intergovernmental Panel on Climate Change (IPCC), *Climate Change 2007: Synthesis Report* (IPCC: 2007). Sea-level-rise projections are discussed in detail in Chapter 2 of this volume.

4 For further discussion of the effects of sea level rise on habitability, see Chapter 2.

2. Sea Level Rise and Maritime Zones: The Baseline Dilemma

2.1. International Law and Maritime Zones

International law relating to entitlement to maritime zones is set out in the 1982 Law of the Sea Convention (LOSC).[5] Under the LOSC, all coastal States are entitled to certain maritime zones: internal waters, a territorial sea not exceeding 12 nautical miles (nm) in breadth, an exclusive economic zone not exceeding a further 188 nm in breadth, a continental shelf out to 200 nm, and, where the geomorphological conditions exist, an extended continental shelf out to a maximum of 350 nm. Within each of these zones States exercise varying degrees of sovereignty. Internal waters are wholly under the jurisdiction and sovereignty of a State and may be equated for present purposes to territory.[6] Within the territorial sea, a coastal State exercises complete sovereignty, subject only to a right of innocent passage for foreign ships.[7] In the exclusive economic zone (EEZ), a coastal State enjoys sovereign rights for the exploration and exploitation of living and nonliving natural resources of the water column,[8] and on the continental shelf a State enjoys sovereign rights for the exploration and exploitation of the natural resources of the seabed and subsoil.[9] Beyond the areas under the national jurisdiction of States lie the "global commons." Here, the high seas water column is subject to an open access regime of equal right of user in which all States enjoy the "freedom of the high seas."[10] The Area, defined as "the sea-bed and ocean floor and subsoil thereof, beyond the limits of

5 United Nations Convention on the Law of the Sea, December 10, 1982, 21 I.L.M. 1245, 1833 U.N.T.S. 397 [hereinafter LOSC]. As of May 31, 2011, 162 States are party to the LOSC.

6 *Id.* at Art. 8.

7 *Id.* at Art. 17.

8 *Id.* at Art. 56.

9 *Id.* at Art. 77.

10 *Id.* at Art. 87.

national jurisdiction,"[11] is governed by the International Seabed Authority (ISA).[12]

Although jurisdictional rights over the territorial sea, EEZ, and continental shelf may differ, the outer boundary of each of these zones is measured from a common baseline. Except where otherwise provided in the LOSC, the normal baseline is the low-water line along the coast, as marked on large-scale charts officially recognized by the coastal State.[13] In the case of islands situated on atolls or those having fringing reefs that are above water at low tide, the baseline is the seaward low-water line of the reef, as shown by the appropriate symbol on charts officially recognized by the coastal State.[14] The use of the low-water line as the baseline in the LOSC is deliberate because it maximizes the maritime area over which States can claim jurisdiction. Where a low-tide elevation[15] is situated wholly or partly at a distance not exceeding the breadth of the territorial sea from the mainland or an island, the low-water line on the elevation may be used as the baseline for measuring the territorial sea. In other words, low-tide elevations within 12 nm from the coast generate a territorial sea, whereas those beyond 12 nm do not.

As an exception to the general rule requiring the use of the low-water line, in certain circumstances, "straight baselines," drawn in accordance with the specific rules set out in Article 7 of the LOSC, may be used. According to Article 7, straight baselines may be used where the

11 *Id.* at Art. 1(1).

12 The ISA was established pursuant to LOSC Art. 156. All States Parties to the LOSC are members of the ISA, which is headquartered in Kingston, Jamaica.

13 LOSC, *supra* note 5, at Art. 5. *See also id.* at Art. 76 (The breadth of the territorial sea is up to 12 nautical miles measured from the baselines. The exclusive economic zone extends for a further 188 nautical miles. The breadth of the continental shelf is up to 200 nautical miles from the baselines or, where an extended continental shelf exists, it may extend up to 350 nautical miles from the baselines).

14 *Id.* at Art. 6.

15 *Id.* at Art. 13(1) ('a low-tide elevation is a naturally formed area of land which is surrounded by and above water at low tide but submerged at high tide').

coastline is deeply indented or where there is a fringe of islands along the coast. In these situations, the LOSC provides that the baseline is created by joining appropriate points, known as basepoints, with a straight line.[16] Basepoints are generally rocky or other prominent points unlikely to suffer from erosion. The straight baseline then forms the baseline from which the maritime zones are measured in the relevant area. Although state practice exists to the contrary,[17] the LOSC makes clear that straight baselines must not significantly depart from the general direction of the coast and that the areas lying within the lines must be sufficiently closely linked to the land to be subject to the internal waters regime.[18] Straight baselines are not to be drawn to and from low-tide elevations unless lighthouses or similar installations that are permanently above sea level have been built, or their use has otherwise been generally accepted.[19]

Given that the normal baseline is the low-water mark, baselines can shift due to coastal realignment. Thus, the LOSC provides for the permanent fixing of baselines in two situations. The first relates to deltas. Pursuant to Article 7(2), straight baselines drawn around deltas or other areas of unstable coastline remain fixed, as a geographical matter, unless changed by the coastal States. The second relates to the outer limits of the continental shelf. Pursuant to Article 76(9), coastal States are to deposit with the Secretary-General of the United Nations charts and relevant information, including geodetic data, that permanently describe the outer limits of their continental shelf. Apart from these exceptions, however, the LOSC does not expressly state whether the outer boundary of maritime zones moves as baselines – or the low-water mark on which they are based – move. A textual interpretation of the LOSC, according

16 *Id.* at Art. 7(1).

17 *See, e.g.*, J. Ashley Roach & Robert Smith, *Straight Baselines: The Need for a Universally Applied Norm*, 31 Ocean Development and International Law 47 (2000); Trent Palmer, *Sea Level Change and Baselines*, *in* Proceedings of the Canadian Hydrographic Conference and National Surveyors Conference, 4 (2008).

18 LOSC, *supra* note 5, at Art. 7(3).

19 *Id.* at Art. 7(4).

to Freestone,[20] Alexander,[21] Caron,[22] and Soons,[23] leads to the conclusion that both baselines and the outer boundaries of the territorial sea, contiguous zone, and exclusive economic zone are ambulatory. The commentators conclude this from the negative implication that, because the convention specifically indicates two situations where baselines are permanently fixed, baselines must otherwise be ambulatory.

2.2. The Baseline Dilemma

The difficulty with ambulatory baselines is, however, immediately apparent in the context of sea level rise: if the baseline moves, then the outer boundaries of the maritime zones also move. A first area of concern relates to low-tide elevations and fringing reefs. Where low-tide elevations are located within 12 nm of land they may be used to generate a baseline.[24] The seaward low-water line of a fringing reef also may be used as a baseline.[25] In both cases, the area of territorial sea claimed will be greatly enhanced. By contrast, permanent inundation of these low-tide elevations and fringing reefs will, for many States, result in significant loss of breadth of their maritime zones. Archipelagic States, a category to which our disappearing States belong, may be particularly affected. An "archipelago" is defined in Article 46(b) of the LOSC to mean "a group of islands, including parts of islands, interconnecting waters and other natural features which are so closely interrelated that such islands,

20 David Freestone, *International Law and Sea Level Rise*, *in* INTERNATIONAL LAW AND GLOBAL CLIMATE CHANGE, 109–25 (Robin Churchill & David Freestone eds., 1992).

21 Lewis Alexander, *Baseline Delimitations and Maritime Boundaries*, 23 VA J. INT'L L.

22 David D. Caron, *When Law Makes Climate Change Worse: Rethinking the Law of Baselines in Light of a Rising Sea Level*, 17 ECOLOGY L, Q, 621–53, 634 (1990); David D. Caron, *Climate Change, Sea Level Rise and the Continuing Uncertainty in Oceanic Boundaries: A Proposal to Avoid Conflict, in* MARITIME BOUNDARY DISPUTES, SETTLEMENT PROCESSES AND THE LAW OF THE SEA (S.Y. Hong & J. Van Dyke eds., 2008).

23 Alfred H. A. Soons, *The Effects of a Rising Sea Level on Maritime Limits and Boundaries*, 37(2) NETHERLANDS INT'L L. REV. 207, 216–218 (1990).

24 LOSC, *supra* note 5, at Art. 13.

25 *Id.* at Art. 6.

waters and other natural features form an intrinsic geographical, economic and political entity, or which historically have been regarded as such." Archipelagic States are defined in Article 47 of the LOSC as States constituted wholly by one of more archipelagos. These States are entitled to draw archipelagic baselines joining the outermost points of their islands and drying reefs (the part of a reef that is above water at low tide but is submerged at high tide) within which a special regime of "archipelagic waters" applies.[26] Archipelagic baselines are subject to a number of criteria, including maximum length constraints and a maximum water-to-land ratio of 1 to 9.[27] These States may find themselves struggling to meet Article 47 requirements if a rise in sea level removes low-tide elevations or drying reefs from the land mass calculations.[28]

Even greater shifts may occur in the case of islands. Although an island can generate all maritime zones, "rocks which cannot sustain human habitation or economic life of their own shall have no exclusive economic zone or continental shelf."[29] Prima facie, once rendered uninhabitable, an island will lose its exclusive economic zone and its continental shelf.[30] Moreover, should the island be reduced to nothing more than a low-tide elevation, it will generate only a territorial sea if it is situated within 12 nm of the mainland or of another island still generating a territorial sea. If the former island is located outside the 12 nm distance or if it is totally submerged, it will lose its territorial sea altogether.[31] The potential effects of this loss may be illustrated by reference to the island

26 *Id.* at Art. 49.

27 *Id.* at Art. 47.

28 Victor Prescott & Clive Schofield, The Maritime Political Boundaries of the World 176 (Leiden: Martinus Nijhoff, 2005).

29 LOSC, *supra* note 5, at Art. 121(3).

30 This issue is not free from doubt. However, Schofield points to the voluntary reclassification of Rockall by the UK from island to rock status and its consequential relinquishment of its claim to some 60,000 square nautical miles of fishery zone as State practice in support of the proposition. *See* Clive Schofield, *Shifting Limits? Sea Level Rise and Options to Secure Maritime Jurisdictional Claim's*, 4 Carbon and Climate L. Rev. 405, 409 (2009).

31 Soons, *supra* note 23, at 216–17.

of Kapingamarangi, the southernmost island in the Federated States of Micronesia (FSM), which is located 300 km south of the nearest island. Should Kapingamarangi become uninhabitable and be reduced to a mere "rock," its 500 inhabitants would lose their home and the FSM would lose in excess of 30,000 square nautical miles of its EEZ.[32] In addition to a significant shift in the location of maritime boundaries and zones, as Khadem puts it, "changes of this magnitude could provide a fertile source of inter-State conflict and spark disputes over navigation rights and, more particularly, sovereignty rights to living and non-living resources."[33]

States may seek to reinforce natural baseline and basepoint features to preserve them from inundation or erosion by wave action. The Netherlands provides the most extensive, but by no means the only, example of shoreline reinforcement and sea defense. In the South China Sea, Vietnam, China, Malaysia, and the Philippines have built up relatively small features such as sand cays and modest rocks into military facilities with runways, barracks, and helipads to substantiate EEZ claims.[34] Okinotorishima, two small rocks less than one meter above sea level on which Japan has constructed extensive concrete and rock breakwaters, is perhaps the most well-known and one of the most controversial examples of reinforcement for the purpose of claiming vast maritime zones.[35] Such artificial conservation of natural coastline basepoint and baseline features for the purpose of generating and maintaining maritime zones appears to be generally accepted under international

32 Stuart Kaye, *Maritime Claims in the Indian and Pacific Oceans*, Canberra: Sea Power Centre Australia (2011).

33 Alain Khadem, *Protecting Maritime Zones from the Effects of Sea Level Rise*, 5(3) IBRU Boundary and Security Bulletin (Autumn 1998); *see also* Jonathan Lusthaus, *Shifting Sands: Sea Level Rise, Maritime Boundaries and Inter-State Conflict*, 30 Politics 113 (2010).

34 David Whiting, *The Spratly Islands Dispute and the Law of the Sea*, 26 Denv. Int'l J. L. & Pol'y 897 (1998).

35 Yann-huei Song, *Okinitorishima: A "Rock" or an "Island"? Recent Maritime Boundary Controversy between Japan and Taiwan/China*, *in* Maritime Boundary Disputes, Settlement Processes and the Law of the Sea (John van Dyke & Seoung-Yong Hong eds., 2009).

law.[36] In other words, reinforced natural features do not appear, by virtue of that reinforcement, to lose their status as basepoints even if the natural feature itself is no longer above water.

The situation is somewhat different in the case of artificial islands, many of which have been built to host major construction works such as airports, harbors, industrial sites, or power plants and, in recent years, to provide housing needs where land is limited. An example particularly germane to the discussion here is the island of Hulhumalé, which was constructed by the government of the Maldives on the Kaafu Atoll next to the capital of Male. The island hosts the Malé airport and other housing, commercial, and industrial developments.[37] As Article 60(8) of the LOSC makes clear, artificial islands do not generate maritime zones. The distinction between the two cases may, however, often be a difficult one to make. For further discussion of the possible utility of physical embankments in preserving marine territories, see Chapter 6.

3. International Law and Statehood

3.1. The Statehood Dilemma

The difficulties posed by the ambulatory nature of baselines are most acute in the case of small low-lying island States. Only States are entitled to claim maritime zones. The traditional international law criteria for statehood include territory and a permanent population.[38] As the

36 For a full discussion of the legality of such efforts, see Chapter 6 in this volume.

37 Grigoris Tsaltas, Tilemachos Bourtzis, & Gerasimos Rodotheatos, *Artificial Islands and Structures as a Means of Safeguarding State Sovereignty Against Sea Level Rise: A Law of the Sea Perspective*, paper presented at the 6th Advisory Board on the Law of the Seas (ABLOS) Conference, Monaco, October 25–27, 2010, *available at* http://www.gmat.unsw.edu.au/ablos/ABLOS10Folder/ABLOS.htm.

38 Montevideo Convention on the Rights and Duties of States, art. 1, 49 Stat. 3097, T.S. 881, December 26, 1933. For a fuller discussion of the criteria for statehood, see Chapter 3 in this volume.

territory of a threatened island State disappears beneath the waves, the criteria of territory will no longer be met. Of course, as noted at the outset, physical disappearance of the land is most likely to be a gradual process, with the territory being rendered uninhabitable and the population having fled long before the territory's total physical disappearance. In that case too, however, the criteria for statehood will cease to be met from the time of evacuation.

Although rules exist in international law relating to the creation, dissolution, and succession of States, none exist that address specifically the question of what happens when a State loses its permanent population and/or physical territory.[39] Arguably there is a presumption in favor of the continuity of existing States, even in situations where their factual claim to continued statehood seems dubious.[40] Indeed in its advisory opinion on the *Legality of the Threat or Use of Nuclear Weapons*,[41] the International Court of Justice concluded that it would be lawful for a State to resort to the use of nuclear weapons in a situation where its very existence is at stake. If a State is entitled to resort to the use of nuclear weapons when its existence is threatened, the extinction of a State is clearly not lightly to be presumed. Nevertheless, as the case of Taiwan exemplifies, claims to statehood are ultimately a matter of acceptance by the international community. Collective nonrecognition by the international community of continued claims to statehood by disappearing States whose territory has been rendered uninhabitable by sea level rise, could ultimately lead to their extinction as a legal matter. The maritime zones of these States would then lapse and either revert to the high seas and "the Area" or, where geographical conditions permit, attach to neighboring States.[42]

39 *See* James Crawford, The Creation of States in International Law (2d ed. Oxford: Oxford University Press, 2006).

40 *Id.* at 34.

41 35 ILM 809 & 1343 (1996).

42 Caron, *supra* note 22, at 650; Soons, *supra* note 23, at 230.

3.2. Resolving the Statehood Dilemma

One potential solution to the statehood dilemma is for the disappearing State to acquire new territory from a distant State by a treaty of cession.[43] Like the Alaska purchase, sovereignty over the ceded land would transfer in its entirety to the disappearing State, which could then relocate its population to the new territorial location, thus securing the continued existence of the State in accordance with the traditional rules of international law. In the same way as any State currently claims maritime zones, the State would maintain its maritime zones, regardless of lack of geographical proximity.

From a legal perspective, acquisition of new territory by cession undoubtedly represents the most straightforward and appealing solution. However, as a practical matter, the political, social, and economic ramifications of ceding territory are likely to exceed the capacities – and courage – of existing governments.

Another option is for the disappearing State to merge, possibly into some form of federation, with another State.[44] Again, this would sustain the preexisting maritime zones. However, these zones would now inure to the "host" State. The disappearing State would cease to exist and, depending on the terms of confederation, would have no further say in the exploitation and management of its former maritime zones.

This, too, may seem like a straightforward, pragmatic, and legally sound solution to the problem of disappearing States. However, the equity and fairness of any solution that requires those States that are least to blame for climate change to relinquish jurisdiction and control over their vast and potentially extremely lucrative maritime zones to States that may well be among those most to blame for climate change must be questioned.

43 Soons, *supra* note 23, at 230. Cession of State territory refers to the "transfer of sovereignty over state territory by the owner-state to another state." *See* R. Jennings & A. Watts (eds.), Oppenheim's International Law (9th ed.), Vol 1., § 224 (1992).

44 Soons, *supra* note 23, at 230; Caron, *supra* note 22, at 650.

The application of equitable principles to ensure avoidance of inherently inequitable results is an established practice in the jurisprudence of the International Court of Justice (ICJ). To echo the ICJ in the *Gulf of Maine* case, "a legitimate scruple lies rather in concern lest the overall result . . . should unexpectedly be revealed as radically inequitable, that is to say, as likely to entail catastrophic repercussions for the livelihood and economic well-being of the population of the countries concerned."[45] The need to determine maritime boundaries to achieve an equitable solution has since been recognized as a principle of customary international law.[46] If maritime boundaries are to be determined on the basis of avoidance of inequity, then arguably so too should the issue of retention of maritime zones. Enforced confederation, in the circumstances of sea-level-threatened disappearing States, would appear to be intrinsically inequitable and contrary to international law.

An alternative that I have previously suggested as an equitable and fair solution would be the recognition in international law of a new category of State, the "deterritorialized State."[47] Such States would continue to exist as international legal entities despite their physical disappearance through population loss or inundation of territory caused by sea level rise, and they would maintain sovereignty over their maritime zones. As I have discussed elsewhere,[48] the concept of the deterritorialized State is neither new nor inconsistent with current international law that already recognizes that sovereignty and nationhood may be separated from territory. It is thus fully capable of recognizing the continuation of sovereign rights and entitlements without further victimizing disappearing States by forcing them to relinquish territory and maritime rights. It is simply a matter of working out the details. A more

45 Delimitation of the Maritime Boundary in the Gulf of Maine Area (Canada/United States of America), Advisory Opinion, 1984 ICJ 246 (January 20).

46 Maritime Boundary in the Area between Greenland and Jan Mayen (Denmark v. Norway), Advisory Opinion, 1993 ICJ 38 (June 14).

47 Rayfuse, *supra* note a. This alternative is explored in greater detail in Chapter 4 of this volume.

48 See the publications by Rayfuse, *supra* note a.

detailed discussion of this option, and how some of these details might be resolved, is available in Chapter 4.

In this respect, there is no reason, in principle, why the management of maritime zones would be more problematic for a disappeared State than for a State in possession of distant offshore territories. Admittedly, the challenges of monitoring, control, surveillance, and enforcement would be great. However, these challenges can be met by the use of sophisticated technologies and through cooperation and coordination within regional fisheries management organizations, the International Seabed Authority, the International Maritime Organization, and other relevant international organizations. From an oceans governance perspective the continued recognition of a deterritorialized State as a State would contribute to the ongoing promotion and achievement of the LOSC's objectives of peace, stability, certainty, fairness, and efficiency in oceans governance.

However, the exercise, protection, and retention of rights presuppose their initial possession. Thus, we must return to the baseline dilemma and its possible resolution.

4. Resolving the Baseline Dilemma and Preserving the Indicia of Statehood

As the saying goes, hindsight is 20/20. With hindsight, it is easy to suggest that the drafters of the LOSC should have considered the effects of sea level rise on the legal regime they were crafting and provided rules to cover its eventuality. Nevertheless, although not considered in the 1970s,[49] the issue is now arguably ripe for consideration in light of changing circumstances and the increasing evidence of both real and projected global sea level rise.

49 David Freestone & John Pethick, *Sea Level Rise and Maritime Boundaries: International implications of impacts and responses*, *in* Maritime Boundaries (Gerald Blake ed., 1994).

Two sets of legal approaches can be taken to deal with the effects of sea level rise on maritime boundaries. The first set relates to actions taken in accordance with existing law. The second set involves the development of new rules of international law permanently fixing baselines.

4.1. Actions in Accordance with Existing Law

With respect to existing rules, a first approach presents itself in the guise of a (re)interpretation and application of Article 5 of the LOSC, which suggests that the practical effect of marking the low-water line on large-scale charts officially recognized by the coastal State may be to "fix" that baseline as against coastal regression and the claims of other States, at least until such time as new charts are produced.[50] As Caron notes, this approach does not resolve the legal question of whether the LOSC intended baselines to be fixed or ambulatory in the case of coastal regression.[51] Moreover, as geographers and cartographers note, coastal regression may result in a disjuncture between the low-water line marked on the chart and the actual low-water line, placing at risk the management and enforcement of maritime zones, as well as the safety of navigation.[52] States seeking to retain their baselines may simply refrain from updating charts, with possibly disastrous consequences. Nevertheless, it is unlikely that States will willingly revise their charts to their detriment so as to relinquish existing baseline and maritime claims. Thus, the agglomeration of state practice may ultimately result in the development of a new rule of customary international law effectively freezing baselines at the location marked on the chart.

A second approach entails States taking greater advantage of the rules on straight baselines. Article 7(2) of the LOSC allows for the use of straight baselines where, because of natural conditions, the coastline is highly unstable. Although originally considered to apply only to deltas,

50 D. Kapoor & Adam Kerr, A Guide to Maritime Boundary Delimitation 31 (1986).

51 Caron, *supra* note 22, at 634.

52 Schofield, *supra* note 30, at 408.

it is open to States to (re-)interpret the criteria of instability to apply in the context of sea level rise. Article 7(5) of the LOSC provides that, where the method of drawing straight baselines is applicable, account may be taken of economic interests peculiar to the region concerned, the reality and the importance of which are clearly evidenced by long usage. There is no question that maritime zones are of economic interest to coastal States, particularly so for small island States whose economies depend in large part on revenues associated with activities in these zones. The emergence of an unstable coastline where once a stable coastline existed may provide new opportunities for States to draw straight baselines and to do so in such a way as to maintain their existing maritime zones. Although this approach raises the often controversial question of whether the rules on straight baselines are applicable in the first place, as a practical matter, States can ensure their ability to draw straight baselines in the future by, for example, building lighthouses and other installations on unstable basepoints or low-tide elevations, thereby enabling their continued use as basepoints for fixed straight baselines long after submergence of the geological feature.

Of particular relevance to island States is the possibility of taking greater advantage of the rules on archipelagic baselines in Article 47 of the LOSC. Article 47 allows archipelagic States to draw archipelagic straight baselines that join the outermost points of the outermost islands and drying reefs of the archipelago, provided that certain criteria are met: the water-to-land ratio of the area included within the baselines must not exceed 9 to 1; the length of the various baseline segments must not exceed 100 nm, although, exceptionally, up to 3% of the total number of baselines may be up to 125 nm in length; the baselines must not depart to any appreciable extent from the general configuration of the archipelago; and the basepoints must not be drawn from low-tide elevations unless either permanent above-water structures have been built on them or they are otherwise within the territorial sea of the nearest island. Although in theory archipelagic States may fail to meet these requirements as a result of sea level rise, the construction of even modest structures will ensure the continued validity of these features as basepoints even after their permanent inundation.

Articles 16, 47, and 75 of the LOSC require coastal States to publicize their straight and archipelagic baseline claims and the outer limit lines of their EEZs and to deposit charts and/or lists of relevant geographic coordinates with respect to those claims with the Secretary-General of the United Nations. Importantly, the ability to stipulate geographic coordinates indicates that baselines and charts may be disaggregated. As such, the charts may subsequently be amended for navigation purposes to reflect the physical reality of a receding coast while the baselines and outer EEZ limits remain fixed. Thus, once declared and publicized, these claims attract a presumption of permanence.

This presumption becomes particularly germane in the case of the Pacific island States whose domestic maritime zone legislation, all prepared with the assistance of the Commonwealth Secretariat during 1970s, follows a common model. In roughly identical terms, provision is made in the legislation of the Cook Islands,[53] Fiji,[54] Federated States of Micronesia (FSM),[55] Kiribati,[56] the Marshall Islands,[57] Nauru,[58] Niue,[59] Palau,[60] Papua New Guinea,[61] Samoa,[62] Solomon Islands,[63] Tokelau,[64] Tonga,[65] Tuvalu,[66] and Vanuatu[67] for the drawing of archipelagic baselines. Where archipelagic baselines are not used, baselines are generally

53 Territorial Sea and Exclusive Economic Zone Act (Act No. 16/1977) (Cook Is.).
54 Marine Spaces Act 1977 (Act No. 18) (Fiji).
55 The Code of the Federated States of Micronesia, Territory, Title 18.
56 Marine Zones (Declaration) Act 1983 (Act No. 7) (Kiribati).
57 Marshall Islands Revises Code, art. 33, ch. 1.
58 Sea Boundaries Act 1997 (Act No. 16) (Nauru).
59 Niue, Territorial Sea and Exclusive Economic Zone Act 1996 (Act No. 220/1997).
60 Palau National Code, Title 27, Ch. I, Subchapter III.
61 National Seas Act 1978 (Act No. 7) (Papua N.G.).
62 Maritime Zones Act 1999 (Act No. 18) (Samoa).
63 Delimitation of Marine Waters Act 1978 (Act No. 32) (Solom. Is.).
64 Territorial Sea and Exclusive Economic Zone Act 1977 (Act No. 125) (Takelau).
65 Territorial Sea and Exclusive Economic Zone Act 1978 (Act No. 30) (Tonga).
66 Maritime Zones (Declaration) Act (Chapter 24A/1983) (Tuvalu).
67 Maritime Zones Act 1982 (Act No. 23) (Vanuatu).

defined as the low-water line along the coast of each island except where an island is situated on an atoll or where it has a fringing reef, in which case the baseline is the seaward low-water line of the reef. Territorial seas, exclusive economic zones, and fisheries limits are all defined as measured from baselines, whereas the continental shelf is defined only by reference to the EEZ, as being the seabed and subsoil thereof contained within the EEZ.

The implications of this legislation are immediately obvious. In the absence of any declaration of straight or archipelagic baselines, should the low-water line move, the baselines will move – not as an effect of the operation of international law but rather as an effect of the operation of these States' own domestic law. Only Fiji, Papua New Guinea, Solomon Islands, and Vanuatu have declared and published straight and/or archipelagic baselines.[68] Nauru has declared the basepoints around the entire island from which its baseline is determined.[69] Kiribati, which claims archipelagic State status, and other Pacific island States have not published their baselines. Thus, even without resolution of the international legal issue relating to the ambulatory nature of baselines, the prudent course would be for the Pacific island States, as a matter of priority, to declare and publish their straight and archipelagic baselines and the outer limits of their EEZs by reference to geographical coordinates.

In addition, the Pacific island States should amend their domestic legislation to stipulate that the breadth of their maritime zones is measured from baselines that are proclaimed by periodic subsidiary regulation. The objectives of such an amendment are twofold. First, it preempts any domestic legislative inference of ambulatory baselines, an inference that could be relied on by other States as evidence of state practice confirming the ambulatory nature of baselines under international law. Second, it allows for more expeditious amendment should the need (or the desire) arise. Unless and until amended, these coordinates will "fix"

68 Maritime Space: Maritime Zones and Martime Delimination, http://www.un.org/Depts/los/LEGISLATIONANDTREATIES/depositpublicity.htm (last visited June 26, 2011).

69 Sea Boundaries Act 1997(Act No. 16) (Nauru).

the location of the baselines as a matter of domestic law. Once filed with the UN Secretary-General these coordinates can be relied on to assert that baselines drawn in accordance with the coordinates have been "fixed" internationally, regardless of the location of the low-water line.

The proposed legislative approach is consistent with that already adopted by many States. Australian legislation, for example, provides that the limits for its maritime zones and the baselines from which they are to be measured are to be declared, from time to time, by Proclamation of the Governor-General.[70] It is highly unlikely that any State adopting this methodology will willingly revise its baseline and outer limits coordinates to its detriment. Thus, by accretion of State practice, baselines may become "fixed" at the geographical coordinates expressed in domestic legislation. This does not entirely resolve disputes over excessively large baseline claims. However, where the location of baselines and the outer limits of the maritime zones are not currently disputed, this approach represents a relatively straightforward precautionary measure that should be taken as a matter of urgency.

A third approach calls for States to permanently establish the outer limits of their continental shelf. When claiming an extended continental shelf beyond 200 nm, coastal States are required by Article 76 of the LOSC to deposit with the UN Secretary-General charts and relevant information, including geodetic data, permanently describing the outer limits of their continental shelf. It is unclear whether this permanent designation was intended to apply to all of the continental shelf or only to that part of the shelf beyond 200 nm. However, according to Hayashi, the implication is that if a State files coordinates relating to the limits of all parts of its continental shelf, then those limits are permanently set for both the 200-nm limit and any extended shelf.[71] Consequently, the recession of baselines due to sea level rise will have no

70 Seas and Submerged Lands Act 1973 (Act No. 161/1973) (Au.).

71 M. Hayashi, *Sea Level Rise and the Law of the Sea: Legal and Policy Options*, *in* PROCEEDINGS OF THE INTERNATIONAL SYMPOSIUM ON ISLANDS AND OCEANS 78–90 (Terashima ed., 2009).

effect on the breadth of the continental shelf. Outer continental shelf submissions, or notices of intention to submit to the Commission on the Limits of the Continental Shelf, have been filed by a number of small island States, including the Cook Islands, Fiji, FSM, Maldives, Mauritius, Palau, Papua New Guinea, Seychelles, Solomon Islands, Tonga, and Vanuatu.[72]

A fourth mechanism States can use to protect many of their maritime entitlements is to definitively delimit their maritime boundaries in bilateral treaties with other States. Under international law generally a treaty may become void where there is a fundamental change of circumstances. However, the operation of that rule is expressly excluded in the case of a treaty that establishes a boundary.[73] Careful negotiation and drafting are essential because the permanence of the boundary may depend on the terms of treaty. Although most treaties establish boundaries as a series of geographical coordinates, if a treaty simply refers to a method of delimitation – such as the use of an equidistance line – the loss of land territory may affect the boundary. For example, the delimitation treaty between Tuvalu and France in respect of Wallis and Fortuna[74] refers only to a method of delimitation. Should some of the southernmost islands of Tuvalu be submerged or be otherwise disqualified from generating an EEZ, precedent exists to support an argument by France in favor of a northward shift in the boundary to the detriment of Tuvalu.[75] To prevent such shifts, boundaries, like baselines, should be expressed as a series of immutable geographical coordinates.

72 For final submissions *see* http://www.un.org/Depts/los/clcs_new/commission_submissions.htm. For preliminary information notifications *see* http://www.un.org/Depts/los/clcs_new/commission_preliminary.htm.

73 Vienna Convention on the Law of Treaties, art. 62, May 23, 1969, 1155 U.N.T.S. 331 8 1. L. M. 679.

74 Exchange of Notes constituting an agreement concerning a provisional maritime delimitation between two countries, France-Tuvalu, August 6, 1985-November 5, 1985, 1506 U.N.T.S.1987.

75 *See* the *Chamizal Case* (Mex. v. U.S.), 11 R.I.A.A. 309 (1911).

4.2. Developing New Law

Over the years, a number of commentators have expressed the need to reject the ambulatory theory of baselines and to adopt new positive rules of international law freezing either baselines or the outer limits of maritime zones. New rules, it is argued, are necessary to counter the potential for economic waste, instability, and conflict that the ambulatory theory implies.

Caron suggests a rule that would permanently fix the boundaries of all maritime zones on the basis of presentably accepted baselines.[76] Such a rule, he argues, would be fair and equitable because it would merely freeze the present division of authority and allocation of maritime entitlements agreed to in the LOSC. Under this rule, no State would gain any additional entitlement even if the baselines were to recede. Judge José Luis Jesus of the International Tribunal for the Law of the Sea agrees that once baselines have been established in accordance with the LOSC they should be seen as permanent, irrespective of any changes in sea level.[77] Judge Jesus would also apply this presumption in the case of newborn islands and the future qualification of rocks as either habitable or uninhabitable. A future rise in sea level would therefore entail neither a loss nor an acquisition of ocean space and of associated jurisdictional rights over maritime spaces and resources. States would maintain their current entitlements, and neither the international seabed nor the high seas commons would be affected. Similarly, Soons proposes a new general rule of international law freezing the outer limits of maritime zones "where they were located at a certain moment in accordance with the general rules in force at the time."[78]

76 Caron, *supra* note 22, at 623, 640–641.

77 José Luis Jesus, *Rocks, New-Born Islands, Sea Level Rise and Maritime Space, in* VERHANDELN FÜR DEN FRIEDEN, NEGOTIATING FOR PEACE 602–602 (J. Frowein et al. eds., 2003).

78 Soons, *supra* note 23, at 225. *See also* Caron, *supra* note 22, at 650.

In assessing these approaches, we should appreciate the different legal ramifications of permanently fixing baselines versus permanently fixing the outer limits of maritime zones. If the baseline is fixed, any new ocean areas created as a result of sea level rise will lie landward of the baseline and become internal waters. The legal status of the newly submerged area will be no different than when it was dry land. Conversely, if only the outer limits of the maritime zones are fixed then, as the sea level rises, the baselines will move and the breadth of the territorial sea or EEZ will expand, diminishing a State's jurisdiction over the newly submerged area. It has been suggested that the former approach is preferable, not only as a matter of fairness but also because it would avoid the need to amend the rules on the breadth of the territorial sea and the EEZ and would, in any event, have the effect of fixing the outer limits of the maritime zones.[79]

A final consideration relates to the date from which baselines are frozen. Various options exist, ranging from the date of entry into force of the LOSC, either in general or for the individual State, to the date on which the established baselines are publicized on the relevant charts and deposited with the Secretary-General of the United Nations as required by Article 16 of the LOSC. Hayashi argues that the latter is optimal because it is both consistent with explicit LOSC obligations and it will encourage States to establish their baselines.[80]

This is not to suggest that the freezing of baselines would resolve all disputes over entitlement to and delimitation of maritime zones. Like the freeze on disputed sovereignty claims in Antarctica, freezing declared baselines would not necessarily imply the resolution of disputed claims. Preexisting disputes over the status of rocks and islands or the location and legitimacy of straight baselines would persist until resolved through the normal processes. However, pending resolution of any such outstanding disputes, a freezing of baselines and the associated outer limits

79 Hayashi, *supra* note 71, at 83.
80 *Id.* at 84.

of all maritime zones would be consistent with, and would significantly assist in, the promotion and achievement of the LOSC objectives of peace, stability, certainty, fairness, and efficiency in ocean governance.[81] As Caron suggests, freezing of baselines would also be a valuable climate change adaptation strategy. Resources could be directed to substantive adaptation needs rather than to the preservation of baselines; new wetlands and coastal ecosystems could be created to replace those lost to rising seas, thereby assisting in the relocation and conservation of species and habitats under threat; and maritime zones, the prime asset of many coastal States, in particular, small island States, would be preserved.[82]

Various procedural options for achieving the adoption of these new rules exist. First, the generation of new rules can occur through the normal process for the development of customary international law. Although generally slow and unwieldy, precedent exists for the rapid generation of customary rules in the context of the law of the sea. Unknown to international law before the 1970s, the concept of the EEZ had gained sufficient support in state practice and *opinio juris* by 1985 to be considered by the International Court of Justice to have crystallized into customary international law.[83] By taking practical steps to declare and publicize baselines, particularly those established by reference to geographical coordinates, and refraining from changing these declarations in the face of sea level rise, States may generate a new custom rebutting the ambulatory interpretation of the LOSC.

Given the vagaries of the process of formation of customary rules, however, other more expeditious or determinative mechanisms should be considered. Hayashi identifies four possible approaches.[84] First, States might consider modification or expansion of the LOSC provisions either

81 LOSC, *supra* note 5, at pmbl.

82 Caron, *supra* note 23, at 642–50.

83 Continental Shelf (Libya Arab Jamahiriya/Malta), Advisory Opinions and Orders, 1985 I.C.J. 29 (June 3).

84 Hayashi, *supra* note 71, at 87–89.

by way of formal amendment or by de facto amendment by a decision of the Meeting of the States Parties (SPOLS). With respect to a formal amendment of the LOSC, Hayashi notes that States have been reluctant to invoke the amendment procedure, which has never been used, and concludes that any such attempt is unlikely.[85] In contrast, de facto amendment by the States Parties has occurred on four occasions. However, Hayashi points out that these changes have all related to procedural provisions establishing certain time limits. Indeed, the ability of SPLOS to deal with substantive matters is a controversial issue among the States Parties to the LOSC.[86] However, as Freestone and Oude Elferink note, no real impediment exists in international law that prevents the States Parties from agreeing on and adopting substantive amendments, including protocols or similar agreements, by way of decision of the SPLOS.[87] Nevertheless, any such amendment would only be binding on the States Parties.

Mechanisms external to the LOSC might also be used. These include the adoption of a supplementary agreement for modification of implementation of the LOSC provisions either by a separate diplomatic conference or by UN General Assembly resolution. As Hayashi notes, the first procedure was used in the negotiation and adoption of the UN Fish Stocks Agreement, whereas the second was used in the adoption of the Part XI Implementing Agreement.[88] These processes provide an additional advantage because they may involve all interested States, not just the States Parties to the LOSC. However, therein also lies their disadvantage, because there appears to be little appetite in the international community for engaging in detailed and possibly protracted negotiations on the matter.

85 *Id.* at 87.

86 *Id.* at 89.

87 David Freestone & Alex Oude Elferink, *Flexibility and Innovation in the Law of the Sea – Will the LOS Convention Amendment Procedures Ever be Used?*, *in* Stability and Change in the Law of the Sea: The Role of the LOS Convention 209 (A. Oude Elferink ed., 2005).

88 Hayashi, *supra* note 71, at 89–90.

5. Conclusion

The right to claim and maintain maritime jurisdictional entitlements is a quintessential hallmark of statehood. Current international law does not adequately address the continued maintenance of these entitlements in the context of sea level rise. Adoption by the international community of a positive rule freezing baselines would address the concerns of all States threatened by sea level rise and would be consistent with the LOSC objectives of peace, stability, certainty, fairness, and efficiency in ocean governance. In the case of small island States whose physical existence may be threatened by sea level rise – our disappearing States – a positive rule freezing baselines would also assuage some concerns as to their future, in a manner fully consistent with the presumption of the continuity of States.

Adoption of a positive rule freezing baselines does not appear to be high on the international agenda at the moment. Nevertheless, coastal States that are threatened by sea level rise have a range of options available to them to help ensure the continued preservation of their maritime jurisdictional interests into the future. Possibly the most straightforward, least costly, and immediately achievable measure is to ensure that their domestic legislation and baseline practice are consistent with their international ambitions. Given sufficient state practice in this regard, these ambitions may one day become law.

Part III RESETTLEMENT PROTECTIONS AND PROPOSED SOLUTIONS

8 Human Rights and Climate Change

Reflections on International Legal Issues and Potential Policy Relevance

Siobhán McInerney-Lankford

1. Introduction

Evidence of the social and human implications of a changing climate attests to actual and potential negative impacts on the full spectrum of human rights. Indeed, it would be difficult to identify a human right that is not potentially negatively affected by climate change, whether from slow-onset change in the natural environment or increasingly frequent sudden environmental disasters and extreme weather events.[1]

1 *See* Intergovernmental Panel on Climate Change [IPCC], *Climate Change 2007: Impacts, Adaptation and Vulnerability: Contribution of Working Group II to the Fourth Assessment Report of the IPCC*, at 82 (2007).

The author thanks David Freestone, Michael Gerrard, Nathan Lankford, and John Knox for comments on earlier drafts of this article, and Behnaz Bonyadian for excellent research assistance. Responsibility for errors or omissions remains with the author. This article is based on a presentation given at Columbia Law School on May 24, 2011, which included findings of a study commissioned by the World Bank for which the author was Task Team Leader: the author acknowledges the work of Mac Darrow and Lavanya Rajamani in that regard. The views expressed in this article are those of the author and do not necessarily reflect the views of the Board of Executive Directors of the World Bank or the governments they represent.

This chapter considers elements of the potential added value[2] of human rights to climate change[3] by tracing the connection between the two from the perspective of public international law. The analysis considers the potential contribution that human rights law[4] may make to current discussions of climate change. However, it does not in any way presume the exclusive relevance of legal approaches, given that rights also encompass moral and ethical claims[5] for prompting action on climate change.[6] Nor does this approach ignore how rights discourse can, more broadly, promote empowerment and mobilize political support for climate advocacy, given that human rights are underpinned by powerful ethical and moral arguments.[7]

2 *See* Siobhán McInerney-Lankford, *Human Rights and Development: A Comment of Challenges and Opportunities from a Legal Perspective*, 1 J. Hum. Rts. Prac. 51 (2009) [hereinafter McInerney-Lankford, *A Comment of Challenges*], for a similar analysis elaborated by the author on the value added of human rights to development.

3 *See generally* Int'l Bank for Reconstruction & Dev. [IBRD], World Bank Grp. [WBG], *A World Bank Study: Human Rights and Climate Change: A Review of the International Legal Dimensions* (2011) (by Siobhán McInerney-Lankford, Mac Darrow, & Lavanya Rajamani).

4 There are numerous authoritative texts on human rights law. *E.g.*, Bertrand G. Ramcharan, The Fundamentals of International Human Rights Treaty Law (2011); International Human Rights Law (Daniel Moeckli et al. eds., 2010); Andrew Clapham, Human Rights: A Very Short Introduction (Very Short Introductions Vol. 163, 2007); Jack Donnelly, Universal Human Rights in Theory and Practice (2d ed. 2003); Thomas Buergenthal, *The Normative and Institutional Evolution of International Human Rights*, 19 Hum. Rts. Q. 703 (1997); Antonio Cassese, Human Rights in a Changing World 9–48 (1990).

5 Theories of Right (Jeremy Waldron ed., 1984); Allen E. Buchanan, Justice, Legitimacy, and Self-Determination: Moral Foundations for International Law (2004); Amartya Sen, *Elements of a Theory of Human Rights*, 32 Phil. & Pub. Aff. 315 (20040; Carlos S. Nino, The Ethics of Human Rights (1991).

6 Sara C. Aminzadeh, *A Moral Imperative: The Human Rights Implications of Climate Change*, 30 Hastings Int'l & Comp. L. Rev. 231 (2007).

7 Edward Cameron, *Development, Climate Change and Human Rights: From the Margins to the Mainstream?* (World Bank Soc. Dimensions of Climate Change Practice Grp., Social Development Papers No. 123, 2011).

Existing work points to the difficulty of applying human rights law to climate change[8] and the limits of human rights litigation in addressing climate change,[9] and this contribution acknowledges those challenges. Its central contention is rather that legal and policy discussions of global climate change are not complete without some consideration of international human rights obligations. Human rights obligations are part of the global legal and normative environment – and "the normative environment cannot be ignored."[10]

The analysis recalls existing characterizations of the human rights impacts of climate change[11] and extrapolates the potential application of human rights law to climate change. The chapter is divided into three parts. The first part offers a theoretical analysis of human rights obligations generated by international human rights treaties and how they may relate to climate change through the framework of obligations to respect, protect, and fulfill. The second part considers the mutual relevance of human rights and climate change obligations under international law and traces areas of distinction and tension, as well as areas of convergence. The third, concluding section offers reflections on potential policy implications of human rights obligations in addressing climate change.

8 John H. Knox, *Climate Change and Human Rights Law*, 50 VA. J. INT'L L. 163 (2009) [hereinafter Knox, *Climate Change*].

9 Eric A. Posner, *Climate Change and International Human Rights Litigation: A Critical Appraisal*, 155 U. Pa. L. Rev. 1925 (2007).

10 Int'l Law Comm'n [ILC], *Fragmentation of International Law: Difficulties Arising from the Diversification and Expansion of International Law: Report of the Study Group of the ILC*, ¶ 419, at 211, U.N. Doc. A/CN.4/L.682 (Apr. 13, 2006) (finalized by Martii Koskenniemi) [hereinafter ILC, *Fragmentation Report*].

11 Int'l Council on Human Rights Policy [ICHRP], *Climate Change and Human Rights: A Rough Guide* (2008) [hereinafter ICHRP, *Rough Guide*]; Ctr. for Int'l Envtl. Law [CIEL], *Climate Change and Human Rights: A Primer* (May 2011), *available at* http://www.ciel.org/Publications/CC_HRE_23May11.pdf; CIEL & Friedrich-Ebert-Stiftung [FES], *Human Rights and Climate Change: Practical Steps for Implementation* (Feb. 25, 2009).

2. What Makes Climate Change a Human Rights Issue?

The multitude of social and human impacts that result from accelerating climate change are clear and well evidenced.[12] The *Fourth Assessment Report of the Intergovernmental Panel on Climate Change*[13] (IPCC AR4) concludes that the global climate system is warming and doing so mainly because of human-made greenhouse gas emissions. It documents the observed and projected changes in weather due to global warming to include the following: the contraction of snow-covered areas and shrinking of sea ice, rising sea levels and higher water temperatures, increased water acidification, heavy precipitation events, and an increase in tropical cyclones and droughts. The IPCC AR4 also documents in stark terms the human impacts of such climate change in areas such as food, water, health, settlement, and society.[14]

Accounts of the impacts of climate change on the physical environment and the lives of peoples around the world can also be understood as accounts of the undermining of human rights, and a body of literature exists characterizing it in this way.[15] But such accounts can also be understood as accounts of the nonrealization, or even violation, of human rights *law*[16] with impacts on civil and political rights, as well as

12 *See generally Social Development: Social Dimensions of Climate Change*, The World Bank, http://go.worldbank.org/SD8MH5HLU0 (last updated June 21, 2011) (demonstrating the extensive work of the World Bank on the social impacts of climate change).

13 IPCC, *Climate Change 2007: Synthesis Report: Contribution of Working Group I, II, and III to the Fourth Assessment Report of the IPCC* (2007).

14 *Id.* at 48–53.

15 *E.g.*, Sumudu Atapattu, *Global Climate Change: Can Human Rights (and Human Beings) Survive this Onslaught?*, 20 Colo. J. Int'l Envtl. L. & Pol'y 35 (2008); Stephen L. Kass, *Integrated Justice: Human Rights, Climate Change and Poverty*, 18 Transnat'l L. & Contemp. Probs. 115 (2009); Int'l Bank for Reconstruction & Dev., *supra* note 3, at 11–19; Office of the United Nations High Comm'r for Human Rights [OHCHR], *Annual Report of OHCHR on the Relationship between Climate Change and Human Rights*, at 8–16, U.N. Doc. A/HRC/10/61 (Jan. 15, 2009) [hereinafter OHCHR, *Annual Report*].

16 OHCHR, *Annual Report*, *supra* note 15, at 8.

economic, social, and cultural rights.[17] These rights include the right to life and self-determination, the right to adequate food, the right to water, the right to health, and the right to adequate housing.[18] In addition, there are cross-cutting equality rights implicated by disproportionate impacts on discrete groups, such as women, children, indigenous peoples,[19] and persons with disabilities. This disproportionate impact is relevant in the climate change context, where climate effects will be felt most acutely by those groups who are already in vulnerable situations.[20]

Among the most prominent attempts to link human rights and climate change[21] through a legal strategy was the 2005

17 *See, e.g.*, Submission of the Maldives to the OHCHR under Human Rights Council Res. 7/23 (Sept. 25, 2008), *available at* http://www2.ohchr.org/english/issues/climatechange/submissions.htm; Submission of the Marshall Islands to the OHCHR under Human Rights Council Res. 7/23 (Dec. 31, 2008), *available at* http://www2.ohchr.org/english/issues/climatechange/submissions.htm.

18 Jessie Hohmann, *Igloo as Icon: A Human Rights Approach to Climate Change for the Inuit?*, 18 Transnat'l L. & Contemp. Probs. 295 (2009).

19 OHCHR, *Annual Report*, *supra* note 15, at 15–18. *See also* Emily Gerrard, *Climate Change and Human Rights: Issues and Opportunities for Indigenous Peoples*, 31 U. New S. Wales L.J., no. 3, 2008 at 941; Marguerite E. Middaugh, *Linking Global Warming to Inuit Human Rights*, 8 San Diego Int'l L.J. 179 (2006).

20 *See* U.N. Human Rights Council Res. 16/11, Human Rights and the Environment, 16th Sess., Feb. 28-Mar. 25, 2011, U.N. Doc. A/HCR/RES/16/11 (March 24, 2011). The Resolution notes that environmental damage can have negative implications, both direct and indirect, for the effective enjoyment of human rights, and it recognizes that environmental damage is most acutely felt by those segments of the population already in vulnerable situations.

21 Well-developed literature exists connecting human rights and the environment, including work on a free-standing right to a clean and healthy environment. *E.g.*, Dinah Shelton, *Human Rights, Environmental Rights, and the Right to Environment*, 28 Stan. J. Int'l L. 103 (1991); Paul W. Gormley, *Legal Obligation of the International Community to Guarantee a Pure and Decent Environment: The Expansion of Human Rights Norms*, 3 Geo. Int'l Envtl. L. Rev. 85 (1990); Iveta Hodkova, *Is There a Right to a Healthy Environment in the International Legal Order?*, 7 Conn. J. Int'l L. 65 (1991); Alan E. Boyle, "The Role of International Human Rights Law in the Protection of the Environment," in Human Rights Approaches to Environmental Protection 43 (Alan E. Boyle & Michael R. Anderson eds., 1998); Dinah Shelton, Environmental Rights, in Peoples' Rights 185 (Philip Alston ed., 2001); Dinah Shelton, "The Environmental Jurisprudence of International Human Rights Tribunals," in Linking Human Rights and the Environment 1 (Romina Picolotti & Jorge D. Taillant eds., 2003);

Inuit[22] Petition to the Inter-American Commission on Human Rights against the United States. It was submitted for violations of Inuit rights protected under the Inter-American Declaration on Human Rights which the Petitioners alleged resulted from the United States' acts and omissions related to climate policy.[23] Central to the petition was the claim that the failure to implement climate change policies by the United States, as one of the world's largest producers of carbon emissions per capita, violated the Inuits' right to the benefits of their culture, their right to use and enjoy their personal property and the lands they have traditionally used and occupied, and their right to the preservation of health. The Commission declined to process the petition on the basis that the information provided did not enable it to determine whether the alleged facts would tend to characterize a violation of rights protected by the American Declaration (for further discussion of sovereign obligations to their people, see Chapter 15).[24]

James R. May, *Constituting Fundamental Environmental Rights Worldwide*, 23 PACE ENVTL. L. REV. 113 (2006). See also G.A. Res. 2398 (XXIII), *Problems of the Human Environment*, U.N. GAOR, 23d Sess., U.N. Doc. A/L.553/Add.1–4 (Dec. 3, 1968) (noting the link between the impairment of the environment and the enjoyment of human rights); United Nations Conference on Environment and Development, Rio de Janiero, Braz., June 3–14, 1992, *Rio Declaration on Environment and Development*, Princ. 1, U.N. Doc. A/CONF.151/26 (Vol. 1), Annex 1 (Aug. 12, 1992) [hereinafter *Rio Declaration*] ("Human beings are at the centre of concerns for sustainable development. They are entitled to a healthy and productive life in harmony with nature.").

22 The NGO that brought the petition is the Inuit Circumpolar Conference (ICC), representing the Inuit people of Alaska, Canada, Greenland, and Russia. For a discussion of the Inuit Petition, *see* Hari M. Osofsky, *The Inuit Petition as a Bridge? Beyond Dialectics of Climate Change and Indigenous Peoples' Rights*, 31 AM. INDIAN L. REV. 675 (2007); Joanna Harrington, *Climate Change, Human Rights, and the Right to be Cold*, 18 FORDHAM ENVTL. L. REV. 513 (2007).

23 Inuit Circumpolar Conference [ICC], Center for International Environmental Law [CIEL] & Earthjustice, *Petition to the Inter-American Commission on Human Rights Seeking Relief from Violations Resulting from Global Warming Caused by Acts and Omissions of the United States* (Dec. 7, 2005), *available at* http://www.ciel.org/Publications/ICC_Petition_7Dec05.pdf.

24 Instead the Commission invited petitioners to present testimony and held a hearing on the relationship between human rights and climate change on March 1, 2007.

In nonlegal strategies, the efforts of the Maldives stand out internationally in advancing the recognition of the connection between human rights and climate change.[25] In November 2007, the Maldives hosted a conference of the Association of Small Island States (AOSIS) that addressed squarely the question of how climate change affects the human rights of citizens of such States. The outcome of the Conference was the *Malé Declaration on the Human Dimension of Global Climate Change*, which asserted unequivocally that "climate change has clear and immediate implications for the full enjoyment of human rights."[26]

During the December 2007 Bali Conference of Parties (COP 13), this theme was taken up by the Deputy High Commissioner for Human Rights, Kyung-wha Kang, in an address that explicitly linked the two agendas and laid out the implications of climate change for human rights. The Deputy High Commissioner drew attention to the need to consider the human dimensions of climate change and noted that a human rights perspective shifts the focus more directly to the individual. Critically, she observed that "[a]long with the human factor, human rights also introduce an accountability framework that is an essential element of the promotion and protection of human rights itself, by holding governments, the duty-bearers accountable to reducing the vulnerability of their citizens to global warming and assisting them in adapting to the consequences."[27] She concluded that "[a]s climate change will inevitably affect the enjoyment of human rights, safeguarding of human rights

25 *See Human Rights and Climate Change*, Ministry of Foreign Affairs, Republic of Maldives, http://www.foreign.gov.mv/v3/?p=menu_item&sub_id=50&submenu=Human%20Rights%20and%20Climate%20Change (demonstrating the efforts by the Maldives).

26 Small Island Developing States [SIDS], Malé, Republic of Maldives, Nov. 13–14, 2007, *Malé Declaration on the Human Dimension of Global Climate Change*, at 2, (Nov 14, 2007), *available at* http://www.ciel.org/Publications/Male_Declaration_Nov07.pdf.

27 Kyung-wha Kang, Deputy High Comm'r for Human Rights, OHCHR, Address at the Convention on Climate Change and its Kyoto Protocol: Climate Change and Human Rights (Dec. 14, 2007), *available at* http://www.ohchr.org/EN/NewsEvents/Pages/DisplayNews.aspx?NewsID=200&LangID=E.

should be a key consideration in efforts to address the impact of climate change. International human rights law imposes several obligations on States that are relevant to addressing human vulnerabilities to climate change."[28]

In 2008, the UN Human Rights Council adopted its first resolution on human rights and climate change[29] sponsored by the Maldives.[30] The resolution recognized that climate change has implications for the full enjoyment of human rights and called on the Office of the UN High Commissioner for Human Rights (OHCHR) to prepare a detailed analytical study on the relationship between climate change and human rights to be submitted to the tenth session of the Council in March 2009. In March 2009, the Human Rights Council adopted a second resolution on human rights and climate change[31] in which it, inter alia, notes that "climate change-related impacts have a range of implications, both direct and indirect, for the effective enjoyment of human rights."[32] It further recognizes that the effects of climate change "will be felt most acutely by those segments of the population who are already in a vulnerable situation."[33] The resolution confirmed that the realization of human rights relies on collaborative action to stem climate change by noting that "effective international cooperation to enable the full, effective and sustained implementation of the United Nations Framework Convention on Climate Change . . . is important in order to support national efforts for the realization of human rights implicated by climate change-related impacts" and affirming that "human rights obligations and commitments have the potential to inform and strengthen international and national

28 *Id.*

29 The Resolution secured eighty co-sponsors and was adopted by consensus. UNHRC Res. 7/23, U.N. Doc. A/HRC/7/78 (Mar. 28, 2008).

30 *See generally* John H. Knox, *Linking Human Rights and Climate Change at the United Nations*, 33 Harv. Envtl. L. Rev. 477 (2009).

31 The Resolution secured eighty-nine co-sponsors and was adopted by consensus. U.N. Human Rights Council Res. 10/4, U.N. Doc. A/HRC/RES/10/4 (Mar. 25, 2009).

32 *Id.* at 1.

33 *Id.* at 1.

policy-making in the area of climate change."[34] In March 2011, the UN Human Rights Council adopted a resolution on *Human Rights and the Environment*.[35] The resolution also calls on the OHCHR to conduct a detailed analytical study on the relationship between human rights and the environment to be submitted to the Human Rights Council prior to its nineteenth session[36] and in time to provide substantive inputs into the Earth Summit of 2012.[37] September 2011 saw the passing of another resolution on human rights and climate change,[38] which affirmed that human rights obligations, standards, and principles have the potential to inform and strengthen international and national policy making in the area of climate change, promoting policy coherence, legitimacy, and sustainable outcomes. That resolution also called for the convening of a seminar to address the adverse impacts of climate change on the full enjoyment of human rights, and promote respect of human rights in all climate change-related actions and policies and identify actions and best practices.

These recommendations resonate with some of the commitments made in the Outcome Document of the July 2012 Rio+20 Conference *The Future We Want.* It reaffirms the importance of all human rights, including the right to development (paragraph 8), as well as the importance of the Universal Declaration of Human Rights and other

34 *Id.* at 2. In Resolution 10/4, the Human Rights Council decided to hold a panel discussion on the relationship between climate change and human rights at its eleventh session in order to contribute to the realization of the goals set out in the Bali Action Plan. *Id.* at ¶ 1.

35 U.N. Human Rights Council Res. 16/11, *supra* note 20.

36 *Id.* at ¶ 1.

37 *See generally Earth Summit* 2012, http://www.earthsummit2012.org/, for more information on the UN Summit on Sustainable Development ("Earth Summit 2012", also called Rio+20. In this connection, OHCHR's key messages for Rio+20 centered on ensuring full coherence between efforts to advance the green economy, on the one hand, and human rights obligations on the other, and that all policies and measures adopted to advance sustainable development must be firmly grounded in, and respectful of, all internationally agreed human rights and fundamental freedoms, including the right to development.

38 Resolution 8/22.

international law instruments relating to human rights, emphasizing States' obligations to respect, protect and promote human rights and fundamental freedoms for all (paragraph 9). The Outcome Document further affirms that green economy policies in the context of sustainable development and poverty eradication should "promote sustained and inclusive economic growth, foster innovation and provide opportunities, benefits and empowerment for all and respect all human rights" (paragraph 58).

Finally, a recent concrete outcome of the UN Human Rights Council initiatives was the establishment of a special procedure.[39] In July 2012, the Human Rights Council appointed the first Independent Expert on the issue of human rights obligations related to the enjoyment of a safe, clean, healthy and sustainable environment.[40]

3. Implications of Characterizing Climate Impacts as Human Rights Impacts

3.1. Value Added from Human Rights Law

What is the specific implication of characterizing climate impacts as human *rights* impacts?[41] In short, it is the focus on obligations. A premise of this chapter is that the consideration of human rights introduces the concept of duty, whether in a moral and ethical sense or a legal sense (as transposed into binding legal obligation). That is because rights do

39 Established by Council's resolution 19/10 of 22 March 2012 (A/HRC/RES/19/10).

40 On July 6, 2012, the UN Human Rights Council named John Knox as the first Independent Expert on human rights and the environment.

41 John Merrills characterized the implication in the following way: "if rights are a good way of ensuring that something is taken seriously, designating an entitlement a human right is even better, on account of the status of this class of rights in legal and moral discourse." John G. Merrills, *Environmental Rights*, in OXFORD HANDBOOK OF INTERNATIONAL ENVIRONMENTAL LAW 663, 666 (Daniel Bodansky, Jutta Brunnee, & Ellen Hey eds., 2007).

not exist independently of duties[42] and because rights require correlative duties[43] and the identification of right-holders and duty-bearers. In a similar way, enforceable legal rights rely on the existence of correlative binding legal obligations. Thus, by introducing the concept of right, it focuses some consideration on the protection of an interest, a freedom, or an entitlement. The concept of duty identifies responsibility for that protection in a particular actor (usually the State). More fundamentally, however, the frameworks of right and duty, or right-holder and duty-bearer, generate a form of accountability for the right or entitlement in question and provide concrete baselines against which to assess processes and outcomes relevant to those rights.

Considering the effects of climate change from the perspective of human rights law locates the analysis in goods or entitlements that are the subject of human rights treaties and the object of binding legal obligations assumed by States.[44] However, the following analysis does not aim to establish that the negative impacts on the enjoyment of human rights result in violations of such rights in a strict legal sense.[45] Rather it aims to highlight the distinct value added, and potential relevance of, human rights law to policy approaches to climate change, notwithstanding some important constraints discussed herein. In this way, the analysis is not intended to demonstrate how an individual might hold a State responsible for harm caused by climate change, as much as how human rights obligations might inform State responses to climate change.

42 *See* Bernard Mayo, *What Are Human Rights?*, *in* Political Theory and the Rights of Man 68, 72 (David D. Raphael ed., 1967) (discussing the "correlativity of rights and duties").

43 Asbjørn Eide, *Economic, Social and Cultural Rights as Human Rights*, *in* Economic Social and Cultural Rights: A Textbook 22 (Catarina Krause, Asbjørn Eide, & Allan Rosas eds., 2d rev. ed. 2001) [hereinafter Eide, *Economic, Social, and Cultural Rights*].

44 *See* OHCHR, *Annual Report*, *supra* note 15, at 23. *See generally* Siobhán McInerney-Lankford, *Climate Change and Human Rights: An Introduction to Legal Issues*, 33 Harv. Envtl. L. Rev. 431, 437 (2009).

45 *See* OHCHR, *Annual Report*, *supra* note 15, at 23.

An essential and distinguishing element of the contribution of human rights to climate change discussions is the principle of accountability through the identification of duties and legally binding obligations under human rights treaties. The following analysis employs the tripartite typology of human rights obligations to analyze the contribution that human rights law may make to climate change policy. Moreover, the analysis focuses on what some commentators refer to as the "vertical axis" of human rights, which governs the relationship between States and their citizens,[46] rather than that between States and citizens outside their jurisdiction.

Several considerations are appropriate to preface this discussion. First, the current analysis focuses primarily on human rights protected under universal human rights treaties, and the two core treaties of the international bill of rights in particular[47]: the International Covenant on Civil and Political Rights (ICCPR)[48] and the International Covenant on Economic, Social, and Cultural Rights (ICESCR).[49] The present analysis considers how existing human rights obligations might be interpreted in a way that has direct relevance for climate change, rather than how provisions protecting a right to a clean and healthy environment in human rights instruments[50] could be interpreted to extend to climate

46 John H. Knox, *Diagonal Environmental Rights*, *in* Universal Human Rights and Extraterritorial Obligations 82 (Mark Gibney & Sigrun Skogly eds., 2010).

47 The analysis does not discount the possible relevance of customary law to the present context. *See* John Lee, *The Underlying Legal Theory to Support a Well-Defined Human Right to a Healthy Environment as a Principle of Customary International Law*, 25 Colum. J. Envtl. L. 283 (2000). However, it confines the analysis to the nomenclature of right and obligation under human rights treaties and those concluded under the aegis of the United Nations in particular. *See generally International Law*, OHCHR, http://www2.ohchr.org/english/law/index.htm.

48 International Covenant on Civil and Political Rights [ICCPR], Dec. 16, 1966, 999 U.N.T.S. 171, 6 I.L.M. 368.

49 International Covenant on Economic, Social and Cultural Rights [ICESCR], Dec. 19, 1966, 993 U.N.T.S. 3, 6 I.L.M. 360.

50 For example, Article 11 of the Protocol of San Salvador (to the American Convention on Human Rights) provides "Right to a Healthy Environment: 1. Everyone shall have the right to live in a healthy environment and to have access to basic public services. 2. The States parties shall promote the protection, preservation, and improvement of the

change.[51] Environmental human rights are often characterized as third-generation or solidarity rights, and the traditional typology of human rights law obligations does not apply quite so neatly to these rights, which, by definition, protects more collective interests and admits more diffuse and expansive definitions of obligation. It is also clear, however, that the obligations to respect, protect, and fulfill could have relevance to a right to a satisfactory environment where such a right is recognized, because the right is potentially characterized as "hybrid" and does not fit neatly into any single category or "generation" of rights, but straddles all three categories.[52]

3.2. International Human Rights Law Obligations

The primary frame of reference for this chapter's analysis of obligations is the structure of obligations to protect, respect, and fulfill[53]

environment." Org. of American States, Additional Protocol to the American Convention on Human Rights in the Area of Economics, Social, and Cultural Rights (Protocol of San Salvador) art. 11, Nov. 17, 1988, O.A.S.T.S. No. 69, 28 I.L.M 156. Article 24 of the African Charter on Human and Peoples Rights recognizes people's right to "a general satisfactory environment." Org. of African Unity, African [Banjul] Charter on Human and Peoples' Rights art. 24, June 27, 1981, OAU Doc. CAB/LEG/67/3 rev. 5, 21 I.L.M. 58 (1982). Other human rights treaties contain more indirect environmental provisions. *See, e.g.*, the Convention on the Rights of the Child provides that States parties shall take appropriate measure to combat disease and malnutrition "through the provision of adequate nutritious foods and clean drinking water, taking into consideration the dangers and risks of environmental pollution." United Nations Convention on the Rights of the Child art. 24, ¶ 2(c), Nov. 20, 1989, 1577 U.N.T.S. 3. *Id.* at art. 24. *See also* Int'l Labor Org., Convention Concerning Indigenous and Tribal Peoples in Independent Countries, June 27, 1989, No. 169, 72 ILO Official Bull. 59, 28 I.L.M. 1382 (protecting the environment of areas that indigenous peoples occupy or use).

51 The right to a satisfactory environment is recognized in a host of domestic jurisdictions but is comparatively rare to find protected in public international law. It is often categorized among third-generation rights' or solidarity rights (along with the right to development).

52 *See* Patricia W. Birnie & Alan E. Boyle, International Law and the Environment 253 (2d ed. 2002).

53 OHCHR, *Frequently Asked Questions on a Human Rights-Based Approach to Development*, at 9, HR/PUB/06/8 (2006).

employed by the UN Committee on Economic, Social, and Cultural Rights (CESCR).[54] These obligations are transversant and potentially apply to all rights, because each category of right may give rise to affirmative duties to take action or to duties of restraint.[55] This is relevant also to the way in which each type of obligation is understood and potentially linked to climate change in the following analysis. Although there are variations on this tripartite typology,[56] it is quite widely accepted as capturing the qualitative differences among types of human rights obligations established under human rights treaties.[57] The three categories

54 CESCR has relied on this typology since 1999. *See* U.N. Econ. & Soc. Council [ECOSOC], Committee on Economic, Social, and Cultural Rights [CESCR], *General Comment No. 12: The Right to Adequate Food*, 20th Sess., Apr. 26-May 14, 1999, U.N. Doc. E/C.12/1999/5 (May 12, 1999) [hereinafter CESCR *General Comment No. 12*]; ECOSOC, CESCR, *General Comment No. 14: The Right to the Highest Attainable Standard of Health*, 22d Sess., Apr. 25-May 12, 2000, U.N. Doc. E/C.12/2000/4 (Aug. 11, 2000) [hereinafter CESCR *General Comment No. 14*]; ECOSOC, CESCR, *General Comment No. 15: The Right to Water*, 29th Sess., Nov. 11–29, 2002, U.N. Doc. E/C.12/2002/11 (Jan. 20, 2003) [hereinafter CESCR *General Comment No. 15*]. For a thoughtful critique of the typology in relation to economic, social, and cultural rights, *see* Ida E. Koch, *Dichotomies, Trichotomies, Waves of Duties*, 5 Hum. Rts. L. Rev. 81 (2005) (discussing whether the typology properly reflects the complex legal issues connected with the implementation and judicial enforcement of economic, social, and cultural rights).

55 Sandra Fredman, Human Rights Transformed: Positive Rights and Positive Duties 68 (2008).

56 The duty to fulfill is defined by some scholars and commentators to include the duty to promote – as a duty to generate awareness about rights among rights-holders.

57 *See* Henry Shue, Basic Rights: Subsistence, Affluence, and U.S. Foreign Policy 51–52 (2d ed. 1980). Shue's classification specifies the duties to avoid depriving, to protect from deprivation, and to aid the deprived. Other human rights law typologies exist, including the obligations to "protect, promote, and provide." *Id.* at 5. Asbjørn Eide added the obligation to '"facilitate" to the standard tripartite typology – that obligation required that States parties "facilitate opportunities by which the rights listed can be enjoyed." Asbjørn Eide, *Universalisation of Human Rights versus Globalisation of Economic Power*, in Rendering Justice to the Vulnerable: Liber Amicorum in Honor of Theo van Boven 111 (Fons Cooman et al. eds., 2000). Van Hoof identifies the obligations to "respect, protect, ensure and promote." G. Van Hoof, *The Legal Nature of Economic, Social and Cultural Rights: A Rebuttal of Some Traditional Views*, *in* The Right to Food 97, 106 (Philip Alston & Katarina Tomaševski eds., 1984). Steiner and Alston identify still others: to respect the rights of others, to create the

are understood here as "levels of obligation,"[58] rather than sharply distinguishable types of obligations, according to positive or negative qualities or categories, strictly correlated with either grouping of right. With respect to the type of general obligation imposed, the ICCPR and the ICESCR contain different textual provisions that are considered in the following analysis.[59]

The following discussion explores the nature of obligations to respect, protect, and fulfill by drawing on examples of specific rights under the ICESCR and ICCPR to illustrate how the constitutive elements of human rights obligations may be brought to bear on climate change.

3.2.1. The Obligation to Respect

The obligation to respect captures the "basic obligation not to interfere or deprive to the extent that this jeopardizes current human rights enjoyment for any individual."[60] The *American Convention on Human Rights* establishes it as a fundamental cross-cutting obligation in a general provision covering all Convention rights.[61] It can be seen as a general obligation governing all ICCPR rights through the terms of ICCPR Article 2, paragraph 1, which provides that "[e]ach State Party to the present Covenant undertakes to respect and to ensure to all individuals within its territory and subject to its jurisdiction the rights recognized in the

institutional machinery essential to the realization of rights, to protect against violations, and to provide goods and services to satisfy rights and to promote rights. HENRY J. STEINER & PHILIP ALSTON, INTERNATIONAL HUMAN RIGHTS IN CONTEXT: LAW, POLITICS, MORALS 182 (2d ed. 2000).

58 Asbjørn Eide, *Economic and Social Rights*, *in* HUMAN RIGHTS: CONCEPT AND STANDARDS 109, 134 (Janusz Symonides ed., 2000).

59 It is noteworthy that the African Charter imposes the same obligations on both types of rights and requires States parties to "recognize the rights" and "undertake to adopt legislative or other measures to give effect to them." Org. of African Unity, *supra* note 50, art. 1.

60 SIGRUN I. SKOGLY, BEYOND NATIONAL BORDERS: STATES HUMAN RIGHTS OBLIGATIONS IN DEVELOPMENT COOPERATION 68 (2006) [hereinafter SKOGLY, BEYOND NATIONAL BORDERS].

61 American Convention on Human Rights art. 1, Nov. 22, 1969, O.A.S.T.S. No. 36, 1144 U.N.T.S. 123.

present Covenant."[62] However, the obligation to respect may also apply to ICESCR rights, particularly in relation to nondiscrimination, which constitutes an obligation with immediate effect in the ICESCR[63] and is central to the ICCPR.[64] In this, it is primarily a negative duty targeting "violations through acts of commission,"[65] requiring that a State refrain from actions that result in direct or indirect violation of the rights within the treaty's coverage[66] or that undermine the level of enjoyment of rights.[67]

"The obligation to respect obliges the State to avoid depriving a human being of this existential status in situations where that status has been attained."[68] Breaches of the obligation to respect human rights

62 ICCPR, *supra* note 48, art. 2.

63 ECOSOC, CESCR, *General Comment No. 3: The Nature of States Parties' Obligations*, 5th Sess., Nov 26-Dec. 14, 1990, UN Doc. E/1991/23, Annex III at 86 (Dec. 14, 1990) [hereinafter CESCR *General Comment No. 3*]. But in respect of the ICCPR, see also UN Human Rights Committee, International Covenant on Civil and Political Rights [ICCPR], *General Comment No. 31: The Nature of the General Legal Obligations Imposed on States Parties to the Covenant*, 80th Sess., ¶ 5, UN Doc. CCPR/C/21/Rev.1/Add.13 (May 26, 2004) [hereinafter Human Rights Committee *No. 31*]. ("The article 2, paragraph 1, obligation to respect and ensure the rights recognized by in the Covenant has immediate effect for all States parties.").

64 UN Human Rights Committee, ICCPR, *General Comment No. 18: Non-discrimination*, 37th Sess. (1989), *reprinted in* Compilation of General Comments and General Recommendations Adopted by Human Rights Treaty Bodies, at 146, UN Doc. HRI/GEN/1/Rev.6 (May 12, 2003).

65 Int'l Comm'n of Jurists, et al., Maastricht Guidelines on Violations of Economic, Social, and Cultural Rights ¶ 6 (1997), *available at* http://www.fao.org/righttofood/KC/downloads/vl/en/details/212297.htm. On the occasion of the tenth anniversary of the Limburg Principles on the Implementation of the International Covenant on Economic, Social, and Cultural Rights (the Limburg Principles), a group of more than thirty experts met in Maastricht to elaborate on the Limburg Principles as regards the nature and scope of violations of economic, social, and cultural rights and appropriate responses and remedies.

66 Human Rights Committee *No. 31*, *supra* note 63, ¶ 6; Frederic Megret & Florian Hoffmann, *The UN as a Human Rights Violator? Some Reflections on the United Nations Changing Human Rights Responsibilities*, 25 Hum. Rts. Q. 314, 319 (2003).

67 Margaret L. Satterthwaite, *Crossing Borders, Claiming Rights: Using Human Rights Law to Empower Women Migrant Workers*, 8 Yale Hum. Rts. & Dev. L.J. 1, 12 (2005).

68 Rolf Kunnemann, *A Coherent Approach to Human Rights*, 17 Hum Rts. Q. 323, 328 (1995).

are probably what are generally understood in ordinary language as "human rights violations,"[69] because the obligation generates limits on the exercise of State power that might threaten people's existing access to rights.[70] In some important respects, therefore, it is an obligation of conduct more than result.

In addition, the obligation to respect can be understood to require that States maintain basic procedural safeguards where their actions have the potential to interfere with the enjoyment of rights. In its General Comment on the *Nature of States Parties' Obligations*, the Committee that monitors the ICESCR noted,

> Among the measures which might be considered appropriate, in addition to legislation, is the provision of judicial remedies with respect to rights which may, in accordance with the national legal system, be considered justiciable. The Committee notes, for example, that the enjoyment of the rights recognized, without discrimination, will often be appropriately promoted, in part, through the provision of judicial or other effective remedies.[71]

In the context of climate change, the obligation to respect would appear to be the most generally applicable of all human rights obligations, because it very clearly embodies a minimum duty to do no harm[72] and requires that a State's actions not accelerate climate change (including harm to the environment that does so) in a way that undermines the

69 Frederic Megret & Florian Hoffman, *supra* note 66, at 320.

70 Special Rapporteur on the Right to Food, *Rep. on the Right to Food*, Comm'n on Human Rights, ECOSOC, ¶ 22, UN Doc. E/CN.4/2006/44 (Mar. 16, 2006) (by Jean Ziegler).

71 CESCR *General Comment No. 3*, *supra* note 63, ¶ 5. *See also* CESCR *General Comment No. 15*, *supra* note 54, ¶ 56; ECOSOC, CESCR, *General Comment No. 4: The Right to Adequate Housing*, 6th Sess., Nov 25-Dec. 13, 1991, UN Doc. E/1992/23, Annex III at 114 (Dec. 12, 1991); ECOSOC, CESCR, *General Comment No. 7: The Right to Adequate Housing: Forced Evictions*, 16th Sess., Nov. 17-Dec. 5, 1997, UN Doc. E/1998/22, Annex IV at 113 (May 5, 1997).

72 The duty to do no harm can, however, be seen to underpin all three constitutive types of human rights obligation – to respect, protect, and fulfill. Moreover, the duty to do no harm can be argued to exist as a free-standing principle that finds expression in both human rights law and environmental law.

existing access to rights.[73] Moreover, the obligation has been understood in general terms to apply to both natural and human-made disasters, and it can therefore be argued to apply to the climate change context regardless of the problems of multiple causes or the difficulty of establishing causation at all.

Given that climate change is generally accepted as being caused or accelerated by the actions of States, such as by emitting pollutants or authorizing their emission, this obligation would require States to desist from such actions insofar as they result in a harmful impact to the realization of rights.[74] Other examples might include avoiding human rights harms related to overexploitation of natural resources,[75]

73 Support for such an obligation may be drawn from environmental human rights jurisprudence under the ECHR. *See* Lopez-Ostra v. Spain, 303 Eur. Ct. H.R. (ser. A) at ¶ 51 (1994), where the Court found a violation of the applicant's rights to home and private and family life (under Article 8 ECHR) based on the inaction of the State and its failure to act in relation to the fumes from a State-subsidized plant treating liquid and solid waste on municipal land twelve meters from the applicant's home. However the Court also made clear that the violation could also have been characterized in relation to the State's positive duty to take reasonable and appropriate measures to secure the applicant's rights or *in terms of an interference by a public authority*, and that in either case the applicable principles are broadly similar. *Id.* at ¶ 51.

74 Of course, arguments are also mobilized in support of developing countries having higher emission levels based on their right to develop, and this finds expression in the UNFCCC. Article 3.4, for instance, recognizes that "[t]he Parties have a right to, and should, promote sustainable development. Policies and measures to protect the climate system against human-induced change should be appropriate for the specific conditions of each party and should be integrated into the national development programmes, taking into account that economic development is essential for adopting measures to address climate change." Such arguments are sometimes based on the right to development itself. *See generally*, Paul Baer, Tom Athanasiou, Sivan Kartha, & Eric Kemp-Benedict, *The Greenhouse Development Rights Framework, Second Edition: The Right to Development in a Climate Constrained World*, November 2008 (Stockholm Environment Institute).

75 Int'l Union for Conservation of Nature [IUCN], *Climate Change and Overexploitation of Natural Resources: Inclusion in the IUCN Program*, Res. 4.074, 4th IUCN World Conversation Congress [WCC4], Oct 5–14, 2008, *in* IUCN, WCC4 RESOLUTIONS AND RECOMMENDATIONS 92 (2009), *available at* http://data.iucn.org/dbtw-wpd/edocs/WCC-4th-005.pdf.

unsustainable fossil fuel extraction, or deforestation. As such, the obligation to respect constrains State action by requiring that the level of access to rights not be reduced due to the actions or omissions of a State Party that contribute to climate change (or to the degradation of the natural environment;[76] that is, measures to address climate change may themselves affect human rights and cannot be assumed to be "human rights neutral," in part because of the deep interconnections between the spheres of climate and human rights, but also because of the inherent and unavoidable need to balance different rights. The obligation to respect could be argued to have relevance for mitigation measures insofar as they aim to prevent dangerous anthropogenic interference with the climate and underpin at least a commitment to work toward reduced emissions of CO_2 and other greenhouse gases. Examples of rights-based approaches to mitigation already exist, illustrating the potential for implementing at least the obligation to respect in climate adaptation.[77]

The obligation to respect may also generate a slightly different set of potential examples, arising from potentially harmful acts taken in response to climate change, and from the obligation's requirement that human rights not be breached by either climate change itself or "as a by-product of a given mitigation or adaptation policy."[78] The instances of unintended potential harm to rights, caused by actions aimed at combating climate change, may concern substantive rights or more process-based rights. Substantive examples include avoiding violations of the

76 The African Commission on Human and Peoples Rights has recognized that environmental degradation can interfere with economic, social, and cultural rights, as well as civil and political rights. *See also* Soc. & Econ. Rights Action Ctr. [SERAC] v. Nigeria, African Comm'n on Human & Peoples Rights, Commc'n. No. 155/96 ¶ 53 (2001), where the Commission confirmed Nigeria's obligations to refrain from violations of human rights.

77 *See* Paul Baer et al., The Greenhouse Development Rights Framework: The Right to Development in a Climate Constrained World (2d ed. 2008), for a recent example of a GHG mitigation model based on the right to development.

78 ICHRP, *Rough Guide*, *supra* note 11, at 18.

right of adequate housing through forced evictions, where individuals have to move from a high-risk zone due to climate impacts;[79] safeguarding against violations of the right to food due to biofuel substitution programs affecting food security or land rights (e.g., palm oil in Indonesia);[80] or protecting the rights of indigenous peoples or forest-dependent communities, in the context of Reducing Emissions from Deforestation and Degradation (REDD), to both participate in and share the benefit of such programs.[81] Process-based rights that should be upheld in adaptation measures include the right of equal access to information, public participation, and freedom of speech and association. In any of the foregoing sets of examples, environmental human rights jurisprudence supports a requirement that States carry out prior assessments to predict and evaluate the effects of actions that might degrade the environmental and thereby harm individuals' rights.[82] This potential entry point

79 In each case, the obligation can be argued to have special relevance to vulnerable groups who are especially dependent on the natural environment or whose livelihoods, culture, or identity is inextricably bound to natural resources. *See, e.g.*, *UNPFII Report Highlights Increased Vulnerability to Climate Change among Indigenous Communities*, Climate Change Policy & Practice News, Jan. 18, 2010, *available at* http://climate-l.iisd.org/news/unpfii-report-highlights-increased-vulnerability-to-climate-change-among-indigenous-communities/. *See also* Secretariat of the UN Permanent Forum on Indigenous Issues [UNPFII], *Climate Change: An Overview* (UNPFII Working Paper, Nov. 2007), *available at* http://www.un.org/esa/socdev/unpfii/en/EGM_CS08.html; Melinda Janki, *Indigenous Peoples Rights and the Environment: Issues and the Future*, (UN Environment Programme & OHCHR Drft. Paper, 2009), *available at* http://www.unep.org/environmentalgovernance/LinkClick.aspx?fileticket=gUAoSyFdNG8%3d&tabid=2046&language=en-US.

80 In its 2007 *Concluding: Observations on Indonesia*, the Committee on the Elimination of Race Discrimination noted concern about how palm oil plantations would affect indigenous peoples and threaten their rights to own their lands and enjoy their culture. Comm. on the Elimination of Race Discrimination [CERD], *Concluding Observations: Indonesia*, 71st Sess., July 30-Aug. 18, 2007, ¶¶ 2, 5, U.N. Doc. CERD/C/IDN/CO/3 (Aug. 15, 2007).

81 *See generally* UN Collaborative Initiative on Reducing Emission from Deforestation and Forest Degradation in Developing Countries [UN-REDD], http://www.un-redd.org/ (last visited July 9, 2011).

82 Soc. & Econ. Rights Action Ctr. [SERAC] v. Nigeria, African Comm'n on Human & Peoples Rights, Commc'n. No. 155/96 ¶ 53 (2001); Taskin v. Turkey, 2004-X Eur. Ct.

for human rights in climate policy is discussed again in the concluding section.

One final, though more controversial, potential application of the obligation to respect to climate change is as an extraterritorial duty to do no harm. In this, States must not cause harm to human rights outside the State territory,[83] interfere with the enjoyment of rights in another State, or undermine other States' ability to fulfill their human rights obligations.[84] This argument may draw support from the environmental and transboundary genesis of the duty to "do no harm" in the 1938 *Trail Smelter Arbitration*[85] and the ICJ's 1996 *Nuclear Weapons* recognition of the duty as "part of the corpus of international law relating to the environment."[86] The duty is also manifest in the Rio Declaration[87] and in Principle 21 of the Stockholm Declaration of the UN Conference on the Human Environment, which provides the following: "States have, in accordance with the Charter of the United

H.R. 179 (2004); Saramaka People v. Suriname, Preliminary Objections, Merits, Reparations, and Costs, Judgment, Inter-Am. Ct. H.R. (ser. C) No. 172 (Nov. 28, 2007). *See generally* U.N. Econ. Comm'n for Europe [UNECE], Aarhus Convention on Access to Information, Public Participation in Decision-making and Access to Justice in Environmental Matters, June 25, 1998, 2161 U.N.T.S. 447, 38 I.L.M. 517 (1999).

83 This would be so, even where the State's level of control falls short of that of an occupying power. John Cerone, *Jurisdiction and Power: The Intersection of Human Rights Law and the Law of Non-International Armed Conflict*, 40 Isr. L. Rev. 396, 423–24 (2007).

84 Skogly, Beyond National Borders, *supra* note 60, at 68–69. *See generally* Theodore Meron, *Extraterritoriality of Human Rights Treaties*, 89 Am. J. Int'l L. 78, 82 (1995). For a thorough discussion of the extraterritorial application of the ECHR, *see* Michael Kearney, *Extraterritorial Jurisdiction of the European Convention on Human Rights*, 5 Trinity C.L. Rev. 126 (2002).

85 *See* Trail Smelter Arbitration (U.S. v. Canada), 3 R.I.A.A. 1911 (1938), 3 R.I.A.A. 1938 (1941), where the tribunal established a general duty of care on the part of a State to protect other States from injurious acts by individuals within its jurisdiction.

86 Legality of the Threat or Use of Nuclear Weapons, Advisory Opinion, 1996 I.C.J. 226, ¶ 29, at 241 (July 8). The Court held that "[t]he existence of the general obligation of States to ensure that activities within their jurisdiction and control respect the environment of other States or of areas beyond national control is now part of the corpus of international law relating to the environment." *Id.* at ¶ 29.

87 *Rio Declaration*, *supra* note 21, Princ. 2.

Nations and the principles of international law, the sovereign right to exploit their own resources pursuant to their own environmental policies, and the responsibility to ensure that activities within their jurisdiction or control do not cause damage to the environment of other States or of areas beyond the limits of national jurisdiction."[88]

The jurisprudence of the International Court of Justice[89] and a number of human rights bodies[90] supports the existence of extraterritorial responsibility for violations of human rights law and humanitarian law in circumstances where the State exercises effective control,[91] thereby disaggregating the notions of territory and jurisdiction in Article 2, para. 1 of the ICCPR. Independent of establishing effective control, there may be potential to argue for a general obligation to respect rights in other States, notwithstanding the limitations on the extraterritorial application of human rights treaties. Matthew Craven has observed of the ICESCR that "while a State's primary responsibility is to the local populace; each State is also required to ensure that it does not undermine the enjoyment of rights of those in foreign

88 United Nations Conference on the Human Environment, Stockholm, Swed., June 5–16, 1972, *Declaration of the United Nations Conference on the Human Environment*, UN Doc. A/CONF.48/14, 11 I.L.M. 1416 (June 16, 1972). *See also* Draft Articles on the Prevention of Transboundary Harm from Hazardous Activities, Rep. of the Int'l Law Comm'n, 53rd Sess, Apr. 23-June 1, 2001, ch. V, ¶ 97, art. 3, 4, UN Doc. A/56/10, Supp. No. 10 (Nov. 30, 2001) (requiring States to inter alia "take all appropriate measures to prevent, or to minimize the risk of, significant transboundary harm").

89 *See, e.g.*, Armed Activities on the Territory of the Congo (DRC v. Uganda), 2005 I.C.J. 116, 220 (Dec. 19); Legal Consequences of the Construction of a Wall in Occupied Palestinian Territory, Advisory Opinion, 2004 I.C.J. 163, ¶ 111 (July 9).

90 Sergio Euben Lopez Burgos v. Uru., UN Human Rights Committee, Commc'n No. R.12/52, at 176, UN Doc. A/36/40, Supp. No. 40 (1981); Coard et al. v. U.S., Case 10.951, Inter-Am. Comm'n H.R., Report No. 109/99, OEA/ser.L/V/II.106, doc. 3 rev. (1999).

91 The Human Rights Committee stated that "a State party must respect and ensure the rights laid down in the Covenant to anyone within the power or effective control of that State Party, even if not situated within the territory of the State Party." Human Rights Committee *No. 31*, *supra* note 63, ¶ 10.

territory."[92] Similarly, in respect of the right to food, the Committee on Economic, Social and Cultural Rights observed that "[t]he extraterritorial obligation to respect the right to food requires States to 'do no harm.' ... It requires states to ensure that their policies and measures do not lead to violations of the right to food for people living in other countries."[93] Following this logic in the context of climate change, the obligation would constrain State actions that undermine access to rights in other States. The study of the OHCHR on the relationship between climate change and human rights appears to support the application of the obligation to respect in the context of climate change by arguing that States have obligations to refrain from interference with the enjoyment of rights in other countries and to ensure that human rights are given due attention in international agreements and that such agreements do not adversely affect human rights (for a litigation strategy for enforcing international climate obligations, see Chapter 12).[94]

3.2.2. The Obligation to Protect

The obligation to protect requires States to regulate the behavior of third parties and protect against human rights violations.[95] It typically entails a range of State actions,[96] including the obligation to prevent, investigate, punish, and ensure redress for the violations caused by abuses of

92 Matthew Craven, *The Violence of Dispossession: Extraterritoriality and Economic, Social, and Cultural Rights*, *in* Economic, Social and Cultural Rights in Action 71 (Mashood Baderin & Robert McCorquodale eds., 2007).

93 Christine Kaufmann & Simone Heri, *Liberalizing Trade in Agriculture and Food Security – Mission Impossible?*, 40 Vand. J. Transnat'l L. 1039, 1052 (2007) (quoting CESCR *General Comment No. 12*, *supra* note 54, ¶ 19).

94 OHCHR, *Annual Report*, *supra* note 15, ¶ 86.

95 Rolf Kunnemann, *supra* note 68, at 328.

96 *See* Int'l Comm'n on Intervention & State Sovereignty [ICISS], The Responsibility to Protect (2001) (discussing the responsibility to protect, so-called R2P in the humanitarian context). Other duties flowing directly from the terms of Article 2 are the duties to promote and fulfill, which are confirmed by the interpretation of the Human Rights Committee in General Comment 31. Human Rights Committee *No. 31*, *supra* note 64, ¶ 7.

human rights by State instrumentalities or by third parties – private individuals, commercial enterprises, or other non-State actors.[97] The State may discharge this obligation through appropriate policies, regulation, and adjudication. "The obligation to *protect* requires States to prevent violations of such rights by third parties. Thus, the failure to ensure that private employers comply with basic labor standards might amount to a violation of the right to work or the right to just and favorable conditions of work."[98] The obligation to protect may be understood to impose affirmative duties on States parties to take (adequate or appropriate) action against infringements, and to constitute an obligation of conduct (as well as result).[99] Conversely, the "international responsibility of the State" arises "because of the lack of due diligence to prevent the violation."[100]

In the context of climate change, the obligation to protect is more positive, because it requires States to *take steps* to protect persons within its jurisdiction or to exercise effective control against the negative impacts of climate change. It may also require that States *take action* to protect persons against the harmful acts of private parties and non-State actors that cause or accelerate climate change.[101] In substantive terms, these obligations could entail elaborating a coherent and comprehensive legal and policy framework to effectively regulate emissions[102] by taking steps toward reaching an international agreement on climate change at the global level or developing adequate mitigation plans at the national

97 Human Rights Committee *No. 31*, *supra* note 63, ¶ 8 (discussing Article 2(2) of the ICCPR). The obligation to protect also targets harm that may result from State *omission* (i.e., harm resulting from a State's failure to act).

98 Int'l Comm'n of Jurists, et al., *supra* note 65, ¶ 6 (emphasis added).

99 CESCR *General Comment No. 3*, *supra* note 63, ¶ 1.

100 Velásquez Rodríguez v. Honduras, Merits Judgment, Inter-Am. Ct. H.R. (ser. C) No. 4, ¶ 172, 176 (July 29, 1988).

101 From the environmental human rights jurisprudence, one may again draw useful analogies to the climate change context. *See, e.g.*, Soc. & Econ. Rights Action Ctr. [SERAC] v. Nigeria, African Comm'n on Human & Peoples Rights, Commc'n. No. 155/96 ¶ 53 (2001), where the Commission underscored Nigeria's obligations to both refrain from violations of human rights and protect its citizens from damaging acts that may be perpetrated by private parties (in this case Shell Oil).

102 *See, e.g.*, Massachusetts v. Envtl. Prot. Agency, 549 U.S. 497 (2007).

level to ensure the protection of economic and social rights, particularly those of vulnerable groups.

In a different way, the obligation to protect may require States to take positive action to secure adequate procedural safeguards and remedies where rights are infringed by third parties. As the CESCR has stated,

> Among the measures which might be considered appropriate, in addition to legislation, is the provision of judicial remedies with respect to rights which may, in accordance with the national legal system, be considered justiciable. The Committee notes, for example, that the enjoyment of the rights recognized, without discrimination, will often be appropriately promoted, in part, through the provision of judicial or other effective remedies.[103]

The requirements to provide redress and to respect certain procedural standards are therefore similar to those described earlier in relation to the obligation to respect human rights, but they operate as safeguards against third-party infringements.

Some commentators have argued that the duty to protect extends beyond the confines of a State's borders and that it may have extraterritorial application.[104] In relation to Article 2, paragraph 1 of the ICCPR, the Committee held "a State party must respect and *ensure* the rights laid down in the Covenant to anyone within the power or effective control of that State Party, even if not situated within the territory of the State Party."[105] Similarly, Louise Arbour has written authoritatively of the responsibility (obligation) to protect as a duty of care in international law and practice in the prevention of genocide.[106] Although anchoring it in

103 CESCR *General Comment No. 3*, *supra* note 63, ¶ 5. *See also* Human Rights Committee *No. 31*, *supra* note 63, ¶ 8.

104 It is clear, however, that the nature and extent of this obligation may be more limited where the effects of a State's action have an impact on the human rights of people located physically outside of the State's territory.

105 Human Rights Committee *No. 31*, *supra* note 63, ¶ 10 (emphasis added).

106 Louise Arbour, *The Responsibility to Protect as a Duty of Care in International Law and Practice*, 34 REV. INT'L STUD. 445, 450 (2008). Despite some points of similarity,

humanitarian intervention, rather than human rights law, Arbour identifies the responsibility as "a notion of due diligence, a concept well understood in international human rights law in relation to the positive obligation of a State to act in response to threats to human rights, notably to the life and security of the person within its own jurisdiction."[107] Arbour opines that because of their preeminence, global reach, and capabilities, other States may also be in a position to act in furtherance of the responsibility to protect. She identifies three elements that may ground that responsibility: influence, proximity, and information.

Nevertheless, "human rights obligations requiring the adoption of affirmative measures may be more limited in an extraterritorial context."[108] A variety of reasons may be put forward for such a limitation: "because a State may not ordinarily exercise authority in the territory of another State, there is no obligation upon it to act to ensure respect for human rights there, whether by the government of that State or by private persons."[109] In addition, "positive obligations are also limited by considerations of fairness and the effectiveness principle itself."[110]

Notwithstanding these limitations and those related to applying human rights jurisprudence on transboundary environmental harm,[111] the obligation to protect could be argued to support a range of more or less onerous obligations, of both a substantive and procedural nature, in the climate context. Such an application could draw on the reasoning of the *Trail Smelter Arbitration (US and Canada)*[112] in which the tribunal

such as that sovereignty is not an absolute bar to intervention, this is a responsibility to be distinguished from doctrines of humanitarian intervention.

107 *Id.* at 452.

108 John Cerone, *supra* note 81, at 441.

109 Louis Henkin, The Age of Rights 45 (1990).

110 Damira Kamchibekova, *State Responsibility for Extraterritorial Human Rights Violations*, 13 Buff. Hum. Rts. L. Rev. 87, 97 (2007).

111 Knox, *Climate Change*, *supra* note 8, at 200–210.

112 Trail Smelter Arbitration, *supra* note 85, at 1965. The case is widely recognized as the source of 'do no harm' as a principle of customary international environmental law. Under the principles of international law ... no State has the right to use or permit the

established a general duty of care on the part of a State to *protect* other States from injurious acts by individuals within its jurisdiction.[113] This obligation could be argued to generate requirements that States take action to protect persons against human rights harm resulting from the acts of third parties under its jurisdiction or control (e.g., corporations), which occurs within *or outside* their territory.[114]

3.2.3. The Obligation to Fulfill

The obligation to fulfill connotes a more positive obligation than either the obligations to respect or protect,[115] because it requires States to create enabling conditions for individuals to fully enjoy their rights.[116] "The obligation to fulfill obliges a State to secure the existential status for human beings in situations where that status has not been attained (i.e., situations of deprivation)."[117] This is apparent in ICESCR Article 2, para. 1, which provides that "[e]ach State Party to the present Covenant undertakes to take steps, individually and through international assistance and co-operation, especially economic and technical, to the

use of its territory in such a manner as to cause injury by fumes in or to the territory of another or the properties or persons therein, when the case is of serious consequences and the injury is established by clear and convincing evidence. *Id.* at 1965.

113 The arbitral agreement recognized the responsibility of a State for the acts of non-State actors in addition to those of the State and its organs. Understood in this way, the duty can be seen to be related to a general international law duty not to harm foreign nationals and to make reparations for breaches.

114 Skogly, Beyond National Borders, *supra* note 60, at 70. *See also* CESCR *General Comment No. 14*, *supra* note 45, ¶ 39 ("States parties have to respect the enjoyment of the right to health in other countries, and to prevent third parties from violating the right in other countries, if they are able to influence these third parties by way of legal or political means"); *see also* CESCR *General Comment No. 15*, *supra* note 47; ECOSOC, CESCR, *General Comment No 19*: *The Right to Social Security* UN Doc. E/C.12/GC/19 (Feb. 4, 2008).

115 Philip Alston & Gerard Quinn, *The Nature and Scope of States Parties' Obligations under the International Covenant on Economic, Social and Cultural Rights*, 9 Hum. Rts. Q. 156, 186 (1987) (noting that the obligation to guarantee represents "the highest rung of state obligation under the Covenant").

116 Margaret L. Satterthwaite, *supra* note 67, at 12.

117 Rolf Kunnemann, *supra* note 68, at 328.

maximum of its available resources, with a view to achieving progressively the full realization of the rights recognized in the present Covenant by all appropriate means, including particularly the adoption of legislative measures."[118]

Because of the inherently affirmative nature of the obligation to fulfill, it is mediated by the principle of progressive realization, giving States some leeway in achieving the realization of human rights because immediate realization is not realistic (in view of resource constraints, as well as contextual and temporal factors). States are nevertheless under an immediate obligation to take "deliberate, concrete, and targeted steps"[119] toward the full realization of human rights. The obligation to fulfill is therefore both an obligation of conduct and result.[120]

In an attempt to counter the "inherent relativism of the programmatic standard of 'progressive realization'"[121] of the ICESCR, the Committee has confirmed that States must uphold a "minimum core obligation to ensure the satisfaction of, at the very least, minimum essential levels of each of the rights."[122] Furthermore, the CESCR has held that "even where the available resources are demonstrably inadequate, the obligation remains for a State party to strive to ensure the widest possible

118 ICESCR, *supra* note 49, art. 2, ¶ 1.

119 CESCR *General Comment No. 3*, *supra* note 63, ¶ 2.

120 *Id.* at ¶ 1.

121 Katherine G. Young, *The Minimum Core of Economic and Social Rights: A Concept in Search of Content*, 33 Yale J. Int'l L. 113, 121.

122 CESCR *General Comment No. 3*, *supra* note 63, ¶ 10 (noting inter alia, that at a minimum, States are required to provide for the basic needs of the population). For arguments that even the "minimum core" socioeconomic rights require sensitivity to national resource constraints, *see* Eide, *Economic, Social, and Cultural Rights*, *supra* note 43; Asbjørn Eide, *Realization of Social and Economic Rights and the Minimum Threshold Approach*, 10 Hum. Rts. L.J. 35 (1989); Craig Scott & Philip Alston, *Adjudicating Constitutional Priorities in a Transnational Context: A Comment on Soobramoney's Legacy and Grootboom's Promise*, 16 S. Afr. J. Hum. Rts. 206, 250 (2000) (positing a distinction between absolute core minimums [applicable to all States] and relative [State-specific] core minimums in which some degree of differentiation is warranted).

enjoyment of the relevant rights under the prevailing circumstances."[123] This obligation requires States to take appropriate legislative, administrative, budgetary, judicial, and other measures toward the full realization of such rights.[124] Thus, the failure of States to provide essential primary health care to those in need may amount to a violation of the rights to health and medical care.[125] Consistent with its "affirmative" and programmatic character, the obligation to fulfill includes duties to facilitate (increase access to resources and means of attaining rights) and provide (ensure that the whole population may realize their rights where individuals are unable to do so themselves).[126]

In addition, the principle of non-retrogression, derived from Article 2 of the ICESCR and its obligation of result,[127] prohibits the dismantling of existing protections of guaranteed rights,[128] the degradation of rights already realized,[129] and any decrease in people's access to a right recognized.[130] "Any deliberately retrogressive measures in that regard would require the most careful consideration and would need to be fully justified by reference to the totality of the rights provided for in the Covenant and in the context of the full use of the maximum available resources."[131]

123 *Id.*

124 Margaret L. Satterthwaite, *supra* note 67, at 12.

125 INT'L COMM'N OF JURISTS, ET AL., *supra* note 65, ¶ 6.

126 CESCR *General Comment No. 12*, *supra* note 54, ¶ 15.

127 ICESCR, *supra* note 49, art. 2, ¶ 1.

128 An analogous principle can be argued to underpin EU rights discourse, such that rights created by the EU (such as EU citizenship rights) are established "in addition to national ones, *whilst existing rights are not reduced*." Dimitris N. Chryssochoou, *Civic Competence and the Challenge to EU Polity-Building*, 9 J. EUR. PUB. POL'Y 756, 765 (2002) (citing Andrew Geddes, *Immigrant and Ethnic Minorities and the EU's Democratic Deficit*, 33 J. COMMON MARKET STUD. 197 (2000)) (emphasis added).

129 Jessica Schultz, *Economic and Social Rights in the United States: An Overview of the Domestic Legal Framework*, *in* 11 HUM. RTS, BRIEF 2 (Am. U. Washington Coll. of Law: Ctr for Human Rights & Humanitarian Law 2003).

130 *See Id.*

131 CESCR *General Comment No. 3*, *supra* note 63, ¶ 9. In respect of the Committee's approach, Katherine Young observed, "Although the Committee described retrogression as a move away from the direction of full realization (rather than a move below

Perhaps even more strongly than the obligation to protect, the obligation to fulfill could be seen to generate or support actions in relation to climate mitigation, and particularly adaptation, which by their nature relate to the conditions necessary for the realization of rights. Because even the most conservative predictions about climate change require collective action, the affirmative nature of the obligation to fulfill is inherently suited to support and inform the types of actions required to address climate change's negative social and human impacts,[132] whether in terms of adaptation or mitigation measures. Consistent with action to tackle climate change, the obligation is a programmatic one, envisaging steps to be taken individually and collectively over longer periods of time to progressively realize human rights.

This obligation requires that States move as expeditiously and effectively as possible toward the goal of progressive realization,[133] and it could be argued to require actions in furtherance of mitigation or adaptation, which must also advance the realization of human rights and be guided by human rights standards. Such affirmative obligations under human rights treaties could be relevant to both underscoring the responsibility of all States to negotiate a solution to climate change[134] and to the design objectives of adaptation measures. Conversely, they would also prohibit actions or omissions related to climate change that reduce the attained level of enjoyment of rights and would further require that any deliberately retrogressive measures connected with climate change be justified by reference to the totality of rights protected under the human rights treaty commitments of the State in question.[135]

any minimum), there are unexplored parallels between a 'ratchet-effect' standard of retrogression and a state-specific minimum core." Katherine G. Young, *supra* note 121, at 121 n.40. It is significant that deliberately retrogressive measures do not constitute a prima facie violation. For a critical view of this, *see* MATTHEW CRAVEN, THE INTERNATIONAL COVENANT ON ECONOMIC, SOCIAL, AND CULTURAL RIGHTS: A PERSPECTIVE ON ITS DEVELOPMENT 131–32 (1995).

132 CESCR *General Comment No. 3*, *supra* note 63, ¶ 9.

133 *Id.*

134 ICHRP, *Rough Guide*, *supra* note 11, at 13.

135 CESCR *General Comment No. 3*, *supra* note 63, ¶ 9.

Like the obligations to respect and protect, the obligation to fulfill can be argued to be a source for *procedural rights*, such that the actions a State may take that result in changes to the climate and environment, which in turn affect the rights of persons, should be subject to certain procedural safeguards and remedies. These might include the right to information, the right to consultation, the right to participate in the decision-making process, and even the right to consent (or withhold consent) for certain types of actions related to climate. In a broader sense, the obligation may require action on more formal (rather than substantive) aspects of rights and include a range of judicial, administrative, and educative measures to be fulfilled.[136] The Human Rights Committee has also made clear that, in addition to effective protection of ICCPR rights, States parties must ensure that individuals also have accessible and effective remedies to vindicate those rights.[137]

The obligation to fulfill can also be argued to have an extraterritorial component because of the provision in Article 2 of the ICESCR requiring that "[e]ach State Party to the present Covenant undertakes to take steps, individually and *through international assistance and co-operation*, especially economic and technical, to the maximum of its available resources, with a view to achieving progressively the full realization of the rights recognized in the present Covenant by all appropriate means, including particularly the adoption of legislative measures."[138] The Committee has emphasized that this provision refers both to resources existing within a State and those available from the international community through international cooperation and assistance,[139] and it underscored

136 Human Rights Committee *No. 31*, *supra* note 63, ¶ 7.

137 *Id.* ¶ 15.

138 ICESCR, *supra* note 47, art. 2, ¶ 1 (emphasis added). Sigrun Skogly also views the provision on international cooperation and assistance as part of the obligation to fulfill. *See* SKOGLY, BEYOND NATIONAL BORDERS, *supra* note 60, at 71.

139 *See, e.g.*, ICESCR, *supra* note 49, art. 22, 23. *See also* ECOSOC, CESCR, *General Comment No. 2: International Technical Assistance Measures*, 4th Sess., UN Doc. E/1990/23, Annex III at 86 (1990).

the essential role of such cooperation in facilitating the full realization of the relevant rights.[140]

In successive General Comments,[141] the Committee on ESCR has relied on this provision to argue for extraterritorial obligations under the ICESCR,[142] although developed countries have resisted the idea that the Covenant generates a legal obligation for them to provide international financial assistance to other States.[143] It is noteworthy, however, that the obligations of international cooperation and assistance have been interpreted as applying to both developed and developing countries. In the latter case, the obligation has been understood to include requirements to "actively seek assistance" where needed, identify technical cooperation needs, ensure that assistance programs are monitored and refrain from obstructing such monitoring, and ensure that assistance prioritizes the minimum core entitlement of rights and the rights of the most vulnerable.[144]

140 CESCR *General Comment No. 3*, *supra* note 63, ¶¶ 13, 14. In this connection the Committee highlighted the 1986 Declaration on the Right to Development. G.A. Res. 41/128, UN Doc. A/RES/41/128 (Dec. 4, 1986).

141 *See generally* CESCR *General Comment No. 3*, *supra* note 63, ¶ 13 ("A final element of article 2 (1), to which attention must be drawn, is that the undertaking given by all States parties is 'to take steps, individually and through international assistance and cooperation, especially economic and technical.' . . . The Committee notes that the phrase 'to the maximum of its available resources' was intended by the drafters of the Covenant to refer to both the resources existing within a State and those available from the international community through international cooperation and assistance.") *See also* CESCR *General Comment No. 12*, *supra* note 54, ¶ 38; CESCR *General Comment No. 14*, *supra* note 54, ¶ 40; CESCR *General Comment No. 15*, *supra* note 54, ¶ 34.

142 *E.g.*, ICESCR, *supra* note 49, art. 11, 15 ¶ 4, 22, 23.

143 ECOSOC, Comm'n on Human Rights, *Rep. of the Open-Ended Working Group to Consider Options Regarding the Elaboration of an Optional Protocol to the IECSCR on its Second Session*, ¶ 76, UN Doc. E/CN.4/2005/52 (Feb. 10, 2005). *See also* Philip Alston, *Ships Passing in the Night: The Current State of the Human Rights and Development Debate as Seen through the Lens of the Millennium Development Goals*, 27 Hum. Rts. Q. 755, 777 (2005).

144 *See also* Magdalena Sepúlveda, *Obligations of 'International Assistance and Cooperation' in an Optional Protocol to the International Covenant on Economic, Social and Cultural Rights*, 24 Neth. Q. Hum. Rts. 271, 287 (2006). *See generally* Casting the Net Wider: Human Rights, Development, and New Duty Bearers (Margot Salomon,

As such, although the Covenant's provision on international assistance and cooperation has been the subject of divergent views,[145] it could be argued, in some broad theoretical sense, to support the provision of resources and technology to support and encourage climate change adaptation in developing countries. From a human rights perspective, this transfer of resources is essential to avert the catastrophic consequences on human rights in terms of immediate threats to basic subsistence. But equally, human rights arguments could support technology transfer and technical cooperation, in the longer term, to address longer term human rights fulfillment, which itself relies on renewable technologies.[146] Climate change adaptation by definition requires collaborative action and specialized technical assistance. Because many forms of environmental damage (including climate change) are transnational in character, effective international cooperation is essential to address such damage and support national effects for the realization of human rights.[147] This basic interpretation is supported in the analysis of the OHCHR study, which contends that States, depending on the availability of resources, have obligations to take steps through international

Arne Tostensen & Wouter Vanderhole eds., 2007), which aims to identify (new) duty-bearers in relation to development and world poverty and to clarify their various obligations in that regard. For an argument about global responsibility for world poverty from a moral and political philosophy perspective, *see* THOMAS W. POGGE, WORLD POVERTY AND HUMAN RIGHTS: COSMOPOLITAN RESPONSIBILITIES AND REFORMS (2002).

145 Sigrun I. Skogly, *The Obligations of International Assistance and Cooperation in the International Covenant on Economic, Social and Cultural Rights*, *in* HUMAN RIGHTS AND CRIMINAL JUSTICE FOR THE DOWNTRODDEN: ESSAYS IN HONOUR OF ASBJORN EIDE 403 (Morten Bergsmo ed., 2003); SKOGLY, BEYOND NATIONAL BORDERS, *supra* note 60; Sigrun I. Skogly, *Global Responsibility for Human Rights*, 29 OXFORD J. LEGAL STUD. 827 (2009); Magdalena Sepúlveda Carmona, *Obligations of 'International Assistance and Cooperation' under the International Covenant on Economic, Social and Cultural Rights. A Possible Entry Point to a Human Rights Based Approach to Millennium Development Goal* 8, 13 INT'L J. HUM. RTS. 86 (2009).

146 *See generally* Int'l Council on Human Rights Policy [ICHRP], *Human Rights and Climate Technology Policy: Using Technology to Protect Rights in a Climate-Constrained World* (Draft Report, Mar. 2011), *available at* http://www.ichrp.org/files/assets/230/138_draft.pdf.

147 U.N. Human Rights Council Res. 16/11, *supra* note 20.

assistance and cooperation to facilitate the fulfillment of human rights in other countries, including disaster relief, emergency assistance, and assistance to refugees and displaced persons.[148]

4. Constraints on Climate-Specific Application of Human Rights Obligations

Despite the potential relevance of human rights obligations to climate change as described earlier in theoretical terms, distinct challenges emerge because of features inherent in the existing international human rights law framework[149] and its relationship to the regimes governing climate change, environment,[150] and development.

Given the foregoing analysis on the potential relevance of human rights obligations to action on climate change, what are the reasons for the absence of human rights discourse and obligations in climate change discussions and the apparent disconnect between the two fields?

From a legal point of view,[151] the two fields reflect a separate evolution[152] and emanate from distinct "treaty-based sub-systems,"[153] which have generated their own sector-specific areas of international law.[154] In this way, they exemplify a broader phenomenon of technically

148 OHCHR, *Annual Report*, *supra* note 15, ¶ 86.

149 These have been written about extensively. *See, e.g.*, Knox, *Climate Change, supra note 8; Sammy Adelman, Rethinking Human Rights: The Impact of Climate Change on the Dominant Discourse, in* Human Rights and Climate Change 159 (Stephen Humphreys ed., 2010).

150 *See generally* Philippe Sands, Principles of International Environmental Law 297–307 (2d ed. 2003).

151 Other commentators have written about this disconnect or "mutual disinterest" in less legalistic terms considering a range of reasons that might explain it, among which they identify "disciplinary path dependence."

152 Joost Pauwleyn, Conflict of Norms in Public International Law: How WTO Law Relates to Other Rules of International Law 9 (2003).

153 *Id.*

154 The limitations identified here resemble the disconnects between human rights law and development policy that coexist in parallel legal and policy spheres, despite connecting and overlapping in a range of ways. Siobhán McInerney-Lankford, *Human*

specialized cooperation networks with global scope and expert cooperation that transgress national boundaries and are difficult to regulate through traditional international law.[155] Moreover, although the logic of separate specialized international regulation may be elusive and normative conflict simply endemic to international law,[156] one cannot assume that such separateness was not in some ways intentional. A background concern in assessing the "horizontal"[157] relationship between treaty frameworks in an increasingly fragmented international law is the rise of specialized rules and rule systems that have no clear relationship to one other.[158] This discussion considers the nature of the relationship between human rights and climate change, and the sources of its disconnects and limitations.

The international bill of rights[159] is a product of the post–World War II era[160] and is oriented toward preventing specific moral harms and preserving a set of substantive and procedural entitlements for all. The climate change regime for its part emerged much later,[161] as the outcome

Rights and Development. Some Institutional Perspectives, 23 Neth. Q. Hum. Rts. 459 (2007). But beyond analogies, they can be seen as connected, because tackling climate change raises questions about access to and use of resources, which have direct bearing on human rights. This links also to the fact that public international law possesses no clear means for evaluating development activities in explicit and comprehensive human rights terms or upholding clear human rights accountability in development policy and activities.

155 ILC, *Fragmentation Report*, *supra* note 10, ¶ 481.

156 *Id.* ¶ 486; Joost Pauwleyn, *supra* note 152, at 8–22.

157 As opposed to the vertical relationship between norms of public international law and national law, sometimes called "conflict of laws" or "private international law."

158 ILC, *Fragmentation Report*, *supra* note 10, ¶ 483.

159 Term to refer collectively to the Universal Declaration on Human Rights, the ICESCR, and the ICCPR.

160 Hersch Lauterpacht, An International Bill of Rights of Man (1945); Louis Henkin, The International Bill of Rights: The Covenant on Civil and Political Rights (1981); Brian Simpson, *Hersch Lauterpach and the Genesis of the Age of Human Rights*, 120 L.Q. Rev. 49 (2004).

161 The two principal instruments regulating international climate change are the United Nations Framework Convention on Climate Change (total of 192 States parties), May 9, 1992, S. Treaty Doc. No. 102–38, 1771 U.N.T.S. 107 [hereinafter UNFCCC], and the Kyoto Protocol to the United Nations Framework Convention on Climate Change

of very different movements and awareness. It is aimed at tackling harms that were not contemplated in the design of human rights frameworks, which are therefore ill equipped to tackle them. The distinctness of their respective goals is apparent in how human rights law and discourse protect specific enforceable entitlements of individuals (and sometimes groups) against the infringements and omissions of duty-bearers. In contrast, climate change instruments are more consensus driven and oriented toward collective solutions and the protection of the "climate system."[162] They employ more opaque concepts like equity associated with deliberation and fairness,[163] in contrast to the human rights terminology of equality, right, violation, and redress. Climate change frameworks are by definition forward-looking and focused on potential future or projected harms of a more general and diffuse sort, whereas human rights frameworks and methodologies tend to look retrospectively on specific infringements and violations of the rights of identifiable individuals or groups that have already occurred.[164]

In more structural and institutional terms too, climate change frameworks, such as the UNFCCC, are horizontal in nature, are concluded between States to regulate the equitable distribution of global burdens with respect to emissions, and promote cooperation to address a collective problem, in contrast to the more formalistic language of rights and

("Kyoto Protocol") (with a total of 176 States parties), Dec. 10, 1997, 37 I.L.M. 22, UN Doc. FCCC/CP/1997/7/Add.1. The UNFCCC originally set a voluntary goal of reducing emissions of developed countries; these voluntary reductions were replaced by binding emission limits in the Kyoto Protocol.

162 The UNFCCC has the long-term objective of stabilizing greenhouse gas concentrations in the atmosphere "at a level that would prevent dangerous anthropogenic interference with the climate system." UNFCCC, *supra* note 161, art. 2.

163 *See* Stephen Humphreys, *Competing Claims: Human Rights and Climate Harms*, *in* Human Rights and Climate Change, *supra* note 149, at 47 [hereinafter Humphreys, *Competing Claims*] (discussing equity and common but differentiated responsibilities).

164 That is not to discount the more forward-looking, programmatic approaches that can be derived from human rights law, which are considered in the final section of this article.

equality, upheld through individual complaints and redress (extraterritorial application versus that within the State's effective control). Human rights treaties represent a more vertical commitment between a State and its citizens and establish formal and substantive norms to govern the behavior of duty-bearers and constrain their exercise of power. Climate change remains a collective global problem[165] with aggregate causes and disparate impacts, whereas the international human rights law framework is structured primarily to address direct entitlements and violations, most often on the individual level, even if some group or collective rights are recognized.[166]

A related problem is the limited jurisdictional reach of international human rights treaty frameworks, as primarily State-centric and underpinned by "vertical" obligations between a government and its citizens, whereas climate harms suffered by individuals may be generated from outside the national jurisdiction and by actors other than their own governments.[167] Despite the potential for the extraterritorial application of human rights law obligations to respect, protect, and fulfill as outlined earlier, it is also true that the application is limited at best[168] and widely contested,[169] which creates very real limitations on its potential to

165 This has sometimes generated the argument that such generalized harms or injuries are best addressed by political rather than legal branches.

166 These limitations have led some commentators to argue in favor of a free-standing right to a sustainable environment or a universal environmental right. *See* Sammy Adelman, *supra* note 140, at 172. *But see* Philip Alston, *Conjuring Up New Human Rights: A Proposal for Quality Control*, 78 Am. J. Int'l L. 607 (1984) (discussing the risks of declaring new human rights).

167 Knox, *Climate Change*, *supra*, note 8, at 210–211.

168 The ICCPR and the Inter-Am. Comm'n on Human Rights [IACHR] limit their application to persons within their jurisdiction: this limits the clear and direct application to territorial jurisdiction and leaves in doubt what form of protection could lie beyond that. Extraterritorial Application of Human Rights Treaties 2 (Fons Coomans & Menno Kamminga eds., 2004). In the case of the ECHR, extraterritorial application is considered exceptional. *See* Michal Gondek, *Extraterritorial Application of the European Convention on Human Rights: Territorial Focus in the Age of Globalization?*, 52 Neth. Int'l L. Rev. 349 (2005).

169 Although the ICESCR would appear to envisage extraterritorial application in its provision on international assistance and cooperation under Article 2, Paragraph 1,

address the global and diffuse phenomenon of climate change.[170] Even if there was scope to recognize the negative obligation to refrain from causing (transboundary) harm in the context of international human rights law, such an application would "treat climate change as a series of individual transboundary harms, rather than as a global threat to human rights."[171] If the obligation to respect would be difficult to transpose into the climate change context, as the duty not to cause transboundary harm, then the application of more affirmative obligations to support the provision of adaptation resources or technology is likely to be met with even more resistance.

Moreover, the correlation of rights and duties that underpins human rights law obligations relies in critical ways on establishing causation between the acts or omissions of the duty-bearer and the impacts on the rights-holder. In the case of climate change, such causation is elusive at best and difficult to establish because it results from a variety of causes and from the acts and omissions of States all over the world. The impacts of climate change contribute in indeterminate ways to aggregate problems of increased emissions, temperature rise, increased acidification of the seas, melting of permafrost, droughts, floods, and extreme weather events, such as cyclones and tsunamis. Furthermore, the harms that result are disparate and multifarious, spread broadly across vast areas, and present additional problems related to the standing of persons to take suit.[172] This makes the harms impossible to trace to particular sources, much less to the actions or omissions of States parties as duty-bearers under binding human rights treaties.

this also remains contested and politically divisive. Nevertheless, almost every General Comment adopted by the Committee on Economic, Social, and Cultural Rights has relied on this provision to set forth extraterritorial obligations. *See* Int'l Bank for Reconstruction & Dev., *supra* note 3.

170 ICHRP, *Rough Guide*, *supra* note 11, at 4.

171 Knox, *Climate Change*, *supra* note 8, at 211.

172 *See, e.g.*, Bradford C. Mank, *Standing and Global Warming: Is Injury to All Injury to None?*, 35 Envtl. L. 1 (2005).

A different limitation attendant to the international human rights framework, noted by Stephen Humphreys, is that the most severe climate change impacts will be catastrophic – drought, floods, vector-borne diseases, famine, and mass migration affecting large numbers of people. In those circumstances, States are entitled to,[173] and typically will, declare an emergency, which in turn allows them to suspend the application of human rights law protections.[174]

The sobering conclusion is, therefore, that human rights law does not display any special capacity to tackle the social and human impacts of climate change and that "the existing weaknesses of the human rights regime appear exacerbated in conditions of climate change, with little obvious sign of renewal or reinforcement in the future."[175]

5. Opportunities for Mutual Reinforcement and Policy Relevance

Despite the formal differences and disconnects between human rights and climate change from an international law perspective, it is also clear that the two areas of law are increasingly brought into contact because of evidence that climate change affects rights protected under international human rights law.[176] Whereas the foregoing discussion highlighted points of tension and disconnect, this concluding section aims to highlight the mutual relevance of human rights and climate change under international law, and it suggests structural and principled arguments that may mediate that relationship and help identify congruence between them.

173 Most human rights treaties and many domestic laws allow for the suspension of certain rights protections in times of national emergencies. *See, e.g.*, European Convention for the Protection of Human Rights and Fundamental Freedoms [ECHR], Nov. 4, 1950, 213 U.N.T.S. 220.

174 Stephen Humphreys, *Introduction: Human Rights and Climate Change*, *in* Human Rights and Climate Change, *supra* note 149, at 6.

175 Humphreys, *Competing Claims*, *supra* 154, at 65.

176 *See* discussion *supra* § 2.

The structural arguments are rooted primarily in concerns about how to approach the "fragmentation of international law,"[177] which, more than creating duplication and possible contradiction, may contribute to the weakening of protection. From a human rights law perspective, the concern is that this fragmentation may lead to uncertainty about the content and protection of human rights, because a difference in views about the content of general law diminishes legal security and "puts legal subjects in an unequal position vis-à-vis each other. The rights they enjoy depend on which jurisdiction is seized to enforce them."[178] Indeed some international legal scholars argue that general international law supports the coherent and mutually reinforcing interpretation of different bodies of law based on a strong presumption against normative conflict.[179]

Given the range of harms that climate change may inflict on the enjoyment of human rights,[180] one may argue for the need to take account of human rights obligations in the context of climate negotiations, and to explore how human rights and climate change law obligations may be viewed in a mutually supportive light,[181] where a State is party to one or both of the human rights Covenants and the United Nations Framework Convention on Climate Change (and Kyoto Protocol). In such cases, one may argue that human rights law obligations are part of the relevant rules of international law applicable in relations between the parties, within the meaning of Article 31 of the Vienna Convention on the Law of Treaties (VCLT),[182] which provides for the general rule of interpretation.[183]

177 ILC, *Fragmentation Report*, *supra* note 10.

178 *Id.* ¶ 52.

179 *See Id.* ¶ 37.

180 OHCHR, *Annual Report*, *supra* note 15.

181 ILC, *Fragmentation Report*, *supra* note 10, ¶ 417.

182 Vienna Convention on the Law of Treaties, May 23, 1969, 1155 U.N.T.S. 331, 8 I.L.M. 679.

183 Article 31, Paragraph 3 of the VCLT provides, "There shall be taken into account, together with the context.... (c) Any relevant rules of international law applicable in the relations between the parties." Vienna Convention on the Law of Treaties, art. 31, May 23, 1969, 1155 U.N.T.S. 331, 8 I.L.M. 679.

The foregoing may support the argument that all States Parties to core human rights instruments have human rights obligations, which are relevant to their obligations under climate change agreements where the membership of these international treaty norms overlaps. The nature of climate impacts described earlier results in an inevitable interaction between the regimes of climate change and human rights, such that the norms governing each may inform one another. This normative interplay can be characterized negatively or positively. In negative terms, it could be cast as a responsibility to ensure that the interpretation of norms related to climate change does not undermine the fulfillment of human rights obligations. From a more positive perspective, it could be argued that human rights obligations can be drawn on to inform the goals set forth in climate agreements and policy.[184]

Furthermore, human rights obligations may have direct, positive import for climate policy and may "inform and strengthen international, regional, and national policymaking in the area of environmental protection, and promoting policy coherence, legitimacy and sustainable outcomes."[185] A key assumption underlying arguments for the policy relevance of human rights is therefore that human rights are inextricably linked with sustainable development.[186] Human rights law obligations

184 *See e.g.*, ILC, *Fragmentation Report*, *supra* note 10, ¶¶ 410–480 (discussing "systematic integration"). *See also* Campbell McLachlan, *The Principle of Systematic Integration and Article 31(3)(C) of the Vienna Convention*, 54 INT'L & COMP. L.Q. 279 (2005).

185 U.N. Human Rights Council Res. 16/11, *supra* note 20, at 2; MARGOT E. SALOMON, GLOBAL RESPONSIBILITY FOR HUMAN RIGHTS: WORLD POVERTY AND THE DEVELOPMENT OF INTERNATIONAL LAW 106 (2007); McInerney-Lankford, *A Comment of Challenges*, *supra* note 2.

186 Dominic McGoldrick, *Sustainable Development and Human Rights: An Integrated Conception*, 45 INT'L & COMP. L.Q. 796 (1996); ORG. FOR ECON. CO-OPERATION & DEV. [OECD], DAC ACTION-ORIENTED POLICY ON HUMAN RIGHTS AND DEVELOPMENT (2007); Third High Level Forum on Aid Effectiveness, Accra, Ghana, Sept. 2–4, 2008, *Accra Agenda for Action*, ¶ 3, *available at* http://siteresources.worldbank.org/ACCRAEXT/Resources/4700790–1217425866038/AAA-4-SEPTEMBER-FINAL-16h00.pdf ("Gender equality, respect for human rights, and environmental sustainability are cornerstones for achieving enduring impact on the lives and potential of poor women, men, and children. It is vital that all our policies address these issues in a more systematic and coherent way.").

may operate as substantive standards or thresholds to provide guidance on the "levels of protection for individual rights that can be regarded as the minimum acceptable outcome under a given policy scenario."[187] In this way, they also mandate a minimum standard of "do no harm," such that the acts and omissions of States with respect to climate change must at least not lower the standard of human rights enjoyment that currently prevails. More positively, these treaty obligations require that States act to protect their citizens from the impacts of climate change, regardless of whether the harms of climate change result from their acts or omissions and regardless of whether the State is itself to blame for climate change. By definition, mitigation policy involves identifying priorities and acceptable outcomes, and human rights obligations may, if not provide a formula for such prioritization, then contribute legally binding standards against which to assess those policies and offer minimum acceptable thresholds for their outcomes.

Approaches that take account of human rights obligations, like human rights approaches in general,[188] emphasize human welfare and people-centered perspectives on climate change and prioritize its human and social dimensions over its purely environmental science, technical, or even economic considerations. They may also ensure that the plight of the most vulnerable (States or communities) are considered, that safeguards exist against unjustified harm to their rights,[189] and that their

187 ICHRP, *Rough Guide*, *supra* note 11, at 18.

188 For a UN definition of the Human Rights Based Approach (HRBA), see United Nations Development Group [UNDG], *The Human Rights Based Approach to Development Cooperation: Towards a Common Understanding Among UN Agencies* (2003), *available at* http://www.undg.org/archive_docs/6959-The_Human_Rights_Based_Approach_to_Development_Cooperation_Towards_a_Common_Understanding_among_UN.pdf.

189 Margot Salomon has put forward similar arguments in the context of economic globalization: "A rights-based approach to globalization seeks to place international human rights standards and principles at the centre of international economic affairs, to have them successfully inform all cooperative endeavors that may impact on their exercise." Margot E. Salomon, *supra* note 174, at 11.

participation and "voice" are adequately protected through information and consultation.

Human rights approaches may help highlight or counter power imbalances that underpin the climate change problem.[190] They may also facilitate a policy focus on the underlying causes of special vulnerability to climate change, because that vulnerability may also relate causally to the denial or violation of rights. As Stephen Humphreys notes,

> The most dramatic impacts of climate change are expected to occur (and are already being experienced) in the world's poorest countries, where rights protections are too often weak for a variety of reasons. But the effect is also causal and mutually reinforcing. Populations whose rights are poorly protected are likely to be . . . less able to lobby effectively for government or international action and more likely to lack the resources needed to adapt to expected alterations in their environmental and economic situation. A vicious circle links precarious access to natural resources, poor physical infrastructure, weak rights protections, and vulnerability to climate-related harms.[191]

In addition, one should recognize the special dependence that certain groups may have on the environment, whether as a means of subsistence or as a central element of their cultural identity and heritage, which contributes a new and often accentuated form of vulnerability.[192]

In more operational terms, recognition of the policy relevance and applicability of human rights obligations to climate change might encourage greater reliance on human rights impact assessments (HRIAs)[193] as a potential tool to integrate human rights legal accountability into the

190 Amy Sinden, *Climate Change and Human Rights* 27 J. Land Resources & Envtl. L. 255 (2007).

191 Stephen Humphreys, *Introduction: Human Rights and Climate Change*, *in* Human Rights and Climate Change, *supra* note 149, at 1–2.

192 Indigenous Peoples and Climate Change (Jan Salick & Anja Byg eds., 2007), *available at* http://www.ecdgroup.com/docs/lib_004630823.pdf; Frances Seymour, *Forests, Climate Change and Human Rights: Managing Risk and Trade-offs*, *in* Human Rights and Climate Change, *supra* note 149, at 207.

193 Although human rights impact assessments (HRIA) are a relatively new tool in human rights practice with no accepted definition, and although it appears they

process of policy formulation.[194] In this context, use of HRIAs would be an explicitly human-rights-based process to help duty-bearers understand the potential human rights impacts of their actions, and to orient climate policy to minimize those effects that are harmful,[195] by implementing a type of "precautionary principle."[196] This process would be particularly apposite in respect of the duties to respect,[197] but possibly also has relevance in relation to the duties to protect and even to fulfill. A HRIA can be characterized as a tool to measure the gap between the commitments made by a State[198] and the ability of individuals, groups,

may fulfill a broad range of functions depending on context and institutional factors, they are recognized to have evolved from social impact assessments (SIAs) and environmental impact assessments (EIAs), and they draw on social scientific analysis and work in fields such as development, environment, business administration, and public policy to respond to specific human rights-related needs and concerns. SIMON WALKER, THE FUTURE OF HUMAN RIGHTS IMPACT ASSESSMENTS OF TRADE AGREEMENTS 5 (Sch. of Human Rights Research Ser. No. 35, 2009). *See generally Guide to Human Rights Impact Assessment and Management*, INT'L FIN. CORP. [IFC], WORLD BANK GROUP (June 25, 2010), http://www.ifc.org/ifcext/sustainability.nsf/Content/Publications_Handbook_HRIA.

194 For examples of the growing body of research on quantitative assessments of human rights, *see* Rep. of the High-Level Task Force on the Implementation of the Right to Development on its 5th Sess., Apr. 1–9, 2009, UN Doc. A/HRC.12/WG.2/TF/2, Human Rights Council, 12th Sess. (June 17, 2009); Edward Anderson & Marta Foresti, *Assessing Compliance: The Challenge for Economic and Social Rights*, 1 J. HUM. RTS. PRAC., no. 3, Nov. 2009 at 469. *See also Assessing the Impact of Human Rights Work: Challenges and Choices*, INT'L COUNCIL ON HUMAN RIGHTS POLICY [ICHRP] (2011), http://www.ichrp.org/en/projects/181 (last visited July 13, 2011).

195 *See, e.g.*, Soc. & Econ. Rights Action Ctr. [SERAC] v. Nigeria, African Comm'n on Human & Peoples Rights, Commc'n. No. 155/96 ¶ 53 (2001).

196 The precautionary principle is now integral to international environmental policy and embodied in Principle 15 of the Rio Declaration. *Rio Declaration*, *supra* note 21, Princ. 15. *See* THE PRECAUTIONARY PRINCIPLE AND INTERNATIONAL LAW: THE CHALLENGE OF IMPLEMENTATION (David Freestone & Ellen Hay eds., 1996); Hannes Veinla, *Precautionary Environmental Protection and Human Rights*, XII U. TARTU JURIDICA INT'L L. REV. 91 (2007), *available at* http://www.juridicainternational.eu/index/2007/vol-xii/precautionary-environmental-protection-and-human-rights/?highlight=precautionary,environmental.

197 *See* discussion *infra*.

198 This process could be considered applicable to any and all parts of the normative framework and could therefore apply to civil and political, as well as economic, social, and cultural rights. Int'l Bank for Reconstruction & Dev., *supra* note 3, at 32.

and communities in a country to enjoy these rights.[199] Others define HRIA as a systematic process to investigate and measure the impact of politics, programs, projects, and interventions on human rights. It is based on the normative framework of international human rights law laid down in such documents as the *Universal Declaration of Human Rights* and the *International Covenants on Civil and Political Rights and on Economic, Social and Cultural Rights*.[200] Although an in-depth analysis of HRIA lies beyond the scope of this analysis, the purpose is merely to flag their potential application as either *ex ante* or *ex post* assessments of climate change policy options[201] or of agreements connected with climate change,[202] while bearing in mind the scientific evidence of direct and indirect climate impacts and the human rights law obligations that govern all spheres.

In a different way, recognition of the relevance of human rights to climate change might also argue for the IPCC assessments,[203] or traditional environmental assessments (EA), to incorporate the human rights obligations and standards, including the respect of international human rights treaties. Indeed, many models of EA incorporate environmental

199 *See Introduction to Human Rights Impact Assessment*, Human Rights Impact Resource Center, http://www.humanrightsimpact.org/introduction-to-hria/hria-tutorial/introduction/ (last visited July 13, 2011).

200 Saskia Bakker et al., *Human Rights Impact Assessment in Practice: The Case of the Health Rights of Women Assessment Instrument (HeRWAI)*, 1 J. Hum. Rts. Prac., no. 3, Nov. 2009 at 436 (quoting Olga Lenzen & Marina d'Engelbronner, Aim for Human Rights, Guide to Corporate Human Rights Impact Assessment Tools 9 (2009), *available at* http://www.aimforhumanrights.org/fileadmin/user_upload/pdf/HRB_Guide_to_corporate_HRIA_2009-def.pdf).

201 Some scholars have attempted to categorize human rights impact assessments into *ex ante* and *ex post*, by distinguishing between the assessment of projects and programs that are either directly or indirectly related to human rights. *E.g.*, Todd Landman, Studying Human Rights (2006).

202 For a discussion of the obligation to respect as applied to climate change and arguments, *see* OHCHR, *Annual Report*, *supra* note 15.

203 Up to now, the IPCC Assessments have not considered human rights explicitly. On the IPCC bridging the science-policy interface, *see generally* Neth. Envtl. Assessment Agency, Assessing an IPCC Assessment: An Analysis of Statements on Projected Regional Impacts in the 2007 Report (2010), *available at* http://www2.lse.ac.uk/CATS/publications/papersPDFs/82_Petersen_IPCCreport.pdf.

and social impacts together in an integrated way,[204] such that including consideration of human rights impacts might also be possible. Moreover, many EA models may be based on environmental impact assessments but allow for elements of other instruments to be included as appropriate.[205] Strategic environmental assessments employed in the context of climate change policy may therefore present entry points for the integration of human rights considerations.[206] This might offer a more ad hoc or piecemeal consideration of specific human rights or principles and could rely more on individual human rights indicators (or sets of indicators), rather than entire human rights impact assessments.[207]

6. Conclusion

The philosopher Simon Caney has said, "[A] human rights perspective has important insights and any account of the impacts of climate change that ignores implications for people's enjoyment of human rights is fundamentally incomplete and inadequate."[208] The key contention of this chapter is that without some account of international human rights obligations, international law and policy on climate change are not complete.

204 *See, e.g.*, World Bank Group [WBG], *Operational Policies 4.01: Environmental Assessment*, ¶ 2, OP 4.01 (Jan. 1999), *available at* http://go.worldbank.org/RUEQVWD550.

205 Mohamed A. Bekhechi & Jean-Roger Mercier, The Legal and Regulatory Framework for Environmental Impact Assessments: A Study of Selected Countries in Sub-Saharan Africa 2 (2002).

206 OECD, DAC Network on Env't & Dev. Co-operation [ENVIRONET], *Strategic Environmental Assessment and Adaptation to Climate Change* (Oct. 30, 2008), *available at* http://www.oecd.org/dataoecd/0/43/42025733.pdf.

207 It is clear, however, that any form of HRIA would rely heavily on a potentially wide range of qualitative and quantitative human rights indicators, and this area has also been the subject of a growing body of work and scholarship. See, for example, *Human Rights Indicators: Indicators to Promote and Monitor the Implementation of Human Rights*, OHCHR, http://www2.ohchr.org/english/issues/indicators/index.htm (last visited July 13, 2011), for a project of the OHCHR on human rights indicators, which has yielded a sophisticated conceptual and methodological framework that is starting to be applied in country and institutional contexts.

208 Simon Caney, *Climate Change, Human Rights and Moral Thresholds*, *in* Human Rights and Climate Change, *supra* note 149, at 90.

Human rights treaty obligations are part of the legal landscape that governs the actions of a majority of States in a wide array of realms, including those connected with climate change. As such they are binding law and should not be ignored.[209]

209 Discussing the idea of human rights thresholds as a policy requirement, the ICHRP *Rough Guide* states, "A policy requirement that basic threshold levels should not be breached – either as an effect of climate change itself or as a by-product of a given mitigation or adaptation policy – ought not to be controversial, especially as such a goal is also a legal requirement for policy actors." ICHRP, *Rough Guide*, *supra* note 11, at 18.

9 Protection of Persons Displaced as a Result of Climate Change

Existing Tools and Emerging Frameworks

Michele Klein Solomon and Koko Warner

Introduction

In the 1990s, the Intergovernmental Panel on Climate Change (IPCC) noted that "one of the gravest effects of climate change may be on human migration."[1] The findings of this first IPCC Assessment Report were substantiated by its Fourth Assessment Report in 2007,[2] which showed that climate change is likely to raise the risk of humanitarian emergencies and trigger population movements due to increasingly intense weather events, sea level rise, and accelerated environmental degradation, including coastal erosion and desertification.

Research has substantiated the fact that environmental change is one of a larger set of factors that affect human migration and displacement worldwide. Processes such as natural disasters and shifts in climate patterns that may bring glacial melt, sea level rise, and desertification are and will increasingly affect migration and displacement. Some of

1 Intergovernmental Panel on Climate Change (IPCC), *Climate Change: The IPCC Impacts Assessment* 5, Working Group II Report (1990); IPCC, *First Assessment Report (FAR)* (1990), *available at* http://www.ipcc.ch/publications_and_data/publications_and_data_reports.shtml.

2 IPCC, *Fourth Assessment Report: Climate Change 2007 (AR4)* (2007), *available at* http://www.ipcc.ch/publications_and_data/publications_and_data_reports.shtml.

The authors are indebted to Gervais Appave, Senior Policy Adviser to the Director General, International Organization for Migration (IOM), for sharing his forthcoming publication on a related subject.

the most vulnerable regions include low-lying islands and deltas, coastal areas, areas dependent on glacial-fed water systems, and areas subject to persistent drought. Field-based research suggests that most environmentally induced migrants will move within their own countries. Some movements will resemble familiar migration and displacement patterns, but other movements will likely occur in emergency circumstances or complex humanitarian crises, particularly where climate change exacerbates natural hazards, such as cyclones. Climatic-related human mobility can also make more acute factors contributing to communal violence and conflict.[3]

Migration is one adaptation and survival strategy employed by populations that experience the negative effects of climate change. In fact, migration has shaped the interaction between humans and their environment for millennia. With climate change exacerbating environmental degradation – resulting in both more intense slow-onset changes (such as desertification and soil and coastal erosion) and more frequent sudden-onset disasters (such as tropical storms, floods and heat waves) – human mobility in response to such environmental events and processes is expected to increase.[4]

What portion of such migration can be attributed specifically or exclusively to climate change (as opposed to other environmental or human factors) is an important empirical question that will be difficult to answer. However, it is already widely understood that most of both climate and environmentally induced movement of people is likely to be of an internal nature and not involve the crossing of an international border. The case of sinking islands will, by definition, raise the prospect

3 *See e.g.*, the findings of the FP7 MICROCON (*A Micro-Level Analysis of Violent Conflict*, http://www.microconflict.eu/) and CLICO (*Climate Change, Hydro-Security, and Human Security*, http://www.clico.org/) projects looking at water-related cooperation and conflict at sub-national scales.

4 Please note that volcanic eruptions, earthquakes and tsunamis are tectonic events. They are not related to climatic factors and climate change. Of course, they often also give rise to displacement of persons.

of cross-border movements, as the territories of certain small Island States become, over time, entirely uninhabitable. As discussed later, this distinction is important in analyzing the legal frameworks and protection afforded under international law to persons affected by climate change.

No international legal instrument specifically and explicitly defines environmental migration or environmental migrants (or climate migration/migrants or refugees), includes provision for cross-border movements induced by the environment, or provides for the protection of persons who move due to environmental factors. Neither the UN Framework Convention on Climate Change (UNFCCC) nor its Kyoto Protocol includes any provisions concerning specific protection of or assistance to those who will be directly displaced by climate change.

The topic of human mobility recently appeared for the first time in a decision of the Conference of the Parties of the UNFCCC. The issue is couched in pragmatic, solution-oriented terms under adaptation, reflecting the realization of Parties following the IPCC's Fourth Assessment report that the impacts of climate change are highly likely and may already be manifesting themselves in different regions of the world. Migration, displacement, and planned relocation are featured in the Cancun Adaptation Framework as technical issues in a part of the text that highlights a list of activities that may qualify for adaptation funding in the future. This chapter explores the potential importance of the first-ever agreed text on migration, displacement, and planned relocation in the UNFCCC climate negotiations process: Cancun Adaptation Framework paragraph 14(f).

This chapter is presented in two parts. The first part examines the nature of the relationship among climate change, environmental degradation, and migration, including the applicability of refugee law in the context of climate-change-related displacement and other migration-related laws and institutions and how they can help. The second part reviews developments within the UNFCCC leading up to and including the Cancun Adaptation Framework paragraph 14(f), and what implications that paragraph has for migration planning efforts going forward.

The chapter gives particular attention to the emergence and shaping of human mobility in the UNFCCC context, which sets it apart from other attempts to raise the political weight of the issue, including those that have focused on using or expanding tools and mandates of existing frameworks such as the 1951 UN Convention relating to the Status of Refugees (which are addressed in this and the following chapter).[5] New opportunities are emerging for moving practical solutions and discussions forward in the UNFCCC process through the Cancun Adaptation Framework (paragraph 14(f)), the Climate Finance and the Adaptation Committee, and the Subsidiary Body on Implementation's (SBI) Work Program on Loss and Damage.[6] These and other policy processes will likely catalyze nationally and regionally driven work on the topic of migration, displacement, and planned relocation in the context of climate change.

5 Approaches from this school include writings such as David Hodgkinson, Tess Burton, Heather Anderson & Lucy Young, *"The Hour when the Ship Comes in": A Convention for Persons Displaced by Climate Change* (2009), *available at* http://www.ccdpconvention.com/documents/Hour_When_Ship_Comes_In.pdf (last visited Aug. 2, 2011); Michel Prieur et al., *Draft Convention on the International Status of Environmentally-Displaced Persons* (2d version, May 2010); Bonnie Docherty & Tyler Giannini, *Confronting a Rising Tide: A Proposal for a Convention on Climate Change Refugees*, 33 Harv. Envtl. L. Rev. 349–403 (2009); Frank Biermann & Ingrid Boas, *Protecting Climate Refugees: The Case for a Global Protocol.* posting to Environment Magazine, http://www.environmentmagazine.org/Archives/Back%20Issues/November-December%202008/Biermann-Boas-full.html (Nov.-Dec. 2008); Frank Biermann, Philipp Pattberg, Harro van Asselt, & Fariborz Zelli, *The Fragmentation of Global Governance Architectures: A Framework for Analysis, in* Global Environmental Politics, November 2009, Vol. 9, No. 4: 14–40; Frank Biermann & Ingrid Boas, *Preparing for a Warmer World: Towards a Global Governance System to Protect Climate Refugees, in* Global Environmental Politics, 60–88 (MIT Press, 2010).

6 The SBI is one of four sub-committees in the UNFCCC climate negotiations process. COP decisions related to preparation of activities for implementation (such as the Clean Development Mechanism and others) are often sent to the SBI for further elaboration and preparation before implementation.

1. The Relationship among Climate Change, Environmental Degradation, and Migration

Drawing in large part on Docherty and Giannini and as further developed in Appave (forthcoming), the relationship among climate change, environmental degradation, and human mobility can usefully be examined along several overlapping axes.[7]

- **Slow-onset versus sudden/catastrophic change.** Slow-onset change includes rises or falls in temperature, rises or falls in sea level, desertification, coastal erosion, and changes in amounts of precipitation. Sudden-onset changes present as disasters in the form of cyclones, tornadoes, floods, landslides, and avalanches. Earthquakes and tsunamis are environmental events, but are of tectonic origin and are, as a general rule, not climate related.[8] In the short term, disasters capture media and policy attention, but slow-onset change – such as that experienced by the so-called sinking island States – is likely to have as significant if not greater impact in the longer term, particularly because such changes are often irreversible.
- **Forced versus voluntary movement.** There can be clear cases of unavoidable, forced movement, as well as clear cases of planned and voluntary movement. However, most current and anticipated movement is likely to fall somewhere in between where it is difficult to determine precisely the point at which movement is no longer optional. In the vast majority of cases, the complexity of environmental migration does not allow for a clear-cut distinction between "forced" and "voluntary" movements. The situation of the sinking

7 Docherty & Giannini, *supra* note 5.

8 There is some scientific evidence that tectonic movements might be related to global warming because of the impact it has on the volume and weight of the water column. *See* Agence France-Presse, *Climate change affects tectonic plate movement, causing earthquakes: study*, posting to The Raw Story, http://www.rawstory.com/rs/2011/04/13/climate-can-drive-seismic-shifts-study/ (Apr. 13, 2011).

islands is a case in point. It is unlikely that the inhabitants will remain at home until the islands actually sink. Rather, over time, individual islanders will decide to leave at a time and in a manner when opportunities to do so present themselves, such as an invitation from a family member abroad to come work and live with them. Whether that movement is "forced" or "voluntary" is less significant than what the needs and rights are of the affected populations.

- **Permanent versus temporary movement.** Historically, people have migrated in response to seasonal changes, particularly in areas where sharp and generally expected variations in rainfall and temperature determine the nomadic behavior of traditional pastoralists. Alterations in habitat due to climate change, such as desertification, may require longer-term changes in residence, but here again it is difficult to establish precisely why and when the option of return is no longer available and the stay away from home turns into permanent relocation.
- **Last-resort survival versus adaptation strategy.** Most people think of migration as a survival measure, especially in cases of impending or acute natural disaster when people have no choice but to leave their homes. However, migration can also be an adaptation strategy developed over a longer period of time. It can help people cope with gradual changes in their environment that are affecting their ways of life. In this case migration is not so much an accident that is best avoided but rather a positive adaptation strategy.
- **Internal versus international mobility.** The overwhelming majority of both climate and environmentally induced movement of people to date has been and is likely to continue to be of an internal nature within the borders of a country, not involving the crossing of an international border. This is a fundamental distinction for questions of legal responsibility. However, flows can also spill across a border into another country, and therefore managing population movements that have both internal and international manifestations, either simultaneously or sequentially, can be particularly challenging. The case of the

sinking islands will, by definition, raise the prospect of cross-border movements, as the territories of certain small island States become, over time, entirely uninhabitable.

It is useful to acknowledge that, to date, there is no global agreement in the climate change or any other context governing the movement of persons that provides criteria or quotas for the admission of the citizens of one State into the territory of another. The international law that exists relevant to the movement of persons concerns, largely, the treatment of foreign nationals once they are in a foreign territory, whether they be refugees[9] or migrant workers,[10] and does not provide a basis for their orderly admission. States continue to regard the admission of foreign nationals to their territories as a matter of national sovereignty, to be determined according to national priorities and criteria. They have resisted subjecting such matters to agreed global criteria not least to permit flexibility and adaptability in their immigration laws to changing national economic, labor market, and demographic circumstances.

Moreover, there is no academic or policy consensus on the construction of a definition of environmental or climate-change-displaced persons. In the absence of such consensus, the International Organization for Migration (IOM) has proposed a working definition that is broad enough to enable inclusive discussion of aspects of human mobility related to environmental factors: "Environmental migrants are persons or groups of persons who, for compelling reasons of sudden or progressive change in the environment that adversely affects their lives or living conditions, are obliged to leave their habitual homes, or choose to do

9 *See, e.g.*, the Convention Relating to the Status of Refugees, July 28, 1951, 189 U.N.T.S. 150 [hereinafter 1951 Refugee Convention] and the Protocol relating to the Status of Refugees, Oct. 4, 1967, 606 U.N.T.S. 26, as well as regional instruments such as the Convention Governing the Specific Aspects of Refugee Problems in Africa, Sept. 6–10, 1969 [hereinafter African Union Refugee Convention of 1969] and the Cartagena Declaration on Refugees, Nov. 22, 1984 [hereinafter Cartagena Declaration].

10 *See, e.g.*, the International Convention on the Protection of the Rights of all Migrant Workers and Members of their Families, Dec. 18, 1990, 2220 U.N.T.S. 3.

so, either temporarily or permanently, and who move either within their country or abroad."[11] The intention behind this definition is to attempt comprehensive coverage of all forms of population movements whether they be temporary or permanent, internal or international, voluntary or forced, or triggered by sudden or incremental change. No legal status is intended to flow from IOM's working definition. The terminological debate will clearly continue for some time to come in both academic and policy circles.

This general reticence by the international community to develop binding international legal definitions, standards, or clear burden-sharing and interstate cooperation mechanisms relevant to the movement of persons is exacerbated in the case of the movement of people related to climate change. There are several reasons for this reticence, put forward by Boncour and Burson.[12]

First, they note the difficulty of identifying or establishing simple, direct causal relationships between environmental events and processes (such as cyclones, floods, desertification, soil erosion, and changing coastlines), climate change, and migratory flows. This is in part because environmental drivers of migration interact with underlying economic, social, and developmental factors that can accelerate or, conversely, mask the effects of climate change. For example, persons who are already vulnerable due to their social or economic status (i.e., indigenous populations in many Andean countries) are likely to be less resilient to and therefore even more vulnerable to the effects of climate change.

Second, climate-change-related migration straddles several policy domains, including migration proper, development, the environment,

11 IOM, Glossary on Migration, at 33 (2nd ed. 2011). For a further discussion of the usefulness of the concept of "environmental migrant" *see* Howard Chang, *The Environment and Climate Change: Is International Migration Part of the Solution or Part of the Problem*? (Paper presented to the 16th Symposium of the Fordham Envtl. L. Rev.).

12 Philippe Boncour & Bruce Burson, *Climate Change and Migration in the South Pacific Region: Policy Perspectives*, *in* Climate Change and Migration: South Pacific Perspectives 5–28 (Bruce Burson ed., 2010).

and humanitarian assistance. Issues of policy or administrative ownership have to be resolved, and effective lines of communication between policy makers and researchers in each of these fields need to be established before serious work can begin.

Third, migration policy makers and program managers tend to prioritize security-related objectives. This means that they are likely to address climate-change-related migratory movements from a defensive perspective, focusing on threats to national security that need to be countered, rather than seeing migration in this context as a potentially adaptive response to climate change.

Finally, further complicating the debate is the uncertainty surrounding how many people have been or are likely to be affected.[13] Estimates based on expert estimates and statements of the numbers of persons displaced already or likely to be displaced by 2050 as a result of climate-change-related effects vary wildly from 25 million persons to more than 2 billion persons.[14] Indeed, although an empirical base is slowly emerging,[15] the lack of basic data related to human mobility and environmental factors make more robust and reliable "global" estimates

13 *See* UK Government Office for Science, Migration and Global Environmental Change: Future Challenges and Opportunities, Foresight Report (2011); Norman Myers, Environmental Refugees: An Emergent Security Issue (2005); Oli Brown, Climate Change and Forced Migration: Observations, Projections and Implications, UNDP Human Development Report, Occasional Paper No. 17 (2007); Kniveton et al., Climate Change and Migration: Improving Methodologies to Estimate Flows, IOM, Migration Research Series No. 33 (2008).

14 Biermann & Boas, *Protecting Climate Refugees: The Case for a Global Protocol*, *supra* note 5; Hodgkinson, et al., *supra* note 5.

15 *See, e.g., Environmental Change and Forced Migration Scenarios*, www.each-for.eu (summarized in Jill Jäger, Johannes Frühmann, Sigrid Grünberger, & Andras Vag, Synthesis Report, *Environmental Change and Forced Migration Scenarios*, Project 64–66 (2009), *available at* http://www.each-for.eu/documents/EACH-FOR_Synthesis_Report_090515.pdf; Koko Warner, Charles Erhart, Alex de Sherbinin, Susana Adamo, & Tricia Chai-Onn, *In Search of Shelter: Mapping the effects of climate change on human migration and displacement* (2009), *available at* http://www.adaptationlearning.net/sites/default/files/clim-migr-report-june09_final.pdf); UK Government Office for Science, Foresight Report, *supra* note 13; Alex de Sherbinin et al., *Preparing for Resettlement Associated With Climate Change*, 334 Science 456 (Oct. 28, 2011).

difficult.[16] The intent of many of these estimates seems more to be one of seeking to raise public awareness of the issue and its implications for potential economic and social disruptions, rather than of reporting with scientific accuracy.[17] Tacoli has cautioned that "alarmist predictions will backfire and result in policies that marginalize the poorest and most vulnerable groups."[18]

Despite these hurdles, recent years have witnessed a burgeoning of research and dialogue on these issues, not least of which that stimulated and led by the institutions with which we are affiliated, namely the United Nations University (UNU) and particularly its Institute for Environment and Human Security, and the International Organization for Migration (IOM). There is now wide consensus that, although climate change on its own does not necessarily lead to population displacements, it does produce and exacerbate existing environmental impacts that make it difficult, and in some cases over time impossible, for some people to stay where they are.[19]

There are several specific circumstances pertaining to migration related to climate change and environmental degradation that are associated with vulnerability and may give rise to the need for protection of and assistance to the persons affected. These cases include persons who are already outside of their country or areas of habitual residence and cannot safely return due to environmental factors and have no legal status in the country or region of stay or have lost legal status over time;

16 Warner et al., *supra* note 15, at 1 (text box summarizing various attempts to make global estimates and associated difficulties).

17 Richard Black, *Environmental Refugees: Myth or Reality?* (UNHCR Working Paper No.34, 2001); Stephen Castles, *Environmental Change and Forced Migration: Making Sense of the Debate* (UNHCR Working Paper No. 70, 2002).

18 Cecilia Tacoli, *Not Only Climate Change: Mobility, Vulnerability and Socio-Economic Transformations in Environmentally Fragile Areas of Bolivia, Senegal and Tanzania* V (Human Settlements Working Paper Series, Rural-Urban Interactions and Livelihood Strategies – 28. London: IIED, Feb. 2011).

19 Migration, Climate Change and the Environment: Assessing the Evidence (Frank Laczko & Christine Aghazarm eds., Geneva: IOM, 2009), *available at* http://publications.iom.int/bookstore/free/migration_and_environment.pdf.

they also include persons who flee from their homes whether internally or across an international border as a result of predominantly environmental factors and are in need of protection and assistance.

As is explored in greater depth in this chapter, in 2010, the Cancun Adaptation Framework (hereinafter "Cancun Framework")[20] to the UNFCCC,[21] for the first time in an international agreement, called explicit attention to climate-change-induced displacement, migration, and planned relocation, in paragraph 14(f). In many respects, paragraph 14(f) is an initial step, a first-time mention of the issue in the context of global environmental change and particularly climate change. It calls attention to the fact that there is and will continue to be human movement as a result of the effects of climate change, but it leaves room for further development and interpretation of specific modalities of interstate cooperation or obligations on States; for example, to protect and assist persons displaced in relation to climatic variability and change. Before considering the development of and potential next steps related to paragraph 14(f), this chapter explores international law relevant to the movement of persons and its applicability or potential applicability to persons moving as a result of climate change, and in particular to the situation of inhabitants of sinking island States.

1.1. Relevance of International Refugee Law

The image of a person or family fleeing a natural disaster or seeking a place to live peacefully and with the prospect of earning a livelihood when their farms are flooded or their earth is scorched after successive years of drought conjures in the public mind the situation of refugees. Indeed, there have been multiple and repeated calls for States to afford

20 United Nations Framework Convention on Climate Change (UNFCCC), Conference of the Parties, Decision -/CP.16, Outcome of the work of the Ad Hoc Working Group on Long-term Cooperative Action under the Convention (2010), *available at* http://unfccc.int/files/meetings/cop_16/application/pdf/cop16_lca.pdf.

21 *Ibid.*

protection and assistance to "climate" or "environmental refugees" pursuant to existing international refugee law or to develop a specific refugee law instrument to address persons displaced as a result of environmental or climate change forces.[22] Terms such as "climate change refugee" or "environmental refugee" are therefore commonly used in the media, but there is a lively debate in the scholarly literature about the precise terminology that ought to be used to describe those who move as a consequence of climate or environmental change, not least of which because of the legal rights and responsibilities that flow therefrom.

22 *See* Jane McAdam, *Climate Change Displacement and International Law: Complementary Protection Standards* (UNHCR Legal and Protection Policy Research Series, May 2011); Susin Park, *Climate Change and the Risk of Statelessness: The Situation of Low-lying Island States* (UNHCR Legal and Protection Research Series, May 2011); Roger Zetter, *Protecting Environmentally Displaced People: Developing the Capacity of Legal and Normative Frameworks* (Oxford Refugee Studies Centre, Research Report, 2010), *available at* www.rsc.ox.ac.uk; Essam El-Hinnawi, *Environmental Refugees*, United Nations Environment Programme, Nairobi (1985); Steve Lonergan, *The Role of Environmental Degradation in Population Displacement*, Environmental Change and Security Project Report, No 4: 5–15 (1998), Koko Warner, *Climate Change Induced Displacement: Adaptation Policy in the Context of the UNFCCC Climate Negotiations*, (May 2011), *available at* http://www.unhcr.org/refworld/docid/4e09a3d32.html; Nansen Conference on Climate Change and Displacement Bulletin: A Summary Report of the Nansen Conference on Climate Change and Displacement in the 21st Century, IISD Reporting Services, May 10, 2011, *available at* http://www.iisd.ca/download/pdf/sd/ymbvol189num1e.pdf; United Nations General Assembly, Annual Report of the United Nations High Commissioner for Human Rights and Reports of the Office of the High Commissioner and the Secretary-General, *Report of the Office of the United Nations High Commissioner for Human Rights on the relationship between climate change and human rights*, Jan. 15, 2009; United Nations Economic and Social Council, Commission on Human Rights, Report of the Sub-Commission on the Promotion and Protection of Human Rights on Its Fifty-Seventh Session, *The legal implications of the disappearance of States and other territories for environmental reasons, including the implications for the human rights of their residents, with particular reference to the rights of indigenous peoples*, Resolution 2005/20, Oct. 17, 2005, *available at* http://www.un.org/terrorism/pdfs/3/G0516031.pdf; UNHCR Expert Roundtable, *Climate Change and Displacement: Identifying Gaps and Responses* (Feb. 2011), *available at* http://www.unhcr.org/refworld/docid/4d7604ec2.html; UNHCR, *Forced Displacement in the context of climate change: challenges for States under international law* (May 19, 2009), *available at* http://unfccc.int/resource/docs/2009/smsn/igo/049.pdf.

Those who favor use of the term "environmental refugee" or "climate change refugee" or some such variant argue, essentially by analogy, that those who flee for climate or environmental change reasons do so in circumstances of such overwhelming distress as to be tantamount to persecution. Biermann and Boas cannot see reason why inhabitants of the Maldives, threatened as they are by a rise in sea levels, should receive less protection than those who fear political persecution.[23] Although the calls for action are laudable and understandable, international refugee law as it now stands is not prima facie well suited to this purpose. Moreover, it is questionable whether it could be adapted in the near term in a meaningful way to serve the aim of protecting and assisting persons fleeing their countries as a result of environmental degradation and/or climate change.

Many scholars and the UN Human Rights Commission (UNHCR) itself, the custodian of the 1951 Convention relating to the Status of Refugees, are skeptical.[24] They point out that the term "refugee" has specific legal meaning that, at best, would have to be uncomfortably stretched if it is to cover those who move as a consequence of environmental or climate change; at worst doing so may risk disturbing refugee status determination procedures and seriously undermining the refugee protection regime.

The United Nations Convention relating to the Status of Refugees of 1951 (hereinafter "Refugee Convention") serves a specific purpose: protection from persecution. The Convention defines a refugee (in

23 Biermann & Boas, *Protecting Climate Refugees: The Case for a Global Protocol*, *supra* note 5.

24 Jane McAdam & Ben Saul, *Displacement with Dignity: International Law and Policy Responses to Climate Change, Migration and Security in Bangladesh* (Univ. N.S.W. Faculty of Law Research Series, Paper No. 63, 2010); Angela Williams, *Turning the Tide: Recognizing Climate Change Refugees in International Law*, 30 Law & Pol'y 502–529 (2008); Bruce Burson, *Protecting the Rights of People Displaced by Climate Change: Global Issues and Regional Perspectives*, *in* Burson, *supra* note 12; UN High Commissioner for Refugees, *Climate Change, Natural Disasters and Human Displacement: A UNHCR Perspective* (Aug, 14, 2009), *available at* http://www.unhcr.org/refworld/docid/4a8e4f8b2.html (last visited Aug. 4, 2011).

Article 1A) as a person who "*owing to a well-founded fear of being persecuted for reasons of race, religion, nationality, membership of a particular social group, or political opinion, is outside the country of his nationality, and is unable to or, owing to such fear, is unwilling to avail himself of the protection of that country*." The concept of a refugee has been expanded by regional conventions in Africa and Latin America to include persons who have fled war or other violence in their home country.[25]

As it currently stands, the 1951 refugee definition would not apply to environmental or climate change migrants per se for the following reasons. First, exposure to climate/environmental factors is not generally considered "persecution" within the meaning of the Refugee Convention. "Persecution" exists when there are serious violations of human rights (which may indeed be present in the case of persons subjected to the prolonged effects of climate change[26]) perpetrated by a "State agent;" that is, through the deliberate policy or practice of a government. This understanding of persecution has been extended in recent years to cover the activities of non-State entities (i.e., rebel groups), yet always requires a concrete identifiable entity perpetrating the harm.

Equally importantly, climate/environmental impacts are indiscriminate. They do not differentiate in terms of whom they affect based on any of the five identified grounds specified in the Refugee Convention. That is, the serious human rights violations must be perpetrated "on account of" the affected person's race, religion, nationality, membership in a particular social group, or political opinion.[27] Exceptionally, a case for persecution may be made if government policies target particular groups reliant on agriculture for survival, use starvation or famine as a political tool, or deliberately withhold disaster aid from a particular social group. Such cases, however, are likely to be rare.

25 African Union Refugee Convention of 1969, *supra* note 9; Cartagena Declaration, *supra* note 9.

26 *See, e.g*, Chapter 8 in this volume.

27 *See* 1951 Refugee Convention, *supra* note 9, art. 1, para. A(2).

Moreover, it is widely acknowledged that the overwhelming number of persons displaced to date as a result of environmental factors have been displaced within the borders of their own country, and that this is expected to continue to be the case. In such instances, by definition, the persons would be internally displaced persons (IDPs, discussed in further detail below) and not refugees (who must have crossed an international border).

Finally, the whole notion of international refugee protection is premised on international protection when there has been a failure of State protection. That is, there is a need for other States to take on the protection functions of a State for persons who have specifically lost the protection of their own State, i.e. when the State turns on them seeking to do them harm. In the case of persons displaced as a result of climate or environmental factors, the State of nationality has not ceded its rights and responsibilities to protect its citizens, but may instead be impeded practically in its ability to afford such protection. The case of the sinking islands is a case in point. It is precisely the governments of such States that are seeking a means to ensure the continued protection of their citizens in the face of their compromised ability to do so themselves at home.

1.2. Amending the Refugee Convention?

One option is bringing those who move for environmental or climate change reasons within the ambit of the 1951 Refugee Convention, possibly through the introduction of an amendment. Biermann and Boas observe that such a proposal was put forward by the Maldives delegation at a 2006 conference the government convened.[28] The authors express doubt about the political feasibility of the idea. They point out that industrialized countries are not likely to take on additional

28 Biermann & Boas, *Protecting Climate Refugees: The Case for a Global Protocol*, *supra* note 5.

protection commitments, given their claim that their capacity to meet current refugee protection needs is vastly overstretched. (Developing countries may also view mass population inflows into their territory from neighboring countries with alarm.) An additional argument by Biermann and Boas against the proposal is that it might, in the end, give rise to issues of priority of treatment between persons currently under within the scope of the Convention and the new category of "climate refugees." For these, and similar reasons, scholarly opinion is generally of the view that the disadvantages of recourse to the 1951 Refugee Convention would outweigh any possible advantages.[29]

Moreover, they do not see reason to believe that the international community will agree to a renegotiation of the 1951 Convention with a view to making it more inclusive. With a regrettably pervasive xenophobic atmosphere in many countries around the world today, where refugees are often subjected to discriminatory treatment and unfairly blamed as an economic burden or source of criminality, efforts to expand the understanding or scope of the definition of refugees cannot be undertaken lightly. Indeed, there is consensus among concerned agencies, including the Office of the United Nations High Commission for Refugees,[30] to avoid terms such as "climate change refugee" or "environmental refugee" for fear that they could potentially undermine adherence to the existing international legal regime for the protection of refugees. In addition, the general consensus among States and agencies is not in favor of a reopening of the 1951 Refugee Convention. Even if the scope of the Refugee Convention were to be expanded to include environmental/climatic factors, it would still exclude from its protections all persons whose migration cannot be strictly classified as forced as well as those who did not cross an international border.

29 For a contrary view, *see* Tiffany T.V. Duong, *When Islands Drown: The Plight of "Climate Change Refugees" and Recourse to International Human Rights Law*, 31 U. Pa. J. Int'l L. 1239–1266 (2010).

30 Summary Record of the 616th meeting of the Executive Committee of UNHCR (A/AC.96/SR.616), at para 26.

1.3. Protection and Assistance Desired

Greater attention, however, will be needed to the question of the nature and purpose of any overarching international framework that might be needed to respond to present and future challenges. That is, beyond the issues discussed earlier regarding classifying persons displaced as a result of climate change as refugees, there is the additional question of whether the protections afforded to refugees would be the principal sorts of protections sought by inhabitants of the sinking islands. Refugee protection consists of two principal elements: (1) in the host State, treatment for persons determined to be refugees equivalent to national treatment (i.e. with regard to access to education, social benefits, and the like, effectively a "nondiscrimination" principle; and (2) a "*non-refoulement*" principle; that is, protection from being sent back to a place of persecution.

In the case of inhabitants of sinking islands, it is likely that both nondiscrimination and "*non-refoulement*" for persons who have already entered the territory of another State would be of interest; that is, not being sent back to a sinking island State and being permitted to remain in the host country lawfully regardless of how they entered.

The evolving interpretation of the *non-refoulement* principle also needs to be taken into account. The protection ensured by the principle of *non-refoulement*, enshrined in the 1951 Refugee Convention relating to the status of refugees, in fact, only relates to people fearing persecution if they would be returned in their country of origin or to a third country. The evolving interpretation promoted by the international bodies verifying respect of the human rights conventions to the principle encompasses also cases in which the life of the individual would be in danger or where he or she would risk being subjected to torture or to inhumane and degrading treatment. The international judicial and semi-judicial bodies have shown a tendency to extend further the scope of this principle by recharacterizing a violation of other rights, for example the right to health, as a form of inhumane treatment.[31]

In a situation of extreme natural disasters or of serious environmental pollution, where the life of affected individuals might be threatened or where individuals would risk suffering from inhumane or degrading living conditions due to the hardships caused by the environmental situation, the *non-refoulement* principle could potentially protect them from being returned to the country they fled.

However, it would not address two principal interests sought by advocates on behalf of the citizens of sinking islands: (1) assistance to permit them to remain in their national territories for as long as possible and (2) a right to and means for orderly migration to a new country where they could live safe and productive lives. As indicated previously, there is no precedent under refugee or other international law requiring States to provide the citizens of another State assistance to permit them to remain at home or guaranteed orderly admission to the territory of another State with the prospect of a long-term stay and integration. These two additional needs may need to be addressed through other, innovative means; possibilities for doing so are addressed later in this chapter as well as in the other chapters in this section.

1.4. Other Migration-Related Laws and Institutions and How They Can Help

Although international law provides an explicit protection paradigm for "refugees," this does not mean that individuals falling outside such definitions have no right to protection. They can benefit from the protection granted by existing international migration law instruments,[32] in

31 *Ibid.*

32 International migration law draws together the norms governing the legal relationships between States and those between States and individuals involved in the migration process. It is an umbrella term for an area of law that has developed over time and indeed continues to develop. *See* IOM, International Migration Law,

particular those relating to human rights, which should be interpreted taking into consideration the specific context of climate change to ensure that these rights are practical and effective and not simply theoretical and illusory.[33] In addition, because a range of instruments, spread across various branches of international law, apply to people on the move,[34] human rights law can be considered both to be at the core of the protection of the rights of all those who migrate and a linkage between different relevant branches of international law.

Those who migrate are subjects of human rights obligations, and States are the duty-bearers of those obligations. The obligation to respect human rights is enshrined in international and regional instruments and is in many cases found in national legislation as well. States have the sovereign power to manage migration flows, but this power must be exercised in full respect of international law, including but not limited to human rights law. Accountability mechanisms exist at the international, regional, and national levels, and remedies can be sought in cases where the human rights of migrants are violated.[35] More attention needs to be devoted to mechanisms, law, and policies to prevent abuses of the human rights of migrants.

Just as with other groups crossing a border, persons moving for reasons of climate change or environmental degradation will not necessarily need special protection; many may be capable of managing on their own. However, when vulnerability arises, persons moving as a result of environmental factors may need protection and to claim their human rights. The moment they are in a vulnerable position their human rights

http://www.iom.int/jahia/Jahia/international-migration-law/lang/en, for key publications and analysis on international migration law.

33 *See* Chapter 8. *See also*, *ex multis*, the position expressed by the European court of Human Rights on this point in the case *Airey v. Ireland*, 32 Eur Ct HR Ser A (1979); [1979] 2 E.H.R.R. 305 (October 9), at para. 24.

34 Other branches of international law relevant to migration include humanitarian and labor law, international criminal law, refugee and statelessness law, environmental law and the law of the sea, to name a few.

35 *See* Chapter 8.

will come into play, and they will have a claim to basic rights and protection.

A much-debated question with few conclusive answers is whether there is a "protection gap" for persons moving as a result of environmental factors. There may be a de facto protection gap for environmental migrants. A de facto gap means that legally protection exists – as opposed to a de jure gap; that is, a situation where a group/issue is not covered by law or insufficiently covered by law – but that for some practical reason the effective respect is very difficult to obtain. There are no simple answers to the question of protection, which is usually defined as all activities aimed at obtaining full respect of the rights of the individual in accordance with the letter and spirit of the relevant bodies of law. Protection is thus a practical activity, but one taking place within a legal framework.

In principle, persons affected by environmental factors and crossing an international border remain under the protection of the State of which they are nationals (for example, through consular protection). This stands in contrast to the situation of refugees, who by definition have lost the protection of their State of nationality and therefore are entitled to international protection. Environmental migrants are also protected under international human rights law by the State of destination that has the duty to extend to them all human rights enjoyed by non-nationals under international law (i.e., nondiscrimination), except with respect to political, voting rights, and those rights and exceptions thereto that flow from the migration status of the individual.[36]

International actors have the responsibility to support States in complying with their protection duties, including by strengthening the capacity of State authorities and institutions. Mainly on State requests, a

36 *See* United Nations General Assembly, Human Rights Council, Report of the Special Rapporteur on the human rights of migrants, Francois Crepeau (Apr. 2, 2012), *available at* http://daccess-dds-ny.un.org/doc/UNDOC/GEN/G12/125/96/PDF/G1212596.pdf?OpenElement.

substitution of the State's role can be considered in cases of imminent or ongoing humanitarian emergency, principally through the UN humanitarian coordinated response system.

Within this framework, the role of IOM, among other international actors, will be significant. Although IOM has no legal protection mandate, its activities contribute to protecting persons involved in migration, and it has a de facto protection mandate. This is recognized in the 2007 IOM Strategy and explained in background paper IC/2007/3, "Protection of Persons Involved in Migration: Note on IOM's Role."[37] This de facto protection mandate applies to all migrants – environmental, economic, or otherwise.

1.5. Persons Displaced Internally within Their Own Countries

Given that most environmental migration will be internal in nature, the UN Guiding Principles on Internal Displacement[38] (a nonbinding or "soft law" instrument) are directly relevant. The Guiding Principles are based on international humanitarian and human rights law and analogous refugee law and are intended as an international standard to guide governments, international organizations, and all other relevant actors in providing assistance and protection to Internally Displaced Persons (IDPs). In the definition of IDPs, natural or human-made disasters are explicitly stated as among the factors that can induce internal displacement.

The Principles identify the rights and guarantees relevant to the protection of the internally displaced in all phases of displacement. They provide protection against arbitrary displacement, offer a basis for protection and assistance during displacement, and set forth guarantees for

37 United Nations High Commissioner for Refugees, International Migration Law and Legal Affairs Department, *Protection of Persons Involved in Migration: Note on IOM's Role* (June 2007), *available at* http://www.unhcr.org/4bf644779.html.

38 Global Database, Guiding Principles on Internal Displacement, http://www.idpguidingprinciples.org/.

safe return, resettlement, and reintegration. Although they are not binding, the Principles are consistent with international law.

The Guiding Principles reflect that IDPs legally remain under the protection of their own State, but that effective protection requires reaching them and providing the necessary assistance. The Guiding Principles could potentially serve as an indication of what rights, protection, and assistance are relevant for persons who have moved due to environmental factors, including when the movement has taken place across an international border. Particularly relevant is Principle 15.d, which states the right to be protected against forcible return to or resettlement in any place where their life, safety, liberty, and/or health would be at risk.

The Guiding Principles define IDPs as "*persons or groups of persons who have been forced or obliged to flee or to leave their homes or places of habitual residence, in particular as a result of or in order to avoid the effects of armed conflict, situations of generalized violence, violations of human rights or natural or human-made disasters, and who have not crossed an internationally recognized State border*" (paragraph 2).

There are several features of the Guiding Principles that distinguish them from other major international human rights instruments. The first is that they are informal in nature and therefore nonbinding on States. The reasons for that choice were largely circumstantial: in the 1990s, governments were (as they still are today) wary of surrendering their sovereign prerogatives and thus unwilling to enter into a legally binding treaty stipulating what they could or could not do within their own territory. There was also a desire to avoid lengthy negotiations over the format of an instrument at a time when international norms in this field were urgently needed. Finally, there was a prevailing view that there was a sufficiently broad, if rather dispersed, set of existing legal norms to meet the general needs of internally displaced persons.[39]

39 Roberta Cohen, *Key Policy Debates in the Internal Displacement Field*, Refugee Watch, 32 (Dec. 2008).

Flowing from this is a second distinguishing feature of the Guiding Principles: they do not set out to create new law. They are, in the main, a restatement and consolidation of existing norms and principles, shedding light on the many vulnerabilities of the internally displaced and pointing to appropriate safeguards and remedies already available within the human rights and humanitarian legal framework, although attention was given to the clarification of gray areas and the bridging of perceived gaps.

A further important defining characteristic lies in the fact that the Guiding Principles are premised on the State's prime responsibility for the protection of internally displaced persons. A comparison with the 1951 Convention relating to the Status of Refugees is instructive. Both documents refer to displacement and a need for protection. In the 1951 Convention, however, the State is viewed as being delinquent in its relationship to the national concerned, who is then entitled to seek international protection. In the Guiding Principles, the locus of responsibility lies unequivocally with the State, so that it becomes, in effect, an integral part of the exercise of sovereignty.[40]

One additional distinguishing feature worthy of note is that the Guiding Principles, in keeping with their informal nature, do not envisage the grant of a legal status. Instead, a descriptive approach is preferred, covering a collectivity of individuals and drawing attention to their general profile of vulnerability, rather than prescribing a status determination procedure that would assign a particular status to each individual.[41]

The specific provisions of the Guiding Principles are laid out in four distinct clusters. The first section sets out a number of general framing principles, including the right of internally displaced persons to enjoy the same rights and freedoms as other persons in their country, the primary responsibility of national authorities in the provision of protection

40 Francis M. Deng, Sadikiel Kimaro, Terrence Lyons, Donald Rotschild, & I. William Zartman, Sovereignty as Responsibility: Conflict Management in Africa (Washington, DC: Brookings Institution Press, 1996).

41 Guiding Principles on Internal Displacement, *supra* note 38.

and humanitarian assistance to internally displaced persons, and, nondiscrimination in the application of the Principles. The second section, while acknowledging implicitly that not all displacement is unjustified, provides for a right of protection from displacement in diverse circumstances (such as ethnic cleansing or armed conflict), as well as minimization of the impact on populations of unavoidable and government-sanctioned displacement. The third and most substantial section covers, in detail, the principles relating to protection during displacement, including the rights to life, liberty, and personal security, but, equally, in terms of an adequate standard of living and access to essential services. The fourth and final section outlines the conditions that have to be met to allow the safe and dignified return of internally displaced person to their places of origin, as well as the measures necessary to their social and economic reintegration.

In sum, the Guiding Principles on Internal Displacement are an innovative instrument of international protection. They provide practical guidance to a wide range of stakeholders, including the Secretary-General, UN Member States, international organizations, and non-governmental organizations in relation to "rights and guarantees relevant to the protection of persons from forced displacement and to their protection and assistance during displacement as well as during return and resettlement and reintegration."[42] They have an important consciousness-raising function and also serve as the basis for policy formulation, yet leave room for further efforts toward the development of more binding instruments.[43] Developments since the submission of the Guiding Principles to the Commission on Human Rights in 1998 have demonstrated that there is indeed plenty of potential for the international community to make use of its "soft" substance to fashion instruments with a harder and sharper legal edge.

42 *Ibid.*

43 Simon Bagshaw, *Responding to the Challenge of Internal Forced Migration: The Guiding Principles on Internal Displacement, in* Internal Migration Law (R. Cholewinski, R. Perruchoud, & E. MacDonald, eds., The Hague: Asser Press, 2007).

1.5.1. Application and Impact of the Guiding Principles

Since the introduction of the Guiding Principles by Professor Francis Deng, then-Special Representative of the Secretary-General on internally displaced persons, with the modest hope that they would be "widely circulated and given practical application in the field,"[44] the international response has greatly exceeded this expectation at the global, regional, and national levels, although these responses have varied significantly in scope and nature.

At the global level the Guiding Principles have received wide endorsement, notably by the 2005 World Summit,[45] General Assembly,[46] and the Human Rights Council.[47] These endorsements have established the Guiding Principles as an important framework for the protection of internally displaced persons; they call for international support to enable States to build capacity toward their implementation; they promote solidarity in action among States in dealing with situations of displacement; and they note with satisfaction that an increasing number of States, United Nations organizations, and regional and nongovernmental organizations are applying them as a standard. Progress in this area has been substantial enough for the Representative of the Secretary-General on the human rights of internally displaced persons to express the view that "there are some indications that the Guiding Principles are emerging as customary law, providing a binding interpretation of the international legal norms upon which they are based."[48]

44 Guiding Principles on Internal Displacement, *supra* note 38.

45 United Nations General Assembly [hereinafter UNGA], Draft Resolution A/60/L.1, http://www.who.int/hiv/universalaccess2010/worldsummit.pdf (last visited July 16, 2011).

46 *See, e.g.*, UNGA Resolution A/62/153, http://www.undemocracy.com/A-RES-62–153.pdf (last visited July 15, 2011).

47 *See, e.g.*, Human Rights Council, Resolution 6/32, http://www.internal-displacement.org/8025747B0037BAC5/%28httpResources%29/F923C89362388DDBC12574A600338165/$file/A_HRC_RES_6_32.pdf (last visited July 17, 2011).

48 UNGA, *Report of the Representative of the Secretary-General on the Human Rights of Internally Displaced Persons*, Doc. A/HRC/13/21, http://daccess-dds-ny.un.org/doc/UNDOC/GEN/G10/101/61/PDF/G1010161.pdf?OpenElement (last visited July 17, 2011).

In addition, the Guiding Principles have been adopted – individually or jointly – by international organizations as an authoritative guide to their interventions in situations of internal displacement. UN agencies such as UNHCR, UNDP, OHCHR, and UNICEF, as well as the IOM, have incorporated the Guiding Principles into their internal displacement policies or operating procedures. The Inter-Agency Standing Committee (IASC) Working Group on Internal Displacement has placed the Guiding Principles at the center of its Framework on Durable Solutions for Internally Displaced Persons.[49] The Representative of the Secretary-General on the human rights of internally displaced persons has also played a very important part in the promotion and dissemination of the Guiding Principles and in the mobilization of public opinion through consultation with governments and the development of training materials.

At the national level, a growing number of States have begun to enshrine the Guiding Principles in their legislative and policy frameworks. Peru adopted a law in 2004 based on the Guiding Principles, establishing material benefits for internally displaced persons. In Colombia, the Constitutional Court has cited the Guiding Principles to support its judgments in favor of IDPs. Burundi, Colombia, the Philippines, Sri Lanka, and Uganda are all reported to have incorporated elements of the Guiding Principles into their national policies on internal displacement.[50]

It is, however, at the regional level that the Guiding Principles are exercising greatest influence and this across several continents. In 2006, for instance, the Council of Europe's Committee of Ministers adopted thirteen recommendations on internally displaced persons that do not simply restate the substance of the Guiding Principles, but underline

49 The Brookings Institution – University of Bern Project on Internal Displacement, IASC Framework on Durable Solutions for Internally Displaced Persons (Apr. 2010), http://ochanet.unocha.org/p/Documents/IASC%20Framework%20DS%20for%20IDPs.pdf.

50 Roberta Cohen, *The Guiding Principles on Internal Displacement: An Innovation in International Standard Setting*, Global Governance, 10, 459–80 (2004).

binding obligations undertaken by the its member States.[51] Independently, the members of the Organization for Security and Cooperation in Europe (OSCE) have acknowledged the usefulness of the Guiding Principles as a framework for the work of the OSCE.[52] In the Americas, the General Assembly of the Organization of American States has urged its member States to consider the Guiding Principles on Internal Displacement in designing public policy.[53]

The most significant advances have been achieved in Africa, which remains the only region that has developed and adopted binding instruments to protect internally displaced persons.[54] It is striking that it is in the Great Lakes Region, a vast area that has been affected by massive flows of internally displaced persons in recent decades, that a group of States committed not only to adopt the Guiding Principles but also to enshrine them in law. The International Conference on the Great Lakes Region was launched as a process in 1999 under the joint auspices of the United Nations and the African Union to create a climate of trust and cooperation against the backdrop of a history of conflict, political volatility, and economic instability. Its major outcome was the signing by eleven participating States, in December 2006, of a Pact on Security, Stability, and Development in the Great Lakes Region. The Pact entered into force in June 2008, after ratification by the requisite eight States.

The Protocol on Protection and Assistance to Internally Displaced Persons[55] is an integral part of this pact. It takes note of the level of

51 Corien Jonker, *Protecting IDPs in Europe*, Forced Migration Review 15–16 (Special Issue, Dec. 2008).

52 Cohen, *supra* note 50.

53 Organization of American States, AG/RES. 2055 (XXXIV-O/04), Internally Displaced Persons (June 8, 2004), http://www.internal-displacement.org/8025708F004CFA06/%28httpKeyDocumentsByCategory%29/853F93BBEFFFF276C12571170051E7BD/$file/IDP_OAS_resolution.pdf (last visited July 17, 2011).

54 UNGA, *Report of the Representative of the Secretary-General on the human rights of internally displaced persons*, *supra* note 48.

55 International Conference on the Great Lakes Region, Protocol on the Protection and Assistance to Internally Displaced Persons (Nov. 30, 2006), http://www.internal-displacement.org/8025708F004BE3B1/%28httpInfoFiles%29/29D2872A54561F66C12572FB002BC89A/$file/Final%20protocol%20Protection%20IDPs%20-%20En.pdf (last visited July 18, 2011).

international recognition enjoyed by the Guiding Principles; provides definitions for the concept of "internally displaced persons;" sets down, as one of its central objectives, the establishment of a legal framework in the Great Lakes Region for ensuring the adoption and implementation by Member States of the Guiding Principles on Internal Displacement; assigns prime responsibility to Member States for the physical and material safety of internally displaced persons; and requires Member States to enact national legislation for the incorporation into law of the Guiding Principles.

The continent-wide African Convention for the Protection and Assistance of Internally Displaced Persons in Africa (Kampala Convention)[56] took much of its inspiration from the Great Lakes Protocol, but is broader in scope.[57] Here also there is acknowledgment of the Guiding Principles as "an important international framework for the protection of internally displaced persons."[58] On the foundation of the Guiding Principles, the Kampala Convention builds, in effect, a binding instrument specifying both rights and responsibilities.

The Kampala Convention deals with the cycle of displacement in a comprehensive manner, prescribing a range of measures to prevent the occurrence of displacement, to ensure appropriate protection and assistance during phases of displacement, and to provide durable solutions for those affected. The overarching goal is the establishment of an African Union legal framework to address all aspects of internal displacement with emphasis both on State responsibility and on mutual support and solidarity among Member States. It is noteworthy that the notion of responsibility for the prevention of displacement and the

56 African Union Convention for the Protection and Assistance of Internally Displaced Persons in Africa (Kampala Convention), http://www.au.int/en/sites/default/files/AFRICAN_UNION_CONVENTION_FOR_THE_PROTECTION_AND_ASSISTANCE_OF_INTERNALLY_DISPLACED_PERSONS_IN_AFRICA_%28KAMPALA_CONVENTION%29.pdf [hereinafter Kampala Convention].

57 Prisca Kamungi, *Beyond Good Intentions: Implementing the Kampala Convention*, 34 Forced Migration Review 53–55 (2010).

58 Kampala Convention, *supra* note 56, at 2.

protection of the internally displaced is extended to armed groups and non-State actors. The role of humanitarian organizations is also underlined, as is their duty to respect principles of "humanity, neutrality, impartiality and independence."

States Parties commit to three broad lines of action: the promulgation of national legislation incorporating their obligations under the Convention, the allocation of budgetary resources for purposes of protection and assistance, and the designation of an appropriate authority for the coordination of national responses. To ensure compliance and accountability on the part of States Parties, the Convention provides for the establishment of a monitoring mechanism in the form of a Conference of States Parties, to be convened regularly and facilitated by the African Union. The Convention has been signed so far by thirty-one AU Member States and ratified by four. Fifteen ratifications are required for its entry into force.

By any reasonable measure of progress, the Guiding Principles have had considerable impact on international thinking and action toward the protection of internally displaced persons. They have been given practical application in a variety of ways across the world and have also anchored the development of regional legal instruments. Further developments can be expected, including helping to shape international legal thinking on cross-border displacement occasioned by natural disasters.

1.6. A New Legal Instrument?

At the global level, there has been in recent years much discussion, exchange, and debate on climate change displacement, driven in large part by the speculation about the imminence and magnitude of these outflows and hence the perceived urgency to act. Several authors have developed and proposed alternatives to the 1951 Refugee Convention, although the Convention generally features as a benchmark or at least as an important point of reference. Proposals by Biermann and Boas, Docherty and Giannini, and Hodgkinson et al. all opt for a convention model.[59] Each proposal has particularities of design and content. For instance, there are noteworthy differences in the definitions they use to define climate or environmental "refugees." Biermann and Boas propose a protocol directly linked to the UNFCCC. Docherty and Giannini acknowledge that it would be theoretically possible for their instrument to be developed as a protocol to the UNFCCC but have doubts about the suitability of such a construction. They would, on balance, favor a stand-alone version, as would Hodgkinson et al. Biermann and Boas and Hodgkinson et al. find scope in their models for both internally and internationally displaced persons, and the latter seek to attend to the specific needs of small islands threatened by climate change; Docherty and Giannini cater exclusively to those who cross international borders. In relation to status determination, Docherty and Giannini allow for both group and individual determination, whereas the others prescribe only group determination. Finally, the three models differ substantially in terms of their legal and administrative machinery. Biermann and Boas envisage an eventual executive committee of the States Parties that would have overall responsibility for issues of recognition, protection, and resettlement. A funding mechanism would complement this set-up, whereas actual implementation of the protocol would be effected through existing UN agencies. Docherty and Giannini's core structure rests on three essential components: a global fund, an independent coordinating agency broadly resembling UNHCR in its mode of functioning, and a body of scientific experts that would inter alia identify at-risk populations. Hodgkinson et al. propose an administrative establishment that is the most complex, with a new Climate Change Displacement Organization at its peak, under which there would be an Assembly, a Council, a Climate Change Displacement Fund, and a Climate Change Displacement Environment and Science Organization. Climate Change

59 Biermann & Boas, *Protecting Climate Refugees: The Case for a Global Protocol*, *supra* note 5; Docherty & Giannini, *supra* note 5; Hodgkinson, et al., *supra* note 5. The strong references to tidal and/or nautical imagery in the literature are noteworthy for their allusion to both impending disaster and the need for rescue.

Implementation Groups and a Permanent Secretariat complete the set-up. The Hodgkinson proposal is discussed in greater detail in the next chapter.

The proposals share several fundamental features. They all draw on and seek to be consistent with existing human rights and humanitarian law. In one way or another, they invoke the notion of shared responsibility between origin and destination countries on the one hand, and between developed and developing countries on the other. They provide for both humanitarian assistance and protection, including strong emphasis on resettlement options, and they imply the formation of complex administrative structures and bureaucracies. But their most striking common trait by far is their more or less close adherence to what McAdam refers to as the "protection paradigm,"[60] emulating, on the basis of international law, the grant of asylum to people fleeing from harm. She acknowledges advocacy of international responses by the governments of certain developing countries but queries whether global treaties will indeed provide the answer. She observes that movement due to climate change is difficult to establish in isolation from other causal factors; she also argues that exclusive emphasis on forced or flight-related movement is unwarranted when there are likely to be a range of patterns including those that are more of a migratory nature, thus requiring adaptive migratory answers rather than remedial refugee solutions. She is unable to detect any readiness on the part of governments to create a new international agreement.

Some proposals have limited the scope of a possible convention to population movements induced by *climate change*. To demonstrate that a particular migration event can be causally linked to climate change (as opposed to more general environmental change) could create an onerous burden of proof, something that in practice will be very difficult. Restricting a convention's applicability to climate change would also exclude

60 Jane McAdam, *Swimming against the Tide: Why a Climate Change Displacement Treaty Is Not the Answer*, 23 Int'l J. Refugee L. 2 (2011).

those moving as a result of nonclimatic phenomena, such as geological/tectonic hazards like earthquakes.

Similarly, most proposals for a convention address cross-border movement. Although this is important, it should not be forgotten that the vast majority of environmental migrants will be internal migrants. Hence, frameworks and measures to address internal environmental migration are also required.

Because of its complexity, the distinction between forced and voluntary migration tends to be very difficult in the context of environmental migration. This could complicate access to the provisions of a convention if it demands that movement be proven to be "forced."

A convention would likely have to choose between a remedial and a preventive approach. Both have their limitations in the context of environmental migration: a purely remedial (post hoc) stance would mean missing a chance to act to avert a worst case scenario; a preventive approach, however, should be careful to avoid suppressing migration at all costs, as doing so may in fact increase the risk of vulnerability and harm.

Saul and McAdam are concerned that a refugee-oriented climate change treaty might, in the end, prove damaging to the cause of refugee protection and give rise to abusive claims and even people smuggling.[61] They instead propose recourse to existing international law and principles that are considered to be sufficiently flexible and adaptable to respond to a variety of scenarios; they also advocate the development of an array of policies to address the various aspects of human mobility related to climate change. On the basis of empirical fieldwork conducted in Bangladesh and the South Pacific, Saul and McAdam find considerable virtue in the Guiding Principles on Internal Displacement as a broad-based tool capable of framing protection and assistance endeavors throughout the mobility continuum – from the pre-displacement phase, through the displacement phase proper, and also taking in return

61 McAdam & Saul, *supra* note 24.

and resettlement. This proposition has certain points in common with Williams' submission of regional approaches built around the Guiding Principles, although the latter envisages linkages to the UNFCCC.[62]

The question of the necessity of a potential new convention may depend on the feasibility of a new convention, which will ultimately be determined by States. Given the experience of past conventions addressed to the movement of people or treatment of the human and labor rights of migrants,[63] the negotiation and ratification processes would likely be lengthy. There may be a risk that protracted and inconclusive negotiations or low numbers of ratifying states following a conclusion may serve as an excuse for inaction and distract from actual, current needs – which in the case of sinking islands might be so pressing that they cannot afford to wait for the conclusion of a lengthy process. As discussed more fully in the remaining chapters in this section, a more targeted regional framework or specific national and/or international regime addressed to the particular needs of the sinking island States may prove both more timely and feasible than including their needs within a broader instrument that will necessarily be more complex and time consuming to develop.

1.7. Statelessness

Article 1 of the *1954 Convention Relating to the Status of Stateless Persons* defines a stateless person as "a person who is not considered a national by any state under the operation of its law." Should a State cease to exist, citizenship in that State would cease, because there would no longer be a State of which a person could be a national. This question, which is

62 Williams, *supra* note 24.

63 *See, e.g.*, International Convention on the Protection of the Rights of all Migrant Workers and Members of Their Families, *supra* note 10, which was negotiated over the course of 10 years, took another 13 years to enter into force with the minimal required number of ratifications, and today, 21 years after it was adopted, has 46 State contracting parties, none of which is a major country of immigration, thereby rendering many of its most important provisions untested and unimplemented.

discussed in Chapters 3 and 4 of this book in detail, could have a significant impact both practically and legally on displaced persons, particularly from small island nations.

1.8. National-Level Responses

National-level responses will of necessity form part of a protection and assistance framework for persons crossing a border and needing protection as a result of environmental degradation and/or climate change. Some countries have already introduced provisions in law or policy to respond to the protection or residence needs of victims of environmental disturbances, especially sudden, unforeseen, and generally unpredictable events such as earthquakes, tsunamis, or volcanic eruptions.

Temporary protection (or granting "Temporary Protection Status" [TPS] /"Temporary authorization to stay") is one measure used to provide protection and assistance to those affected by environmental factors. In past State practice,[64] TPS has been granted to prevent the return of individuals to their home countries when a natural disaster had occurred in that country. It should be noted that there is no single definition of temporary protection mechanisms, and practices vary from State to State.[65] In Finland and Sweden, someone who left his or her country and is unable to return due to environmental disaster can qualify as a person in need of protection. The United States may authorize the grant of "Temporary Protection Status" to persons already in the country and unable to return home as a result of environmental disaster; this was applied, for example, after Hurricane Mitch in Honduras and Nicaragua in 1998. The U.S. TPS designation is discretionary and applies only to persons already within the territory of the United States at the time

64 Some of the countries which have applied temporary protection in various forms are: Australia, Bahamas, Canada, Denmark, Ecuador, Finland, France, Netherlands, Sweden, Switzerland, the United Kingdom, and the United States.

65 For more information, *see* Susan F. Martin, *Managing Environmentally Induced Migration*, *in* Laczko & Aghazarm, *supra* note 19.

that the disaster strikes. In the EU, the "Temporary Protection Directive" allows for temporary protection under certain circumstances, when people are suddenly displaced in large numbers and it is not feasible to deal with their cases on an individual basis. After the 2004 Indian Ocean tsunami, Canada, Switzerland, and the United Kingdom temporarily suspended the removal of nationals of affected States. Efforts tend to be ad hoc and reactive. None of these measures is mandatory (or therefore grant rights to individuals), nor do any provide a basis for admission to the territory of the host State.

TPS also needs to be looked at in the context of the broader migration management regime in which it is applied. In some cases, TPS has been used in conjunction with adjustments to immigration procedures aimed at facilitating the entry of persons affected by environmental factors (usually sudden-onset hazard events). Fast-tracked family reunification procedures are one measure used in combination with TPS, but there are also examples of fast-tracking visitor visa applications.

Current practice in granting temporary protection is ad hoc and leaves room for broad State discretion. Efforts to promote a more systematic approach to temporary protection would be helpful. In addition to unilateral measures, temporary protection could also be developed through bilateral or regional consultations between and among affected origin and destination States of environmental migrants, based on existing standards of humanitarian and human rights laws. Consideration should be given to the fact that, in the case of sudden-onset natural disasters, the affected State would rarely be in a position to defend the interests of its nationals abroad. Lengthy negotiations of an agreement would be contrary to the need to provide timely humanitarian assistance. Furthermore, similar discussions could be introduced in the UNFCCC context, if States agree. However, in that case there is a high risk of discrimination against those victims of natural disasters that are not climate-related and who have so far accounted for the majority of the situations where a temporary authorization to stay was granted.

1.9. Assisted Voluntary Return and Reintegration

Where return is feasible and when migrants who do not fear any possibly persisting environmental threat are willing to return to the country they fled, bilateral voluntary return arrangements can be useful. Sustainable voluntary returns will also require comprehensive reintegration initiatives, coupled with timely reconstruction and rehabilitation, including the provision of health and education services. Such initiatives are most effective if the affected communities are included from the outset in the design and preparation of the move.

Where return is not viable, arranging for permanent stay abroad may require exploration and development of innovative partnerships at the bilateral and/or regional level.[66] In the case of the sinking island States, given their small number as well as the comparatively small number of individuals affected (see subsequent chapters in this section for a more detailed analysis of the likely numbers of persons to be affected), targeted country-specific solutions may prove to be both the most expedient and effective means to ensure protection and assistance to islanders who cannot remain at home. For example, bilateral arrangements are currently being explored for the purchase by one or more of the small island States of territory within the jurisdiction of another country for the potential relocation of their population.[67] In addition, consideration is being given to enhanced and/or dedicated education and immigration authorization for the citizens of one or more small island States to gain access to the territory and labor markets of other States. These types of targeted measures negotiated bilaterally or regionally are likely to be concluded far more quickly and result in more immediate and concrete

66 IOM, *Discussion Note: Migration and the Environment*, MC/INF/288 (Nov. 1, 2007), para. 35, *available at* http://www.iom.int/jahia/webdav/shared/shared/mainsite/about_iom/en/council/94/MC_INF_288.pdf.

67 *See Fearing climate change could wipe out their entire Pacific archipelago, the leaders of Kiribati are considering an unusual backup plan: moving the populace to Fiji*, THE AUSTRALIAN, Mar. 10, 2012.

results that benefit the populations of small island nations than what would inevitably be more protracted and potentially inconclusive negotiation of broader instruments targeted to persons displaced as a result of climate change more generally, although they need not be mutually exclusive.

2. Migration, Displacement, and Planned Relocation in the UNFCCC

The information presented in this part of the chapter relies on personal experiences and conversations in the context of the UNFCCC climate negotiations, UN Observer organizations, and empirical research on environmental change and migration. We refer to negotiating texts from the UNFCCC, and the positions taken reflect our analysis and understanding and do not represent official Party or Observer statements. Differences in perspective from others who have also experienced the discourse described in this chapter firsthand will naturally exist.

2.1. Significance of Paragraph 14(f)

Before addressing the emergence of human mobility in the context of discussions on climate change, it is useful to explain why this framing in the UNFCCC climate negotiations is important:

- Mobility in the context of the UNFCCC is acknowledged as having a link to climatic change and is framed as a *phenomenon to be managed*. The mobility discussion provides a "technical"-level (rather than controversial political dialogue) stepping-stone for transitions between immediate-term use of existing approaches to necessary longer term paradigm changes about population shifts, governance of borders and mobility, livelihood viability, planning in certain regions, and the like.
- Paragraph 14(f) couches human mobility within the realm of adaptation to climate change and subtly introduces the thought that

adaptation may require longer-term societal transformations. This suggests that adaptation may be understood as more than marginal changes in the way people live in certain locations.

- Paragraph 14(f) frames human mobility as part of a *wider range of measures that can be funded* under the emerging climate finance regime to help vulnerable countries adjust to current and expected climatic changes. Depending on how Parties articulate their adaptation needs, human-mobility-related activities will be eligible for climate finance (ranging from managing migration, preventing or reducing displacement, and where appropriate undertaking planned relocation).
- Paragraph 14(f) has significance for *implementation*. As the institutional arrangements for adaptation continue to be shaped, human mobility (and the other areas mentioned in the Cancun Adaptation Framework) will expand from a topic for discussion toward a topic for policy and operations. This will have meaning for development cooperation (particularly around livelihoods), humanitarian and disaster risk-reduction work, urban and rural planning, and the like.
- Finally, Paragraph 14(f) provides an opportunity to further articulate *policy options* at appropriate levels (subnational, national, regional, international) and along the spectrum of human mobility. The presence of human mobility in one policy forum (UNFCCC) has and will continue to influence discussions in other arenas, including the UN Security Council, the Global Forum on Migration and Development (GFMD), the 2013 UN General Assembly High Level Dialogue on International Migration and Development, and regional fora, among others.

These last two points are important because, arguably, few other arenas emphasize discussion, action/planning, and financial resources for implementation as the UNFCCC does. As described earlier and in other literature,[68] existing institutional arrangements to manage human mobility related to natural disasters are limited. Few coordination or planning mechanisms are in place to address relocation related to environmental or climate change (most are development-project related). Paragraph 14(f) provides initial inroads into these policy areas.

2.2. Research Role in Bringing Migration and Displacement to the UNFCCC

The entry point of migration and displacement as a specific topic into the UNFCCC climate negotiations was, arguably, science and advocacy driven; the IPCC Fourth Assessment Report, the Stern Review, and other key literature around 2006 and 2007 forged the case for making adaptation a strong part of the institutional and financial discussions in the UNFCCC. This section argues that two groups – research and Observer organizations with an institutional interest in positioning the issue – harnessed the science-supported call for adaptation (as reflected in the IPCC AR4 and other scientific literature). Researchers and other Observer organizations supported delegate dialogue and the positioning of a subparagraph on human mobility in what became the Cancun Adaptation Framework.

Specific references to environmental change and migration began appearing in the scientific literature several decades ago, with occasional papers in the 1970s and 1980s[69] and then more regular mentions throughout the 1990s.[70] Major scientific reviews such as the Intergovernmental

68 Koko Warner, *Global Environmental Change and Migration: Governance Challenges*, Global Environmental Change 20 (2010) 402–13, *available at* http://www.ehs.unu.edu/file/get/8646.pdf.

69 El-Hinnawi, *supra* note 22.

70 Lonergan, *supra* note 22; Thomas Homer-Dixon, *Thresholds of Turmoil: Environmental Scarcities and Violent Conflict*, *in* Contested Grounds: Security and Conflict in the New Environmental Policies (Daniel H. Deudney & Richard A. Matthew, eds., Albany: State University of New York Press, 1999).

Panel on Climate Change (IPCC) mentioned environmentally induced migration since its first assessment report in 1990. The empirical base of investigations began to accelerate in the mid- to late 2000s as a crop of systematic investigation and case studies on environmental change and migration began to be published.[71] These studies were complemented by methodological and conceptual development, as well as analyses of policy implications.[72] Today several projects in different phases of

71 Douglas S. Massey, William G. Axinn, & Dirgha J. Ghimire, *Environmental Change and Out-Migration: Evidence from Nepal*, Report 07-715 (Population Study Center, University of Michigan, Institute for Social Research, 2007), *available at* http://www.psc.isr.umich.edu/pubs/pdf/rr07-615.pdf; Kniveton et al., *supra* note 13; Jäger, Frühmann, Grünberger, & Vag, *supra* note 15; Koko Warner, Marc Stal, Olivia Dun, & Tamer Afifi, *Researching Environmental Change and Migration: Evaluation of EACH-FOR Methodology and Application in 23 Case Studies Worldwide*, *in* Laczko & Aghazarm, *supra* note 19, at 197–244.

72 WGBU, *Welt im Wandel: Sicherheitsrisiko Klimawandel*, Hauptgutachten 2007, Wissenschaftlicher Beirat der Bundesregierung Globale Umweltveränderungen (Springer-Verlag, Berlin, Heidelberg, 2007); McAdam, *supra* note 22; McAdam, *supra* note 60; Roger Zetter, *Legal and Normative Frameworks*, Forced Migration Review 62–63 (No. 31, 2008); Etienne Piguet, *Climate Change And Forced Migration: How Can International Policy Respond to Climate-Induced Displacement?* (UNHCR Evaluation and Policy Analysis Unit, Geneva, 2008), Etienne Piguet, *Climate and Migration: A Synthesis*, *in* Environment, Forced Migration and Social Vulnerability 73-85 (Tamer Afifi & Jill Jäger eds., Heidelberg: Springer, 2010); Laczko & Aghazarm, *supra* note 19, at 197–244; Michelle T. Leighton, *Climate Change and Migration: Key Issues for Legal Protection of Migrants and Displaced Persons* (Background Paper for the Transatlantic Study Team on Climate Change and Migration, German Marshall Fund of the United States, 2010); Sarah Collinson, *Developing Adequate Humanitarian Responses* (Background Paper for the Transatlantic Study Team on Climate Change and Migration, German Marshall Fund of the United States, 2010), *available at* http://www.gmfus.org/galleries/default-file/Collinson_MAH_EditsV2.pdf (last visited Feb. 1, 2011); Philip Martin, *Climate Change, Agricultural Development, and Migration* (Background Paper for the Transatlantic Study Team on Climate Change and Migration, German Marshall Fund of the United States, 2010); Susan F. Martin, *Climate Change and International Migration* (Background Paper for the Transatlantic Study Team on Climate Change and Migration, German Marshall Fund of the United States, 2010); Andrea Warnecke, Dennis Tänzler, & Ruth Vollmer, *Climate Change, Migration, and Conflict: Receiving Communities under Pressure* (Background Paper for the Transatlantic Study Team on Climate Change and Migration, German Marshall Fund of the United States, 2010); A. Narusova, D. Kniveton, & K. Warner, "Climate Change, Environmental, and Migration: Frequently Asked Questions" (Summary for

completion continue to expand the knowledge base and provide a more refined understanding of how environmental factors, including current weather extremes and the potential for significant longer-term changes in climatic systems, affect migration and displacement. The IPCC plans a subchapter on the topic in its fifth assessment report, which will appear in 2014. The topic will also be addressed in several other chapters, particularly regional chapters, as a crosscutting issue. Other work has documented frequently asked questions about migration and displacement in the context of environmental change, reflecting the current state of knowledge.[73] Similarly, other research has documented some of the major gaps in knowledge and possible ways to address these gaps.[74]

Research findings based on field observations began to be published more widely from 2007 onward. During the course of 2007 and 2008, the Environmental Change and Forced Migration Scenarios Project (EACH-FOR) supported by the European Commission conducted twenty-three case studies worldwide.[75] The EACH-FOR and related research findings were first reported in a submission to the UNFCCC in August 2008 at the Accra, Ghana, session of the Ad Hoc

decision makers at the UNFCCC Climate Negotiations in Cancun (COP16), Climate Change, Environment and Migration Alliance (CCEMA), 2010); Warner, *supra* note 68.

73 Narusova et al., *supra* note 72.

74 *See e.g.*, Piguet, *Climate and Migration: A Synthesis*, *supra* note 72; MIGRATION AND CLIMATE CHANGE (Etienne Piguet, Antoine Pecoud, & Paul de Guchteneire eds., UNESCO Publishing/Cambridge University Press, 2011); Marc Stal & Koko Warner, *The Way Forward: Researching the Environment and Migration Nexus*, Research Brief based on the Outcomes of the 2nd Expert Workshop on Climate Change, Environment, and Migration, 23–24 July 2009 (Munich, Germany: United Nations University, October 2009); Warner, et al., *supra* note 71.

75 Other research projects during this time frame are also of interest. The French Program "Management and Impacts of Climate Change" (in French, "Gestion et Impacts du Changement Climatique" or "GICC") is one of the research programs supported by the French Board for Economic Studies and Environmental Evaluation, launched in 1999. *See* http://www2.gip-ecofor.org/publi/page.php?id=261&rang=0&domain=38&lang=en_GB. One of the CRPs (Call for Research Proposal) from 2008 is "EXCLIM – Managing the Displacement of People due to Extreme Weather" (meaning *Exile climatique*). *See* http://www.reseau-terra.eu/article912.html (French).

Working Group on Long-Term Cooperative Action (AWG-LCA).[76] Around this time, States Parties were gearing up under the AWG-LCA to approach a variety of adaptation issues. This position made Parties receptive to research-supported submissions and supported them in bringing questions about migration and displacement into the early informal discussions in Accra. Research on region-specific case studies and related analysis allowed decision makers to see the relevance of the issue for their own country, further underscoring the importance of including migration and displacement in the draft negotiating text. The research spiked interest and then provided a base of evidence to sustain that interest after the Fourteenth Conference of the Parties (COP 14) when the issue had already been included in the assembly text.

From 2009 onward, new developments (in addition to those already discussed) helped sustain the message that migration and displacement were important topics for adaptation, including the emergence of new review efforts such as the UK Foresight project, a host of workshops and international conferences on the topic, the approval of new field research projects funded by governments and foundations, and a high-level panel on climate change and migration at the Global Forum on Migration and Development.[77]

2.3. The Role of the Humanitarian Community and Advocacy

The humanitarian community played a particular role following COP 13 in Bali; for example, through the activities of the Inter-Agency

76 United Nations University (UNU), *Climate Change and Migration: Impacts, Vulnerability, and Adaptation Options* (Submission by the UNU to the UNFCCC 3rd session of the Ad Hoc Working Group on Long-Term Cooperative Action under the Convention (AWG-LCA 3), Accra, Aug. 21-27, 2008, submitted Aug. 18, 2008); UNFCCC, AWG-LCA, FCCC/AWGLCA/2008/MISC3.

77 Susan Martin & Koko Warner, *Impact of Climate Change on Migration and Development*, Paper for the Civil Society Days, Global Forum on Migration and Development, Mexico 2010, Roundtable 3: Migration and Development: Tools and Evidence for Policy and institutional Coherence, Session 3.2. Civil Society Days (Puerto Vallarta, Mexico, Nov. 8-9, 2010), *available at* http://gfmd2010mexico.hsplatform.com/archivos/portal/doc_2_103_ImpactMartinWarner32.pdf.

Standing Committee (IASC) Task Force on Climate Change. The task force, which included organizations involved in implementation, research networks, and advocacy, became active in the course of 2008. Its efforts soon bridged applied research with the advocacy activities of the IASC on behalf of those organizations with a mandate to address migration and displacement, and more broadly the needs of vulnerable people. This brought an additional element of legitimacy: the humanitarian community sent a coordinated signal to States Parties that the findings of research on migration and displacement were relevant to their policy concerns and that operational organizations could support States Parties with expertise and experience in managing migration and displacement.

The IASC informal group on climate change and displacement coordinated wider participation by a spectrum of UN, international, and civil society organizations concerned with migration and displacement. The coordinator of the IASC Task Force on Climate Change provided outreach to interested organizations from the humanitarian and disaster risk management community, as well as a wide array of civil society organizations involved with the UNFCCC process. The IASC's coordination and an active subgroup dedicated to the theme ensured that joint statements, submissions to the UNFCCC,[78] joint activities outside of the UNFCCC,

78 UNU, *supra* note 76; International Labour Office (ILO), Submission on the Bali Action Plan to the AWG-LCA (Sept. 30, 2008), *available at* http://unfccc.int/resource/docs/2008/smsn/igo/016.pdf; IOM, Munich-Re Foundation, UNEP, & UNU-EHS, 2nd Expert Workshop on Climate Change, Environment, and Migration, July 23-24, 2009, Munich-Germany, with support from the Rockefeller Foundation (workshop syllabus *available at* http://unfccc.int/resource/docs/2008/smsn/igo/031.pdf); UNHCR, *Forced Displacement in the Context of Climate Change: Challenges for States under International Law* (Paper submitted in cooperation with the Norwegian Refugee Council, the Representative of the Secretary General on the Human Rights of Internally Displaced Persons and the United Nations University to the 6th session of the Ad Hoc Working Group on Long-Term Cooperative Action under the Convention (AWG-LCA 6), Bonn, Germany, June 1-12, 2009), *available at* http://www.humanitarianinfo.org/iasc/downloaddoc.aspx?docID=4874&type=any; UNHCR, *Climate Change and Statelessness: An Overview* (Submission supported by the IOM and the Norwegian Refugee Council (NRC) to the AWG-LCA 6 under the UNFCCC, Bonn, Germany, June 1-12, 2009); UNHCR & IASC, Letter to Yvo de Boer on the humanitarian impacts of climate change (Apr. 30, 2009), available at: http://unfccc.int/resource/docs/2009/smsn/igo/053.pdf; IASC, *Addressing the Humanitarian*

and networking supported Parties in their questions about migration and displacement. The Special Representative on the human rights of internally displaced persons[79] played a special high-level role both in providing technical support on the negotiating text and raising awareness subsequently.[80]

2.4. From Bali to Poznan: Establishing the Topic, Supporting the Delegates with Research Findings

The chapter now returns to the inclusion of migration, displacement, and planned relocation in the UNFCCC process since late 2007. From COP 13 to COP 16, a shift in emphasis occurred from compensation and liability (early "climate refugee" references) toward technical discussions focusing on adaptation both in situ and via migration.

Adaptation was firmly established as a focus of the UNFCCC climate negotiations by the Conference of the Parties held in 2007 in Bali, Indonesia (COP 13). This session created the Bali Action Plan, which laid out the elements of adaptation that might be considered in an international climate agreement. For example, COP 13 resulted in the adoption of the Bali Road Map, which included the Bali Action Plan.[81]

Challenges of Climate Change, Regional and National Perspectives, Case Studies on Climate Change Adaptation, *available at* http://www.humanitarianinfo.org/iasc/downloaddoc.aspx?docID=5094&type=any.

79 At the time, Professor Walter Kälin, and as of 2011, his successor, Chaloka Beyani, who is following similar lines of thought.

80 For example, after his visit to the Maldives in July 2011, Chaloka Beyani reaffirmed his action in favor of environmentally displaced persons on island States (*see* Climate Change Adaptation: A Human Rights Based Framework for Internal Displacement Necessary, Statement by the Special Rapporteur on the Human Rights of Internally Displaced Persons, http://www.ohchr.org/EN/NewsEvents/Pages/DisplayNews.aspx?NewsID=11260&LangID=E) and will produce a special report on this issue in March 2012 (*see, e.g.*, Christel Cournil, La mise à l'agenda scientifique et politique des migrations environnementales, Revista Interdisciplinar da Mobilidade Humana, No. 36 (Jan./June 2011), at 75–93, *available at* http://iris.ehess.fr/document.php?id=316).

81 UNFCCC, Report of the Conference of the Parties, Part Two: Action Taken by the Conference of the Parties at Its Thirteenth session, FCCC/CP/2007/6/Add.1 (Bali, Indonesia, Mar. 14, 2008).

The Bali Action Plan charted the course for a new negotiating process designed to tackle climate change, with the aim of completion by 2009.[82] It created the AWG-LCA, a subsidiary body intended to prepare the ground for a successful climate agreement to succeed (or complement) the Kyoto Protocol.

For the Poznań session (COP 14, Dec. 2008), AWG-LCA Chair Michael Zammit Cutajar compiled an assembly text from all the submissions between Bali (COP 13) and September 30, 2008.[83] The assembly document mentioned migration for the first time, reflecting submissions by applied research and the humanitarian community.[84] Migration maintained its presence in the draft negotiating text in the ongoing negotiation process.

In the busy and high-pressure COP 15 negotiations in Copenhagen, technical-level UNFCCC delegates to the AWG-LCA continued working on elements of a broader adaptation framework, which included "migration and displacement." Most areas of divergence were contained or resolved during these discussions, and the text got a further boost when industrialized countries committed to providing fast start finance of up to US$30 billion by 2012 and US$100 billion or more per year by 2020. This provided a high degree of legitimacy to the topic, because it meant it would be included in draft negotiating text coming out of Copenhagen. Meanwhile, the fact that textual issues were largely resolved allowed then-AWG-LCA Chair Margaret Mukahanana-Sangarwe to focus on consolidating consensus and preparing for COP 16, at Cancun and not open up the adaptation text in 2010 for any redrafting or discussion.

82 UNFCCC, Conference of the Parties, Bali Action Plan 3, FCCC/CP/2007 (Bali, Indonesia, 2007), www.unfccc.int/resource/docs/2007/cop13/eng/06a01.pdf.

83 UNFCCC (2009a). Ad Hoc Working Group on Long-term Cooperative Action under the Convention (AWG-LCA). FCCC/AWGLCA/2009/8. United Nations Framework Convention on Climate Change. Available at: http://unfccc.int/resource/docs/2009/awglca6/eng/08.pdf.

84 UNFCCC (2009b). Ad Hoc Working Group on Long-term Cooperative Action under the Convention (AWG-LCA). FCCC/AWGLCA/2009/16/Rev.1. Available at: http://unfccc.int/resource/docs/2008/awglca4/eng/16r01.pdf.

2.5. Structure of Paragraph 14(f) and Possible Actions

Finally, at COP 16, Parties decided to accept the draft text containing several key elements for adaptation in the Cancun Adaptation Framework, including paragraph 14(f) on migration and displacement. Paragraph 14(f) reads as follows:

> *14. Invites all Parties to enhance action on adaptation under the Cancun Adaptation*
>
> *Framework, taking into account their common but differentiated responsibilities and respective capabilities, and specific national and regional development priorities, objectives and circumstances, by undertaking, inter alia, the following:*
>
> ....
>
> *(f) Measures to enhance understanding, coordination and cooperation with regard to climate change induced displacement, migration and planned relocation, where appropriate, at national, regional and international levels;*

The final wording of paragraph 14(f) of the Cancun Adaptation Agreement reflected a more nuanced view on climate-change-induced displacement than in earlier iterations. By comparison, in December 2008 at the Poznan talks, the assembly text referred only to "migration and displacement," and the June 2009 AWG-LCA draft text used the term "climate refugees." By Copenhagen, the terminology reflected the understanding that human mobility occurs on an adaptation continuum, and it placed the issue within a larger adaptation framework (rather than a stand-alone or transboundary issue, which may have caused divergence in views among Parties). Finally, the text's more differentiated form featured different levels (national, regional, international) of mobility, different types of mobility (displacement, migration, planned relocation), and different kinds of measures (research, coordination, cooperation).

What kinds of activities can emanate from the inclusion of migration and displacement in the Cancun Adaptation Framework? The text itself outlines kinds of activities (enhanced understanding), types of human mobility (displacement, migration, planned relocation), and levels of addressing the issue (national, regional, and international). This results in a matrix with twenty-seven permutations. This section explores some of the likely combinations of measures, types of movement, and levels that may emerge in the near term. It also explores the kinds of entities that might be involved in future interventions and how they might be funded.

Enhanced understanding will likely be an activity happening at national, regional, and international levels. If the current pattern continues, it is likely that funding for building the empirical base (research, case studies, and the like) will come from bilateral sources as individual countries call for specific studies and dialogue (conferences, meetings). This has already occurred at the Nansen Conference hosted by Norway in June 2011 (focused on climate-induced displacement) and with several research projects supported by various European and Asian governments (looking mostly at migration and displacement). Regional and international dialogue about research findings is also expected, particularly with publication of the 2014 Fifth Assessment Report of the IPCC, which will feature at least one chapter reviewing migration and displacement in the context of adaptation to climate change.

It is expected that coordination efforts on displacement will continue along current humanitarian organizational lines, through the UN cluster approach and under the auspices of disaster risk management. These efforts will likely continue to be financed through humanitarian assistance channels, at least in the short term. Coordination efforts will happen at all levels, but particularly at the regional and national levels. Coordination on voluntary migration is less clear but may begin to be discussed at international and regional levels. Individual countries may choose to address whether tools like TPS may be broadened or altered to include a variety of environmental processes beyond natural disasters (typically rapid-onset events). Planned relocation is possibly the least developed area of coordination at this point, but will likely emerge in the medium term as countries begin thinking through potential consequences of mitigation and adaptation projects that may require population relocation. These kinds of coordination measures will be needed at regional and national levels, and at the international level guiding principles may be needed (such as those now available for development project-related relocation). International expert discussions have already begun on the topic; for example, at two Bellagio roundtables in 2010 and 2011. Moreover, in the context of the IOM governing bodies, more attention is being paid to migrants in crisis situations, with a possible view toward the development of concerted international approaches.

In the medium to longer term, when human mobility related to climatic change is expected to grow, *operational cooperation* will be needed at the national and regional level to manage the flows of people. Where movements (displacement, migration, relocation) are internal, cross-ministry national cooperation and capacity building may be needed. These kinds of activities may be funded through existing bilateral channels or through the emerging climate finance architecture. Where movements occur in border areas, regional cooperation may be necessary. Examples of regional labor migration agreements may be models for the future, but may take some time to design and implement.

2.6. Gaps: Need for Longer-Term Thinking about Human Mobility and Climate Change

The emerging dialogue around human mobility in the context of climate change is focusing on climate variability and the use of existing tools. There is a risk that it may highlight needs that are not addressed by existing tools and institutions. For countries that emphasize using current institutional frameworks, the risk is that these frameworks are already insufficient and may become more stressed in the future. Several examples of policy frameworks addressing this issue are available, such as TPS in the United States and Europe or principles and soft laws for protecting people who have been displaced by environmental events, as discussed earlier. Similarly, existing humanitarian responses are likely to be overwhelmed by growing and potentially permanent population changes associated with climate change. In addition, these gaps also exist outside the humanitarian context. Complex and slow-onset events could pose a major challenge to legal and governance frameworks, in part because responsibility and temporal limits are difficult to assign. Moreover, various institutions that deal with different issues related to the impacts of climate change may have a tendency to operate in "silos" and may approach issues within narrow sectoral perspectives.

These challenges underscore the need for new thinking about managing and planning for the impacts of climate change on human mobility, ranging from migration to displacement to relocation. It will be necessary to develop approaches that consider shifts in the baseline situation of many regions and to incorporate long-term time horizons (or "climate foresight") as opposed to simple "impact/vulnerability" mapping (that provides short-term "coping" strategies) in adaptation planning. The Adaptation Committee and managers of climate finance should take these longer-term perspectives into account when recommending or funding adaptation-related activities (in this case those that relate to mobility).

For example, typical community-based adaptive activities in coastal Bangladesh include providing assistance to vulnerable communities in low-lying areas with house-plinth raising to keep the houses above flood level and rainwater harvesting to provide clean drinking water because of salinity intrusion into surface and groundwater supplies. A longer-term strategy would include empowering, training, and building the skills of rising generations (e.g., children) in those communities to be in a position to adjust to change (not only variability). This may include development-related resilience-building efforts such as making them capable of obtaining climate-appropriate and/or better paying jobs (resilience includes social, financial, and environmental components) in nearby towns (over the next decade) and then be able to move the rest of their families to join them as survival in their present location may become gradually more and more difficult. This kind of longer term "empowered relocation" strategy (as opposed to "forced migration" or even "planned relocation") would locate decisions on when, how, where and who moves within the household/community, rather than from a "top-down" authority.[85]

2.7. Core Principles

In general, human-rights-based strategies and multipronged legal approaches offer a way forward in protecting and assisting environmental migrants. Such approaches would draw on existing relevant bodies of laws and instruments, incorporating both hard and soft law, such as humanitarian, human rights, and refugee law and instruments on internal displacement, disaster management, and legal migration.

Several core principles underpin the legal instruments related to disaster relief activities: humanity; neutrality; impartiality; nondiscrimination; cooperation; sovereignty and non-intervention; and prevention, mitigation, and preparedness. The protection of persons in disasters is fundamental. Existing international human rights obligations lie at the core of the content of protection in the immediate aftermath of the disaster and during the recovery phase in the long term. Several non-binding instruments suggest the existence of the "right to humanitarian assistance."

Yet the protection of those involved in migration and the right to humanitarian assistance cannot be dissociated from the duty to prevent situations that produce victims. We could say that prevention comes before the right to humanitarian assistance, which, by definition, comes

85 Ul Huq, S., Warner, K., Loster, T., Rhyner, J., & Harmeling, S. (2011). Personal communication about community-based adaptation, short-term adaptation vs. longer-term transformational adaptation, Dhaka, July 21, 2011.

after damage has already been inflicted. Humanitarian assistance cannot be disassociated from other measures that have or should be taken to eradicate the causes of dramatic situations. The provision of such aid cannot be used as an alibi for political inaction or to enable States to believe that they are not responsible for eradicating the root causes of problems such as environmental degradation and climate change. Emergency assistance to population whose lives have been devastated by disasters has traditionally focused on the delivery of life-saving aid such as shelter, food, and medical care. Nevertheless, protection of the rights of affected population must inform all stages of the rehabilitation process. To this effect, IOM has, for example, actively participated in the drafting of the Operational Guidelines on Human Rights and Natural Disasters in 2006.

It should be underlined that protection is the prime responsibility of States: a State has the duty to protect all persons under its jurisdiction, be they its own nationals or not. Each State has also the right – and the duty – to defend and protect its nationals abroad and to allow other States to protect their nationals residing on its territory. International mechanisms would only be subsidiary to this national duty to protect and assist. International actors support States in their duty to protect. IOM provides de facto protection for the human rights of migrants through many of its activities and plays a key supporting role in achieving the effective respect of the human rights of migrants.

To attempt to intervene in the increase of environmental-induced migration, a paradigm shift calls for moving from reactively responding to frequent humanitarian crises to successful corrective action that requires, inter alia, better understanding, early action and planning, mitigating causes through environmental management and climate change mitigation and adaptation, better management of movements within the rule of law, and including migration in development strategies.[86]

86 Examples of IOM's activities in migration, climate change, and the environment can be found in COMPENDIUM OF IOM'S ACTIVITIES IN MIGRATION, CLIMATE CHANGE AND THE ENVIRONMENT (2009), http://publications.iom.int/bookstore/index.php?main_

A more systematic approach to temporary protection as a way for States to protect non-nationals who cannot go back to their place of residence due to environmental factors may be appropriate. In addition to unilateral measures, such arrangements could also be developed through bilateral or regional consultations between and among affected origin and destination States of environmental migrants, based on existing standards of humanitarian and human rights law. The development of awareness of regional and international judicial or semi-judicial bodies on this issue, both during the examination of State reports and on the occasion of deciding on an individual complaint, could also help develop further the existing legal tools and to adapt them to the specific situation of environmental migrants. The IOM and others could play a role in drawing the attention of these bodies to this matter and in suggesting possible ways of interpreting the existing instruments.

Whether or not there is a convention, shared responsibility and respect for human rights are central in managing migration, including environmental migration. Although IOM's approach to environmental migration does not pursue normative objectives, its activities are based on human rights law and provisions contained in other branches of international migration law.[87] From IOM's perspective, it is important to support all environmental migrants, in particular displaced persons who have specific assistance and protection needs; this support should be independent of the type of environmental trigger – whether it is climate related (e.g., a drought or a flood) or geological (e.g., an earthquake). IOM's approach to environmental migration focuses on protection and assistance to affected mobile populations, reduction of vulnerabilities, and long-term management of risks, with the aim to reduce forced migration and to harness the potential of migration for adaptation and development. Its involvement in this issue dates back to the early 1990s, and it continues to be discussed in IOM's governing bodies in 2007

page=product_info&products_id=540. *See also* IOM, *supra* note 19 (in particular, the last chapter).

87 IOM, *supra* note 33.

(94th Council Session), 2008 (Third Session of the Standing Committee on Programmes and Finance), and 2011 (International Dialogue on Migration workshop).

One of IOM's principal roles is the provision of humanitarian assistance in response to displacement in natural disasters as well as disaster risk reduction and climate change adaptation activities. Its ability to perform this role is strengthened by IOM's participation in the international community's system for humanitarian response, its commitment as global cluster lead for Camp Coordination and Camp Management in natural disasters, and its active engagement in the Logistics, Emergency Shelter, Protection, Health, and Early Recovery clusters.

2.8. "Sinking Islands" and the Potential Relocation of Entire Communities

Because there is no precedent for the loss of an entire territory or the relocation of an entire population, the legal debate surrounding the future of small and low-lying island States that are threatened by the impacts of climate change (especially sea level rise) is contentious and ongoing. Action to assist potentially threatened States and communities should not wait until there is legal clarity regarding a State's "disappearance." A range of environmental hazards in small islands, such as natural disasters or salinization of drinking water, are likely to jeopardize livelihoods long before the possible submergence of the territory. Rather, as advocated by IOM and others, it is important to manage the migration process in a timely manner and help the affected governments make arrangements to ensure a viable solution well before the actual disappearance of the State. The hosting State should, in this context, ensure that the population affected is able to obtain naturalization in a shorter time period than other migrants and should consider conferring its nationality to children born on its territory. IOM can also help prepare populations for their relocation and the host State in developing policies targeting the relocated and the receiving communities.

2.9. Future Discussions of Human Mobility in the Context of Climate Change[88]

In the longer run, some have speculated that a new UN process may be needed to address the full spectrum of issues related to loss and damage and adaptation – including migration and displacement. In 2012 the Kyoto Protocol´s first commitment period – one mechanism to help Parties track and reduce greenhouse gas emissions – is expiring. The UNFCCC discussions leading up to December 2012 are intensely focusing on ways to preserve at least the most important parts of that agreement. However, regardless of the legal format, vigorous commitment to the reduction of greenhouse gases is needed in this decade if concentrations of greenhouse gas emissions are to be kept to levels consistent with even a 2.0 Celsius average global temperature increase. Some scientists believe this threshold may already have been crossed.

In scenarios of the world beyond 2 degrees, the impacts of climate change combined with other megatrends such as world population growing to a projected 9 billion by 2050, changes in technology, and other unforeseen shifts in society may be so profound as to require a fully new approach or forum for particular discussions such as migration and displacement. At the current time, however, there is appears to be little appetite for notions like an international convention to protect "climate refugees," as these require commitments, may imply liability, etc. This is another reason why paragraph 14(f) is a modest but significant milestone in addressing climate-change related human mobility. No other forum internationally or regionally has created a space recognizing a range of issues and possible activities, and linking this to the upcoming climate finance regime. There is a precedence for a UN forum or more informal intergovernmental process giving rise to other more specific processes. Notably, the Rio Earth Summit in 1992 created three new forums

88 This section is based on excerpts from Susan F. Martin & Koko Warner, *Climate Change and Migration: The UNFCCC Climate Negotiations and Global Forum on Migration and Development, in* Laczko & Aghazarm, *supra* note 19, at ch. 7.

for addressing environmental and development concerns: the Convention on Biodiversity, the Convention to Combat Desertification, and the Framework Convention on Climate Change (UNFCCC). Similarly, the High Level Dialogue on Migration and Development in the General Assembly spurred establishment of the GFMD.

It is possible that further policy on climate-induced displacement may develop within the UNFCCC context. The key will be to align Party appetite and needs with a range of appropriate and politically feasible "asks" – as noted earlier there is sensitivity around issues of liability and compensation and assignment of blame or historical responsibility. Research and operational organizations should avoid asking for overly complex arrangements or for actions that require Parties to use large amounts of political capital to achieve. Calls for new large international agreements on "climate refugees" may seem difficult to achieve at this point. Research suggests that the complex forms of migration and displacement will mix internal and cross-border movements, as well as raise questions when people cannot return to their places of origin because of environmental reasons (sea level rise, desertification, water issues, and so on). Alternatively, a focus on dialogue, building regional understanding and cooperation, and helping States understand potential impacts of migration and displacement on their current institutional frameworks would likely have more resonance. One idea that has particular currency is the development of a set of Guiding Principles around Climate Induced Displacement, based on the positive experience with the Guiding Principles for Internally Displaced Persons (IDPs).

It is hoped that guiding principles will begin to emerge to help States prepare for the expected impacts of climate change on migration and displacement. In the meantime, a structured and inclusive discussion will contribute to progress under paragraph 14(f). Effective policy development and implementation for migration and displacement will require multidisciplinary and multi-stakeholder cooperation at the national, regional, and international levels. The GFMD roundtable came to similar conclusions. Population movements for environmental reasons are generally found to be primarily local and will be the responsibility of governments. However, the involvement of other stakeholders such as international organizations, NGOs, civil society, the private sector, and development partners is also critical. Experts and organizations from development, humanitarian, environment, and climate, as well as migration and displacement communities, need to be included in these discussions.

Perhaps most importantly, the affected communities (both of origin and destination) and migrants themselves need to be actively involved in the planning and implementation of policies to address the challenges and take advantage of the opportunities of environmental migration. This is equally the case in the context of the UNFCCC discussions as in the context of bilateral, regional, or multilateral discussions and planning measures designed to enhance the protection and assistance afforded to vulnerable populations of sinking island States negatively affected by environmental degradation exacerbated by climate change. Although awareness raising remains critically important, the time for action to reduce human suffering and to improve resilience and well-being of affected populations is now.

10 "In the Face of Looming Catastrophe"

A Convention for Climate-Change-Displaced Persons

David Hodgkinson and Lucy Young

1. Introduction

The coastal region of Bangladesh – particularly vulnerable to frequent storms, cyclones, tidal surges, and sea level rise – is home to more than 40 million people. In late 2009, a climate poverty hearing panel addressing the interaction of climate and poverty in this coastal region, and organized by an environmental non-governmental organization, heard from "climate witnesses" that "[t]housands of people [have] become Climate Refugee[s] and [have been] forced to migrate [to]... nearby cities and in the neighboring countries illegally."[1] In the same year James Hansen,

1 Shaheed Hadis Park, *Declaration on Climate Change from the Coastal People of Bangladesh*, posting of Gouranga Nandy and Hasan Mehedi to Save the Coastlines, http://savethecoastlines.blogspot.com/2010/01/climate-change-voice-of-victims-from.html (Nov. 14, 2009).

This chapter is based on research conducted over the past three years and relies on and builds on our previous journal articles on a convention for climate-change-displaced persons. *See* David Hodgkinson et al., *"The Hour When the Ship Comes in": A Convention for Persons Displaced by Climate Change*, 36 Monash U. L. REV. 69 (2010); Hodgkinson et al., *Copenhagen, Climate Change "Refugees" and the Need for a Global Agreement*, 4 Public Policy 155 (2009); Hodgkinson et al., *Towards a Convention for Persons Displaced by Climate Change*, 2008 New Critic 8. For more information on our ongoing climate-change-displaced persons convention project, *see* http://www.ccdpconvention.com. We especially thank Tess Burton for her research and her significant contribution to this project, and Heather Anderson for her work on threatened island nations.

one of the world's preeminent climate scientists, wrote that, in terms of climate change, "[t]he consequences for a nation such as Bangladesh, with 100 million people living within several meters of sea level, are too overwhelming." He also warned of the consequences of climate change for island nations that are near sea level and expressed his hope that "those nations responsible for the changing atmosphere and climate will provide immigration rights and property for the people displaced by the resulting chaos."[2]

In Bangladesh, in small island nations, and around the world, millions of people may be displaced as a result of climate change.[3] We propose a treaty or Convention for Climate-Change-Displaced Persons (CCDPs) to address this displacement problem. In this chapter we outline the scale and nature of the climate-change-displacement problem and explain why we propose a multilateral governance framework. We then set out the detail of our proposed Convention – with specific reference to small or "threatened" island nations – and raise the possibility of such an international agreement forming part of a climate change "regime complex." We also address arguments against a new treaty for climate change displacement.

2 James Hansen, Storms of my Grandchildren: The Truth about the Coming Climate Change Catastrophe and Our Last chance to Save Humanity 258 (2009).

3 In the Maldives the government announced in November 2008 that a portion of its annual budget would be invested in a fund to buy a new homeland. The Maldives is a fragile chain of 1,200 islands and coral atolls; more than 80% of the total land area is less than 1 meter above sea level. Projected rises in sea levels could make much of the country uninhabitable. Target countries being examined by the Maldives for resettlement include Sri Lanka and India, given their similar cultures to the Maldives, and Australia, "because of the amount of unoccupied land available." The president of the Maldives stated that "[w]e do not want to leave ... but we also do not want to be climate refugees living in tents for decades." Randeep Ramesh, *Paradise Almost Lost: Maldives Seek to Buy a New Homeland*, The Guardian, November 10, 2008, at 1. Throughout the Pacific generally, sea-level rise has "generated plans for the relocation of populations ... *in the face of looming catastrophe*." Stuart Beck & Michael K. Dorsey, *At the Water's Edge: Climate Justice, Small Islands and Sustainable Development*, *in* Climate Change and Energy Insecurity 127, 138 (Felix Dodds et al. eds., 2009) (emphasis added).

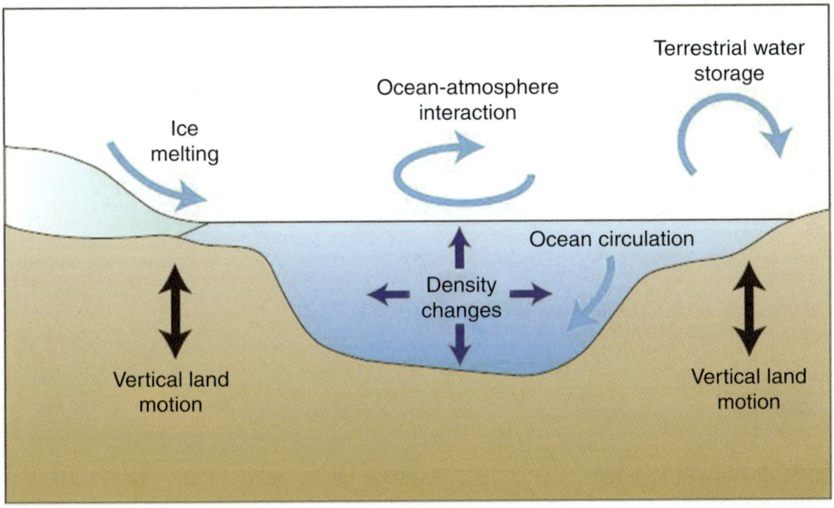

Plate 1. The factors that contribute to changes in sea level. Reproduced from Glenn A. Milne, W. Roland Gehrels, Chris W. Hughes, & Mark E. Tamisiea, *Identifying the Causes of Sea Level Change*, 2 NATURE GEOSCIENCE 471, 472 (2009). Reproduced with permission from Nature Publishing Group.

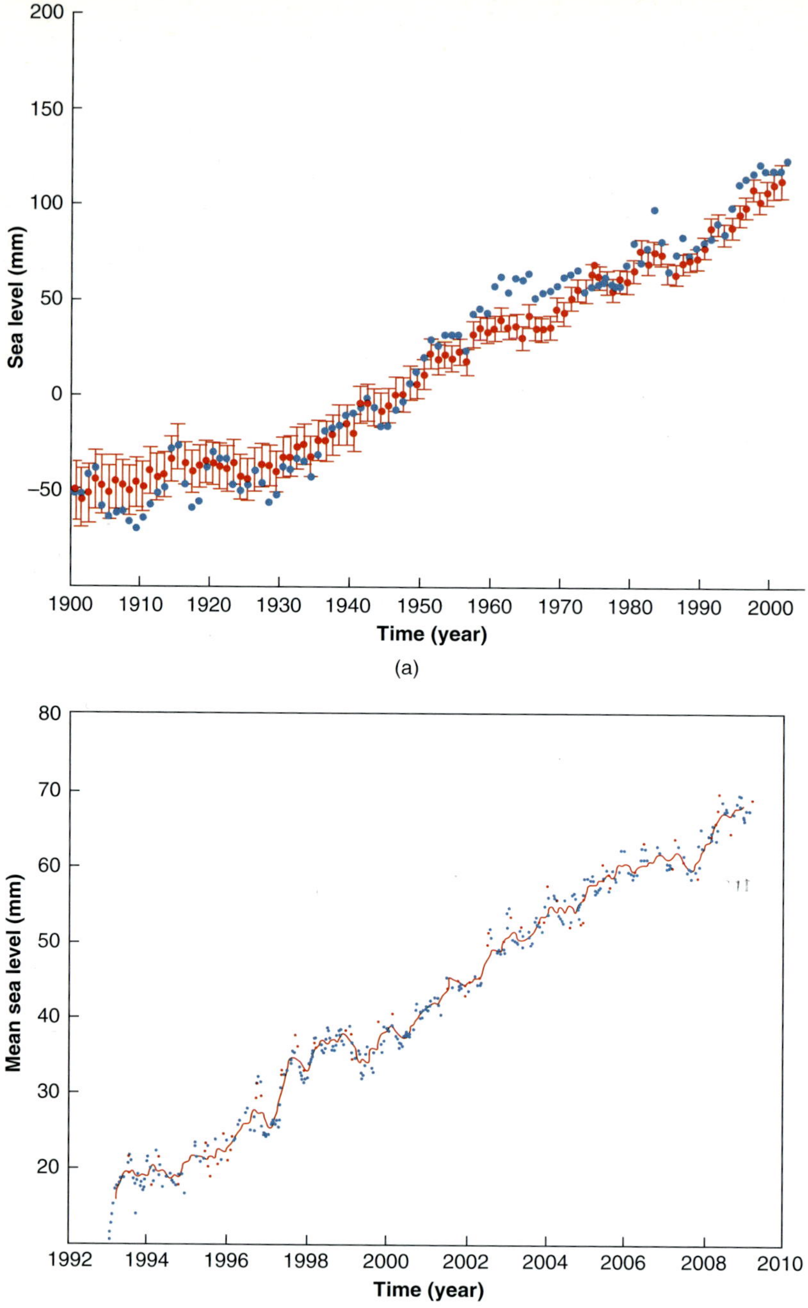

(b)

Plate 2. Observational record of sea level (a) from tide gauges since 1900 (p. 147); (b) from satellite altimeter since 1993 (p. 149). Reproduced from Anny-Cazenave and William Llovel, *Contemporary Sea Level Rise*, 2 Annual Review of Marine Science 145, 147–49 (2010). Annual Review of Marine Science, Copyright 2010 by Annual Reviews, Inc. Reproduced with permission of Annual Reviews, Inc. in the format Other book via Copyright Clearance Center.

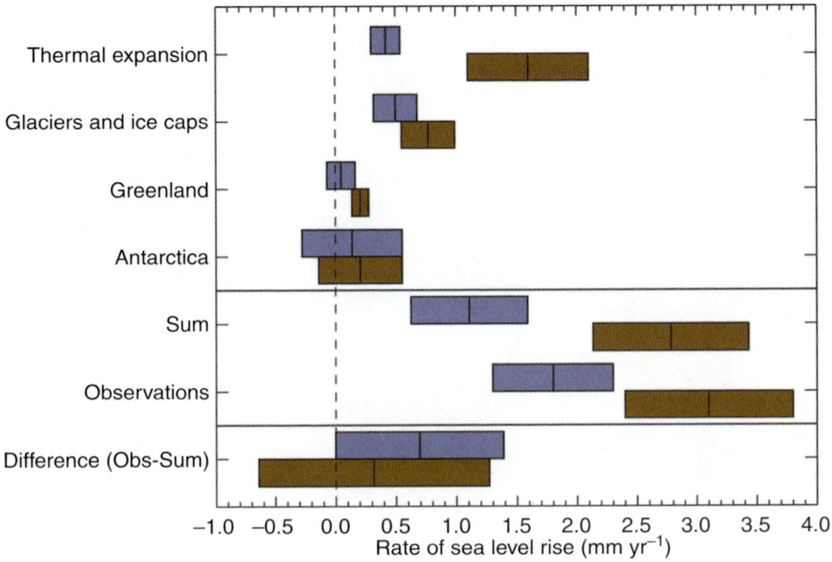

Plate 3. The budget of changing sea level. The bars represent the 90% error range, with blue entries covering 1961–2003 and brown entries covering 1993–2003. The upper four entries are estimates of the contributions to the sea-level-change budget, the middle two sum these contributions with the observed rate of rise, and the lowest bar is the observed rate less the sum of contributions. Reproduced from Nathaniel L. Bindoff et al., *Observations: Oceanic Climate Change and Sea Level*, *in* CLIMATE CHANGE 2007: THE PHYSICAL SCIENCE BASIS. CONTRIBUTION OF WORKING GROUP I TO THE FOURTH ASSESSMENT REPORT OF THE INTERGOVERNMENTAL PANEL ON CLIMATE CHANGE 385, 419 (Solomon, S., D. Qin, M. Manning, Z. Chen, M. Marquis, K. B. Averyt, M. Tignor and H. L. Miller, eds., 2007).

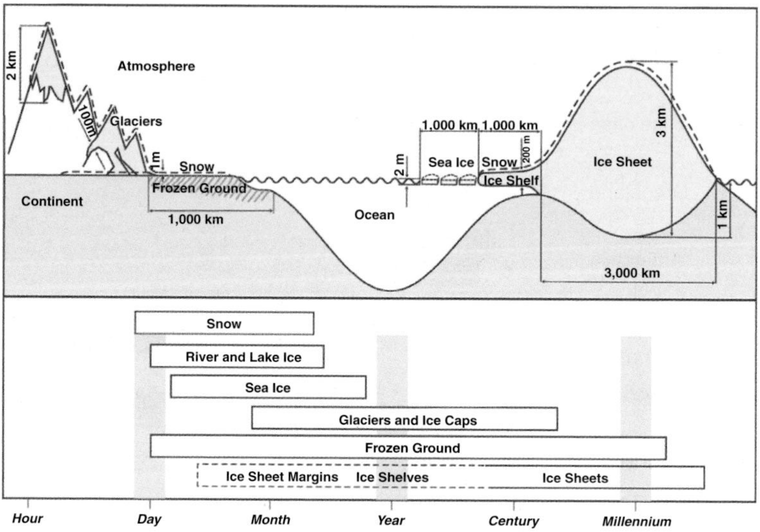

Plate 4. Schematic representation of the cryosphere: snow, river and lake ice, sea ice, mountain glaciers, ice shelves, ice sheets, and frozen ground. Only ice on land can affect sea level. Reproduced from Peter Lemke et al., *Observations: Changes in Snow, Ice, and Frozen Ground*, *in* CLIMATE CHANGE 2007: THE PHYSICAL SCIENCE BASIS. CONTRIBUTION OF WORKING GROUP I TO THE FOURTH ASSESSMENT REPORT OF THE INTERGOVERNMENTAL PANEL ON CLIMATE CHANGE 337, 363–65 (Solomon, S., D. Qin, M. Manning, Z. Chen, M. Marquis, K. B. Averyt, M. Tignor and H. L. Miller, eds., 2007).

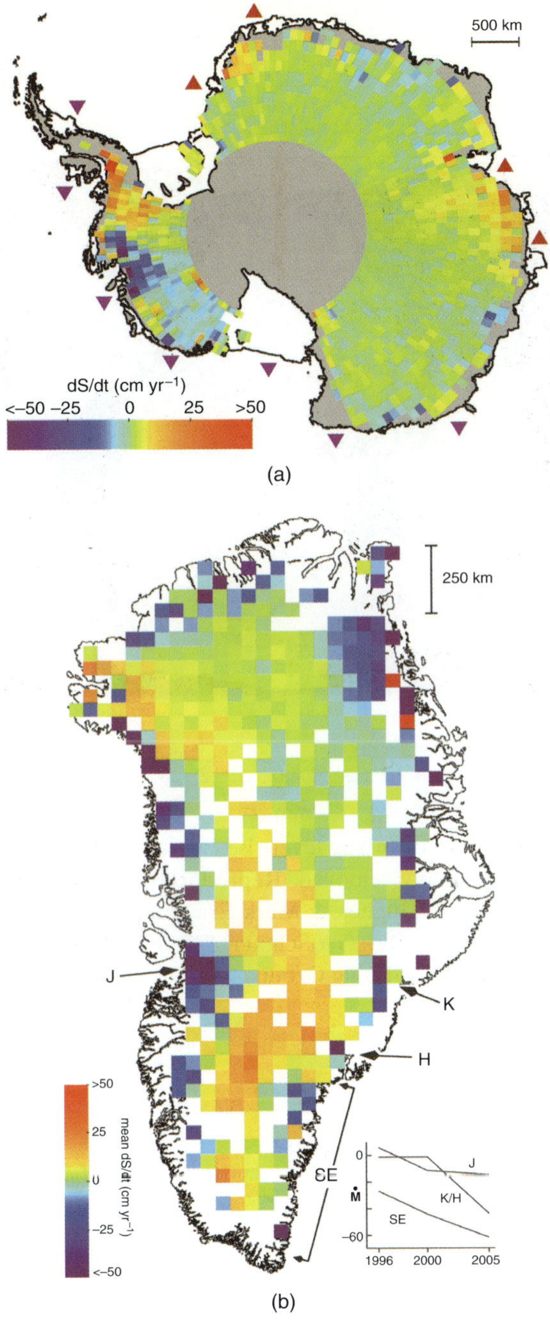

Plate 5. Rate of change in surface elevation in (a) Antarctica, from 1992 and 2003. Red triangles indicate locations of thickening of more than 30 cm per year, whereas purple triangles show thinning of the same magnitude, p. 365; and (b) Greenland, from 1996–2005, p. 363. Reproduced from Peter Lemke et al., *Observations: Changes in Snow, Ice, and Frozen Ground*, *in* CLIMATE CHANGE 2007: THE PHYSICAL SCIENCE BASIS. CONTRIBUTION OF WORKING GROUP I TO THE FOURTH ASSESSMENT REPORT OF THE INTERGOVERNMENTAL PANEL ON CLIMATE CHANGE 337, 363–65 (Solomon, S., D. Qin, M. Manning, Z. Chen, M. Marquis, K. B. Averyt, M. Tignor and H. L. Miller, eds., 2007).

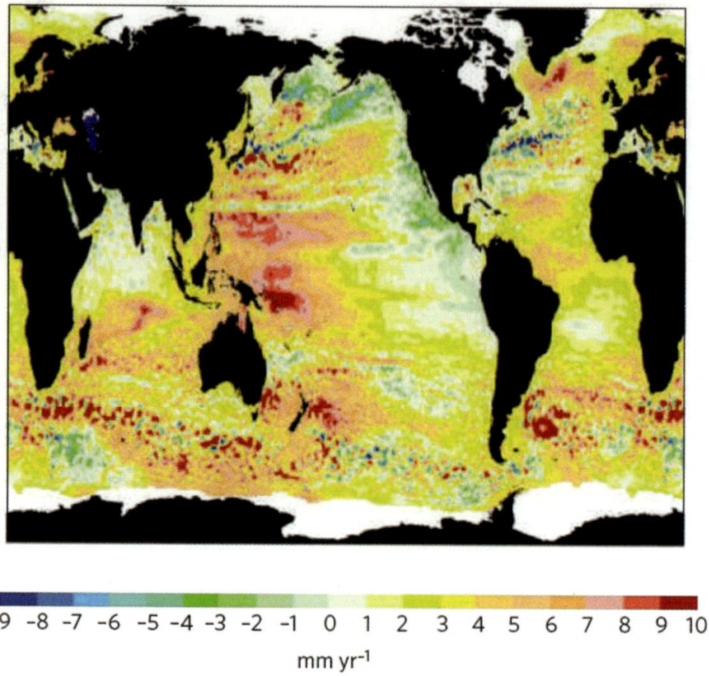

Plate 6. The spatial pattern of sea level rise between 1993 and 2007, with the global average of 3.1 mm/y removed. Reproduced from Glenn A. Milne, W. Roland Gehrels, Chris W. Hughes, & Mark E. Tamisiea, *Identifying The Causes of Sea Level Change*, 2 Nature GeoScience 471, 472 (2009). Reproduced with permission from Nature Publishing Group.

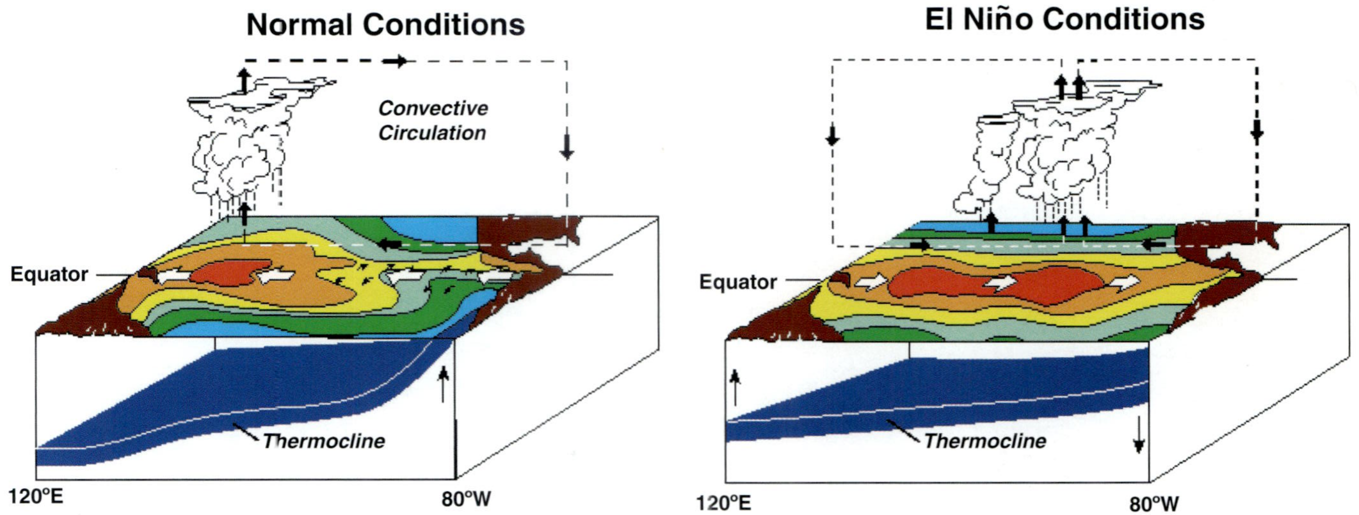

Plate 8. Schematic of normal and El Niño conditions. Reproduced from the website of Pacific Marine Environmental Laboratory (PMEL) of the National Oceanic and Atmospheric Administration (NOAA), http://www.pmel.noaa.gov/tao/elnino/nino_normal.html (last visited June 2, 2011).

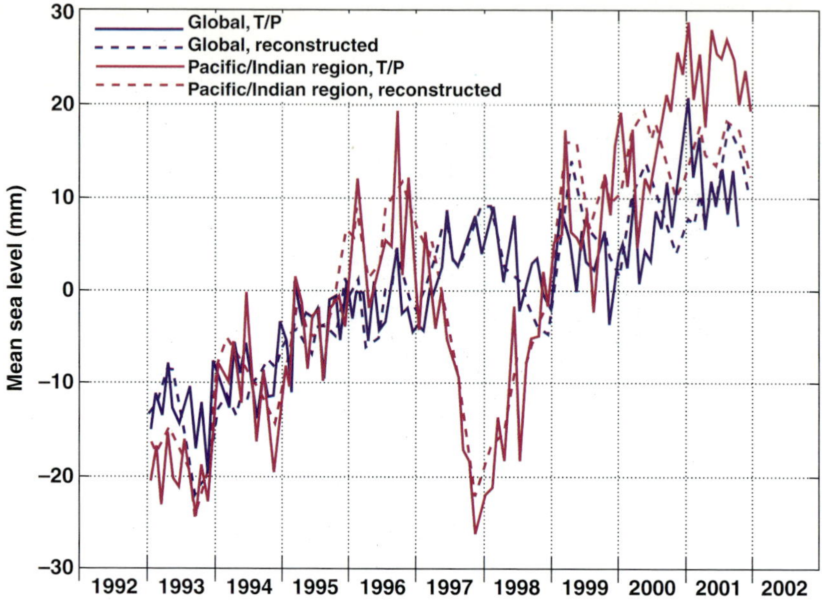

Plate 9. Tide gauge and altimeter records for the western Pacific and Indian tropical oceans from 100E to 120W and from 40S to 40N. This region is the same one depicted in Figure 11. Reproduced from John A. Church et al., *Sea Level Rise at Tropical Pacific and Indian Ocean Islands*, 53 GLOBAL AND PLANETARY CHANGE 155, 160 (2006), with permission from Elsevier.

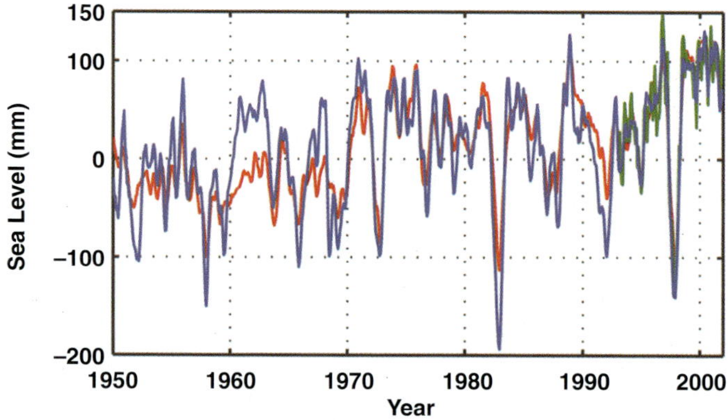

Plate 10. Tide gauge (blue) and altimeter (green) sea level record at Kwajalein (8.4N 167.4E) from 1950–2000, with annual and semi-annual signals removed. The reconstructed sea level indicates ENSO sea level variability exceeding 10 cm. Reproduced from Nathaniel L. Bindoff et al., *Observations: Oceanic Climate Change and Sea Level*, *in* CLIMATE CHANGE 2007: THE PHYSICAL SCIENCE BASIS. CONTRIBUTION OF WORKING GROUP I TO THE FOURTH ASSESSMENT REPORT OF THE INTERGOVERNMENTAL PANEL ON CLIMATE CHANGE 385, 414 (Solomon, S., D. Qin, M. Manning, Z. Chen, M. Marquis, K. B. Averyt, M. Tignor and H. L. Miller, eds., 2007).

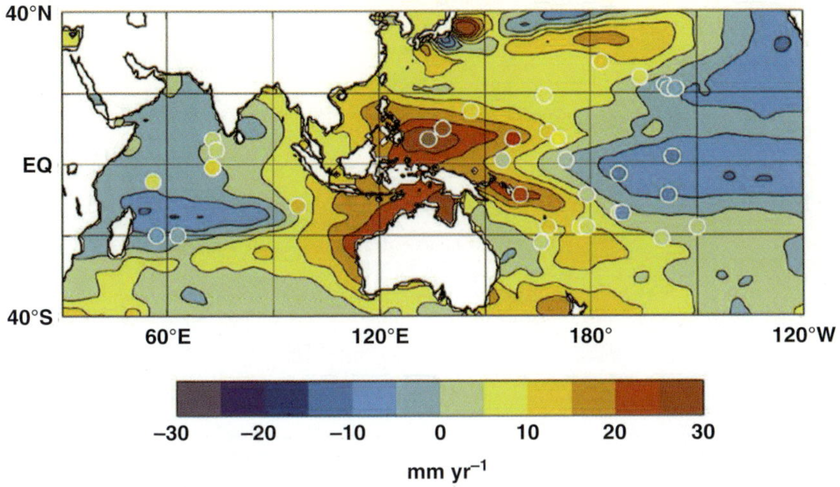

Plate 11. Tide gauge and altimeter data from 1993 to 2001 for the tropical Pacific and Indian Oceans. Reprinted from John A. Church et al., *Sea Level Rise at Tropical Pacific and Indian Ocean Islands*, 53 Global and Planetary Change 155, 159 (2006), with permission from Elsevier.

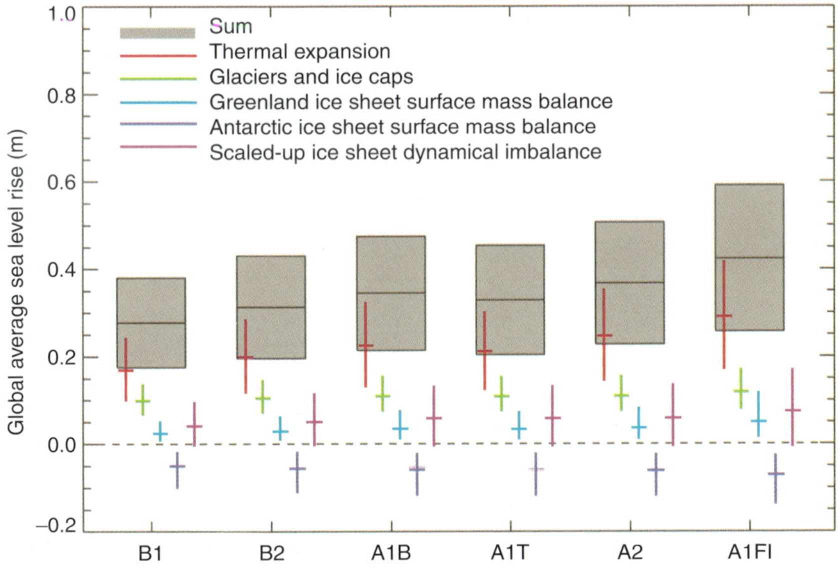

Plate 12. Future projections of sea level rise. B1, B2, A1B, A1T, A2, and A1F1 are different emission scenarios; the range of values corresponds to output from different models. The crosses indicate the range of values corresponding to thermal expansion, melting glaciers, and polar ice sheets. The contribution of polar ice sheets was held constant in time and for each scenario. Reproduced from Gerald A. Meehl et al., *Global Climate Projections*, *in* Climate Change 2007: The Physical Science Basis. Contribution of Working Group I to the Fourth Assessment Report of the Intergovernmental Panel on Climate Change 747, 821 (Solomon, S., D. Qin, M. Manning, Z. Chen, M. Marquis, K. B. Averyt, M. Tignor and H. L. Miller, eds., 2007).

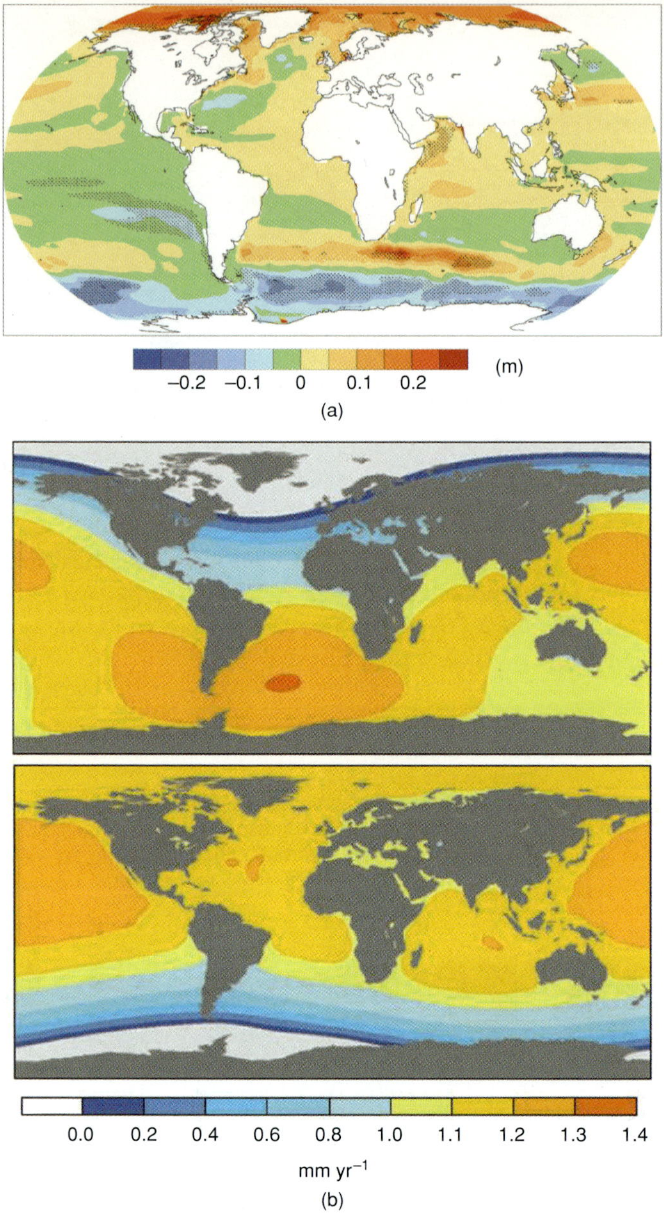

Plate 13. Spatial pattern of projected sea level rise: (a) projected in 2100 relative to a global average rise, calculated as the difference between the global averages for 1980–99 and 2080–99; and (b) expected from the melting of the ice sheets in Greenland (top) and West Antarctic (bottom). (a) reproduced from Gerald A. Meehl et al., *Global Climate Projections*, *in* CLIMATE CHANGE 2007: THE PHYSICAL SCIENCE BASIS. CONTRIBUTION OF WORKING GROUP I TO THE FOURTH ASSESSMENT REPORT OF THE INTERGOVERNMENTAL PANEL ON CLIMATE CHANGE 747, 813 (Solomon, S., D. Qin, M. Manning, Z. Chen, M. Marquis, K. B. Averyt, M. Tignor and H. L. Miller, eds., 2007). (b) reproduced from Glenn A. Milne, W. Roland Gehrels, Chris W. Hughes, and Mark E. Tamisiea, *Identifying the causes of sea level change*, 2 NATURE GEOSCIENCE 471, 472 (2009). Reproduced with permission from Nature Publishing Group.

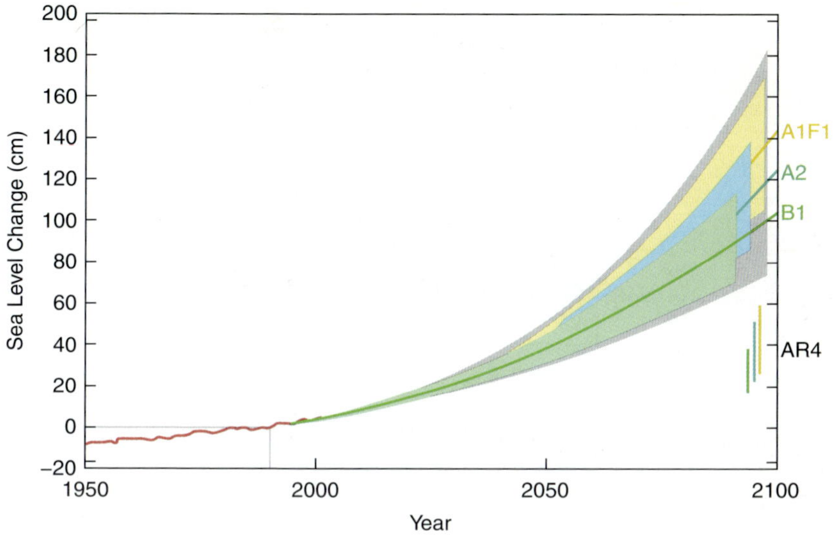

Plate 14. Projected sea level rise from 1990 to 2100, estimated from a statistical relationship between temperature and sea level, applied to nineteen models for each of three different emission scenarios: B1, A2, and A1F1. The red line shows sea level change for the observed temperature record. The green (B1), blue (A2), and yellow (A1F1) lines refer to the standard carbon cycle setting under each scenario, while the corresponding uncertainty range encompasses the high and low settings for the carbon cycle. The gray shading is an additional uncertainty associated with the statistical relationship between temperature and change in sea level rise. The projected sea level change for the three scenarios from AR4 is shown as vertical lines. Reproduced from Martin Vermeer and Stefan Rahmstorf, *Global Sea Level Linked to Global Temperature*, 106, Proceedings of the Nat'l Acad. of Sci. 21527, 21531 (2009).

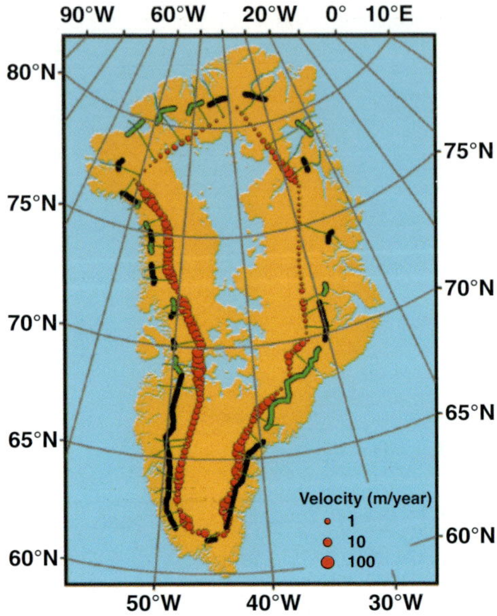

Plate 15. Map showing outlet glacier gates in Greenland: gates that are below sea level are shown as dark green and those above sea level as black. Reproduced from W. T. Pfeffer, J. T. Harper, and S. O'Neel, *Kinematic Constraints on Glacier Contributions to 21st-Century Sea Level Rise*, 321 Science 1340, 1341 (2008). Reprinted with permission from the American Association for the Advancement of Science.

2. Why a Convention?

2.1. Scale and Nature of the Problem

The Intergovernmental Panel on Climate Change (IPCC)[4] and the Stern Review,[5] among many other studies, warn that the effects of climate change – including rising sea levels, heavier floods, more frequent and severe storms,[6] drought, and desertification – will cause large-scale population movements. Climate change displacement presents an urgent problem for the international community (for more information on anticipated impacts from climate change, see Chapter 2).

The existence and scale of climate change displacement are often established by reference to the likely numbers of displaced people; figures range from 50 million to 1 billion. The most cited estimate, that of Norman Myers, is 200 million climate change migrants by 2050, or one person in every forty-five on the planet will be displaced.[7] However, Myers' estimate is contested, and as Brown observes, "nobody really knows with any certainty what climate change will mean for human population distribution."[8] Nonetheless, there is a consensus in the literature

4 *See* WORKING GROUP I, IPCC, 4TH ASSESSMENT REP., CLIMATE CHANGE 2007: THE PHYSICAL SCIENCE BASIS (2007); WORKING GROUP II, IPCC, 4TH ASSESSMENT REP., CLIMATE CHANGE 2007: IMPACTS, ADAPTATION AND VULNERABILITY (2007); WORKING GROUP III, IPCC, 4TH ASSESSMENT REP., CLIMATE CHANGE 2007: MITIGATION OF CLIMATE CHANGE (2007).

5 NICHOLAS STERN, THE ECONOMICS OF CLIMATE CHANGE: THE STERN REVIEW (2007).

6 With particular reference to rising sea levels, heavier floods, and more frequent and severe storms, *see* SUSMITA DASGUPTA ET AL., THE IMPACT OF SEA LEVEL RISE ON DEVELOPING COUNTRIES (World Bank, Working Paper No. 4136, 2007); Gordon McGranahan et al., *The Rising Tide: Assessing the Risks of Climate Change and Human Settlements in Low Elevation Coastal Zones*, 19 ENV'T & URBANIZATION 17 (2007); NICK BROOKS ET AL., GERMAN ADVISORY COUNCIL ON CLIMATE CHANGE, SEA LEVEL RISE: COASTAL IMPACTS AND RESPONSES (2006).

7 *See* NORMAN MYERS, ENVIRONMENTAL REFUGEES: AN EMERGENT SECURITY ISSUE (2005). *See also* IPCC, CLIMATE CHANGE 2001: THIRD ASSESSMENT REPORT (2001). Stern describes Myers' assumptions as "conservative." STERN, *supra* note 6, at 91–2.

8 OLI BROWN, UNDP, HUMAN DEV. REP. OF. OCCASIONAL PAPER NO. 17, CLIMATE CHANGE AND FORCED MIGRATION: OBSERVATIONS, PROJECTIONS AND IMPLICATIONS 5

that climate change will lead to "major forced displacements" over time (for more information on current displacement in small island nations, see Chapter 11).[9]

Any proposal for addressing the issue of climate change displacement must account for the different contexts and forms in which such migration is likely to occur.[10] There are three kinds of climate change impacts that are likely to have the greatest consequences in terms of human movement: sea level rise, increasingly severe weather events, and drought and desertification.[11] The IPCC has highlighted threatened island nations, the continent of Africa, mega-deltas (particularly those in Asia), and the polar regions as areas most exposed to climate change impacts.[12]

The latest projections of global sea level rise are discussed in Chapter 2 of this book. It is likely that, for unmitigated emissions, by the end of the century sea level rise will be at least one meter, with two meters being possible. Furthermore, "several meters of sea level rise must be expected over the next few centuries."[13] Because of the irreversibility of sea level rise, migration is a significant potential consequence.

(2007). A number of publications in the area of climate change displacement have accepted that migration is likely to occur as a consequence of climate change, but urge caution in relation to estimating the scale and extremity of the movements. *See, e.g.*, Kniveton et al., Climate Change and Migration: Improving Methodologies to Estimate Flows (Int'l Org. for Migration, Research Series No. 33, 2008); Vikram Odedra Kolmannskog, Future Floods of Refugees (2008); Etienne Piguet, Climate Change and Forced Migration: How Can International Policy Respond to Climate Change Induced Displacement?, UNHCR, Research Paper No. 153, 2008.

9 Piguet, *supra* note 9, at 8.

10 *See* Walter Kälin, Speech at the ECOSOC Panel on Disaster Risk Reduction and Preparedness: The Climate Change-Displacement Nexus (Jul. 16, 2008), *available at* http://www.brookings.edu/speeches/2008/0716_climate_change_kalin.aspx.

11 Bonnie Docherty & Tyler Giannini, *Confronting a Rising Tide: A Proposal for a Convention on Climate Change Refugees*, 33 Harv. Envtl. L. Rev. 349, 355 (2009).

12 Kolmannskog, *supra* note 9, at 23.

13 Ian Allison et al., The copenhagen Diagnosis, 2009: Updating the World on the Latest Climate Science 9 (2009).

Large delta systems and small island nations are particularly vulnerable to sea level rise. IPCC calculations have indicated that a rise in sea level of 45 centimeters would displace 5.5 million people, submerging more than 10% of Bangladesh.[14] A study commissioned for the Stern Review estimates that 146 million people live at an altitude of less than one meter, with South Asian and East Asian populations being predominantly at risk.[15] However, due to their lack of elevation above sea level, island States such as Tuvalu and Kiribati are perhaps the most threatened,[16] and they also face the prospect of saltwater intrusion, frequent flooding, and the like. It is a real possibility that small island nations in their entirety may become uninhabitable as a consequence of climate change.[17]

The impact of drought and desertification on Africa and Asia will be particularly harsh. The IPCC's 2007 report estimates that yields from rain-fed agriculture in sub-Saharan Africa could fall by up to 50% by 2020. Crop yields in central and south Asia could also fall by 30% by the middle of the twenty-first century.[18]

2.2. Inadequate Assistance and Protection Provisions in Refugee Law

Current protections under international law do not adequately provide for a number of the categories of person likely to be displaced by climate change.[19] International refugee lawyers generally agree that

14 Working Group II, IPCC, 3rd Assessment Rep., Climate Change 2001: Impacts, Adaptation and Vulnerability 569 (2001).

15 David Anthoff et al., Global and Regional Exposure to Large Rises in Sea Level 8 (Tyndall Ctr. For Climate Change Research, Working Paper No. 96, 2006).

16 Piguet, *supra* note 9, at 8.

17 Docherty & Giannini, *supra* note 12, at 355.

18 Working Group II, IPCC, 4th Assessment Rep., Climate Change 2007 – Impacts, Adaptation and Vulnerability 13 (2007).

19 Docherty & Giannini, *supra* note 12, at 358; Office of the High Commissioner for Human Rights, *Report of the Office of the United Nations High Commissioner for Human Rights on the Relationship between Climate Change and Human Rights*, at

CCDPs would not be the subject of protection under the 1951 Convention Relating to the Status of Refugees [hereinafter the "Refugee Convention"].[20] The Refugee Convention, the most comprehensive articulation of refugee rights and State obligations, relies on a restrictive definition of a refugee as someone with a "well-founded fear of being persecuted for reasons of race, religion and nationality, membership of a particular social group or political opinion", and who is "outside the country of his nationality."[21] It would be difficult to establish that a person displaced by climate change has been "persecuted," as required by the Refugee Convention and defined by the existing jurisprudence (for more discussion of the role of human rights law in climate change, see Chapter 8).[22]

Another concern with seeking protection for CCDPs under the Refugee Convention is that doing so would risk devaluing current protections for refugees.[23] Furthermore, to conflate the term "refugee" such that it includes both CCDPs and traditional refugees obscures fundamental differences of experience between the groups; most problematically, as a general rule the nexus between CCDPs and their States will not have been severed through persecution. Such protections would also

19, UN Doc. A/HRC/10/61 (Jan. 15, 2009); Kälin, *supra* note 11; Walter Kälin, *Displacement Caused by the Effects of Climate Change: Who Will Be Affected and What Are the Gaps in the Normative Framework for Their Protection?* (2008), *available at* http://www.brookings.edu/papers/2008/1016_climate_change_kalin.aspx. For a fuller consideration of the current international law regimes that may be applicable to climate change displacement, *see* Angela Williams, *Turning the Tide: Recognizing Climate Change Refugees in International Law*, 30 L. & POL'Y 502 (2008). For a discussion of the appropriateness of the United Nations Security Council as a forum for addressing climate change displacement *see* FRANK BIERMANN & Ingrid Boas, *Preparing for a Warmer World: Towards a Global Governance System to Protect Climate Refugees* 21 (Global Governance, Working Paper No. 33, 2007).

20 *See* Williams, *supra* note 20, at 507–8.

21 Convention Relating to the Status of Refugees art 1A(2), Apr. 22, 1954, 189 U.N.T.S 150.

22 JANE MCADAM, CLIMATE CHANGE "REFUGEES" AND INTERNATIONAL LAW 4–5 (2007).

23 David Keane, *The Environmental Causes and Consequences of Migration: A Search for the Meaning of "Environmental Refugees,"* 16 GEO. INT'L ENVTL. L. REV. 209, 214–17 (2004).

be insufficient; as set out at Sections 3.1 and 3.2, most displacement due to climate change will be internal, meaning most CCDPs will not cross national borders.

2.3. The UNFCCC and Displacement

2.3.1. Limitations as a Framework

Proposals for some kind of legal instrument designed to address the problem of climate change displacement seek, in various ways and for various reasons, to link that instrument with the United Nations Framework Convention on Climate Change (UNFCCC).[24] Yet neither the UNFCCC process nor its Kyoto Protocol[25] contemplates or addresses the issue of displacement. In our view, the UNFCCC – notwithstanding its expansion toward adaptation as well as mitigation – has limitations as a framework for dealing with climate change displacement. Displacement is not its focus; its concerns lie instead in the promotion of climate mitigation activities. Its structure and institutions are not designed to address displacement and the issues associated with it. Moreover, as the 2010 Cancun and 2011 Durban climate change conferences reveal, the UNFCCC cannot easily be altered to accommodate CCDPs; dealing with *existing* provisions is already problematic.

These arguments are put succinctly by Docherty and Giannini. Although they observe that the UNFCCC does apply directly to climate change, they also note that it has legal limitations for dealing with climate change displacement. As an environmental law treaty, the UNFCCC

> primarily concerns state-to-state relations; it does not discuss duties that States have to individuals or communities, such as those laid out in human rights or refugee law. It is also preventive in nature and less

24 United Nations Framework Convention on Climate Change, Mar. 21, 1994, 1771 U.N.T.S. 107. *See generally* Biermann & Boas, *supra* note 20; Williams, *supra* note 20.

25 Kyoto Protocol to the United Nations Framework Convention on Climate Change, Feb. 16, 2005, 2303 U.N.T.S 148.

focused on the remedial actions that are needed in a refugee context. Finally, although the UNFCCC has an initiative to help States with adaptation to climate change, that program does not specifically deal with the situation of climate change refugees. Like the refugee regime, the UNFCCC was not designed for, and to date has not adequately dealt with, the problem of climate change refugees.[26]

2.3.2. Focus Only on International/Cross-Border Displacement

Furthermore, to the extent that the UNFCCC does consider displacement effects from climate change, its focus is strongly on international movement. A UNFCCC "non-paper" is a text used by States Parties as a starting point for work on negotiating texts, as well as a vehicle for comments and proposing revisions. Non-paper 41 on climate change adaptation, produced after the seventh session of the Ad Hoc Working Group on Long-Term Cooperative Action (AWG-LCA) under the UNFCCC in Barcelona in October 2009, refers to migration and displacement as follows:

> All Parties [shall] [should] jointly undertake action under the Convention to enhance adaptation at the international level, including through . . . (b) Activities related to migration and displacement or planned relocation of persons affected by climate change, while acknowledging the need to identify modalities of interstate cooperation to respond to the needs of affected populations who either cross an international frontier as a result of, or find themselves abroad and are unable to return owing to, the effects of climate change.[27]

26 Docherty & Giannini, *supra* note 12, at 358. They also note that attaching any climate change refugee protocol to the UNFCCC has three significant shortcomings: "the limits of the UNFCCC's mandate, which is not focused on remedies; the historical reluctance to incorporate human rights issues explicitly into environmental treaties; and the UNFCCC's track record of inaction." *Id.* at 394.

27 Ad Hoc Working Group on Long-Term Cooperative Action under the Convention, *Contact Group on Enhanced Action on Adaptation and its Means of Implementation* 3–4 (Non-paper No. 41, Draft Text, 2009), *available at* http://www.stakeholderforum.org/fileadmin/files/Non-Paper%2041.pdf. *See also* Norwegian Refugee Council, Climate Changed: People Displaced 23 (2009).

Although the Norwegian Refugee Council argues that "[i]t is important that the reference [in the non-paper] to the international level is not interpreted to mean that only cross-border movements are addressed,"[28] it is clear that, for the AWG-LCA, the emphasis is on international displacement. Yet, as we argue in this chapter, most displacement will likely occur within national borders.

2.4. Existing Proposals to Address the Displacement Problem Are Not Complete or Comprehensive

Proposals for a new instrument providing for people displaced by climate change have been advanced in varying levels of detail by Docherty and Giannini,[29] Biermann and Boas,[30] Williams,[31] Bétaille et al.,[32] the authors of this chapter and their colleagues,[33] and others.[34] All of the proposals agree that CCDPs do not fall within the scope of the existing refugee regime created by the Refugee Convention.[35] However, they differ as to the most appropriate instrument to tackle that problem and the scope and detail of that instrument. Although we do not provide a detailed summary of the various proposals, we do acknowledge at various points in this chapter the contributions made by these authors and the extent to which we have considered and incorporated aspects of their proposals in our own. In our view, however, no proposal has offered a comprehensive, global approach to addressing the displacement

28 *Id.*

29 Docherty & Giannini, *supra* note 12.

30 BIERMANN & BOAS, *supra* note 20.

31 Williams, *supra* note 20.

32 Julien Betaille et al., *Draft Convention on the International Status of Environmentally-Displaced Persons*, 4 REVUE EUROPÉENNE DE DROIT DE L'ENVIRONNEMENT 395 (2008).

33 *See* text accompanying note 2.

34 *See* ENVTL. JUSTICE FOUND., NO PLACE LIKE HOME: WHERE NEXT FOR CLIMATE REFUGEES? (2008); Md Shamsuddoha & Rezaul Karim Chowdhury, *Climate Change Migrants*, 74 TIEMPO 1, 3 (2010).

35 Docherty & Giannini, *supra* note 12, at 359; BIERMANN & BOAS, *supra* note 20, at 17–21; Williams, *supra* note 20, at 507–10.

problem; the proposal presented here, which builds on our earlier 2008, 2009, and 2010 studies, attempts to provide such an approach.

Any convention designed to deal with climate change displacement on a comprehensive basis must include both internal displacement – that which occurs within State borders and is the most likely form of displacement – and international (or transborder) displacement. Our Convention, although it necessarily distinguishes between internal and international displacement, provides for both.[36] For those displaced within State borders it institutes a mechanism for the provision of principled, nondiscriminatory assistance. It also provides an original definition of "climate-change-displaced persons" and a "climate change event."

The Convention attempts to prioritize methods for addressing the climate-change-displacement problem without affecting instruments not designed to deal with that problem, such as the UNFCCC and its Kyoto Protocol. Our Convention proposal provides specifically for the populations of threatened island nations that may become uninhabitable due to the effects of climate change and differentiates such States from others that may be affected by large-scale displacement.

2.5. No "Coherent Multilateral Governance Framework" Exists

As summarized earlier, there is "no coherent multilateral governance framework for environmental migration . . . [regulation] is extremely fragmented and disparate."[37] Neither existing climate change law nor refugee law adequately provides for CCDPs. There has been no coordinated response by governments to address human displacement due to climate change – whether domestic or international, temporary or permanent. And given the nature and magnitude of the problem that climate change displacement presents, ad hoc measures based on existing domestic regimes may lead to inconsistency, confusion, and conflict.

36 "The issue of climate change IDPs [internally displaced persons] is beyond the scope" of Docherty and Giannini's article, *supra* note 12, at 360.

37 Jane Mcadam, Environmental Migration Governance 5 (Univ. N.S.W. Faculty of Law Research Series, Research Paper No. 1, 2009).

The international community has an obvious interest, on security and economic grounds at the very least, in addressing the problem of human displacement in a coordinated fashion. This includes assisting in the provision of preemptive resettlement to those most at risk from impacts of climate change. Our Convention, in addition to providing a general framework for CCDP assistance – its main objective – would address gaps in the international regime of legal protections and humanitarian assistance currently applicable to CCDPs. It would do so by incorporating a mechanism for the provision of nondiscriminatory assistance to internally displaced CCDPs, setting out a framework for the protection of those persons displaced across international borders, and identifying specific principles that should apply to the resettlement of persons from threatened island nations that may become uninhabitable.

3. A Convention for Climate-Change-Displaced Persons

3.1. Introduction: Scope of Convention

We propose a single, multilateral, stand-alone Convention to address the problem of climate change displacement, the scope of which – like the problem, both in terms of causation and consequences – is global; parties to the Convention would include both developed and developing States.

Although our Convention is a stand-alone instrument, it does draw on and adapt provisions of other instruments to adequately provide for, assist, and protect those displaced by climate change. Adopting a multifaceted, cooperative, and international approach to providing for, assisting, and protecting CCDPs, our Convention encompasses those displaced internally *and* those who cross international borders. Migration experts state that most persons displaced by climate change will be unlikely to cross an international border.[38]

38 Kniveton et al., *supra* note 9, at 29; Christian Aid, Human Tide: The Real Migration Crisis 6 (2007); Brown, *supra* note 9, at 13; German Advisory Council on Global change, Climate Change as Security Risk 18 (2007).

Developing-State Parties to the Convention – with economies dependent on the natural environment, but without resources to mitigate and adapt to the effects of climate change – will be those States most in need of displacement assistance. As the International Council on Human Rights Policy notes, "the most dramatic impacts of climate change are expected to occur in the world's poorest countries;" indeed, these countries already experience such impacts.[39]

Finally, the proposed convention would provide for both temporary and permanent relocation. The need for relocation assistance and protection arises whether the relocation is temporary or permanent.

3.2. Convention Provisions

Our Convention proposal is based on the following principles and provides for the following matters, all of which are more fully set out in this chapter:

- The Convention would provide a general framework for assistance to CCDPs and would address gaps in current human rights, refugee, and humanitarian law protections for CCDPs.
- Most people made homeless by climate change are expected to stay within their home countries. The Convention would encompass those displaced within States and those who cross international borders.
- Persons displaced within State borders would be subject to a framework of assistance in which obligations would be shared between the home State and the international community. In the case of CCDPs who have migrated across State borders, the Convention would outline the rights and obligations of the CCDP and the home and host States, while making it clear that the Convention itself would not oblige host States to accept CCDPs (see Section 3.5.2).
- The poorest countries in the world are likely to experience the most severe impacts of global warming. The Convention would provide for

39 Int'l Council on Human Rights Policy, Climate Change and Human Rights 1 (2008).

contributions to a fund for climate change displacement by developed States Parties based on the principle, at least as a starting point, of "common but differentiated responsibilities."[40]

- Recognizing that current levels of scientific knowledge create causation issues regarding the extent to which climate change contributes to a particular weather event or population movement, the Convention would adopt a "very likely" standard.
- Rather than assigning rights and protections on the basis of the individual satisfaction of definitional criteria, as in the Refugee Convention, we believe that *en masse* designation of CCDP status is more appropriate to the characteristics of climate change migration.
- Because of the necessity of integrating complex issues of causation and evolving science into decision making with respect to climate change migration, our proposal calls for the creation of an institutional architecture for designating members of a particular population as CCDPs.
- The very real prospect of entire States becoming uninhabitable differentiates the plight of threatened island nations from other regions in which there is likely to be large-scale displacement, and it requires specific consideration. We propose to apply the principles of proximity, self-determination, and the safeguarding of intangible culture to guide bilateral displacement agreements between threatened island nations and host States. Such agreements would be negotiated by States under the aegis of the Climate Change Displacement Organization, the organization established under the Convention.

The Convention contemplates the collaborative provision of preemptive assistance (and, if necessary, resettlement) to those most at risk from the impacts of climate change. Provision of assistance under the Convention could thus be described as "anticipatory adaptation."

These matters are set out in more detail later.

40 *See* discussion *infra*.

3.3. Causation

Several issues arise with respect to the provision of assistance and protection for persons displaced by climate change. The first is the extent to which climate change *causes* the event giving rise to the displacement. It is not possible at present for science to determine whether some environmental events were caused by climate change.[41] However, the IPCC, for example, identifies (a) increased incidence of extreme high sea level (excluding tsunamis), (b) intense tropical cyclone activity increases, and (c) areas affected by drought increases as "likely" to be a result of climate change; that is, with more than 66% probability.[42] Climate change science continues to evolve. Any instrument that seeks to address migration induced by climate change events must be based on scientific evidence but also must be sufficiently flexible to reflect developments in scientific understanding over time.

A second issue is the extent to which humans contribute to particular climate change events. Docherty and Giannini argue that science can determine the likelihood that humans "contributed to a type of disruption." Their definition references the IPCC's "likelihood ranges" and adopts the IPCC's "more likely than not" standard (a probability greater than 50%).[43]

The Convention would recognize problems with (a) determining the extent to which climate change causes an event giving rise to displacement and (b) establishing the extent to which humans contribute to particular or general climate change events. The Convention would adopt a "very likely" standard (greater than 90% probability) to identify "climate change events" (see Section 3.4) as caused by human-induced

41 Docherty & Giannini, *supra* note 12, at 370.

42 IPCC, Climate Change 2007: Synthesis Report 53 (2007). "Virtually certain" means a probability greater than 99%; "extremely likely" means a probability greater than 95%; and "very likely" means greater than 90%. *Id.* at 27.

43 Docherty & Giannini, *supra* note 12, at 371.

climate change. This higher standard provides increased certainty and targeted resource allocation, vital in the context of a convention that could potentially apply to hundreds of millions of people and would be applied by a proposed climate-change-displacement organization (see Section 3.5).

By adopting the "very likely" standard, and in light of the current state of climate change science, we anticipate that requests from States Parties attracting the operation of our Convention would overwhelmingly concern slow-onset, gradual displacement, which is more likely to be proven as induced by anthropogenic climate change than a sudden disaster.

Issues of causation also arise with respect to the question of whether climate change necessarily leads to migration. Adaptive capacity is integral to displacement decisions – and both adaptation and displacement are determined as much by available resources as by environmental factors. The Convention would largely operate prospectively: assistance to CCDPs would be based on an assessment of their environment becoming uninhabitable due to a "climate change event" such that resettlement measures and assistance were necessary.

Because of the necessity of integrating complex issues of causation and evolving science into decision making in respect of climate change migration, our proposal involves the creation of a modest institutional architecture for designating a particular population as CCDPs. Although the scope of our Convention includes those persons displaced as a result of sudden climate change events (or impacts), as a practical matter the proposed Convention's machinery may not be suited to immediately reacting to an unforeseen disaster. Attempts to apply the Convention in such situations may prevent the operation of existing and more effective disaster relief and management programs. Multiple channels already exist to provide aid, assistance, and protection to those displaced by sudden environmental impacts.

3.4. Definition and Designation of CCDPs

Proposals for definitions of "climate change displaced persons"[44] have advanced a range of strategies for addressing the question of how to distinguish forced from voluntary migration in cases of gradual environmental degradation. In our view, prospective migration based on the likely consequences of climate change (that is, "voluntary migration") is as coerced as migration in response to climate change impacts that immediately render a particular area uninhabitable. In other words, population movements based on the conclusion that a region will no longer be habitable in the future also constitute "forced" migration. The important question *then* becomes institutional rather than definitional and is focused on constructing and administering a set of processes to determine the likely contribution of climate change to both prospective and responsive climate change movements.

Rather than assigning rights and protections on the basis of the individual satisfaction of definition-based criteria, we propose more workable (in our view) *en masse* designations of the status of CCDPs through a process of request and determination by States and Convention institutions. Such an approach would nevertheless require a definition of CCDPs because, as Castles observes, "we cannot get around definitional categories...easily, for definitions are crucial in guiding the policies of governments and international agencies towards mobile people."[45]

We propose, then, the following CCDP definition:

> *CCDPs are groups of people whose habitual homes have or will become temporarily or permanently uninhabitable as a consequence of a climate change event.*

44 Biermann & Boas, *supra* note 20, at 8; Docherty & Giannini, *supra* note 12, at 361; Williams, *supra* note 20, at 522–23.

45 Stephen Castles, Environmental Change and Forced Migration 9 (UNHCR, Working Paper No 70, 2002).

A "climate change event" is defined as "sudden or gradual environmental disruption that is very likely caused by human-induced climate change."

Under our Convention, assistance and protection would not be triggered solely by fulfilling the requirements of a definition, but rather through an international process of status designation informed by scientific studies, affected communities, States, and international institutions.

3.5. Summary of the Convention's Operation

3.5.1. Organization

Although other proposals[46] describe discrete agencies to administer their respective conventions, no proposal has attempted to set out in detail the operation of a climate-change-displacement organization and its constituent bodies, and how those bodies would work together and interact. Our proposal attempts to remedy this gap by putting forward a modest climate-change-displacement organizational structure and, within that structure, setting out the roles and obligations of Convention participants. Although we envisage a single, stand-alone convention with a global scope, the Convention's governance structure also contemplates a role for regional committees and multidisciplinary collaborations across developed and developing States, including government and nongovernment organizations.

Our Convention proposal would create a Climate Change Displacement Organization (CCDO) that would comprise a Council and an administrative body, with links to or direct associations with climate change science research organizations. Climate-change-displacement implementation groups would also be formed to facilitate resettlement.[47]

46 DOCHERTY & GIANNINI, *supra* note 12, at 384–91; BIERMANN & BOAS, *supra* note 20, at 28–29; Williams, *supra* note 20, at 519.

47 Further detail on CCDO institutions, participants, and obligations can be found in Hodgkinson et al., *supra* note 2, at 91–97, from which this section draws. Note, however, that this chapter substantially modifies the CCDO structure set out in that article.

3.5.2. States Parties: Offers of and Requests for Assistance, and Designations

The Convention framework contemplates that States providing displacement assistance funding and States receiving such funding would be parties to the Convention. The Convention limits the provision of assistance to developing States Parties. A developing State Party would make a request to the CCDO for internal or international resettlement assistance. That State Party would, at the time of making a request for assistance, also request or apply for *en masse* designation of the status of CCDPs by the CCDO. In the case of international resettlement assistance, a "host" State request for assistance would reference the relevant "home" State request. It would also be open to the CCDO (through the CCDO Council) to offer assistance to developing States Parties without a request from those parties.

Bilateral displacement agreements would be concluded under the aegis of the Convention between home and host States. The Convention would not compel States to either sign such agreements or to accept CCDPs.

Developing States Parties may also make financial and other requests for assistance in fulfilling their Convention obligations. These obligations would include collecting data on climate change displacement, integrating protection and assistance for CCDPs into future planning programs, and consulting with potential or actual CCDPs and their representative organizations in the collection of data and the development and implementation of policy.

Again, then, the Convention would largely operate prospectively: assistance to CCDPs would be based on a bottom-up/top-down assessment of the likelihood of their environment becoming uninhabitable due to events consistent with anthropogenic climate change, such that resettlement measures and assistance were necessary. In other words, displacement is viewed as a form of adaptation that creates particular vulnerabilities requiring protection as well as assistance through international cooperation. As Mastrandrea and Schneider note,

> [s]uccess in adaptation to climate change will come from the mating of top-down and bottom-up assessment. Scientific projections are *most useful* when joined with the intimate knowledge of existing vulnerabilities that stakeholders possess... Detailed bottom-up studies provide understanding of the structural, institutional, psychological, financial, legal, and cultural frameworks of affected sectors [and]... communities.[48]

3.5.3. Regional Committees

Williams[49] outlines the clear virtues of regional initiatives and cooperation in addressing the plight of persons displaced by climate change, including the ability to assess "current vulnerabilities to climate change impacts, along with information exchange on traditional coping practices, diversified livelihoods, and current government and local interventions."[50] A regional structure, Williams notes, would allow "for various levels of engagement and development by States, depending on the individual capacity of each country involved and the (perceived) severity of the problem in that area."[51] McAdam observes that responses on a regional basis might "be more appropriate and culturally sensitive... taking into account the particular features of the threatened population."[52]

Accordingly, we propose the establishment of regional committees that would inform CCDO decision makers about regional perspectives and enable the unique situation of threatened island States to

48 Michael D. Mastrandrea & Stephen H. Schneider, Preparing for Climate Change 87 (2010) (emphasis added).

49 Williams, *supra* note 20, at 517–23.

50 *Id.* at 519.

51 *Id.* at 521.

52 McAdam, *supra* note 38, at 7–8.

be addressed, in part through the conclusion of bilateral displacement agreements.

3.5.4. Financing

We propose that developed States Parties to the Convention make mandatory financial contributions to a climate-change-displacement fund (supervised by the CCDO), and that such contributions are made on the basis that, at least as a starting point, States and States Parties to the Convention have common but differentiated responsibilities.[53] This principle of common but differentiated responsibilities "recognizes historical differences in the contributions of developed and developing States to global environmental problems, and differences in their respective economic and technical capacity to tackle these problems."[54]

State practice and legal precedent support the general principle of common but differentiated responsibilities, which underpins the obligations of States Parties to the Convention. Our proposal that funding should be apportioned on the basis of the common but differentiated responsibilities principle also finds support in other proposals for climate-change-displacement instruments.[55]

53 The principle of "common but differentiated responsibilities," the basis on which developed State Parties make contributions to the Fund, "recognises historical differences in the contributions of developed and developing states to global environmental problems, and differences in their respective economic and technical capacity to tackle these problems." As the McGill Centre for International Sustainable Development Law notes, the principle has two basic elements: "the common responsibility of states for the protection of the environment, or parts of it, at the national, regional and global levels" and "the need to take into account the different circumstances, particularly each State's contribution to the evolution of a particular problem and its ability to prevent, reduce and control the threat." CTR. FOR INT'L SUSTAINABLE DEV. LAW, THE PIRNCIPLE OF COMMON BUT DIFFERENTIATED RESPONSIBILITIES (2002), *available at* www.cisdl.org/pdf/brief_common.pdf. This has a number of implications, including the imposition of environmental obligations that differ between States.

54 *Id.*

55 BENITO MÜLLER, AN FCCC IMPACT RESPONSE INSTRUMENT AS PART OF A BALANCED GLOBAL CLIMATE CHANGE REGIME (2002), *available at* www.oxfordclimatepolicy.org/publications/documents/iri.pdf; Docherty & Giannini, *supra* note 12, at 386; BIERMANN & BOAS, *supra* note 20, at 26, 28–29. The inhabitants of the Alaskan island

We recognize various limitations to the principle, however, and the hard issue of the level of specific State Party contributions to the Fund would largely be determined during negotiations. We also recognize, of course, that adaptation funding is difficult to come by, as the UNFCCC process has revealed – whether through US$30 billion in "fast start financing" (FSF) from developed States or through a developed-State commitment to provide US$100 billion each year by 2020. FSF has not met even modest expectations, and the question of where to find US$100 billion per year has proven to be divisive. Furthermore, the Copenhagen Accord is vague on climate finance sources for adaptation.[56]

3.6. Assistance and Protection

3.6.1. Introduction

The vulnerability of displaced people has long been recognized as a serious issue by the international community and is reflected in the Refugee Convention, the Guiding Principles on Internal Displacement

of Shishmaref face displacement, and their relocation dilemma highlights some of the problems associated with adaptation funding: "The future of the Shishmaref community has yet to be determined, but two possible solutions have emerged. The first involves moving it to small towns around Nome and Kotzebue, 200 miles to the south and east, respectively, to take advantage of their urban infrastructure. With an estimated cost of $100 million . . . this is the less expensive alternative and the one favoured by the state of Alaska and federal funding agencies. The second alternative, estimated to cost $200 million, would involve relocating the village to the mainland only 12 miles from Sarichef. The village would be fully recreated in an uninhabited area called Tin Creek that would be safe from erosion. 'The state will never pay the extra cost,' says Jonathon Weyiouanna. . . . 'We'll have to find other funding. If we fail, we'll disappear. Our special culture, our community traditions like sharing and respect for our ancestors, our subsistence economy – everything that makes us a unique community will perish in a city like Nome, which is foreign to Inupiaq culture.' The community's struggle can be summed up in these words: relocate to Tin Creek or disappear." Guy-Pierre Chomette, *Alaska, the Kigiqtaamiut in Jeopardy*, *in* CLIMATE REFUGEES 22–23 (Collectif Argos eds., 2010).

56 CLIMATE ACTION NETWORK FRANCE ET AL., CLIMATE CHANGE NEGOTIATIONS: TIME TO FILL THE AMBITION DEFECIT (2011).

[hereinafter the "Guiding Principles"],[57] international humanitarian law, human rights law, and customary law. According to the International Council on Human Rights Policy, climate change is likely to exacerbate existing human rights vulnerabilities and resource inequities. Our proposed Convention not only provides a framework for the provision of adaptive assistance to CCDPs but also addresses gaps in existing protections, as well as identifying principles that should apply to the resettlement of persons from threatened island nations.

Again, most climate change displacement is likely to occur within State borders. The Guiding Principles provide a coherent statement of the matrix of human rights and humanitarian protections that are applicable to Internally Displaced Persons (IDPs). However, it is our view that because climate change is a global problem, the international community has an obligation to provide assistance to CCDPs regardless of whether their movement has a transborder dimension. We therefore propose a Convention that incorporates a mechanism for the provision of aid and assistance to internal CCDPs.

Although our Convention applies to both internal and transborder displacement, it would take account of the clear distinction drawn in international refugee law between refugees and the internally displaced. The distinction arises primarily from the principles of State sovereignty and non-interference, which are fundamental to the international legal system.[58] Our Convention, then, would distinguish among specific rights and protections on the basis of internal and international displacement. Certain ambit provisions, however, should be applicable to persons displaced both internally and across recognized State borders.

The Convention would provide a framework for the provision of assistance to and protection of persons *internally displaced* due to climate change, in which obligations are shared between the home State

57 UNHCR, *Guiding Principles on International Displacement*, U.N. Doc. E/CN.4/1998/53/Add.2 (1998).

58 Catherine Phuong, The International Protection of Internally Displaced Persons 22 (2004).

and the international community. In the case of those displaced *across* borders, our Convention would outline rights of the CCDP and obligations of the host State. Rather than distinguishing between temporary and permanent displacement, the Convention we envisage would allow rights to be gradually accrued based on the duration of the displacement.

Finally, because of their unique circumstances, persons from threatened island nations should be accorded treatment based on a further set of principles that include proximity, self-determination, and the preservation of their culture. These principles are set out at Section 3.7.

3.6.2. Internal Displacement

Protection and assistance provisions for those internally displaced are set out in the Guiding Principles rather than in a binding convention or United Nations Security Council resolution. The Guiding Principles apply to "persons or groups of persons who have been forced or obliged to flee or to leave their homes or places of habitual residence" for reasons including "natural or human made disasters" and "who have not crossed an internationally recognized State border."[59] The Guiding Principles recognize that national authorities have the primary duty and responsibility to provide protection and assistance to IDPs.

Our Convention proposal draws on a range of international law frameworks and precedents, including the Guiding Principles. It adopts a model in which the primary responsibility for CCDPs rests with their own State. It is founded on the recognition in international and human rights law that, although "different States have different capacities," States are responsible for caring for their own people, and it perpetuates this understanding.[60]

Climate-change-induced displacement is also a matter of global responsibility, however, regardless of whether such displacement occurs

59 UNHCR, *supra* note 58. *See also* INT'L LAW ASS'N, LONDON DECLARATION OF INTERNATIONAL LAW PRINCIPLES ON INTERNALLY DISPLACED PERSONS (2000), *available at* www.unhcr.org/refworld/docid/42808e5b4.html.

60 Docherty & Giannini, *supra* note 12, at 380.

across national borders. The international community should therefore be obliged to cooperate in the provision of assistance and protection to internally displaced CCDPs. On request by the home State and in accordance with Convention provisions, other parties to the Convention would be required to provide assistance in cases of internal displacement.

Although the emphasis of a global CCDP instrument should be less on the protection of persons outside their States, and more on supporting governments, local communities, and agencies in assisting those within their own territory,[61] displacement does make people vulnerable to abuses of their human rights.[62] Docherty and Giannini identify the Convention on Cluster Munitions as providing a model framework that requires the provision of "tangible assistance as well as protection of abstract rights."[63] Based on the Convention on Cluster Munitions precedent, State parties would be required to provide CCDPs under their jurisdiction with a practicable level of age- and gender-sensitive assistance, including emergency services, evacuation and relocation, medical assistance, housing, food, measures necessary for social and economic inclusion, property restitution where possible, and the facilitation of family reunion. Moreover, in fulfilling their obligations under the Convention, each State Party would be precluded from discriminating against or among CCDPs.

The operation of our Convention would, in many ways, be prospective. States Parties to the Convention would be obliged to collect reliable and relevant data on climate change displacement; adopt a general policy that integrates assistance and protection for CCDPs into future comprehensive planning programs; disseminate information on likely climate change displacement and State policy toward potential or actual CCDPs; and closely consult with and actively involve potential or actual CCDPs

61 Biermann & Boas, *supra* note 12, at 26.

62 *See* Nils Geissler, *The International Protection of Internally Displaced Persons*, 11 Int'l J. Refugee Law 451, 452 (1999); Phuong, *supra* note 59, at 43.

63 Docherty & Giannini, *supra* note 12, at 378.

and their representative organizations in the collection of data, the development of policy, and its implementation. The Convention would, therefore, provide a forum for the provision of preemptive adaptive resettlement to populations most vulnerable to the impacts of climate change.

As a result, assistance under the Convention would have an adaptive quality, rather than simply facilitating the provision of humanitarian aid; such an approach correlates with the insight that migration is best understood as a form of adaptation to climate change.[64] Situating displacement within the spectrum of adaptive possibilities enables assistance efforts to address the needs of those most vulnerable to climate change, who may otherwise lack the resources to remove themselves from climate change threats.

3.6.3. International Displacement

Both the Refugee Convention and customary international law recognize that, in certain circumstances, the international community is responsible for the provision of rights and assistance to persons displaced across international borders and unable to rely on their own nation for protection. Again, however, there currently exists no provision in international law for most persons migrating across borders as a consequence of climate change.

Docherty and Giannini identify the Refugee Convention as a useful model for what kinds of human rights protections to include in a new instrument for CCDPs, because it provides the "most comprehensive codification of the rights of refugees yet attempted on the international level."[65] We endorse Docherty and Giannini's argument that any treaty for CCDPs should be premised on the rights and protections that States *have already agreed to accord to traditional refugees*. As such, our Convention would guarantee a range of civil, political, economic,

64 Brown, *supra* note 9, at 25.
65 Docherty & Giannini, *supra* note 12, at 376.

social, and cultural rights, based on a principle of nondiscrimination. Further, CCDPs should be guaranteed a minimum standard of treatment at least equivalent to aliens in the host country. However, analogously to the Refugee Convention, certain rights should be afforded to CCDPs at the same level as those enjoyed by nationals in the host country, and in some cases, rights should be afforded based on an absolute standard, rather than being contingent on existing rights in host nations. Rights relating to movement are especially significant to CCDPs, and in particular, CCDPs should enjoy the right to *non-refoulement*, a core principle of refugee law under Article 33 of the Refugee Convention. In the context of the new instrument, *non-refoulement* would prohibit the forcible return of a refugee to a situation in which "climate-induced environmental change would threaten the refugee's life or ability to survive."[66]

Following the Refugee Convention, the rights of CCDPs displaced across international borders should expand on an incremental basis, with further rights accruing the longer that CCDPs remain in a host nation. Adoption of the Refugee Convention model of a gradually deepening set of rights enables the Convention to flexibly adapt to changing environmental conditions and scientific knowledge. As proposed by Docherty and Giannini, those displaced across borders would remain eligible for assistance until they acquired a new nationality, voluntarily returned to their home country, or refused to return when it was safe for them to do so.[67]

Any instrument for CCDPs should ensure that basic survival needs are met. Again, a duty of international cooperation and assistance, based on the principle that climate change is a global problem, is equally applicable to CCDPs who cross borders *and* to those who remain within their own States.

66 *Id.* at 377.
67 *Id.* at 369.

3.7. Threatened Island Nations

Rising sea levels and the prospect of island nations becoming uninhabitable are among the most visible consequences of climate change. McAdam suggests that "their small physical size, exposure to natural disasters and climate extremes, very open economies and low adaptive capacity make them particularly susceptible, and less resilient, to climate change."[68] The populations of such threatened island nations may not only be displaced but may also see the effective disappearance of their homelands. As a result, although they may amount only to a fraction of the total number of likely CCDPs, the interests and expectations of the populations of these island nations have a high profile in the international system.

The possibility of effective loss of homelands differentiates threatened island nations from other regions in which there is likely to be large-scale displacement, and it requires specific consideration. It may also signify the end of unique ways of life that are intimately connected to precarious physical landscapes. Such a scenario is unprecedented, and existing legal regimes do not adequately articulate the rights that should be accorded to CCDPs from threatened island nations in order to recognize this loss. We propose that the principles of (a) proximity, (b) self-determination, and (c) the safeguarding of intangible culture should be applicable to regional bilateral displacement agreements between threatened island nations and host States, negotiated under the Convention framework.

3.7.1. Proximity

It has been suggested that a climate change treaty should be implemented with recourse to a principle of proximity that requires the least separation of persons from their cultural area.[69] Proximate

68 McAdam, *supra* note 38, at 10.
69 Docherty & Giannini, *supra* note 12, at 358.

resettlement of threatened island nationals may be particularly appropriate given their strong connections to both land and seascapes. A principle of proximate resettlement should *guide* bilateral agreements between threatened island nations (home States) and host States in the relevant region.

As an alternative to the most proximate location for resettlement, small island populations may wish to migrate to a State with a sizable or established community of migrants from their nation, such as the population of Kiribati Islanders in New Zealand.

3.7.2 Self-Determination

A second guiding principle that could inform resettlement is self-determination. Self-determination is enshrined in the United Nations International Covenant on Economic, Social and Cultural Rights. Article 1 states that "[a]ll peoples have the right of self-determination. By virtue of that right they freely determine their political status and freely pursue their economic, social and cultural development."[70]

This articulation of self-determination is particularly useful in the context of threatened island nations. We consider self-determination relevant to two distinct elements of relocation: (a) *when* people will abandon their territories and (b) *where* people choose to resettle.[71] First, the proposed Convention would recognize that citizens of threatened island nations may want to remain in their home States for as long as practicable. Adaptation strategies could be funded through the capacity of the proposed Convention's fund to provide prospective resources for persons likely to be displaced by climate change.

70 United Nations International Covenant on Economic, Social and Cultural Rights art. 1, Jan. 3, 1976, 993 U.N.T.S 3.

71 It is probable that the former would operate on an individual basis, whereas the latter would be on a population basis.

In this regard we note the work of Barnett and Campbell in *Climate Change and Small Island States*:

> [F]ocusing on large-scale adjustments in social systems or large-scale engineering works is likely to be a less effective and efficient approach to adaptation...than multiple small-scale processes to build adaptive capacity at the community...local government...and national government levels....These approaches do not mean that the international community [is]...absolved of responsibility for adaptation....[A]daptation responses...are likely to have major costs, little if any of which should be borne by communities forced to take adaptive action.[72]

The principle of self-determination should also apply to the second aspect of relocation: the destination to which threatened island nation CCDPs relocate. Although proximity is also one factor here, collective self-determination should inform agreements between threatened island nations and host States. Threatened island nations' preferences may relate to existing migration patterns or proximity and should, to the extent possible, be supported or respected.

Our aim, above all, is to enable people to make the choice themselves whether to remain in their homes for as long as possible or to seek out new, more stable communities elsewhere, while continuing to plan for future potential resettlement activities by larger groups of people (for a discussion of self-determination options under existing law, see Chapters 3 and 4).

3.7.3. Preservation of Intangible Culture

Finally, our Convention would recognize the effects of climate change on populations, as reflected in our inclusion of *en masse* CCDP designations. Finding, as Docherty and Giannini do, that "group status

72 Jon Barnett & John Campbell, Climate Change and Small Island States 182–83 (2010).

determinations . . . increase opportunities to formulate solutions that would keep the integrity of a group intact, which could help preserve cultures and national identities,"[73] our Convention emphasizes the importance of the maintenance of the CCDP group's social and cultural cohesion in their host country.

For CCDPs from threatened island nations, the protection of cultural autonomy is of particular concern given that their State may effectively cease to exist. Although there exists no explicit protection of refugees' cultural autonomy in current international legal regimes, the protection of social and cultural rights in the Refugee Convention may be invoked as a useful precedent. Together with the Refugee Convention, the Convention for the Safeguarding of the Intangible Cultural Heritage is also relevant. Its purpose, as set out in Article 1, is to (a) safeguard intangible cultural heritage; (b) ensure respect for the intangible cultural heritage of the communities, groups, and individuals concerned; (c) raise awareness of the importance of intangible cultural heritage; and (d) provide for international cooperation and assistance.[74]

4. Is "Advocacy for a New Treaty" Misplaced?

McAdam argues that "advocacy for a new treaty to address climate-related movement is presently misplaced for a number of reasons."[75] This section summarizes the arguments that McAdam makes and briefly responds with reference to our proposed Convention. In our view, the structure and emphasis of our Convention mean that a number of McAdam's arguments against treaties generally do not apply. Further arguments against a new convention are articulated by Katrina Wyman, in Chapter 11.

73 Docherty & Giannini, *supra* note 12, at 375.

74 Convention for the Safeguarding of the Intangible Cultural Heritage art. 1, Apr. 20, 2006, 2368 U.N.T.S 3.

75 Jane McAdam, *Swimming against the Tide: Why a Climate Change Displacement Treaty Is Not the Answer*, 23 INT'L J. REFUGEE LAW 2, 4 (2011).

4.1. Empirical Evidence on Movement

McAdam states that

- "Treaty proposals are premised on certain assumptions about climate change and human movement that are not borne out in the empirical studies…which show that movement is likely to be predominantly internal and/or gradual, rather than in the nature of refugee 'flight.'"[76]
- "The rationale behind a treaty is to address international movement."[77]
- If a treaty is thought desirable, "it is important that it be viewed as one of a number of mechanisms that may respond to climate-induced displacement, rather than as *the* solution (which is often how it is posited)."[78]

The assumptions that McAdam claims for treaty proposals in general are not those that underlie our proposed Convention. As we make clear, any proposal for addressing the issue of climate change displacement must take account of the different contexts and forms that such migration is likely to take.[79] Thus, although our Convention encompasses both internal and international displacement, our proposal is based on empirical consensus that most movement will occur internally, rather than across international borders.[80] The rationale behind our treaty thus avoids McAdam's complaint about treaties as solely addressing international movement. Indeed, a focus of our proposed Convention is on regional responses, action, and agreements. Further, our proposal anticipates that requests from States Parties to the Convention would mostly concern slow-onset, gradual displacement and, indeed, should not be

76 *Id.* at 8.
77 *Id.*
78 *Id.*
79 *Id.*
80 *Id.*

treated like "refugee 'flight.'"[81] In response to McAdam's third criticism, our convention provides a general framework for CCDP assistance as one element of a wider international climate change "regime complex."[82]

4.2. Climate Change as a Cause of Movement

McAdam also notes that "it is conceptually problematic and empirically flawed in most cases to suggest that climate change alone *causes* migration" and that "it would seem both practically impossible and conceptually arbitrary to attempt to differentiate between those displaced people who deserve 'protection' on account of climate change and those who are victims of 'mere' economic or environmental hardship."[83]

As we state, our proposed Convention recognizes problems with determining the extent to which climate change causes an event giving rise to displacement and so adopts a "very likely" standard.[84] Our Convention creates an institutional architecture for designating a particular population as CCDPs because of the necessity of integrating issues of causation and evolving science into decision making in respect of climate change migration.[85] We acknowledge that causation will be a difficult issue, but note that similarly complex questions of causation have been resolved in other contexts; complexity of a pursuit does not necessarily render that pursuit any the less worthwhile.

We also note that the UNFCCC appears to have resolved issues of practicality and arbitrariness in its focus on adaptation assistance or "protection" to address the effects of climate change; States Parties agree specifically to cooperate "in preparing for adaptation to the impacts of climate change."[86]

81 *Id.*
82 *See* discussion *infra*.
83 McAdam, *supra* note 76, at 12, 13.
84 *Id.*
85 *Id.*
86 UNFCCC, *supra* note 25, Art. 4.1(e); *see also id.* Art. 4.8, 4.9.

More so than other worldwide global economic and environmental problems, there is some international recognition that responsibility for the effects of human-induced climate change is global and, in particular, that it rests with developed countries. Despite the difficulties[87] to date in concluding and then giving effect to the UNFCCC and its Kyoto Protocol, it may be that the international community is more likely to provide assistance to communities displaced by climate change rather than by other economic and environmental factors. Although this provision of aid may be arbitrary in some cases, it represents an opportunity to provide assistance to affected populations that should not be ignored.

4.3. Political Obstacles to a New Treaty

McAdam notes that "States presently seem to lack the political will to negotiate a new instrument requiring them to provide international protection to additional groups of people" and asks why States would "be willing to commit to, and realize protection for, people displaced by climate change" (a "compelling" answer provided by McAdam is that "an instrument would provide a basis for calling for shared responsibility").[88] Again, the assumptions that McAdam claims for treaty proposals in general are not those of our Convention; the underlying basis of our proposal is *not* "to provide a rights-based framework for people forced to move when the impacts of climate change render life and livelihoods at home impossible."[89] Our Convention would not require States to provide protection to additional groups of people; it contemplates above all the collaborative provision of preemptive *adaptation* assistance, in part because evidence strongly suggests that most displacement will take place internally, within a State. Provision of protection to "additional

87 Difficulties that could be addressed through development of a climate change "regime complex." *See id.*

88 McAdam, *supra* note 76, at 15–16.

89 *Id.* at 7.

groups" – or, put another way, States' acceptance of those *internationally* displaced – would be effected through bilateral displacement agreements between home and host States, concluded under the aegis of the Convention, and which host States would not be compelled to sign. No treaty requirement on a host State to accept CCDPs would exist.

McAdam further notes – and we agree – that "even if a treaty text could be agreed upon, its ratification, implementation and enforcement could not easily be compelled."[90] We also agree that "[a]ny treaty is necessarily an instrument of compromise and, even once achieved, States must demonstrate sufficient political will to ratify, implement and enforce it,"[91] and that "[i]f a treaty is nonetheless thought desirable, it is important that it be viewed as one of a number of mechanisms that may respond to climate-induced displacement, rather than as *the* solution."[92]

It should also be noted that there are clear benefits to be found just in the process of discussing or negotiating a treaty; such conversations emphasize the urgency of the displacement problem and global responsibility. They also, of course, do not rule out other ways of addressing the displacement problem.

Finally, a further concern noted by McAdam is that

> if a treaty becomes the *main* focus of international policy development, attention may shift from... alternative and additional responses that may enable people to remain in their homes for as long as possible (which is the predominant wish among affected communities), or to move safely within their own countries, or to migrate in a planned manner over time.[93]

Our aim, of course, through a treaty, is just that: to enable people to remain in their homes for as long as possible, to move safely within their own countries, or to migrate in a planned manner over time.

90 *Id.* at 17.
91 *Id.* at 25.
92 *Id.* at 8.
93 *Id.* at 5.

5. A "Regime Complex" or "Building Blocks" Approach

Japan, Russia, and Canada[94] have said they will not extend emissions cuts beyond December 31, 2012 (beyond the Kyoto Protocol's first commitment period), unless all major emitters sign on to a binding deal.[95] As the Climate Group notes,

> The rejection by Japan, Russia and Canada of any further Kyoto commitments indicates that the likelihood of a comprehensive and environmentally effective post-2012 Kyoto outcome is diminishing. The EU and other industrialised Kyoto parties, such as Australia, are likely to find it difficult to justify to domestic audiences further Kyoto commitments in the absence of these three key countries. As the continuation of the [Kyoto] Protocol remains [important]... for developing countries, the future of Kyoto will remain a flash point for disagreement.[96]

The current UNFCCC negotiating process is unlikely to make significant progress unless something changes in the positions of major developed and developing States and the relationship between them regarding climate change. Moreover, even if a deal were possible among "large" emitters, the UNFCCC's consensus decision-making model allows a relatively small group of countries to block agreement. One alternative approach

> to moving forward would be to split the climate change problem up into different pieces and address... the pieces in more specialized forums [this happens already with aviation]... more progress may be possible in specialized forums, which have long traditions of cooperation, than in the UNFCCC.[97]

94 All parties to the Kyoto Protocol.

95 *Extra UN climate talks set for April in Bangkok*, REUTERS, Jan. 29, 2011, http://www.reuters.com/article/2011/01/28/us-climate-talks-idUSTRE70R1WK20110128.

96 THE CLIMATE GROUP, POST-CANCUN ANALYSIS 6 (2011), *available at* http://www.theclimategroup.org/_assets/files/Post-Cancun-Analysis_1.pdf.

97 DANIEL BODANSKY, THE INTERNATIONAL CLIMATE CHANGE REGIME: THE ROAD FROM COPENHAGEN, 5 (2010), *available at* http://belfercenter.ksg.harvard.edu/files/Bodansky-VP-October-2010–3.pdf.

As Bodansky notes, "since an agreement among the major emitters is unlikely anytime soon, we should seek progress where we can, through whatever means and in any forums that are available."[98]

Robert Keohane and David Victor, after stating that "[t]here is no integrated, comprehensive regime governing efforts to limit the extent of climate change," propose a "climate change regime complex" – a loosely coupled set of specific regimes. They say that efforts to "build an effective, legitimate, and adaptable comprehensive regime are unlikely to succeed" and argue that a climate change regime complex has advantages in terms of adaptability and flexibility, characteristics that are "particularly important in an environment of high uncertainty, such as in the case of climate change where the most demanding international commitments are interdependent yet governments vary widely in their interest and ability to implement such commitments."[99]

It is an argument put forward by others – a group from the London School of Economics and Political Science and Keele University, for example, refers to a "building blocks approach" that develops different elements of climate governance in an incremental fashion and embeds them in an international framework.[100] Although outside the scope of this chapter, it may be that any climate-change-displacement treaty could form one element of such a climate governance approach.

6. A Treaty for Climate-Change-Displaced Persons?

In Bolivia, the Chacaltaya glacier has lost 80% of its surface area since 1982. In the Andes generally, if warming trends continue, many tropical glaciers may disappear within twenty years; among other things, the water supplies for more than seventy million people would be

98 *Id.* at 6.

99 Robert O. Keohane & David Victor, The Regime Complex for Climate Change 1 (Harvard Project on Int'l Climate Agreements, Discussion Paper 10–33, 2010).

100 Robert Falkner et al., *International Climate Policy after Copenhagen: Towards a "Building Blocks" Approach*, 1 Global Pol'y 252 (2010).

threatened. Residents of the Bolivian town of Palca, reliant on glacier meltwater for survival, are migrating to the city of El Alto and other locations. However, the water in El Alto is also running dry. Although, at present, "there is not a major problem in El Alto because the additional glacial melt has compensated for the demand, providing more water flow," the Bolivian vice minister of water resources states, "We're going to begin to have problems."[101] They are problems that "even money won't completely solve," Pablo Solón, Bolivia's ambassador to the United Nations, has said.

Ambassador Solon asks this question: "What do you do when your glacier disappears or your island is under water?"[102] We propose in this chapter a treaty for climate-change-displaced persons that sets out a framework for the collaborative provision of preemptive adaptation assistance – and, if necessary, relocation – *before* glaciers melt and *before* islands are under water.

101 Carolyn Kormann, *Retreat of Andean Glaciers Foretells Global Water Woes* posting to Yale Environment 360, http://e360.yale.edu/content/feature.msp?id=2139 (Apr. 9, 2009).

102 Elisabeth Rosenthal, *In Bolivia, Water and Ice Tell of Climate Change*, N.Y. TIMES, Dec. 14, 2009, at A1.

11 The National Immigration Policy Option

Limits and Potential

Katrina M. Wyman

1. Introduction

Kiribati, the Maldives, the Republic of the Marshall Islands (RMI), and Tuvalu are small island nations whose existence may be threatened by climate change and whose citizens therefore may have to relocate to other countries.[1] There are concerns, however, that climate change refugees from these and other nations will lack a right of refuge under existing international law and the domestic immigration laws of likely destination countries. These concerns recently have prompted proposals for a new multilateral instrument to give climate-displaced persons a right of refuge as a matter of international law (for an articulation of one of these proposals, see Chapter 10).[2]

1 Nicholas Schmidle, *Wanted: A New Home For My Country*, N.Y. TIMES MAG., May 10, 2009, at 38; Neil MacFarquhar, *Refugees Join the List of Climate-Change Issues*, N.Y. TIMES, May 29, 2009, at A4.

2 Frank Biermann & Ingrid Boas, *Preparing for a Warmer World: Towards a Global Governance System to Protect Climate Refugees*, GLOBAL ENVTL. POL., Feb. 2010, at 60; Bonnie Docherty & Tyler Giannini, *Confronting a Rising Tide: A Proposal for a Convention on Climate Change Refugees*, 33 HARV. ENVTL. L. REV. 349 (2009); A CONVENTION FOR PERSONS DISPLACED BY CLIMATE CHANGE, http://www.ccdpconvention.com; Michel Prieur et al., *Draft Convention on the International Status of Environmentally-Displaced Persons* (Le Centre International de Droit Comparé de l'Environnement, Second Version, 2010), *available at* http://www.cidce.org/pdf/Draft%20Convention%20on%20the%20International%20Status%20on%20environmentally%20displaced%20persons%20%28second%20version%29.pdf. For proposals for regional responses, possibly under an international umbrella,

This chapter argues that there is more potential for using the domestic immigration laws of likely destination countries to respond to climate migration from threatened island nations than is commonly assumed. Although likely destination countries do not have a category of admission specifically for climate-displaced persons, some have immigration provisions that could be the basis for accommodating climate-related migration from some of the most threatened island nations. Policymakers should recognize the potential utility of building on the existing immigration policies of likely destination countries and think of the option of a new multilateral climate displacement treaty as a backstop rather than a top priority.[3]

The starting point for the argument in favor of building on existing national immigration policy is that there already is noticeable outmigration from Kiribati, Tuvalu, and the RMI, the three poorest of the four small island nations commonly thought to be most vulnerable to extinction due to climate change.[4] Moreover, outmigration mainly occurs in a relatively orderly way to two highly developed countries (New Zealand and the United States) that in combination with a third (Australia) could afford to increase the number of migrants they admit from Kiribati, Tuvalu, and the RMI.

As currently structured, the existing outmigration channels would not be sufficient to deal with large-scale migration from even Kiribati, Tuvalu and the RMI, let alone the Maldives or other countries.

see Angela Williams, *Turning the Tide: Recognizing Climate Change Refugees in International Law*, 30 Law & Pol'y 502 (2008); Benoît Mayer, *The International Legal Challenges of Climate-Induced Migration: Proposal for an International Legal Framework*, 22 Colo. J. Int'l Envtl. L. & Pol'y 357 (2011).

3 For other arguments against a climate displacement treaty, *see* Graeme Hugo, *Climate Change-Induced Mobility and the Existing Migration Regime in Asia and the Pacific*, *in* Climate Change and Displacement: Multidisciplinary Perspectives 9 (Jane McAdam ed., 2010), which also argues for examining the potential of existing migration policy to deal with climate change migration; Jane McAdam, *Swimming against the Tide: Why a Climate Change Displacement Treaty Is Not the Answer*, 23 Int'l J. Refugee L. 2 (2011).

4 On the relative wealth of the four threatened island nations, *see infra* note 23.

However, these channels could be expanded gradually to deal with the need to facilitate migration by large numbers of people if climate change requires it. Furthermore, there are important reasons for threatened island nations and likely destination countries to pursue the policy option of gradually expanding existing immigration channels.

This chapter focuses on using domestic immigration policy in likely destination countries to deal with climate migration from the most threatened small island nations. However, its point about the utility of national immigration policy also likely extends to cross-border climate migration from other countries. Consider Bangladesh, a country often mentioned as likely to generate large numbers of climate migrants, but not likely to become completely uninhabitable.[5] It is already one of the top source countries for international migrants.[6] Like the most threatened island nations, Bangladesh also may be well served by exploring the expansion of existing outmigration channels to deal with additional cross-border migration required by climate change.[7]

2. The Existing Outmigration

Discussions of the fate of Kiribati, the Maldives, the RMI, and Tuvalu in a warming world often overlook that three of these nations already have considerable experience with outmigration and that this experience could provide options for dealing with possible future climate migration.

Table 1 provides background population and migration information about Kiribati, the Maldives, the RMI, and Tuvalu. It indicates by country the total population, generally for 2009; the estimated net migration rate for 2011; and the leading destination country for migrants, where identified. Included in the column identifying the leading destination

5 Biermann & Boas, *supra* note 2, at 70 (Table 1: Climate Change Impacts by Region) (sea level rise estimated to create 26 million climate refugees in Bangladesh by 2050).

6 The World Bank, Migration and Remittances Factbook 2011, at 3 (2d ed. 2011).

7 McAdam, *supra* note 3, at 22–25 (Bangladesh's approach to climate change migration).

Table 1. *Population and Migration Data for Kiribati, Maldives, the RMI, and Tuvalu*

	Population (2009)	Net Migration Rate (2011)[8]	Leading Destination Country
Kiribati	98,000[9]	−2.85	Maybe New Zealand (1,116 i-Kiribati in New Zealand in 2006, measured by ancestry/ethnicity)[10]
Maldives	300,000[11]	−12.62	N/A
RMI	61,000[12]	−5.19	United States[13] (22,434 Marshallese in United States in 2010, measured by race)[14]
Tuvalu	9,730 (2008)[15]	−7.02	New Zealand[16] (2,625 Tuvaluans in New Zealand, measured by ancestry/ethnicity)[17]

country is an estimate of the diaspora population in that destination country, again where it is determinable.

Table 1 suggests several points. First, the population figures indicate that the four island nations commonly mentioned as existentially threatened by climate change have relatively small populations,

8 CIA, The World Factbook, https://www.cia.gov/library/publications/the-world-factbook.

9 The World Bank, Factbook, *supra* note 6, at 151.

10 *Bedford & Bedford*, *supra* note 21, at 89, 108 (Table 5.6: Kiribati and Tuvalu populations in Australia and New Zealand (2006) and approvals for residence (2004–2007)).

11 The World Bank, Factbook, *supra* note 6, at 171.

12 *Id.* at 174.

13 *See e.g.* Pac. Inst. of Pub. Policy, *The Micronesian Exodus* (Discussion Paper 16, 2010), *available at* http://wp.pacificpolicy.org/wp-content/uploads/2012/05/D16-PiPP.pdf.

14 Data on the RMI population in the United States come from the 2010 U.S. decennial census, as reported in U. S. Gov't Accountability Office, GAO-12–64, Compacts of Free Association: Improvements Needed to Assess and Address Growing Migration 65 (2011) (Appendix IV: 2010 Decennial Census Data on Micronesians by Race).

15 *Bedford & Bedford*, *supra* note 21, at 97 (Table 5.1: Populations of Kiribati and Tuvalu and their urban areas). The population estimate for Tuvalu is for 2008. *Id.*

16 *Id.* at 104.

17 *Id.* at 108.

totaling 468,730. This indicates that resettling the populations of these island nations would be a manageable proposition if it becomes necessary due to climate change.[18] By way of context, the United States accepted 1,107,100 permanent migrants in 2008, Australia accepted 205,900, and New Zealand accepted 51,700.[19] In other words, in 2008 alone, these three destination countries together admitted as permanent migrants almost three times the combined population of Kiribati, the Maldives, the RMI, and Tuvalu.

Second, the data in the table about net migration rates indicate that Kiribati, Maldives, the RMI, and Tuvalu are all currently losing population due to outmigration. The net migration rate is the number of migrants per 1,000 population, taking into account both persons migrating into and out of the country. A negative net migration rate, which the table indicates for all four countries, implies the country is losing population to migration after taking into account emigration and immigration.[20] There is considerable support from other sources that Kiribati, Tuvalu, and the RMI are experiencing net outmigration.[21] However, other sources provide strong reasons for doubting the data in the table that the Maldives is experiencing net negative migration.[22] In

18 *See also* Ross Garnaut, The Garnaut Climate Change Review, Final Report 149 (2008).

19 N.Z. Dep't of Labour, Int'l Migration, Settlement & Emp't Dynamics, Migration Trends & Outlook – 2009/2010, at 11 (2010) (Table 2.1: Comparison of immigration and population statistics in selected OECD countries).

20 For the CIA's definition of the net migration statistic, *see* CIA, The World Factbook, https://www.cia.gov/library/publications/the-world-factbook/rankorder/2112rank.html.

21 Richard Bedford & Graeme Hugo, N.Z. Dep't of Labour, Labour and Immigration Research Ctr., Population Movement in the Pacific: A Perspective on Future Prospects 14 (2012) (Table 5: Some Demographic Rates, Pacific Populations, 2000s); Brian Opeskin & Therese MacDermott, *Resources, Population and Migration in the Pacific: Connecting Islands and Rim*, 50 Asia Pac. Viewpoint 353, 359 (2009); Richard Bedford & Charlotte Bedford, *International Migration and Climate Change: A Post-Copenhagen Perspective on Options for Kiribati and Tuvalu, in* Climate Change and Migration: South Pacific Perspectives 89, 108 (Bruce Burson ed., 2010).

22 For example, the World Bank publishes data indicating that in 2009 remittances flowing out of the Maldives exceeded remittances coming into the country by US$113

fact, anomalous among the four island nations mentioned as existentially threatened by climate change, the Maldives is better considered a labor-receiving rather than a labor-sending country.[23] The lack of existing outmigration from the Maldives may provide an argument for the Maldives to take a strong position in favor of a new multilateral protection instrument, because the country does not seem to have any existing outmigration channels on which it could seek to expand if large-scale relocation becomes necessary due to climate change.[24] Another option for the Maldives, more consistent with the approach this chapter advocates, would be to negotiate bilateral outmigration arrangements for its citizens with possible destination countries, which the Maldives then could attempt to expand if its citizens must relocate due to climate change.[25]

Third, Table 1 provides information about the countries to which i-Kiribati, Marshallese, and Tuvaluans are migrating. It indicates that there are currently two leading destination countries for migrants from

million. THE WORLD BANK, FACTBOOK, *supra* note 6, at 171. The Maldives publishes statistics on expatriate employment in the Maldives that indicate that in 2009, 70,259 expatriates were employed in the Maldives, a very large number given the Maldives' population of roughly 300,000. REP. OF MALDIVES, DEP'T OF NAT'L PLANNING, STATISTICAL YEARBOOK OF MALDIVES 2010, *available at* http://planning.gov.mv/yearbook2010/yearbook.html (Table 5.2 Expatriate Employment by Industry and Nationality, 2009).

23 Deshal de Mel & Suwendrani Jayaratne, Inst. of Policy Studies Sri Lanka, *Migration, Remittances and Development Nexus in the Maldives* (Int'l Conference on Migration, Remittances and Dev. Nexus in S. Asia, May 4–5, 2009), *available at* http://www.ips.lk/events/workshops/4_5_09_development_nexus/session3.pdf. The Maldives likely is the wealthiest of the four small island nations commonly discussed as existentially threatened by climate change, which may help explain its labor-receiving status. Per capita gross national income in US$ in 2009 was $1,890 for Kiribati, $3,870 for the Maldives, and $3,060 for RMI. THE WORLD BANK, FACTBOOK, *supra* note 6, at 151, 171, 174. The World Bank does not report statistics for Tuvalu. As a rough proxy, a 2002 estimate of its per capita gross domestic product (PPP) was $1,600. THE WORLD BANK, FACTBOOK, *supra* note 6.

24 "[T]he Maldives in 2006 proposed amending the 1951 Refugee Convention to extend the definition of a 'refugee' in article 1A(2) to include 'climate refugees.'" *McAdam*, *supra* note 3, at 6.

25 In 2008, the then President of the Maldives, Mohammed Nasheed, proposed to respond to climate change by buying land in another country to resettle Maldivians. *Schmidle*, *supra* note 1.

these three countries: New Zealand and the United States. Currently New Zealand is home to the largest number of Tuvaluans outside of Tuvalu,[26] although a fairly sizable share of Tuvaluans in New Zealand were born in New Zealand.[27] When measured by ancestry or ethnicity, the population of Tuvaluans in New Zealand represents about 27% of the population living in Tuvalu, a sizable diaspora. Citizens of Kiribati seem less inclined than Tuvaluans to migrate to New Zealand, and there are fewer i-Kiribati in New Zealand in absolute terms than Tuvaluans.[28] Nonetheless, New Zealand currently may be home to the largest number of i-Kiribati outside of Kiribati. Measured by ancestry or ethnicity, the number of i-Kiribati in New Zealand amounts to approximately 1% of the population living in Kiribati. There also are significant communities of i-Kiribati in Fiji and the Solomon Islands due to mass relocations several decades ago.[29]

The United States is home to the largest community of Marshallese outside the RMI.[30] Numbering 22,434 people, the U.S.-based population of Marshallese is approximately 37% of the population in the RMI, also a sizable diaspora.

A word is in order about the reasons for the existing outmigration from Tuvalu, the RMI, and Kiribati. There is little evidence that people are currently leaving these island nations in large measure because of climate change. As a historical matter, migration is not a new phenomenon for Pacific Islands.[31] Tuvalu, Kiribati, and the RMI are not

26 *Bedford & Bedford*, *supra* note 21, at 97 (Table 5.1: Populations of Kiribati and Tuvalu and their urban areas).

27 *Id.* at 106 (Table 5.5: New Zealand's Tuvalu-born and Tuvaluan populations, 1986–2006).

28 *Id.* at 104.

29 John Campbell, *Climate-Induced Community Relocation in the Pacific: The Meaning and Importance of Land*, *in* CLIMATE CHANGE AND DISPLACEMENT: MULTIDISCIPLINARY PERSPECTIVES 57, 71–78 (Jane McAdam ed., 2010); BEDFORD & HUGO, *supra* note 21, at 28.

30 *See supra* note 23 and accompanying text.

31 MICHAEL J. LEVIN, U.S. DEP'T OF THE INTERIOR, OFFICE OF INSULAR AFFAIRS, THE STATUS OF MICRONESIAN MIGRANTS IN THE EARLY 21ST CENTURY: A SECOND STUDY OF THE IMPACT

economically prosperous, and significant economic development is unlikely in the future.[32] Evidence suggests that people are migrating because of the prospect of better employment and educational opportunities and access to higher quality health care in their destination countries, as well as networks of relatives and friends in these countries that can provide information and support.[33]

3. The Mechanisms

Many of the existing proposals for new multilateral instruments to provide climate-displaced persons with a right of refuge assume that people from the affected countries have little ability to resettle elsewhere under existing law. Analysts tend to discount the utility of existing law to deal with climate migration because destination countries do not usually allow persons to immigrate permanently because they have been displaced from their home countries for environmental reasons.[34]

of the Compacts of Free Association based on Censuses of Micronesian Migrants to Hawaii, Guam and the Commonwealth of the Northern Mariana Islands 6 (2007); *Bedford & Bedford*, *supra* note 21, at 93–107; *Opeskin & MacDermott*, *supra* note 21, at 353, 358–359, 369; Bedford & Hugo, *supra* note 21, at 24–25.

32 *McAdam*, *supra* note 3, at 8–9; U.S. Dep't of the Interior, Office of Insular Affairs, Report to Congress on the Compacts of Free Association with the Federated States of Micronesia (FSM) and the Republic of the Marshall Islands (RMI) for Fiscal Year 2006, *available at* http://www.doi.gov/oia/reports/PDF/CompactReport2006(OMB%20Final)8.24.07.pdf; Emil Friberg et al., *US Economic Assistance to Two Micronesian Nations: Aid Impact, Dependency and Migration*, 47 Asia Pac. Viewpoint 123 (2006); Opeskin & MacDermott, *supra* note 21.

33 Colette Mortreux & Jon Barnett, *Climate Change, Migration and Adaptation in Funafuti, Tuvalu*, 19 Global Envtl. Change 105 (2009); Levin, *supra* note 31, at 45; Bedford & Hugo, *supra* note 21, at 22, 31.

34 Susan Martin, Int'l Org. for Migration, Background Paper WMR 2010: Climate Change and International Migration 10 (2010), *available at* http://publications.iom.int/bookstore/free/WMR2010_climate_change_migration.pdf. However, Sweden's and Finland's immigration legislation might allow persons displaced by climate change to resettle permanently. *Id.* at 9. Briana Dema, *Sea Level Rise and the Freely Associated States: Addressing Environmental Migration under the Compacts of Free Association*, 37 Colum. J. of Envtl. L. 177, 197–98 (2012). National laws sometimes include temporary protection provisions. Martin, *supra* note 31, at 8–10.

Moreover, the definition of "refugee" in the international Refugee Convention, which many countries use in deciding whether to grant refugee status under their domestic immigration laws, likely will not cover most people who might be displaced by climate change.[35]

However, lost amid the generalizations about the inadequacy of existing law are relatively little-known policies in the United States, New Zealand, and, to a lesser extent, Australia that currently enable citizens of small island nations to migrate. Although inadequate as currently structured to deal with climate migration, these provisions could and should be expanded to accommodate such migration.

3.1. Analytical Framework

To evaluate the potential of immigration policy in likely destination countries to adequately deal with climate change migration, it is necessary to have a sense of what climate change migration might require of that policy. As an initial matter, for national immigration policy to accommodate climate migration, it will need to have at least two characteristics: (1) enable large numbers of citizens of threatened island nations to gain admittance to destination countries and (2) permit permanent, not just temporary, migration.

For national immigration policy to respond to climate migration, the policy of one or of a group of likely destination countries must allow large numbers of citizens of threatened island nations to gain admittance, although not necessary all at once. For the benefit of migrants as well as the receiving countries, it is hoped that climate migration would occur gradually to facilitate planning and adaptation.[36] No one destination country's immigration policy has to be sufficiently capacious on its own to admit all the citizens of all the threatened island nations.

35 Jane McAdam & Ben Saul, *An Insecure Climate for Human Security? Climate-Induced Displacement and International Law*, *in* Human Security and Non-Citizens: Law, Policy and International Affairs 357 (Alice Edwards & Carla Fertsman eds., 2010).

36 *Bedford & Bedford*, *supra* note 21, at 125.

The threatened island nations and likely destination countries could divide up responsibility among likely destination countries for migration from threatened island nations. Alternatively, such a division of responsibility could emerge informally from the bottom up by likely destination countries working bilaterally with specific threatened island nations to improve migration opportunities.

For large numbers of citizens to be able to migrate from threatened island nations, immigration policies in likely destination countries collectively must not numerically cap migration from the threatened island nations at levels insufficient to permit the gradual absorption of their populations, taking into account population growth. In addition, national immigration policy in likely destination countries must not implicitly restrict migration from threatened island nations by limiting migration opportunities to skilled labor. Facilitating the flow of unskilled labor is vital because many Pacific islanders lack the skills necessary to migrate as skilled labor.[37] National immigration policy in likely destination countries also would need to allow for the entry of persons from threatened island nations unable to participate in the workforce, such as elderly and young persons, and persons with severe disabilities.[38]

Provided the national immigration policies of likely destination countries collectively permit the absorption of the populations of the threatened island nations, there is room for destination countries to pursue different policy approaches. For example, individual destination countries could numerically cap the number of persons they admit from individual threatened island nations, providing again that the caps do not frustrate the ultimate goal of absorbing large numbers of citizens of threatened island nations. Destination countries also could pursue different strategies in relation to skilled and unskilled labor, again provided that the

37 Charles W. Stahl & Reginald T. Appleyard, Austl. Agency for Int'l Dev., Migration and Development in the Pacific Islands: Lessons from the New Zealand Experience (2007), *available at* http://www.ausaid.gov.au/Publications/Documents/migration.pdf.

38 Martin, *supra* note 31, at 12 (criticizing New Zealand's Pacific Access Category).

strategies collectively permit the absorption of large numbers of citizens of threatened island nations. New Zealand and the United States historically and to the present have offered preferential immigration mechanisms that facilitate the flow of unskilled labor from the Pacific islands. Historically Australia has not offered preferential policies for unskilled Pacific islanders, although it recently established a temporary horticultural worker program for Pacific islanders.[39]

A second point to consider in assessing the potential of immigration policy in likely destination countries to adequately handle climate migration is whether that policy facilitates permanent immigration and the acquisition of citizenship. If climate change renders island nations uninhabitable, their citizens will need to permanently migrate to other countries, which will require permanent migration options. Such options will need to provide the opportunity to live elsewhere in perpetuity and likely to acquire citizenship in the destination country. Climate migrants might need to acquire the citizenship of the destination country because they could become stateless persons if their island nations cease to be considered States under international law due to uninhabitability.[40]

Notwithstanding the need for permanent migration options to deal with climate migration, temporary migration programs nonetheless could play a role in responding to climate migration. Temporary

39 On the historical policy differences between Australia, and New Zealand and the United States, *see e.g.*, Stahl & Appleyard, *supra* note 37, at v, 21–22; Bedford & Hugo, *supra* note 21, at 52, 54; E.J. Moore & W. Smith, *Climatic Change and Migration from Oceania: Implications for Australia, New Zealand and the United States of America*, 17 Population & Env't: J. Interdisc. Stud. 105, 109–10 (1995); *Opeskin & MacDermott*, *supra* note 21, at 365–67.

40 However, there are arguments that island nations might continue to be recognized as States in international law even if they are uninhabitable. Jane McAdam, *'Disappearing States', Statelessness and the Boundaries of International Law*, *in* Climate Change and Displacement: Multidisciplinary Perspectives 105 (Jane McAdam ed., 2010); Jenny Grote Stoutenberg, *When Do States Disappear? Thresholds of Effective Statehood and the Continued Recognition of "Deterritorialized" Island States* (see Chapter 3 in this volume).

migration programs could help migrants acquire skills and contacts that may enable them to move permanently to the country where they work temporarily. Past experience in New Zealand suggests that temporary work can be a path to permanent residence for Pacific islanders. A noticeable share of i-Kiribati and Tuvaluans who have acquired New Zealand permanent residency first came to the country on temporary work permits.[41] For temporary migration programs to serve as a path to legal permanent residence in the destination country, however, that country must provide persons who initially come temporarily with a legal route to permanent residence.[42]

In addition to the two characteristics outlined earlier, it would be desirable for national immigration policy in likely destination countries to have other features. For example, persons who need to relocate due to climate change may require financial help for resettlement, and therefore it would be desirable for the mechanisms under which they migrate to include financial assistance.[43] Such funding might not be a minimum requirement, however, because there may be sources of funding other than the destination countries, such as the migrants' home governments, nongovernmental organizations, or donor agencies from States other than the destination or sending countries. Another feature that could be even more desirable to incorporate into migration mechanisms is guaranteed access to necessary health and social services in the destination

41 Bedford & Hugo, *supra* note 21, at 67–68.

42 For destination countries, there is always a risk that temporary worker programs could facilitate illegal permanent migration. Ultimately, persons who illegally overstay may become legal permanent residents under legalization programs. Donald M. Kerwin, Migration Policy Inst., More Than IRCA: US Legalization Programs and the Current Policy Debate (2010). New Zealand previously has passed legislation enabling overstayers to become lawful permanent residents. Stahl & Appleyard, *supra* note 37, at 24. New Zealand's Recognised Seasonal Employer policy, discussed later, was designed to limit the potential for overstaying and appears to be successful in doing so. N.Z. Dep't of Labour, Int'l Migration, Settlement & Emp't Dynamics, Final Evaluation Report of the Recognised Seasonal Employer Policy (2007–2009) xiv (2010).

43 *Dema*, *supra* note 34, at 199.

countries.[44] It also would be desirable for migration mechanisms to be embedded in a legal framework that makes them difficult for political actors in destination countries to cut back easily in response to anti-immigration sentiment. Migration mechanisms embedded in legislation, or better yet a combination of legislation and a bilateral agreement, likely will be less susceptible to anti-immigration politics than mechanisms grounded in policy that the executive can change unilaterally.[45] Overall, it is best to consider the two requirements identified earlier as the barebones requirements for national immigration policy in likely

44 Jon Barnett & Michael Webber, *Accommodating Migration to Promote Adaptation to Climate Change* 25 (The World Bank, Dev. Econ., Office of the Senior Vice President and Chief Economist, Background Paper to the 2010 World Dev. Report, Policy Research Working Paper 5270, 2010) (refugees and internally displaced persons "are able to pursue the livelihood strategies that best suit their skills and values when their entitlements are maximized, and outcomes are most equitable when those entitlements are equal to their local hosts"). Marshallese migrants to the United States likely have been adversely affected by their ineligibility for Medicaid since the Personal Responsibility and Work Opportunity Reconciliation Act of 1996, Pub. L. No. 104–193, 110 Stat 2105 (1996). U.S. General Accounting Office, GAO-02–40, Migration from Micronesian Nations Has Had Significant Impact on Guam, Hawaii, and the Commonwealth of the Northern Mariana Islands (2001).

45 However, even immigration provisions embedded in legislation and bilateral agreements could be vulnerable to shifting political winds in destination countries. For example, RMI's immigration privileges with the United States are embedded in a bilateral agreement and U.S. legislation, Compact of Free Association Amendments Act of 2003, Pub. L. No. 108–188, 117 Stat. 2720 (2003). However, under the terms of the COFA, it could be terminated unilaterally by the United States, the United States could regulate the immigration privileges, and the United States could administratively reduce migration from the RMI through exclusions and deportations. *See also Dema*, *supra* note 34, at 186–189, 199–201; *Friberg et al.*, *supra* note 32, at 131. Recently, three members of Congress and six senators wrote to Secretary of State Hillary Clinton and Secretary of the Interior Kenneth Salazar expressing concerns about the costs of providing health care and other social services to migrants from the RMI and other freely associated States. Letter from Lisa Murkowski, U.S. Senator, et al. to the Honorable Hillary Rodham Clinton and the Honorable Kenneth L. Salazar, May 12, 2011 (on file with author). The costs of migration from the freely associated States are analyzed in U.S. Gov't Accountability Office, *supra* note 44. Efforts to restrict migration from the freely associated States are reported in Giff Johnson, *Inouye: US Entry Screening a Must*, Marshall Islands J., Aug. 5, 2011, at 2; *US to Limit Entry?*, Marshall Islands J., Sept. 16, 2011, at 1.

destination countries to facilitate climate change migration, not an exhaustive list of the features that national immigration policy should include to respond adequately to climate change migration.

3.2. Analysis

Analysis of the mechanisms through which i-Kiribati, Marshallese, and Tuvaluans currently migrate to New Zealand, the United States, and, to a much lesser extent, Australia, suggests that the national immigration policies of these likely destination countries could, with changes, be used to permanently resettle large numbers of citizens of the three threatened island nations. Current practice suggests a possible allocation of responsibility for resettling the populations of Kiribati, Tuvalu, and the RMI under which the United States would have responsibility for resettling the population of the RMI, New Zealand for Tuvalu, and New Zealand and Australia for the population of Kiribati. Presently, however, the three threatened island nations are differentially positioned to deal with the possible need to resettle large numbers of their citizens due to climate change.

As mentioned earlier, the United States currently is the principal destination country for Marshallese migrants, reflecting the two countries' historical and ongoing ties. The RMI is very well positioned, relative to Kiribati and Tuvalu, to respond to any climate-related need to relocate permanently large numbers of its citizens because of the immigration privileges its citizens enjoy under the country's Compact of Free Association (COFA) with the United States.[46] Marshallese entering under

46 On the history of the COFA, *see, e.g.*, Chimène I. Keitner & W. Michael Reisman, *Free Association: The United States Experience*, 39 TEX. INT'L L.J. 1, 33–38, 45 (2003). Palau and the Federated States of Micronesia (FSM) have similar immigration privileges under COFAs with the United States. Compact of Free Association Amendments Act of 2003, Pub. L. No. 108–188, 117 Stat. 2720 (2003) (Federated States of Micronesia); Compact of Free Association, Pub. L. No. 99–658, 100 Stat. 3672 (1986) (Palau). For a perceptive analysis of the adequacy of the RMI's immigration privileges under the COFA for dealing with climate change displacement, *see Dema*, *supra* note 34.

these privileges enjoy a special status in U.S. immigration law, with more privileges than standard nonimmigrants but fewer than permanent resident aliens.[47]

These privileges enable Marshallese to enter the United States and its territories "to lawfully engage in occupations, and establish residence as a nonimmigrant."[48] Unlike many other foreigners entering the United States for work, RMI citizens do not have to meet labor certification requirements and do not require a visa to enter.[49] Marshallese admitted under the COFA provisions are entitled to work legally in the United States,[50] and there is no limit on the time that they can remain in the United States.

Persons can enter regardless of their skill level. Indeed there is no requirement that RMI citizens have an offer of employment or even be employable to rely on the immigration provisions. An individual admitted potentially could be deported if he or she "cannot show that he or she has sufficient means of support in the United States,"[51] but a summary of the immigration provisions distributed by the Compact Negotiations Office of the RMI states that "execution of an affidavit of support by a family member or other person with sufficient funds will prevent the RMI citizen from being removed for 'public charge' reasons."[52]

Provided the privileges remain largely in their current form, the provisions would enable large numbers of Marshallese to relocate to the United States permanently if climate change requires it.[53] The

47 GAO, *supra* note 44, at 8.

48 COFA of 2003 § 141(a). The eligibility requirements for claiming the privileges are slightly more complicated than mentioned earlier. *Dema*, *supra* note 34, at 186

49 COFA of 2003 § 141(a).

50 *Id.* § 141(d).

51 *Id.* § 141(f)(1).

52 Rep. of the Marshall Islands, Ministry of Foreign Affairs, Compact Negotiations Office, A Brief Summary of the Immigration Provisions Under the Compact of Free Association, As Amended 2, *available at* http://www.rmiembassyus.org/Compact/Issue%20Paper%20Immigration%20Summary.doc (Immigration Summary).

53 On the possible threats to the immigration privileges in the COFA, *see supra* note 45.

provisions, however, do not create a special path to U.S. citizenship for Marshallese migrants, although Marshallese can become U.S. citizens through the regular U.S. immigration streams.[54] Presumably because of the difficulty of acquiring U.S. citizenship through the regular immigration streams, few Marshallese migrants in the United States are citizens,[55] even though many regard themselves as living permanently in the United States.[56] The U.S.-born children of Marshallese migrants are U.S. citizens, however, because the United States grants citizenship based on birth in the United States regardless of the immigration status of the parents.[57] There are different ways that the lack of a special path to U.S. citizenship for Marshallese migrants could be addressed. The RMI might seek to negotiate a new political relationship with the United States that includes U.S. citizenship for Marshallese. Alternatively, without changes to the COFA, through special country-specific legislation, the United States might grant permanent resident status and a path to citizenship to Marshallese migrants within the United States as of a certain date, essentially on humanitarian grounds. There are precedents for country-specific legislative grants of permanent resident status and, through it citizenship, for groups of migrants who arrived in the United States in other capacities, although these ad hoc legislative measures tend to be rare and controversial.[58]

As mentioned earlier, New Zealand is the principal destination for Tuvaluans and perhaps also for i-Kiribati. New Zealand currently has a preferential immigration policy – the Pacific Access Category

54 COFA of 2003 § 141(h); *see also Dema*, *supra* note 34, at 188–189.

55 Levin, *supra* note 31, at 145 (according to 2000 U.S. decennial status, about 25% of the 6,320 Marshallese by birth in the United States were U.S. citizens).

56 In a 2003 survey of RMI post-Compact migrants in Hawaii, "73 percent" of Marshallese "self-defined themselves as permanent residents," referring to their intention to remain, rather than their legal status in the United States. *Id.* at 45.

57 Stephen H. Legomsky & Cristina M. Rodríguez, Immigration and Refugee Law and Policy 1290–1294 (5th ed. 2009).

58 *Id.* at 167–168, 252, 1178–1179 (Cubans, Haitians, Mexicans); *see also* Kerwin, *supra* note 32.

(PAC) – for Tuvaluans and i-Kiribati to gain permanent residence status and citizenship.[59] The PAC "allows up to 75 citizens of Kiribati, 75 citizens of Tuvalu and 250 citizens of Tonga to be granted residence in New Zealand each year."[60] Applicants are chosen by ballot and must meet a number of requirements to qualify for residence, including being age 18 to 45, having "an acceptable offer of employment" or a partner with such an offer, and satisfying "minimum income," English-language proficiency, and "health and character" requirements.[61] Successful applicants may bring partners and dependent children, who also count against the country quotas.[62] The PAC is the main channel through which citizens of Kiribati and Tuvalu currently obtain permanent residence in New Zealand, because they usually lack the skills to enter as skilled migrants or the eligibility for family sponsorship.[63]

For the PAC to become a vehicle for climate migration, the requirements for entering under it would need to be addressed. Currently, the employment requirements in particular prevent some i-Kiribati and Tuvaluans who are selected through the ballot from taking up the option of New Zealand residence under PAC.[64] One possibility would be to forge stronger links between the PAC and New Zealand's seasonal worker program (the Recognised Seasonal Employer policy, discussed later). The idea is that employers of i-Kiribati and Tuvaluans entering for

59 Under the Trans-Tasmanian Agreement, New Zealand citizens, including persons who originally came under the PAC, are eligible to work in Australia. Email from David Hammond, Acting Manager, Gov't Exec. and Ministerial Support, N.Z. Dep't of Labour, to author (June 14, 2011) (on file with author).

60 N.Z. Dep't of Labour, Immigration N.Z., *Summary of Terms, Quota under Pacific Access Category*, http://www.immigration.govt.nz/migrant/stream/live/pacificaccess (number of citizens window).

61 N.Z. Dep't of Labour, Immigration N.Z., *How Do I Qualify for Residence under the 2012 Pacific Access Category?*, http://www.immigration.govt.nz/migrant/stream/live/pacificaccess/residence/default.htm.

62 Hammond, *supra* note 59.

63 *Bedford & Bedford*, *supra* note 21, at 108–9.

64 Stahl & Appleyard, *supra* note 37, at 31–32; *Bedford & Bedford*, *supra* note 21, at 126; N.Z. Dep't of Labour, Migration Trends & Outlook, *supra* note 19, at 49.

temporary work might provide permanent job offers that could enable i-Kiribati and Tuvaluans to qualify for the PAC.[65]

In addition, for the PAC to be used to absorb large numbers of Tuvaluans and i-Kiribati, the annual quota for Tuvaluans and i-Kiribati would need to be increased. Richard Bedford and Charlotte Bedford raise the prospect of New Zealand gradually increasing the PAC quotas for Tuvalu and Kiribati to facilitate emigration from these countries.[66] They also offer an analysis that sheds light on the actual size of the increases that might be necessary to reduce the population of Tuvalu and Kiribati, taking into account expected population growth.[67] Bedford and Bedford argue that New Zealand alone could absorb the population of Tuvalu or at least a significant proportion of it without significantly affecting New Zealand immigration policy.[68] The current population of Kiribati (98,000) is much larger than that of Tuvalu (9,730), and therefore New Zealand likely would require assistance from Australia to absorb the population of Kiribati without significant changes to New Zealand's immigration flow.[69] Increasing the quotas for Tuvaluans and i-Kiribati and changing the PAC in other ways to facilitate additional migration would require a political will in New Zealand that has been lacking to date, but that might emerge as the effects of climate change become more tangible in the Pacific.[70]

It is worth noting that Australia currently is funding a scholarship program that may facilitate migration from Kiribati through standard

65 *Bedford & Bedford*, *supra* note 21, at 126–27.

66 *Id.* at 127.

67 *Id.* at 113–23.

68 *Id.* at 115, 128.

69 *Id.* at 122. Bedford and Bedford state that the immigration levels required to gradually reduce the population of Kiribati "will pose a challenge for neighbouring countries that might receive the migrants," implying that assistance might be required from countries in addition to Australia and New Zealand. *Id.*

70 *Id.* at 115, 128; N.Z. Ministry of Foreign Affairs and Trade, *New Zealand's Immigration Relationship with Tuvalu*, http://www.mfat.govt.nz/Foreign-Relations/Pacific/NZ-Tuvalu-immigration.php.

immigration streams.[71] Funded by AusAid, the Australian foreign aid agency, the Kiribati-Australia Nursing Initiative (KANI) is enabling a small number of i-Kiribati to train in Australia to become nurses.[72] Funding is available under KANI for up to ninety students, and seventy-five currently are enrolled at Griffith University in Australia.[73] Although KANI itself does not include any immigration privileges, it provides enrollees with skills that may enable them to acquire permanent residence in Australia or elsewhere. To significantly assist in the absorption of i-Kiribati, however, Australia would need to fund the upgrading of the skills of many more i-Kiribati than the number funded by KANI.[74] It also probably would have to establish a preferential admissions policy such as New Zealand's PAC, because skills upgrading is unlikely to be a feasible means of facilitating migration of very large numbers of i-Kiribati, due to the large expenditures that would be required. Additional funding for skills upgrading and a preferential policy would require political will in Australia, an ingredient that has not been evident to date but

71 Stahl & Appleyard, *supra* note 37, at vii, distinguish two strategies for promoting migration from Pacific Islands: training Pacific Islanders to acquire sufficient credentials to gain admittance under standard immigration provisions, and offering opportunities for unskilled workers.

72 On KANI, *see* Austl. Agency for Int'l Dev., *Kiribati*, http://www.ausaid.gov.au/countries/pacific/kiribati/Pages/home.aspx; Press Release, Solveig Roeyrvik, Asia-Pac. Assoc. for Int'l Educ., Educational Life-line – The Kiribati Australia Nursing Initiative (June 23, 2010), http://www.apaie.org/resources/press_read.php?uid=109&page=1; Brian Reed, *Preparing for Sea Level Rise, Islanders Leave Home*, Nat'l Pub. Radio, Feb. 17, 2011, http://www.npr.org/2011/02/17/133681251/preparing-for-sea-level-rise-islanders-leave-home.

73 Email from Yadira Galarza, Commc'ns Officer, Media Section, Austl. Agency for Int'l Dev., to author (Jun. 7, 2011) (on file with author).

74 The Lowy Institute for International Policy in Australia has recommended an expanded effort to facilitate migration from Kiribati and Tuvalu to Australia that would start with "increased numbers of scholarships" being made available to Tuvaluans and i-Kiribati "to study in Australia." Austrl. Senate, Foreign Affairs, Defence and Trade References Comm., Volume II: Security Challenges Facing Papua New Guinea and the Island States of the Southwest Pacific 105 (2010) (quoting the Lowy Institute for International Policy).

that also might emerge as the effects of climate change become more tangible.[75]

Given the possibility for temporary migration programs to assist with climate change migration, it is worth recognizing that New Zealand, Australia, and the United States each have temporary worker programs open to citizens of some or all of Kiribati, the RMI, and Tuvalu, among other countries. New Zealand's Recognised Employer (RSE) policy admits workers to work in orchards and vineyards.[76] It is open to citizens of Kiribati, the RMI, and Tuvalu, among other countries. Tuvaluans and i-Kiribati also are eligible to work in Australia under the Seasonal Worker Program announced in 2011.[77] A horticultural worker program open to citizens of nine countries, it replaces Australia's Pacific Seasonal Worker Pilot Scheme (PSWPS), which was modeled on New Zealand's

75 The idea of Australia establishing a preferential immigration quota for Pacific islanders has been proposed in the past but not adopted to date. ADRIENNE MILLBANK, PARLIAMENT OF AUSTL., DEP'T OF PARLIAMENTARY SERVS., A SEASONAL GUEST-WORKER PROGRAM FOR AUSTRALIA? 5–8 (2006); *Moore & Smith*, *supra* note 39, at 109–110; *Opeskin & MacDermott*, *supra* note 21, at 366.

76 The RSE policy was implemented in 2007. On the RSE, *see* Sankar Ramasamy et al., *The Recognised Seasonal Employer Policy: Seeking the Elusive Triple Wins for Development through International Migration*, PAC. ECON. BULLETIN, no. 3, 2008, at 171; N.Z. DEP'T OF LABOUR, FINAL EVALUATION REPORT, *supra* note 42; David McKenzie & John Gibson, *The Development Impact of a Best Practice Seasonal Worker Policy* (The World Bank, Dev. Research Grp., Fin. and Private Sector Dev. Team, Policy Research Working Paper 5488, 2010), *available at* http://www-wds.worldbank.org/external/default/WDSContentServer/WDSP/IB/2010/11/30/0001583 49_20101130131212/Rendered/PDF/WPS5488.pdf; Rochelle Ball, *Australia's Pacific Seasonal Worker Pilot Scheme and its Interface with the Australian Horticultural Labour Market: Is It Time to Refine the Policy?*, PAC. ECON. BULLETIN, no. 1, 2010, at 114; N.Z. DEP'T OF LABOUR, IMMIGRATION N.Z., *Recognised Seasonal Employers (RSE) Work Category*, http://www.immigration.govt.nz/migrant/stream/work/hortandvit/rse/default.htm.

77 AUSTL. DEP'T OF EDUC., EMP'T AND WORKPLACE RELATIONS, *Pacific Seasonal Worker Pilot Scheme Home*, http://www.deewr.gov.au/Employment/Programs/PSWPS/Pages/default.aspx; Press Release, Ministers' Media Ctr., Educ., Emp't and Workplace Relations portfolio, Pacific and East Timor Workers Helping Australian Farmers and Tourism Industry (Dec. 18, 2011), http://ministers.deewr.gov.au/shorten/pacific-and-east-timor-workers-helping-australian-farmers-and-tourism-industry.

RSE policy.[78] Since 2011, the U.S. H-2A agricultural worker and H-2B nonagricultural worker programs also have been open to citizens of Kiribati and Tuvalu, as well as many other countries.[79] For the New Zealand, Australian, and U.S. programs to become part of the policy toolbox for dealing with climate migration, the numbers of citizens of threatened island nations working under the programs would need to be higher to enable more people to enter and develop the networks that facilitate the transition to permanent residence. Few i-Kiribati and Tuvaluans and no Marshallese have worked in New Zealand under the RSE policy.[80] Few i-Kiribati worked in Australia under the PSWPS.[81] No i-Kiribati

78 The PSWPS was implemented in 2008 as a pilot project to last until 2012. On the PSWPS, *see* AUSTL. DEP'T OF IMMIGRATION AND CITIZENSHIP, FACT SHEET: THE PACIFIC SEASONAL WORKER PILOT SCHEME, *available at* http://www.immi.gov.au/skilled/pacific-seasonal-worker/_pdf/factsheet-pswps.pdf; *Ball*, *supra* note 76; TNS SOCIAL RESEARCH, INTERIM EVALUATION OF THE PACIFIC SEASONAL WORKER PILOT SCHEME, EXECUTIVE SUMMARY, PREPARED FOR AUSTL. DEP'T OF EDUC., EMP'T AND WORKPLACE RELATIONS (2010), *available at* http://www.deewr.gov.au/Employment/Programs/PSWPS/Documents/PSWPSEvaluationExecSummary.pdf.

79 Press Release, U.S. Citizenship and Immigration Servs., New Countries Eligible to Participate in H-2A and H-2B Programs (Jan. 14, 2011), http://www.uscis.gov/portal/site/uscis/menuitem.5af9bb95919f35e66f614176543f6d1a/?vgnextoid=88404e9c7c08d210VgnVCM100000082ca60aRCRD&vgnextchannel=3381c0ed71f85110VgnVCM1000004718190aRCRD; Danielle Hay, *New Pacific Seasonal Workers Scheme*, DEV. POL'Y BLOG (Jan. 31, 2011), http://devpolicy.org/new-pacific-seasonal-workers-scheme.

80 N.Z. DEP'T OF LABOUR, FINAL EVALUATION REPORT, *supra* note 42, at 84 (Appendix C: Number of Recognised Seasonal Employer Workers in First Two Seasons: Table 23: Number of Recognised Seasonal Employer (RSE) workers in first seasons); N.Z. DEP'T OF LABOUR, *RSE Workers Arrival Into New Zealand, RSE Financial Year 2009/2010*, http://dol.govt.nz/initiatives/strategy/rse/information/arrival-2009–2010.asp. The New Zealand government's evaluation of the RSE policy discusses the low numbers of workers from Kiribati and Tuvalu and efforts to increase the numbers of workers from these countries. N.Z. DEP'T OF LABOUR, FINAL EVALUATION REPORT, *supra* note 42, at 51–52. The number of workers from the two countries has increased. BEDFORD & HUGO, *supra* note 21, at 70–72.

81 AUSTL. DEP'T OF IMMIGRATION AND CITIZENSHIP, *supra* note 78, at 2 (Table 1: Special Program (subclass 416) – Pacific seasonal worker visas granted). Tuvalu was not included in the PSWPS until the fall of 2011. *Id.* The lack of uptake under the PSWPS generally has been the subject of commentary. *Ball*, *supra* note 76; TNS SOCIAL RESEARCH, *supra* note 78. After an evaluation of the program, Australia made

and Tuvaluans had worked in the United States under the H-2A or H-2B programs as of June 29, 2011.[82] In addition, the options for legally transitioning to permanent residence in the destination countries would have to be expanded. As mentioned earlier, New Zealand's PAC has tight caps on the number of eligible i-Kiribati and Tuvaluans. Neither Australia nor the United States has a preferential policy for admitting i-Kiribati and Tuvaluans.

4. The Case for Prioritizing Bilateral Responses

There are important reasons for threatened island nations and likely destination countries to improve the existing outmigration channels and develop new mechanisms, while keeping the option of a new multilateral instrument covering climate migration on the back burner.

Turning first to island nations' perspectives, although the literature on climate migration tends to regard migration as undesirable, there is an alternative literature on the relationship between migration and development that tends to portray migration as potentially beneficial for sending countries (as well as receiving countries and the migrants themselves).[83]

changes to the PSWPS to increase the number of workers coming under the program. AUSTL. DEP'T OF EDUC., EMP'T AND WORKPLACE RELATIONS, FACTSHEET: IMPORTANT CHANGES TO THE PILOT PROVIDE MORE FLEXIBILITY, *available at* http://www.deewr.gov.au/Employment/Programs/PSWPS/Documents/PSWPSFactsheet.pdf; Nic Maclellan, *Australia Revamps Troubled Scheme*, DEV. POL'Y BLOG (Feb. 15, 2011), http://devpolicy.org/australia-revamps-troubled-scheme.

82 Personal communication from U.S. Dep't of State, Visa Issuance Div., to Rachel Jones (June 29, 2011) (on file with author).

83 *Hugo*, *supra* note 3, at 29; BEDFORD & HUGO, *supra* note 21, at 6; *Barnett & Webber*, *supra* note 44, at 3, 19, 22. Useful introductions to this literature include JORGEN CARLING, PEACE RESEARCH INST. OSLO, MIGRANT REMITTANCES AND DEVELOPMENT COOPERATION (2005); THE WORLD BANK, GLOBAL ECONOMIC PROSPECTS: ECONOMIC IMPLICATIONS OF REMITTANCES AND MIGRATION (2006); *Barnett & Webber*, *supra* note 44, 22–26; AARON TERRAZAS, MIGRATION POLICY INST., MIGRATION AND DEVELOPMENT: POLICY PERSPECTIVES FROM THE UNITED STATES (2011), *available at* http://www.migrationpolicy.org/pubs/migdevpolicy-2011.pdf. Sources discussing the impact of migration in the Pacific include JOHN CONNELL & RICHARD P.C. BROWN, ASIAN DEV. BANK, REMITTANCES IN THE PACIFIC: AN OVERVIEW (2005), *available at*

This literature emphasizes that the consequences of migration are not uniform and vary depending on an array of factors.[84] Research suggests that, for countries of origin, international migration may yield numerous economic benefits, principally through the remittances that migrants may send back.[85] Remittances may help reduce poverty, "smooth household consumption," "diversify" household incomes, and provide funds that can be used to invest in education and health and thereby improve human capital.[86] If migration opportunities are available to low-income earners, remittances also may promote greater socioeconomic equality,[87] although the evidence about the impact of remittances "on inequality is mixed."[88] In addition to the benefits from remittances, international migration of low-skilled workers may have favorable labor market impacts, such as lowering the unemployment rate for remaining low-skilled workers and potentially boosting the incomes of the diminished pool of workers.[89] Migration and the attendant remittances and

http://apgml.org/frameworks/docs/8/ADB_remittances-pacific_Mar2005.pdf; The World Bank, At Home & Away: Expanding Job Opportunities For Pacific Islanders through Labour Mobility (2006); Bedford & Hugo, *supra* note 21, at 6–8, 32–36.

84 Carling, *supra* note 83, at 1; The World Bank, Global Economic Prospects, *supra* note 83, at xi, 57; Terrazas, *supra* note 93, at 8; *Barnett & Webber*, *supra* note 44, at 30.

85 The World Bank, Global Economic Prospects, *supra* note 83, at xiii; *Barnett & Webber*, *supra* note 44, at 22. For origin countries, remittances also may have macroeconomic effects, which are not discussed in this chapter. The World Bank, Global Economic Prospects, *supra* note 83, at 99–105.

86 The World Bank, Global Economic Prospects *supra* note 83, at xiii; *id.* at 117–27; Carling, *supra* note 83, at 30–31; *Barnett & Webber*, *supra* note 44, at 22. For analysis of the benefits of remittances for countries of origin focusing on Pacific island nations, *see* Connell & Brown, *supra* note 83; The World Bank, At Home & Away, *supra* note 83.

87 The World Bank, At Home & Away, *supra* note 83, at ix; Carling, *supra* note 83, at 31.

88 The World Bank, Global Economic Prospects, *supra* note 83, at 118; *id.* at 121–22; Connell & Brown, *supra* note 83, at 36–37; Carling, *supra* note 83, at 31.

89 The World Bank, Global Economic Prospects, *supra* note 83, at xiv, 64. Remittances also may create employment for persons who remain by providing a source of funds for activities such as construction. Connell & Brown, *supra* note 83, at 1.

labor market impacts may be particularly helpful for Pacific small island nations given their limited domestic development potential, due to factors such as limited natural resource endowments, small domestic markets, and distance to larger markets.[90]

Kiribati, Tuvalu, and the RMI are well aware of the potential benefits of migration, quite apart from as a mechanism for dealing with climate change. "Kiribati and Tuvalu and to a lesser extent the Marshall Islands . . . have actively encouraged international labor migration."[91] Kiribati and Tuvalu are "partially remittance-dependent economies."[92] In contrast, remittances to the RMI from abroad are minimal,[93] and anomalously among the three countries, remittances from the Marshallese in the RMI to migrants abroad may exceed remittances from migrants back to the RMI.[94] However, remittances from Marshallese in the United States might increase in the future as migrants become more established there.[95]

In addition to the benefits that expanding existing migration generally may afford small island nations, expanding these options offers such nations a mechanism for dealing with climate change, a point

90 Connell & Brown, *supra* note 83, at 8; *Opeskin & MacDermott*, *supra* note 21, at 354–361.

91 Connell & Brown, *supra* note 83, at 43–44; *id.* at 40. There also may be arguments for greater outmigration from the Maldives, irrespective of climate change. de Mel & *Jayaratne*, *supra* note 23.

92 Connell & Brown, *supra* note 83, at 10. Bedford & Hugo, *supra* note 21, at 32, 34, and 35 (provides data on the importance of remittances for Kiribati and Tuvalu).

93 Bedford & Hugo, *supra* note 21, at 35 (Table 10: Contribution of trade, aid and remittances to GDP in Pacific countries, 2003) (remittances represented 0.6% of RMI GDP in 2003).

94 Michael R. Ogden, MIRAB and the Marshall Islands, 2 *ISLAND J. Micr. Stud.* 237, 261 (1994); Jim Hess, *Wave and Reflection: Charting Marshallese Participation in Globalizing Processes*, *in* Globalization and Culture Change in the Pacific Islands 182, 193 (Victoria S. Lockwood ed., 2003).

95 *Ogden*, *supra* note 94, at 263. For possible explanations of the Marshallese pattern of remittances, some of which cast doubt on the idea that remittances will increase as Marshallese in the United States become more established, *see Ogden*, *supra* note 94, at 259; *Hess*, *supra* note 94, at 193.

not lost on national governments.[96] Kiribati and Tuvalu have sought additional migration opportunities to New Zealand and Australia specifically to help deal with climate change.[97] Gradually increasing migration would enable threatened island nations to expand their existing diasporas in likely destination countries. These diasporas in turn could facilitate the resettlement of subsequent migrants using the knowledge, experience, and assets that they have acquired in the destination countries.[98] As a legal matter, earlier generations of migrants also might be able to sponsor subsequent migrants, relying on family reunification policies in destination-country immigration policies.

To be sure, greater outmigration is not a risk-free proposition for small island nations. A population drain of highly skilled workers that would negatively affect island economies and potentially the governance of island nations is among the risks.[99] In addition, if the populations of island nations drop to very low levels due to large-scale migration, their legal status as States may become vulnerable, because a permanent population is one of the criteria for statehood in international law.[100]

96 *See also Barnett & Webber*, *supra* note 44, at 22–26 (arguing that migration can help countries of origin, as well as migrants and destination countries, adapt to climate change); *id.* at 46 (proposing the development of "bilateral and unilateral migration benefits agreements and regional migrant agreements").

97 Connell & Brown, *supra* note 83, at 12; McAdam, *supra* note 3, at 20; *see also Reed*, *supra* note 72 (discussing Kiribati President Anote Tong's "migrating with dignity" strategy, which KANI is consistent with). Fiji also might be a destination country. Paul Chapman, *Sea Levels Force Kiribati to Ask Fijians for New Home*, Sydney Morning Herald, Mar. 9, 2012 (reporting that Kiribati is in negotiations to buy land from Fiji to relocate i-Kiribati and referring to President Tong's "'education for migration' program"). Fiji has suggested that it would accept 'climate change refugees' from Tuvalu and Kiribati in the future. Bedford & Hugo, *supra* note 21, at 83 (quoting *Bedford &* Bedford, *supra* note 21, at 90).

98 The World Bank, Global Economic Prospects, *supra* note 83, at 61–62; Bedford & Hugo, *supra* note 21, at 65. *See* also McAdam, *'Disappearing States,' supra* note 40, at 109 (Tuvalu and Kiribati are attempting to create diasporas).

99 The World Bank, Global Economic Prospects, *supra* note 93, at xiv, 58, 67–71; Bedford & Hugo, *supra* note 21, at 8, 29–30.

100 *Stoutenberg*, *supra* note 40.

Yet remittances from migrants could help sustain "a population nucleus" or "caretaker population" in small island nations and therefore help preserve their legal recognition as a State.[101] It would be in the legal interest of migrants to preserve the status of statehood because perpetuating this status would protect migrants from becoming stateless persons.[102] Although the loss of highly skilled migrants might be particularly detrimental, these workers also might be expected to remit higher amounts than lower skilled persons and therefore make a disproportionate contribution to the legal maintenance of the State.[103]

There also are reasons for likely destination countries to expand the migration options available to the citizens of threatened island nations. For example, high-income destination countries including Australia, New Zealand, and the United States face a demographic problem.[104] Because of the combined effects of the graying of the "baby boomers" and stable or lower fertility rates, high-income countries are expected to face labor shortages that could be alleviated by additional immigration.[105] The age structures of the populations of the United

101 *Id.*

102 *Id.*

103 Connell & Brown, *supra* note 83, at 29 ("there is some relationship between income levels and the amount of remittances"); *see also McAdam*, *supra* note 3, at 21 (remittances might finance climate change adaptation). If small island nations become dependent on remittances from international migrants to help sustain claims to statehood, these nations might need to worry about whether remittances will decline over time. There is considerable research on whether remittances decay over time. Connell & Brown, *supra* note 83, at 23–29; Carling, *supra* note 83, at 18–19; Bedford & Hugo, *supra* note 21, at 35–36. However, migrants from island nations, and possibly even their overseas-born children, might feel a special obligation to continue to remit for a long time to support a continued claim to statehood. The island nations also might attempt to develop policies that reinforce a sense of obligation to contribute among their diasporas. *See* Carling, *supra* note 83, at 42.

104 The World Bank, Global Economic Prospects, *supra* note 83, at 29. Greater migration might have other benefits for high-income countries. *Id.* at 12, 13.

105 *Id.* at 29; The World Bank, At Home & Away, *supra* note 83, at 18–19 (Australia and New Zealand); Bedford & Hugo, *supra* note 21, at 6 (high-income countries).

States, New Zealand, and Australia are not identical,[106] but the aging of the population is an issue for all three countries – as well as other high-income countries[107] – that could provide a motivation for increasing immigration levels in the future, including from threatened island nations.

Policy makers in high-income countries that may require additional immigrants in the future to meet labor market needs might ask themselves why they should admit citizens from threatened island nations in particular, rather than boosting immigration levels on an entirely nationality-neutral basis.[108] One reason for high-income countries to have preferential policies facilitating immigration from threatened island nations is that, if climate change renders these nations uninhabitable, there is an argument that the high-income countries will be partially responsible because of their history of greenhouse gas emissions. Opening their doors to migrants from the threatened island nations thus could be regarded as a means by which they partially discharge a moral debt owed to threatened island nations for contributing to the climate change.[109] The term "partially" is used advisedly because it would be difficult to maintain that admitting migrants from the threatened

106 Org. for Econ. Co-operation and Dev., Society at a Glance 2006: OECD Social Indicators 43 (2006) (GE2.1. The age-dependency ratio will increase by around one-fourth over the next 50 years).

107 *Id.*

108 The World Bank, At Home & Away, *supra* note 83, at 20. The report suggests that Australia and New Zealand should consider preferential policies for admitting Pacific islanders because of the potential for migration to significantly improve "Pacific livelihoods . . . and also enhance regional integration and stability." *Id.*

109 Sujatha Byravan & Sudhir Chella Rajan, *Providing New Homes for Climate Change Exiles*, 6 Climate Pol'y 247 (2006); Sujatha Byravan & Sudhir Chella Rajan, *The Ethical Implications of Sea-Level Rise due to Climate Change*, 24 Ethics & Int'l Aff. 239 (2010); *see* also *Opeskin & MacDermott*, *supra* note 21, at 369. Elsewhere I have questioned the use of a corrective justice framework to address climate change migration. Katrina Miriam Wyman, *Sinking States*, *in* Property in Land and Other Resources 439 (Daniel H. Cole & Elinor Ostrom eds., 2012).

island nations represents full compensation for the loss of a way of life.[110]

Another reason why it may be in the interests of Pacific nations such as Australia and New Zealand, and perhaps the United States, to expand migration options for threatened island nations is the possibility that they may face a sudden mass influx of refugees from vulnerable island nations if they do not implement mechanisms that facilitate gradual outmigration.[111] The probability of a mass influx is likely low, but it could be triggered by sudden events such as extreme weather events or flooding.[112] A sudden mass influx likely would be economically and politically costlier for the destination countries to accommodate than gradual migration.[113]

Notwithstanding the arguments for building on the existing outmigration channels, there are nonetheless grounds for putting a degree of effort into developing a new multilateral protection instrument. Advocacy for a new instrument might assist efforts to bilaterally expand migration options by raising public awareness in destination countries of the possibility of migration due to climate change. Such advocacy also could help respond to the argument that there is no reason for any country to take the lead in dealing with climate migration because the problem was created by countries around the world.[114] Supporters might use the advocacy to argue that momentum is developing for an

110 John Connell, *Environmental Change, Economic Development, and Emigration in Tuvalu*, *in* Globalization and Culture Change in the Pacific Islands 260, 271 (Victoria S. Lockwood ed., 2004).

111 Alan Dupont & Graeme Pearman, Lowy Inst. for Int'l Policy, Heating Up the Planet: Climate Change and Security 55–58 (2006); *see* also *Opeskin & MacDermott*, *supra* note 21, at 369–370.

112 Dupont & Pearman, *supra* note 121, at 58.

113 The World Bank, Global Economic Prospects, *supra* note 83, at 49; *Bedford & Bedford*, *supra* note 21, at 125.

114 *See McAdam*, *supra* note 3, at 16–17 (arguing that "the most compelling" reason for a new treaty may be that it "would provide a basis for . . . shared responsibility" and discussing the response of the Australian Labor Party, once in office, to a proposal for protection visas for climate refugees).

international response to climate change migration and that taking the lead might induce others to contribute. In contrast, opponents might respond that destination countries acting on their own in anticipation of climate change migration may reduce, rather than increase, the pressure for an international framework that fairly allocates responsibility for climate change migration.

A second, perhaps more convincing, argument for devoting effort to developing a new multilateral instrument is that there may not be sufficient political will in New Zealand, Australia, and the United States to amend their domestic immigration laws to facilitate the resettlement of citizens of threatened island nations.[115] Advocacy for a multilateral instrument might enlist other countries to become destination countries for citizens of threatened nations, even if the advocacy does not yield a new agreement, by drawing attention to the possible need to resettle persons due to climate change. Other possible destination countries for Pacific islanders are Canada, which already has a sizable community of Fijians, as well as European countries such as the United Kingdom, France, and Germany, which have small populations of Pacific islanders.[116]

In spite of the potential utility of advocacy for a new protection instrument, it is important to remember that developing and implementing a new international instrument would require changes in national immigration legislation similar to the expansion of the existing outmigration mechanisms. To succeed, the advocates for a new international protection instrument will need to engage in the immigration politics of

115 As mentioned earlier, Australia and New Zealand have rejected calls to increase migration from Kiribati and Tuvalu and there are concerns in the United States about the costs of the immigration privileges enjoyed by the RMI and other freely associated States. Concerns in industrialized countries about admitting greater numbers of immigrants are summarized in THE WORLD BANK, AT HOME & AWAY, *supra* note 83, at 21. *See also* Robert Suro, *America's Views of Immigration: The Evidence from Public Opinion Surveys, in* MIGRATION, PUBLIC OPINION AND POLITICS 52 (Bertelsmann Stiftung & Migration Policy Inst. eds., 2009).

116 BEDFORD & HUGO, *supra* note 21, at vii, viii, 63–64.

destination countries, because ultimately these countries must consent to an international agreement to effectively protect the citizens of threatened island nations.[117] Indeed, advocates likely would place an especially high priority on convincing the United States, New Zealand, and Australia – the same three countries that could be lobbied to improve their existing outmigration mechanisms – given their current role in accepting migrants from countries widely considered existentially threatened by climate change.

Moreover, the domestic advocacy required to achieve a new international protection agreement likely will be more complicated than the advocacy required to improve the existing outmigration mechanisms in New Zealand, the United States, and Australia. Advocates will have to make the case for a new protection instrument that will ask destination countries to assume responsibilities for protecting a category of climate-displaced persons, whose numbers will be hard to predict because of uncertainty about the effects of climate change on human settlement and the ability of humans to adapt to climate change without relocating. By comparison, it should be easier for destination countries to forecast the consequences of expanding existing outmigration mechanisms that grant admission based on nonclimate criteria, such as existing nationality. The advocacy for a new multilateral instrument also will be complicated by the need to engage in the domestic immigration politics of many more countries than would be required to improve the existing outmigration mechanisms of selected destination countries.

International agreements sometimes are portrayed as a means by which national elites secure changes in domestic policy by locking in their countries to certain policies at the international level. However, the international legal protection option does not offer an escape from domestic politics about immigration in destination countries. In game-theoretic terms, international law is a two-level game where action is required in the international and the domestic spheres for an

117 *Mayer*, *supra* note 2.

agreement to take effect.[118] Ultimately, the strongest reason for prioritizing the expansion of existing outmigration channels rather than the development of a new international protection framework is that any such framework will require domestic legislative changes in destination countries to offer meaningful protection.

5. Conclusion

This chapter offers a corrective to the argument that a new multilateral protection instrument is required to address climate change migration from threatened island nations. Drawing attention to the existing outmigration from several threatened island nations, it argues that the immigration mechanisms facilitating this outmigration could be improved and expanded through bilateral arrangements. Other nations whose citizens may need to relocate to other countries because of climate change also might expand or develop bilateral arrangements with potential destination countries. Developing such arrangements will require political will in destination countries to boost immigration from countries threatened by climate change, an ingredient that cannot be assured. Yet any new multilateral protection also will require greater political openness in destination countries to increased immigration. By grappling with climate change migration through existing immigration policy, policy makers may be able incrementally to develop arrangements from the bottom up that would facilitate large-scale migration if climate change requires it.

118 Robert D. Putnam, *Diplomacy and Domestic Politics: The Logic of Two-Level Games*, 42 Int'l Org. 427 (1988).

12 Domestic Law for Resettlement of Persons Displaced by Climate Change

Leslie A. Stein

1. Introduction

When the people of a nation are displaced due to rising sea levels and another country is prepared to accept them as migrants, myriad novel questions arise. When there are migrations arising from conflicts and natural disasters, those displaced are either resettled internally or in a host nation with the hope that they can return one day. There are, however, no precedents where the land of a nation is disappearing and its people must be relocated to a new country with no possibility of return.

For the purpose of this chapter, those displaced are referred to as climate-change-displaced persons (CCDPs).[1] This chapter addresses only one of the many conundrums that arise in the case of CCDPs: if

1 They have also been referred to as climate change or environmental refugees, the latter being a term coined in 1985: E. El-Hinnawi, *Environmental Refugees* (United Nations Environmental Programme 1985). This is a controversial use of the term "refugee" because they do not fall within the existing definitions under the *United Nations Convention Relating to the Status of Refugees* as enacted by domestic laws: Jessica B. Cooper, *Environmental Refugees: Meeting the Requirements of the Refugee Definition*, 6 *N. Y. U. Envtl. L.J.* 480 (1998) and R. Black, *Environmental Refugees; Myth or Reality?*, U. N. High Commissioner for Refugees Working Paper p. 1134 (2001). The UN Human Rights Council has suggested this phrase not be used because it confuses the rights of displaced persons: Human Rights Council: *Report of the Office of the United Nations High Commissioner for Human Rights on the Relationship between Climate Change and Human Rights*, U.N. Doc. A/HCR/10/61, para 58 (Jan. 15, 2009).

accepted by a host country, what is the appropriate form of domestic law to guide their resettlement?

If CCDPs enter a country on the basis of a refugee intake under existing arrangements, no new domestic laws are necessary, because the onus is on individuals arriving to assimilate into communities, obtain social services, and, if possible, become citizens of the host country. That method may indeed be appropriate for CCDPs, because there may not be a single, mass migration from particular places; sea level rise is gradual and may lead to intermittent or delayed migration. However, there may be migration in groups as the sea level rise makes land uninhabitable (for more information on current displacement in small island nations, see Chapter 10).

Unlike immediate crisis migration from conflicts and natural disasters, planning for resettlement of CCDPs as a group facing gradual displacement can be the subject of a strategy to be refined over time.[2] The nature of this strategy is influenced by the fact that CCDPs may feel that they are victims of excess by developed countries[3] and are thus particularly entitled to human rights in the form of adequate housing.

1.2. Qualifications

This chapter offers an *idealized* solution for the creation of domestic laws for the resettlement of CCDPs arriving as a group in a host country. This solution is based on three assumptions that will not necessarily occur. The first assumption is that the displaced persons will need to be

2 The United Nations High Commissioner for Refugees Deputy Director, Jean-François Durleux, at the Bonn Climate Talks in April 2009, emphasized the need for *long-term* adaptation strategies as a critical aspect of confronting this problem. Adaptation was suggested as the main tool in long-term planning in the 6th session of the Ad Hoc Working Group on *Long-Term Cooperative Action under the Convention* (AWG-LCA 6).

3 The emergence of that sentiment is explored in M. Grubb, *Seeking Fair Weather: Ethics and the International Debate on Climate Change*, 71 INT'L AFF. 463 (1995).

resettled as a group. The second assumption is that a host country will accept the displaced persons and, more importantly, create laws for their resettlement. The third is that the resettlement should not destroy the culture of the displaced persons.

It has been stated that such an idealized solution may be an impossible undertaking:

> The possibilities for international relocation outside the region (as opposed to other forms of migration) are extremely bleak. It is unlikely that any country, including those in the Pacific, would cede sovereignty over a part of their territory to a relocated group, and it is equally unlikely that a relocated group could sustain its "way" in a foreign land that did not accept or understand many of their cultural beliefs or practices, including their bond to the land.[4]

This chapter therefore proceeds on the assumptions that, at some point, migration will be in groups and that long-term strategies can be created by a host country for the resettlement of CCDPs by domestic laws. This is reasonable to assume if a host country accepts CCDPs as a class of migrants.

It is the thesis of this chapter that the solution to domestic laws for resettlement of CCDPs is land use planning regulation. Unlike other domestic laws, land use planning regulation contains the necessary structure to enunciate the goals that are appropriate to provide housing for CCDPs and can create communities that can lead to the effectuation of those goals.

This chapter is divided into two sections. The first examines existing instruments and policies that may form the basis for domestic laws for resettlement of CCDPs. The second examines the tools of land use planning as the basis for domestic laws.

4 J. Campbell, *Climate-Induced Relocation in the Pacific: The Meaning and Importance of Land, in Climate Change and Displacement: Multidisciplinary Perspectives* (J. McAdam ed., Hart Publishing, 2010).

2. The Legal Obligation to Provide "Adequate Housing" as a Basis for Resettlement

2.1. International Law Obligations for Adequate Housing

A "refugee" is defined as a person who is displaced because of "persecution or a well-founded fear of persecution on account of race, religion, nationality, membership in a particular social group, or political opinion."[5] A refugee and a CCDP are not the same in terms of the cause of displacement, yet bear similarities in terms of the effect in seeking the right to suitable accommodation in a host country to which they have migrated.

There are many international laws that require adequate housing for refugees who are displaced by conflict or natural disasters to another country,[6] but no such laws exist for CCDPs. The underlying thesis of the laws applying to non-CCDPs as enshrined in the *Universal Declaration of Human Rights*[7] is that refugees have the right to eventually return home[8] and the obligation of the host nation is to provide housing *until* the refugees can return (for a discussion of the role of human rights protections in climate-related protections, see Chapter 8).[9] As a result, in the development of international law doctrines relating to refugees, there

5 Immigration and Nationality Act, 8 U.S.C. § 208(a) (1952).

6 Set out in: S. Leckie, Housing, Land, Property and Restitution Rights of Refugees and Displaced Persons: Laws, Cases and Materials (Cambridge: Cambridge University Press, 2007). *See also* Centre on Housing Rights and Evictions, *Sources No. 7: Housing and Property Restitution for Refugees and Internally Displaced Persons: International, Regional and National Legal Resources*, (Geneva 2001).

7 Article 13. Also, General Comment No 27: Freedom of Movement, CCPR/C/21/Rev.1/Add 9, as another example.

8 Analyzed in S. Leckie, The Right of IDPs to Return Home and Property Restitution in Returning Home, Housing and Property Restitution Rights of Refugees and Displaced Persons (Transnational Publishers, 2003).

9 "The right to return to one's own country is increasingly seen as linked with the right to adequate housing." Global Consultation on International Protection, *Voluntary Repatriation*, EC/GC/02/5 (April 25, 2002), quoted in *id*.

is little emphasis on the nature of the housing that should be supplied because it is of a temporary nature.

Host nations are obligated under international law to provide "adequate housing" for displaced persons who are conflict or natural disaster refugees until they return home. The phrase "adequate housing" is contained in all relevant international laws.[10] The source of the phrase is Article 11 of the *International Covenant on Economic, Social and Cultural Rights*,[11] which recognizes the right of all to "an adequate standard of living…including adequate food, clothing and housing, and to the continuous improvement of living conditions." The UN Committee on Economic, Social and Cultural Rights in General Comment No. 4,[12] added a longer list of housing concerns: some degree of security of tenure; availability of services, materials, facilities and other infrastructure that provide for health, security, comfort, and nutrition; affordability and habitability with adequate space and protection from hazards; accessibility for those who are disadvantaged; a location that offers employment and other social facilities; and "cultural adequacy."

Article 8.1 of the *Pinheiro Principles*[13] proposes the need for domestic laws in a host country to implement "adequate housing" for conflict refugees: "States should develop a legal framework for protecting the right to housing, land and property restitution which is clear, consistent

10 A list of instruments that mention "adequate housing" is set out in UN Habitat Fact Sheet No. 21/Rev. 1. These instruments do not apply to refugees from internal conflicts or disasters who have not crossed international borders; here restitution but not "adequate housing" is mentioned: Commission on Human Rights, *Guiding Principles on Internal Displacement*, U.N. Doc. E/CN.4/1998/53/Add.2 ; International Law Association, *The Declaration of International Law Principles on Internally Displaced Persons* 12 Int'l J. Refugee L. 672. (2000).

11 G. A. Res. 2200A (XXI), 21 U.N. GAOR Supp. (No. 16) at 49, U.N. Doc. A/6316 (1966), 993 U.N.T.S. 3 (entered into force Jan. 3, 1976).

12 UN Doc. E/1992/23 (1992), (adopted Dec. 13, 1991).

13 *Principles on Housing and Property Restitution for Refugees and Displaced Persons*, endorsed by the U.N. Sub-Commission on the Promotion and Protection of Human Rights (2005).

and, where necessary, consolidated in a single law."[14] The *Pinheiro Principles* are based on the premise[15] of a return of refugees to their country of origin. As a consequence, there are no provisions regarding the nature of housing or the form of resettlement that can be applied to CCDPs.

2.2. Adequate Housing as a Human Right

In referring to the right to "adequate housing," the *Handbook on Housing and Property Restitution for Refugees and Displaced Persons to Implement the Pinheiro Principles*[16] states,

> Governmental obligations derived from this right include duties to take measures to confer security of tenure (and consequent protection against arbitrary or forced eviction and/or arbitrary confiscation or expropriation of housing), to prevent discrimination in the housing sphere, to equality of treatment and access vis-à-vis housing and protection against racial discrimination, guaranteeing housing affordability, regulating landlord-tenant relations and securing access to and provision of housing resources suited to the needs of marginalized and/or vulnerable groups, such as women-headed households, persons with disabilities, the chronically ill, migrant workers, the elderly and refugees and internally displaced persons.[17]

The *Handbook* refers to "adequate housing" as a human right.[18] As a right, it is interlaced with other rights. An example of this interlacing

14 Other documents that reinforce the need for "adequate housing" are Article 21, *International Convention Relating to the Status of Refugees* (1951), adopted by U.N.G.A. Res. 429(V) (Jul. 28, 1951) (entered into force on Apr. 22, 1954); Section III(8), *Vancouver Declaration on Human Settlements* (1976), adopted by the U.N. Conference on Human Settlements in 1976; Agenda 21, adopted by the *UN World Conference on Environment and Development* (UNCED) held in Rio de Janeiro in June 1992; Point 13, *Global Strategy for Shelter to the Year 2000.*

15 Section IV of the Report.

16 Food and Agriculture Organization of the United Nations, Rome (2007) [hereinafter "FAO Rome"].

17 *Id.* at 48.

18 The EU has considered it a "fundamental right" (*Report on Housing and Regional Policy of 28 March 2007* (A6–0090/2007)), similar to the right to a decent standard of living in the *European Social Charter*: CETS No.: 035, in force February 26, 1965.

can be seen in the treatment of Article 21 of the Indian Constitution that provides a "right to life." The Indian Supreme Court stated that this includes the right to shelter and that

> [f]or a human being [the right to shelter] has to be a suitable accommodation which would allow him to grow in every aspect-physical, mental and intellectual.... A reasonable residence is an indispensable necessity for fulfilling the constitutional goal in the matter of development of man and should be taken as included in "life" in article 21.[19]

The *Report of the Office of the United Nations High Commissioner for Human Rights on the Relationship between Climate Change and Human Rights*[20] was concerned with the question of whether emissions from developed nations were violations of human rights. It addressed a claim by the Republic of Maldives that purported to attribute liability for inundation to countries of the North. In the analysis of liability, the report concluded that climate change *may* have an impact on rights related to property, although it provided no further analysis.[21]

The UN Resolution of March 26, 2008,[22] affirmed that the impact of climate change considerations on human rights should be subject to further detailed analysis. The March 2009 Resolution[23] reaffirmed the relationship between human rights and climate change but did not provide the reasons for that conclusion. The UN Human Rights Council Report of January 2009 made a specific link among housing, human rights, and climate change:

Its terms are interpreted liberally: European Court of Justice in *INPS v. Bruno and Pettini*, C-395/08; C-396–08, 31 (June 10, 2010).

19 *Shantstar Builders v. Narayan Khimali Tatome et al* (1) SC 106, Civil Appeal No. 2598 of 1989 (JT 1990); upholding *Olga Tellis v. Bombay Municipal Corporation* 1985 (3)SCC 545.

20 U.N. Doc A/HRC/10/61 (Jan. 15, 2009).

21 The influence of the Report is discussed in J.H. Knox, *Linking Human Rights and Climate Change*, 33 Harv. Envtl. L. Rev. 477 (2009).

22 G.A. Res. 7/23, U.N. Doc. A/HRC/7/78.

23 G.A. Res. 10/4, U.N. Doc. A/HRC/10/L.11 (Mar. 31, 2009).

> While the universal human rights treaties do not refer to a specific right to a safe and healthy environment, the United Nations human rights treaty bodies all recognize the intrinsic link between the environment and the realization of a range of human rights, such as the right to life, to health, to food, to water, and to housing.[24]

Thus it can be said that the right to "adequate housing" is established as the equivalent of other human rights.[25] The question is whether this connection can be used as the basis for providing adequate housing for CCDPs.

2.3. Obligations arising from Adequate Housing as a Human Right

The UNHCR has stated[26] that the standard arising from the right to adequate housing is that of the "local standard." No specific guidelines are provided to interpret this standard.

The *Handbook on Housing and Property Restitution for Refugees and Displaced Persons to Implement the Pinheiro Principles* addresses whether the right to adequate housing requires a host State to provide housing for all refugees:

> Human rights law does not require States to build housing for everyone who may request it or need it. Rather, housing rights provisions require States to create conditions within society – through law, policy, budgetary allocations and so forth – to ensure that everyone can access housing that is affordable, habitable and fully 'adequate' in accordance with international standards. This can – and should – include the direct financing for the construction of new housing stock and budgetary allocations towards this end, to the maximum of a country's resources, in accordance with international housing rights

24 Human Rights Council, *Report of the Office of the United Nations High Commissioner for Human Rights on the Relationship between Climate Change and Human Rights*, UN Doc A/HCR/10/61, para 18 (Jan. 15, 2009).

25 The right is conflated with the right to self-determination. *Id.* at Part 6.

26 Conclusion No. 101 on Legal Safety Issues in the Context of Voluntary Repatriation of Refugees, para (j).

> provisions such as those enshrined in article 11(1) of the Covenant on Economic, Social and Cultural Rights.[27]

Several of these conclusions are derived from the *UN Settlements Programme Report on Housing Rights Legislation* in 2002,[28] which commented on the human rights aspects of adequate housing.[29] That report references the 1995 Report of the Special Rapporteur, Justice Sachar,[30] which indicated that the correspondence of "adequate housing" with human rights does not imply:

(a) That the State is required to build housing for the entire population;
(b) That housing is to be provided free of charge by the State to all who request it;
(c) That the State must necessarily fulfill all aspects of this right immediately upon assuming duties to do so;
(d) That the State should exclusively entrust either itself or the unregulated market to ensuring this right to all; or
(e) That this right will manifest itself in precisely the same manner in all circumstances and locations.

Justice Sachar suggested that if a State does accept the obligation to provide "adequate housing"

(a) That once such obligations have been formally accepted, the State will endeavour by all appropriate means possible to ensure everyone has access to housing resources adequate for health, well-being and security, consistent with other human rights;

27 FAO Rome, *supra* note 16 at 49.

28 United Nations Housing Rights Programme, *Report No.1 Housing Rights Legislation – Review of International and National Legal Instruments*, Nairobi (2002).

29 Perhaps first set out comprehensively in Commission on Human Settlements resolution 14/6: *The Human Right to Adequate Housing*, adopted May 5, 1993 (paragraph 6). See: U.N. Doc. A/48/8.

30 Sub-Commission on the Prevention of Discrimination and Protection of Minorities, U.N. Doc. E/CN.4/Sub.2/1995/12, paragraphs 4–5.

(b) That a claim or demand can be made upon society for the provision of or access to housing resources should a person be homeless, inadequately housed or generally incapable of acquiring the bundle of entitlements implicitly linked with housing rights.[31]

2.4. Implementing the Right to Adequate Housing in Domestic Law

The UN *Conference on Human Settlements* in Istanbul in 1996 promoted a "Habitat Agenda" that relied on the "right to adequate housing":

> We reaffirm our commitment to the full and progressive realization of the right to adequate housing.... In this context, we recognize an obligation by governments to enable people to obtain shelter and to protect and improve dwellings and neighbourhoods.[32]

The *Conference for the implementation of the Habitat Agenda by a Global Plan of Action* in Hangzhou, China, in 2000 suggested that the means to provide shelter was through land use planning: "Land-use patterns should minimize transport demand, save energy and protect open and green spaces to avoid the unbalanced, unhealthy and the unsustainable growth of human settlements."[33]

The requirement that adequate housing be achieved by land use planning is set out in the *Guidelines for Localizing the Habitat Agenda in Asia and the Pacific*.[34] These *Guidelines* are significant in that they suggest the preparation of an annual report on the *State of Cities*. These *State of Cities* reports[35] follow the *Guidelines*, indicating that land use planning is the means by which adequate housing can be provided.

31 *Id.* at para 12.

32 Para 39.

33 In 2001, the General Assembly in "Istanbul+5" adopted this principle in the *Declaration on Cities and Other Human Settlements in the New Millennium*.

34 U.N. Economic and Social Commission for Asia and Pacific, (2001), ST/ESCAP/2154.

35 More recently, *State of the World's Cities 2010/2011 – Cities for All: Bridging the Urban Divide*.

If land use planning laws can be the means of implementing the right to adequate housing, the question arises as to what factors are relevant to include in land use planning process and laws that relate to CCDPs.

2.5. Assessing Economic, Social, and Cultural Dimensions of Adequate Housing

Adequate housing for CCDPs is not solely concerned with shelter but also with the economic, social, and cultural aspects of communities as expressed in a different context in the ILO's *Indigenous and Tribal Peoples Convention*:[36]

> (they) shall have the right to decide their own priorities for the process of development as it affects . . . the lands they occupy or otherwise use, and to exercise control, to the extent possible, over their own economic, social and cultural development.

There is also a psychological dimension concerning the reaction of CCDPs and refugees to what is provided by way of housing:

> Unless they are irrevocably broken by trials, they will soon begin to explore their surroundings, assess the attitudes of the host, and endeavour to find a niche for themselves in which they can feel consistent with their background and with their gradually changing expectations. In doing so, the nature of the country of resettlement and its population will be of vital importance.[37]

The association of settlement patterns with human rights suggests that there are ethical and moral dimensions to consider in domestic laws for

36 Article 7(1) (No. 169), (1989). This is relevant because the rights of climate change refugees have been compared to indigenous rights: F. Hampson, *Expanded Working Paper on the Human Rights Situation of Indigenous Peoples in States and Other Territories Threatened with Extinction for Environmental Reasons*, Economic and Social Council Working Paper (2005).

37 E.F. Kunz, *Exile and Resettlement: Refugee Theory*, 15 Intl. Migration Rev. 42 (1981).

CCDPs.[38] These dimensions are not confined to CCDPs and have been suggested as relevant in the case of refugees who are perceived as victims of persecution or natural disasters.[39] Scott Leckie, commenting on the limits set out by Justice Sachar on the obligation of the State to provide "adequate housing," states,

> While the duty to *respect* housing rights essentially implies a series of limits of state action, the obligation to *promote* compels governments to recognise the multifaceted human rights dimensions of housing and to take steps to ensure that no measures are taken with the intention of deliberately eroding the legal and practical status of this right. Moreover, comprehensive legislative review should take place, with any existing legislation or policies negatively affecting the exercise of the right to housing subjected to repeal or alteration. The promotion function additionally requires States to place sufficient legal and policy emphasis on the full realisation of housing rights, through a series of active measures including national and/or local legislative recognition of the right, the incorporation of housing rights imperatives into housing and related policies, and the identification of discernable 'benchmarks' towards the full enjoyment of housing rights by all sectors of society.[40]

A difficulty in legislative recognition of what Leckie calls the "full enjoyment of housing rights" is that the important economic, social, and cultural dimensions of that right are not easily expressed. When there is an attempt to clarify these dimensions, as in land use planning laws, there are sub-issues that cannot be explained by statutory drafting, such as cohesion of the migrated CCDP group, access to services, or cultural forms of housing.

It is the case that the only legal structure for the integration of social, moral, and cultural norms in a community appears to be a regulatory

38 The relationships are the subject of much critical legal theory. An instructive summary is in: J.M. Baretto, *Ethics of Emotions as Ethics of Human Rights: A Jurisprudence of Sympathy in Adorno, Horkheimer and Rorty* 17 LAW & CRITIQUE 73 (1990).

39 This theme is explored in: D.E. Anker, *Refugee Law, Gender and the Human Rights Paradigm*, 15 HARV. HUM. RTS. J. 133 (2002).

40 United Nations Development Program, *U.N.D.P. Development Report 2000: Housing Rights*, 19.

framework. As an example, liquor laws are a consolidation of social attitudes toward drinking, acceptability of certain drinking behaviors, cultural norms, and economic, business dimensions. However, the regulatory system is defective in not being able to explain or reiterate the underlying norms: the economic, social, and cultural dimensions.[41]

2.6. The Relevance of Cultural Cohesion in Determining Adequate Housing

It is not clear if adequate housing for CCDPs requires relocation in separate, cohesive communities or if cultural assimilation is to be an aspect of domestic laws. This issue is complex because CCDPs express that they do not desire to be assimilated into the population of a host country:

> Pacific Islanders point out that addressing climate change can avert migration as many do not want to be absorbed into populations already struggling to interact with ethnic minorities. Tarita Holm of the Palauan Ministry of Resources and Development notes that "[i]t is about much more than just finding food and shelter.... It is about your identity."[42]

Jane McAdam reports the reaction of the Republic of Nauru to an offer to assimilate into Curtis Island, off the coast of Australia, on the condition that they took on the responsibilities of an Australian citizen: "Your terms insisted on our becoming Australians with all that citizenship entails, whereas we wish to remain as a Nauruan people in the fullest sense of the term even if we were resettled on Curtis Island."[43]

Renos Papadopoulos, a psychologist, explains that the concepts of "home" and territoriality are fundamental human instincts: "the beginning of all journeys and that which confers the *telos*."[44] Home is also the

41 The classic discussion is in L. Lessig, *Regulation of Social Meaning*, 62 U. Chi. L. Rev 943 (1995).

42 E. Burleson, *Climate Change Displacement to Refuge*, 25 J. Envtl. L. & *Litig*. 19 at 27 (2009).

43 J. McAdam, *Disappearing States* in J. McAdam, *supra* note 4 at 125.

44 R. Papadopolous (ed.), Therapeutic Care for Refugees: No Place like Home (London: Karnac Books, 2003), 11.

basis for psychological adjustment and personal space and the foundation of the cultural milieu. Regardless of cultural differences, he explains that loss of a home creates fear that is a universal phenomenon of displacement. There are also, as further consequences, alienation, sadness, and multiple psychological problems. Even if not expressed by CCDPs, there is a desire to maintain social cohesion as an aspect of the cultural connection with the lost home.

Whether CCDPs are assimilated or relocated into communities, it has been found that their integration into a host country depends on the creation of "social capital" that consists of all factors that promote social coherence and integration, such as friendship, kinship, marriage, and the social support of the cultural group. The creation of social capital requires that CCDPs connect with their own and re-form these connections in the host country: "It is I argue, their characteristics as "social capitalists" which assist significantly in the issue of their longer term adjustment . . . and government policies which ignore or disrupt such processes inflict additional penalties upon them."[45]

Accordingly, it is an aspect of the right to adequate housing for CCDPs that they must relocate in a manner that encourages social capital formation in the host country. If the assumption is correct that there must be social capital, it is critical that CCDPs maintain their social coherence.

The only caution in seeking cohesion is that research has shown that relocating a group into an economically deprived area is deleterious to the formation of social capital because of the social problems that follow, such as crime and poor access to work.[46] The *World Congress of the*

45 P. Loises, *Are Refugees Social Capitalists*, in S. Baron, J. Field, & T. Schuller (eds.), Social Capital: Critical Perspective (New York, Oxford University Press, 2000), p. 126, quoted in C. Garland, F. Hume, S. Majid, *Remaking Connections: Refugees and the Development of 'Emotional Capital' in Therapy Groups*, in Papadopolous, *supra* note 44 at 72.

46 G. Bolt, R. van Kempen, & M. van Ham, *Minority Ethnic Groups in the Dutch Housing Market: Spatial Segregation, Relocation Dynamics and Housing Policy*, 45 Urban Stud. 1359 (2008).

International Geographical Union[47] examined the relationship between the ethnic identity of migrants and the negative effect of segregation, concluding that segregation clearly arises in areas where there is competition for jobs by low-skilled migrants and an increase in their dependency on social services.[48] This finding has resulted in public policy shifts in many countries away from treating migrants as a cohesive social group and toward a migrant absorption policy focused on migrants as individuals who should assimilate into society.[49]

In the case of CCDPs, a policy of migrant cohesion cannot simply be overlooked because of the important connection to the human rights dimension of climate migration.[50] There is, in this debate, a precedent for the use of "identity politics" to override migrant absorption policies in appropriate cases. Consequent policies for individual groups rely on the needs of those particular ethnic groups in specific situations[51] and could therefore modify absorption policies in the case of CCDPs. In any event, there is always an argument for a policy of initial cohesion or settlement in a group on the basis that assimilation is a natural process that will take place eventually in respect of new migrants in a host country.[52]

47 *International Migration and Ethnic Segregation; Impacts on Urban Areas*, 35 Urban Stud. 367 (1998).

48 W.A.V. Clark, *Mass Migration and Local Outcomes: Is International Migration to the United States Creating a New Urban Underclass?* 35 Urban Stud. 371, 380 (1998).

49 R. Brubaker, *The Return of Assimilation? Changing Perspectives on Immigration and its Sequels in France, Germany, and the United States*, in Toward Assimilation and Citizenship: Immigrants in Liberal Nation-States 39 (C. Joppke & E. Morawska (eds.) New York, Palgrave, 2003). Also discussed in I.M. Young, *Polity and Group Difference: A Critique of the Ideal of Universal Citizenship* 263 in The Citizenship Debates. A Reader (G. Shafir ed., University of Minnesota Press, 1998).

50 As for example, in J. Habermas *Struggles for Recognition in the Democratic* Constitutional State, in Multiculturalism 107 (Amy Gutmann ed., Princeton University Press, 1994).

51 The politics are analyzed in W. Kymlicka, *Identity Politics in Multination States*, in European Commission for Democracy through Law: Venice Commission, *State Consolidation and National Identity*, Strasbourg: Council of Europe Publishing, Volume 4, 45 (2005).

52 R. Alba, *Rethinking Assimilation Theory for a New Era of Immigration* 31 Int'l Migration Rev. 826 (1997).

2.7. The Effect of Urban Policies on Resettlement

The movement of migrants into the United States is initially to gateway cities with a subsequent outflow to other domestic areas.[53] The same pattern of migration to cities has been reported for the United Kingdom and other European countries.[54] In this sequence[55] there is an initial clustering of migrants with common backgrounds that leads to population growth from births and low deaths, as well as an influx of new migrants attracted to areas already settled. These areas then become too populated, leading to pressure to disperse because of crowding and therefore to find better accommodation, especially in the case of economically successful members of the group. This often results in new clusters of ethnically similar populations in other areas.

There are no predictors of the settlement patterns of new migrants in a metropolitan area.[56] It has been argued that in many cases migrant groups tend to disperse in multiple groups throughout large cities without the clustering effects that lead to segregation.[57]

In the formulation of urban policy, there is a bias against treating a migrant group as a single entity so as to prevent the creation of low-income neighborhoods, the emergence of an underclass, and the concentration of health, employment, and welfare issues that are all caused

53 W. Frey, *Three Americas: The Rising Significance of Regions* 68 J. Am. Planning Assn. 349 (2002). In Canada, F. Hou, *Spatial Assimilation of Racial Minorities in Canada's Immigrant Gateway Cities*, 43 Urban Stud. 1191 (2006). The inflow/outflow is reasonably balanced: Brookings Institute Research Report, *The Great American Migration Slowdown: Regional and Metropolitan Dimensions* (Dec. 2009).

54 J. Stillwell & O. Duke-Williams, *Ethnic Population Distribution, Immigration and Internal Migration in Britain: What Evidence of Linkage at the District Scale?*, in *Proceedings of the British Society for Population Studies Annual Meeting* (Canterbury Sep. 2005).

55 Analyzed in L. Simpson, V. Gavalas, & N. Finney, *Population Dynamics in Ethnically Diverse Towns: The Long-Term Implications of Immigration* 45 Urban Stud. 163, 167–68 (2008).

56 K.B. Newbold & J. Spindler, *Immigrant Settlement Patterns in Metropolitan Chicago* 38 Urban Stud. 1903 (2001).

57 W. Zelinsky & B.A. Lee, *Heterolocalism: An Alternative Model of the Sociospatial Behavior of Immigrant Communities*, 4 Int'l J. Pop. Geo. 281 (1998).

by segregating a group from the general populace.[58] In managing urban growth, policies in all countries therefore are directed to encouraging mixed neighborhoods based on the land use planning theory that mixing low-income migrants in areas with higher quality housing and skilled jobs creates an incentive to advance.[59]

There are no effective methods for controlling the dispersal of population arising from immigration once the refugee is in the host country; to do so would violate basic human rights doctrines.[60] Attempts to separate one group within a country from another have been met with direct criticism, such as with the Bantustans of South Africa.[61] In China, a system of housing registration, the *hukoa* system, controls movement between places, thus providing for the distribution of population to serve higher order state policy.[62] The UN has criticized the Chinese system:

> The (Committee) notes with deep concern the de facto discrimination against internal migrants in the fields of employment, social security, health service, housing, and education that indirectly result[s], inter alia, from the restrictive national household registration system which continues to be in place despite official announcements regarding reforms.[63]

58 An example of "social mixing" is found in J. Uitermark, *'Social Mixing' and the Management of Disadvantaged Neighbourhoods: The Dutch Policy of Urban Restructuring Revisited* 40 URBAN STUD. 531 (2003). The debate is discussed in R. Andersson & S. Musterd, *Social Mix and Social Perspectives in Post-War Housing Estates*, in RESTRUCTURING LARGE HOUSING ESTATES IN EUROPE Chapter 7 (R. van Kempen et. al. eds., Policy Press 2005).

59 The effect of this mix has been doubted in a pilot study and theoretical analysis in Amsterdam: W. Ostendorf, S. Musterd, & S. De Vos, *Social Mix and the Neighbourhood Effect: Policy Ambitions and Empirical Evidence*, 16 HOUSING STUD. 371 (2001).

60 Art. 13(1) *Universal Declaration of Human Rights* (1948); Art. 12(1) *International Covenant on Civil and Political Rights* (1966).

61 B. ROGERS, DIVIDE & RULE: SOUTH AFRICA'S BANTUSTANS (London: International Defence and Aid Fund, 1980).

62 FEI-LING WANG, ORGANIZING THROUGH DIVISION AND EXCLUSION (Stanford University Press, 2005).

63 United Nations Committee on Economic, Social, and Cultural Rights, *Concluding Observations of the Committee on the Initial Report of the People's Republic of China (including Hong Kong and Macao)*, 3 (May 13, 2005).

Indirect means of controlling entry into an area include freezing local developments, which, in addition to reducing housing development, also prevents any increase in population density.[64] However, short of placing CCDPs in a reservation, those formulating urban land use policy must accept that there is nothing to prevent their free movement, as they gravitate to opportunities and to urban conglomerations.[65]

The United Nations Human Settlements Program *Global Report on Human Settlements 2009*[66] focuses on the governance tool of land use planning as the means to resolve the difficulties arising from urbanization and random settlement patterns. It argues that government should provide a planning system that addresses climate change, poverty, poor quality housing, safety, and the other consequences of increasing population pressures.[67] As the immigration of CCDPs will be into cities primarily, this report establishes a direct connection between immigration and land use planning as the only means to address the issues of their placement and accommodation.

3. Domestic Law: Land Use Planning as the Means of Relocation

3.1. Land Use Planning and the New Urbanism

In its early development, land use planning embraced the notion that the regulatory process that produced a map and zones could somehow provide a better way of life. However, increased poverty and crime in the 1960s caused a movement away from physical planning (a map and zones) as the panacea for urban ills, toward its being accepted as merely a regulatory process to separate out deleterious uses in an

64 N. Levine, *The Effects of Local Growth Controls on Regional Housing Production and Population Redistribution in California*, 36 URBAN STUD. 2047 (1998).

65 For example in China and India: M. Dalton et al, *Demographic Change and Future Carbon Emissions in China and India Papers of the Population Association of America Annual Meeting*, New York (2007).

66 *Planning Sustainable Cities – Global Report on Human Settlements 2009.*

67 Key findings: p. xxvii.

efficient manner;[68] it became apparent that there was no bridge between an aspirational set of goals and an ordinance that operated by a map and the restriction of land uses in zones. However, in the last ten years, there has been a clear movement back toward land use planning as a means to deliver social goals.

There are many reasons for this shift,[69] but the current, revived use of planning as having an effect on the social good can probably be traced to two sources. The first is the New Urbanism movement that postulates that a sense of community can be developed by design.[70] The design process is used in New Urbanism to create more compact and bounded communities with an emphasis on well-defined open spaces, diverse land uses and mixed housing types (the opposite of single-use residential zones), an interconnected system of streets and walkways, close proximity to all services and transportation in a "walkable" distance, and other principles that encourage face-to-face interactions, a spatially coherent sense of place, affordable housing, and sustainable development.

New Urbanism originated as a system of a rigid design guidelines that were the minimum requirements to develop a walkable, coherent community. There was substantial objection to these guidelines[71] on the basis that they were difficult to create and inefficient for developers. Yet New Urbanism has become, and is used in this chapter, as more than a device to create better communities as a matter of design; it is now a framework for the consideration and incorporation of social, cultural, and economic functions in land use planning and has revitalized the concept that planning is able to be used to create a social benefit by means of design.[72]

68 This history is traced in L.A. STEIN, PRINCIPLES OF PLANNING LAW (Melbourne, 2009), chapter 1.

69 A brief account of some issues is found in A. von Hoffman, *Housing and Planning: A Century of Social Reform and Local Power*, 75 J. AM. PLAN. ASS'N 231 (2009).

70 An analysis of the background of New Urbanism is in R. Gindroz, *New Urbanism and Smart Growth, City Life and New Urbanism*, 29 FORDHAM URBAN L. J.1419 (2002).

71 For instance, P. Gordon, H.W. Richardson, *Are Compact Cities a Desirable Planning Goal?*, 63 J. AM. INST. PLAN. 95 (1997).

72 Discussed in E. TALEN, NEW URBANISM AND AMERICAN PLANNING: THE CONFLICT OF CULTURES, 278–79 (Routledge, 2005).

It is used as an *example* of how land use planning can embrace nuances that are not derived solely from separating out discordant uses, but rather draw from and promote a quality of life.

The acceptance of New Urbanism ideas as a matter of public policy to realize social goals is also related to its connection to the current imperative of encouraging environmentally sustainable communities with efficient buildings and less dependence on cars. As "sustainable growth" becomes a primary criterion for all new urban development,[73] New Urbanism has gained acceptance as a relevant subset of principles that reduce car trips, increase building efficiency, and allow for a housing mix that includes affordable housing – all of which have been contained under the concept of "Smart Growth," used as the touchstone to create better communities at the urban fringe.[74]

In recent years, the New Urbanism agenda has been subsumed, as mentioned, under the heading of "sustainability," but it has also changed to include transportation options to reduce greenhouse gas emissions, alternative technologies, and the use of mixed use around public transportation nodes to fulfill the dual requirements of creating a sense of community and reducing trips to work.[75]

The second reason for the acceptance of land use planning as carrying out social goals is the capacity to translate the New Urbanism agenda to land use regulation in what is called "Traditional Neighborhood Design (TND)." Pennsylvania and Wisconsin[76] were the initial implementers of this form of regulation, which seeks to provide a checklist to which ordinances must conform, such as the patterning of interconnecting streets

73 P.R. Burke & M.M. Conroy, *Are We Planning for Sustainable Development?*, 66 J. Am. Plan. Ass'n 21 (2000).

74 Reviewed in A. Haines, *Evaluating Smart Growth – Implications for Small Communities*, 27 J. Plan. Educ. & Res. 49 (2007).

75 A review of California initiatives in this regard is in R.H. Freilich & N.M. Popowitz, *The Umbrella of Sustainability: Smart Growth, New Urbanism, Renewable Energy and Green Development in the 21st Century*, 42 Urban Law 1 (2010).

76 These provisions are discussed in R. J. Sitkowski & B. W. Ohm, *Enabling the New Urbanism*, 34 Urban Law. 935 (2002).

and, most important, the fixing of the distance from the center to the edge of the neighborhood. The significance of TND regulation is that it proves that a regulatory framework can bridge the gap between aspirational goals and legal devices.

3.2. Social Welfare Goals in Planning

Robert Erikson analyzes the efficacy of including social welfare issues as land use planning goals.[77] He defines "welfare" as the sum of all of the factors that provide the fulfillment of essential physical needs and a sense of security. It is not, he asserts, possible to assess the presence of these factors from the subjective viewpoint of individuals, because the fulfillment of physical needs and a sense of security depend on their personal aspirations, nor can political welfare concepts be used because they are normative judgments that may not accord with the actual state of affairs. The best option for evaluation of welfare factors, Erickson concludes, is the actual ability of an individual to mobilize the resources at his or her command, "with whose help he/she can control and consciously direct his/her living conditions."[78] Accordingly, the effectuation of welfare goals depends on a relationship between the availability of resources and their ability to be accessed and controlled.

One of the determinants of an individual's ability to access resources is the "design of and conditions in the dwelling unit and ... the housing area."[79] Erickson's study concludes therefore that social welfare considerations are an aspect of land use planning:

> One of the purposes of planning is to make different jobs, services and other benefits available to the inhabitants of the community. Physical planning seeks to bring about accessibility, while other planning seeks to make the benefit in question attainable.[80]

77 R. Erikson & R. Feichtner, *Welfare as a Planning Goal*, 17 ARTS SOCIOLOGICA 273 (1974).

78 *Id.* at 275.

79 *Id.* at 278.

80 *Id.* at 281.

Erickson's work has led to the development of "social indicators" used in the evaluation of the EU's "Social Europe" and the UN Millennium Development Goals.[81] The underlying theme of both documents is to eradicate poverty, and in that respect, the Millennium Goals include the "Social Method of Coordination,"[82] which itself consists of three overarching goals and twelve agreed indicators for nations to measure the decrease in poverty. Three of the indicators are relevant:

- to promote the active inclusion in the society and the labor market of the most vulnerable groups
- to ensure decent housing for everyone
- to overcome discrimination and increase the integration of people with disabilities, ethnic minorities and immigrants, and other vulnerable groups

These Millennium Goals make "welfare" or social well-being dependent on and an aspect of land use planning. This is applicable to the social welfare needs of migrants who have been displaced involuntarily because they may face a lower quality of life on resettlement.[83]

It is appropriate for land use planning goals to include statements of ideology, such as social welfare goals. The details of an ideology need not be fully explained for it to be included in a planning ordinance; for example, a statement of the goal may be a "safer community" or "aesthetic development." Despite this imprecision, land use planning is a major regulatory device for the incorporation of ideologies, and it is assumed

81 The history is discussed in A.B. Atkinson, *Social Indicators, Policy and Measuring Progress*, in, ANALYZING INEQUALITY: LIFE CHANCES AND SOCIAL MOBILITY IN COMPARATIVE PERSPECTIVE, chapter 4 (S. Svallfors ed., Stanford University Press, 2007).

82 Adopted in 2006 and based upon the European Commission Communication *Working Together, Working Better: A New Framework for the Open Coordination of Social Protection and Inclusion Policies in the European Union* COM/2005/0706 final (2005).

83 As examined in M. Kontos, *Biographical Assessment of Quality of Life Processes in Migration*, in CHALLENGES FOR THE QUALITY OF LIFE IN THE CONTEMPORARY WORLD: ADVANCES IN QUALITY OF LIFE STUDIES, THEORY AND RESEARCH, chapter 10 (W. Glatzer, S. von Below, & M. Stoffregan eds., Springer 2007).

in planning theory that it will be able to change to subsume new issues as they arise, such as the welfare of CCDPs:

> What I am certain of is that as long as those who regard themselves as planners are content to remain within the confines of the dominant model of the day, which is to say, its dominant ideology, and define planning solely with respect to that ideology be it civic, social, or economic, our profession will always be exposed to catastrophic changes in meaning, such as those we have witnessed. History will see to that. In this case there can be no such thing as planning, only plannings.[84]

3.3. Appropriate Form of Land Use Regulation for CCDPs

A domestic law for resettlement of CCDPs based on land use planning and its ability to incorporate social welfare goals and ideologies is possible and can take the form of a land use ordinance that incorporates goals and the means of their implementation.

A land use ordinance consists of zones and uses that are permitted within those zones. A CCDP "zone" does not accord with historical zoning conventions that describe a land use, such as "residential" or "commercial" zone. However, the techniques of land use planning have changed,[85] and there are recognized forms that can be used to expressly provide for an area or for criteria for planning for CCDPs.

The basic form for the establishment of a zone is to explain the existing qualities of the locality, the opportunities and constraints, the social and physical objectives, the policies that the plan is intended to implement, and then the physical design to effect those goals. This is the traditional "comprehensive plan" that has as its basis that all of the strands of policy will be gathered together in the one format. Policies are incorporated into the plan but also can exist extraneously in other forms. One such example is a strategy usually referred to as a "Land

84 A. Z. Guttenberg, *Planning and Ideology*, 8 J. PLAN. HIST. 287 at 293 (2009).

85 Some other devices are described in E. J. Kaiser & D. Godschalk, *Twentieth Century Land Use Planning: A Stalwart Family Tree*, 61 J. AM. PLAN. ASSN. 365 (1995).

Classification Plan," which identifies areas available for development in the future.[86] In addition, a "Verbal Policy Plan" expresses goals more generally, not in a regulatory format, but as policies, thereby creating a policy framework.[87] A Policy Plan is often employed by state governments[88] or regional planning authorities to express goals not included in zoning ordinances, which may reflect only local, parochial interests.

"Form-based zoning" is a concept that arises from the precepts of New Urbanism and provides a means to directly address those land use elements that promote community interaction. In this type of zoning, the physical forms of development are addressed in detail, as are street types, location character, and design elements. Because form-based zoning is derived from New Urbanism, it encourages the interaction of uses, as in the creation of walkable neighborhoods, enhanced streetscapes, and open space.

The first form-based code was developed by the firm Duany Plater-Zyberk & Company[89] as a prescriptive regulatory instrument to enable development of a particular built form rather than to provide prohibitions on use and development, as is the case with conventional zones. It therefore is more detailed in its approach and provides a "regulating plan" that establishes the detail of development plans, building envelope standards, and specific requirements for housing elements, such as doors and porches. It also sets out the architectural design standards and range of requirements for different forms of buildings, building elements, and

86 *Id.*

87 Explained in R.T. Legates & F. Stout (eds.), City Reader 410–413 (5th ed., Routledge, 2011).

88 E.M. Hamin, M. O. Steere, & W. Sweetser, *Implementing Growth Management: The Community Preservation Act*, 26 J. Plan. Educ. & Res. 53 (2006).

89 Leading to Congress for the New Urbanism, Planning Association Planning Advisory Service Rep. no. 526, *Codifying New Urbanism: How to Reform Municipal Land Development Regulations*, 12–15 (2004). The New Urbanism movement is attributed to Adreas Duany arising from this work: A. Duany, E. Piater-Zyberk, & J. Speck, Suburban Nation: The Rise of Sprawl and the Decline of the American Dream (North Point Press, 2000).

finishes by providing a range of options from which a developer can choose.

There do not appear to be any legal restrictions against implementing form-based zoning in a zoning ordinance, even when the Standard State Zoning Enabling Act has been adopted.[90] There are specific instances, if fact, where legislative authority has been granted for form-based codes, Maryland being the first in 1997 with a "Smart Growth" law. Although there is debate as to the effectiveness of New Urbanism, it is highly relevant as a method of land use planning that is focused on the goal of creating a better way of life by physical design, accompanied by a regulatory form to enforce the concepts of community interaction and cultural coherence.

3.4. Incorporating CCDPs' Aspirations in Zoning Ordinances

The Standard State Zoning Enabling Act requires that goals be enunciated in the comprehensive plan. This is a necessary element, because projecting land use into the future requires an identifiable path forward, which itself necessitates an enunciation of goals.

A goal can be expressed positively or negatively in an ordinance, be general or specific, and explicit and clear or implicit in its wording. In land use planning, most goals are expressed in a manner that does not allow objective assessment by indicators or other measures; instead they are framed as sentiments, ideologies, or hopes. The goals are, in all cases, either an expression of existing states, with the implication that they remain the same, or aspirational goals as to how some elements may improve. The goals as to future states or aspirations are not expressed in detail because their categorization is often unclear; for example, a "vibrant" residential area.

90 D. W. Owens, *Local Government Authority to Implement Smart Growth Programs: Dillon's Rule, Legislative Reform, and the Current State of Affairs in North Carolina*, 35 Wake Forest L. Rev. 671 (2000).

The importance of goal setting in a land use context is threefold. First, it permits inhabitants to express their views for the locality and encourages participation in framing aspirations. Thus it is possible for CCDPs to examine their hopes and aspirations before they occupy any land. Second, it allows the free-form expression of goals that can include human rights statements, such as "The right to adequate housing of a local standard is a human right that should be carried forward in this Ordinance." Statements such as "The goal of this Ordinance is to provide for a cohesive community setting for those persons affected by climate change" are acceptable; that is, in fact, an important statement, because it creates an overall concept of what can be built and how it can be conceived. Third, the goals serve as a focal point, more than just being regulatory provisions, to determine the nature of the community to be established.

To understand the goals and aspirations of CCPDs with respect to housing, community, culture, and the various family structures that constitute a shared way of life, it is important to elicit their responses before setting goals in an ordinance. The New Urbanism methodology offers assistance in the form of a "charette," which is a forum for the exchange of ideas and goal formulation.[91] It is important that CCDPs participate in goal formulation to express their concepts of land use.

This goal setting is a practical method in all countries, because land use planning is not a foreign concept in any place. As an example, it is understood as a concept in the Pacific islands[92]:

> Forms of interactive community decision making regarding use and development have been practised at a traditional level for centuries in the Pacific Islands. Life is steeped in the concept that all activities

91 R. Grindroz & K. Levine, The Urban Design Handbook: Techniques and Working Methods, chapter 3 (W.W. Norton, 2003). The charette has been used as a prelude to development, for example, in the Haiti, Port-au-Prince Renewal Plan: http://portauprincerp.posterous.com, last accessed on March 20, 2011.

92 Matt McIntyre & Associates, commissioned by the Pacific Islands Forum Secretariat, *Review of Land Administration Systems for Land Administration, Land-Use Planning and Management*, Land Management and Conflict Minimisation Sub-Project 3.2 (2009).

> affect each other, given the very small landmasses, the interdependencies between resources and the huge upheavals that can be caused by natural disaster. The concept of competing land-use pressures (i.e. the choice of whether to use land for community, gardens, plantation, living or church) are [sic] also well appreciated at the family, village and community levels.

There is planning legislation in almost all the island nations,[93] although it has been suggested that the planning process is not effective.[94]

This inquiry, the charette, leads to a general set of aspirations and a detailed description of community needs, forms of housing, networks of communication, mixes of uses, means of transportation, and all the other factors that would support a new location providing social capital and cultural familiarity.

3.5. Incorporating Cultural Matters in the Zoning Ordinance

Expressing goals and aspirations in an ordinance does not guarantee consistency or continuity of culture for those displaced. It is, however, possible to re-create some of the important cultural markers of CCDPs' lifestyle by spatial recognition in housing arrangements, building styles, forms of services, means of internal communication, and community hierarchies. "Urban morphology"[95] is a method used for the examination of the physical pattern of streets and blocks, the building fabric, and patterns of land use in the host country. Conzen, the originator of this technique, believes that these aspects can be understood to

93 Fiji: Town Planning Act (Cap 139); Solomon Islands: Town and Country Planning Act (Cap 154); Samoa: Planning and Urban Management 2004; Kiribati: Land Planning Ordinance; Vanuatu: Physical Planning Act (Cap 193); Marshall Islands: Planning and Zoning Act 1987.

94 Secretariat of the Pacific Communities, Land Resources Division, Policy Brief 3/2008, *Land Use Planning in the Pacific*.

95 The theoretical positions and its nature are discussed in P. Gauthier & J. Gilliland, *Mapping Urban Morphology: A Classification Scheme for Interpreting Contributions to the Study of Urban Form*, 10 Urban Morphology 41 (2006).

explain the pattern of the historical development of communities and "the way they objectivate the spirit of that society."[96] His opinion is based on the concept that physical form is the synthesis of the aspirations related to community structure and amenities of the built environment, and consequently it explains the "sense of place." His analysis does not include community power structures, behavioral codes, institutional structures, family structures, social stratification, land tenure arrangements, the growth of modernity, how identity is established, and those other artifacts of culture that are difficult to ascertain. At most, urban morphology can extract the indicia of urban form in the country of origin by showing how physical space is ordered in terms of housing types and networks of open space.[97]

New Urbanism does *not* require a particular Western, bounded, spatial layout of community structure or housing or a specific type of form. It has accordingly been employed in non-Western cultures, such as Managua, Nicaragua; Valparaiso, Chile; Petit Paradis, Haiti; Antigua;Guatemala; Cambinda; Angola; and in the Port-au-Prince Reconstruction Plan in Haiti.[98] New Urbanism, as used in this chapter and as properly employed, is the means to translate a desire for a sense of community into a housing and community structure form emerging from urban morphology. It is thus flexible in terms of the goals it serves, the patterning of the community, and the types of housing.

Form-based zoning coupled with a general narrative on the goals that are being sought can provide for the full expression of these goals and aspirations, a palette of house forms, the network of roads and walkways,

96 M.R.G. Conzen *Geography and Townscape Conservation*, in *The Urban Landscape: Historical Development and Management* (J.W.R. Whitehand ed.) Institute of British Geographers Special Publication, No. 13, 1981, at 57 quoted in, where it is also applied: K. Gu, *Urban Morphological Regions and Urban Landscape Management: The Case of Central Auckland, New Zealand*, 15 Urban Design Intl. 148 (2010).

97 The classic analysis is in L.H. Schneekloth, Ordering Space: Types in Architecture and Design (Van Nostand Reinhold, 1994).

98 Database of the Congress for the New Urbanism: http://www.cnu.org/search/projects?page=18 (last visited March 9, 2011).

the intermingling of services that are expected, and the social interconnections that are needed for cohesion of the group.

A true re-creation of the nuances of the original location is impossible because of the obvious cultural differences and the host country's own building standards, land tenure laws, and other relevant regulations. It is not intended that there be a match but rather an approximation that can yield meaningful possibilities to allow for some integration of CCDPs into a host country that will conform to human rights criteria and that can bridge that gap between cultures that are unique.

The form-based codes can take into account some of the design interactions of the original housing form of the CCDPs, with regard to climate differences. Some of these may be functional, such as the relative proportions of living spaces, and others could be merely decorative but would be familiar. Offering the CCDPs a chance to define these forms and transposing them to new forms will at least provide some continuity of the built environment. It will also allow improvements of some practices that have developed to the detriment of the CCDPs in their host country, such as the common problem of poor ventilation leading to indoor air pollution from cooking fuels.[99]

The housing forms and arrangements, to be acceptable, must comply fully with domestic regulations because the buildings will not necessarily be occupied indefinitely by CCDPs as they disperse over time. As well, a building has a reduced life expectancy if its design or technology is of a lower quality than that of adjoining buildings.[100] Maintaining building quality will be especially important in terms of financing and attraction of industry and services to the area.

A form-based zoning code, resulting from a New Urbanism approach, may be able to re-create in a host country the juxtaposition

99 The positive effect of changes is set out in S. Dasgupta, D. Wheeler, M. Huq, & M. Khaliquizzaman, *Improving Indoor Air Quality for Poor Families: A Controlled Experiment in Bangladesh*, 19 Indoor Air 22 (2009).

100 A. Ashworth, *Estimating the Life Expectancies of Building Components in Life-Cycle Costing Calculations*, 14 Structural Survey 4 (1996).

of uses, the interweaving of social services, the networks of communication, and housing forms that allow some congruence with forms that were left behind. This approach adds substance to the concept of adequate housing.

There are, of course, arguments against such an approach. It could be said that establishing a community based on particular goals will fail if the CCDPs decide to live elsewhere. This may occur if housing costs are too high because these projects were not publically financed. It has been shown in Israel that there is a rapid acceptance of what has been built for migrant groups only if it is public housing.[101] It may also depend on the siting of the area in terms of transportation, services, jobs, and infrastructure.

Assuming the locational issues and the costs can be mediated, the use of form-based zoning and goal setting, which approximate New Urbanism principles, provides perhaps the only forum for CCDPs to enunciate aspirations, have those aspirations translated into a regulatory framework, and achieve some reflection on the social structures that are inherent in their cultural values.

3.6. Location of CCDPs

Rebelo proposes a methodology for migrant land use location in metropolitan areas. He sets out the fundamental criteria that dictate how immigrants arrange themselves:[102]

> Sociological concentration is mainly explained by the spatial differentiation provided by the characteristics of urban economies, the performance of the housing markets, and the characteristics of inhabitants themselves, whereas the factors that better explain spatial dispersion are socio-economic mobility and acculturation processes, that are rooted in the perception that urban economies arrange

101 B.A. Portnov, *The effect of Housing on Migration in Israel*, 11 J. Pop. Econ. 379 (1998).

102 E.M. Rebelo, *A Methodology to Approach Immigrants' Land Use in Metropolitan Areas*, 27 Cities 137, 139 (2010).

> resources and opportunities unevenly in space, which induces different immigrant groups to be displaced in order to increase their social and employment attainment levels.

Immigrants choose locations primarily based on housing that is affordable, job opportunities, and the promotion of social capital. It is incumbent on any planning process to attract immigrants to areas that can facilitate their engagement in economic production and provide good dwelling and working conditions as well as access to transport, services, and other relevant infrastructure. Economic considerations are accordingly the prime factors in immigrant settlement patterns. The Rebelo study suggests a model based on neural networks or hierarchical decision trees to discover ways to keep ethnically diverse immigrants together, recognizing that they concentrate in urban areas and that those living in areas with language and cultural dissimilarities assimilate less easily. It is not necessary to explore the Rebelo model, but it is important to indicate that its basis is the crucial nature of economic opportunities in creating social capital.

In a metropolitan area, two options flow from this analysis. The first is that areas can be developed from greenfield sites in accordance with New Urbanism and form-based zoning that are in proximity to employment lands. Designating new employment lands that have yet to attract business is not necessarily the correct solution. If this model is an accurate snapshot of immigrant behavior, there must be some nexus with established employment opportunities. If the areas that are chosen are in existing metropolitan areas, then the employment needs are fulfilled by implication.

As defined in this chapter and in its current usage, New Urbanism has applicability to the inner city where communities can be developed by infill of existing areas, by selective siting of available sites such as brownfield land, or by density increases in particular areas. The hallmark of such a project would be high-quality urban design based on the aspirations of the future residents. One study of using New Urbanism in Pittsburgh stated, "Design issues emerged from the bottom up, from the goals of residents working toward revitalizing their own neighborhoods, with the designers assisting the community to visualize, articulate and implement their ideas."[103]

Thus, the application of New Urbanism to inner city development would not have the capacity to create a physical pattern that would reflect the results of morphology of the country of origin. It would, however, have the capacity to state goals and aspirations and reflect, to the extent possible, some aspects of the built form. This would suggest that the ideal location for CCDPs is a greenfield area, which would allow the whole range of devices because it is undeveloped and capable of an imprint that reflects the cultural markers of the community structure and built form. This approach depends on the ability to finance new development, because it is inevitably more expensive to convert vacant land to a new community than to provide for infill.

3.7. Financing and Its Consequences

The choice of a location that is capable of being developed for CCDPs depends on the provision of infrastructure. When state or local budgets cannot afford the infrastructure, funding relies on exactions from developers. There is a legal requirement that, for the exactions to be imposed, there must be a nexus[104] between the consequence of the development and the exactions. Accordingly, regional infrastructure unrelated to a development is not included and becomes an on-budget expenditure for a government agency.

The method of calculating the cost of infrastructure for a large, undeveloped area is to draw on the investment plans of the service providers to determine the stream of future costs, deduct the cost of running the

103 S. Deitrick & C. Ellis, *New Urbanism in the Inner City: A Case Study of Pittsburgh*, 70 J. Am. Plan. Ass'n 426, 428 (2004).

104 A summary of the recent cases is in M. Barker, *Much Ado about Nollan/Dolan: The Comparative Nature of the Legislative-Adjudicative Distinction in Exactions*, 42 Urban Law 171 (2010).

infrastructure as it is owned by the agencies when complete, and then also deduct that which would have taken place but for the development amplifying the timing and need. This method is not straightforward and depends on the hypothetical demand and supply generated from the development. Accordingly, it requires complex calculations as to the timing of full take-up of capacity amortized over a reasonable period and then applying the cost to a unit – usually several – in a single development. It is a different matter from applying an exaction for a single subdivision or a single headwork's charge and is complex and becomes a political issue or one subject to negotiation: the English *Community Infrastructure Levy* has been said to be brokered in a clandestine way in smoke-filled rooms.[105]

The effect on development for CCDPs is that, at a time of rising house and land prices, these impact fees can be "passed back" and absorbed in the price. However, in a time of falling prices or when the land and housing to be provided are low cost, these fees create a profound barrier to development. In the case of CCDPs who are arriving as migrants, there is little doubt that the developer could not afford the linkage fees for a greenfield site. An early study on the impact of New Urbanism on home prices indicated that there is an increase in cost that will be absorbed only by buyers who are willing to pay for additional amenities over and above conventional housing.[106]

The manner in which housing for CCPDs would be subsidized will therefore dictate the extent to which the possibilities for cohesive and responsive communities can be realized. If government pays for public housing, this may limit the extent of innovation in design and layout that is possible.[107] The degree of development in accordance with

105 J. Corkindale, *Planning Gain or Missed Opportunity? The Barker Review of Land Use Planning*, 27 Eco. Aff. 46, 48 (2007).

106 C.C. Tu & M. J. Eppli, *Valuing New Urbanism: The Case of Kentlands*, 27 Real Estate Econ. 425 (1999).

107 This is the summation of the articles in 33 Issue 3 of the J. Urban Hist. (2007) that analyzes public housing programs in North and Latin America.

New Urbanism principles will also depend on whether any subsidy is on the supply side in terms of government construction or personal private partnerships, or on the demand side in the form or low-interest mortgages, rent subsidies, or price controls such as capped mortgages or price ceilings. Governments tend to support demand-side funding for public housing by providing rental assistance or subsidies.[108]

In the United States, the U.S. Department of Housing and Urban Development (HUD) provides funding mechanisms for public housing that can be employed to aid with housing of CCPDs. The *Community Development Block Grants* for low- and medium-income families provide the means to support local initiative and funding of these communities, even though it is argued that the process is slowed down by the requirements of the 2009 *American Recovery and Reinvestment Act* that provides funds to stimulate the economy.[109] The Congress of New Urbanism, in partnership with the Natural Resources Defense Council and Green Building Council, developed the LEED ND system as a green neighborhood rating system for developments that use environmentally sound locations (e.g., not wetlands but abandoned sites), reduce driving by encouraging walking, employ mixed uses, provide open space, and conserve resources. HUD has adopted the LEED ND system[110] for use in assessing grants, thus providing a dovetailing between New Urbanism and the possibility of Community Development Block Grants for housing for CCDPs.

As to the demand side of financing, the 2005 UN Habitat report, *Financing Urban Shelter – Global Report on Human Settlements*,[111] recommends financing alternatives for housing in developing countries.

108 Reviewed in Latin America in T. Klak, *Contextualizing State Housing Programs in Latin America: Evidence from Leading Housing Agencies in Brazil, Ecuador, and Jamaica*, 25 Envtl. Plan. Ass'n 653 (1993).

109 *See* M.B.J. Pavlick, *A Look at the Recovery Act and its Effect on the Community Development Block Grant Entitlement Program*, 62 Admin. L. Rev. 523 (2010).

110 Address by the S. Donovan, Secretary of Housing and Urban Development to the Congress of New Urbanism, Atlanta (May 21, 2010).

111 UN-Habitat, Report HS/752/)SE.

It indicates a trend away from state funding for housing to the use of the mechanisms of subsidies and deregulation of markets. It argues that it is important to extend mortgage financing for lower income buyers by a reduction of interest rates and direct capital grants to reduce mortgage size. As well, it suggests that the level of repayments should be varied over time and be raised only when income improves. It also provides that there should be short-term, small-scale loans and that these are more effective for financing than long-term, large-value loans.

3.8. The Basic Form and Substance of a Model CCDP Ordinance

A "floating zone" is a longstanding device that creates a zone but leaves its exact location to be determined in the future as the need arises.[112] The device is not controversial because it is consistent with the general power to create zones.[113] It is an appropriate device for the establishment of a CCDP zone that could be created in advance and applied when necessary.

Another possible device is an "overlay" zone that leaves the underlying zoning intact and provides variations for particular situations. An overlay can be conditional in the sense that it applies only when there are certain triggers. However, restricting the uses in the underlying zone to prevent it from becoming inconsistent with the future uses defined in the overlay may amount to a taking.[114]

There is no restriction on the form of a floating zone, and therefore it can take account of the aspects previously discussed as the basis for its content. Most importantly, it can express goals and aspirations and define the community structure and those housing elements that are relevant in enhancing the goals and aspirations of CCDPs.

112 Perhaps analyzed for the first time in R. R. Reno *Non-Euclidean Zoning: The Use of the Floating Zone*, 23 Md. L. Rev. 105 (1963).

113 *Sheridan v. Planning Board*, 159 Conn. 1, 17, 266 A.2d 396 (1969).

114 *Vacation Village, Inc. v. Clark County, Nevada*, 497 F.3d 902 (9th Cir. 2007).

The emphasis of the floating zone is on regulating and expressing the urban form rather than on describing land use. It sets out the requirements that will lead to a predictable physical pattern in terms of building placement, public open space, and interconnected street networks and relates these elements to specific locations on a plan. The test of its form will be whether or not the plan reflects the specific goals and aspirations of CCDPs and is consistent with the local policies and regulations that control development. This process requires an understanding that the plan will enhance the cultural goals, social interaction, and the daily needs of the residents.

The SmartCode[115] model for form-based zoning uses a "transect" approach that establishes a cross-section of habitats in a defined space around which the built form can be arranged. Using urban morphologies, it is useful in understanding the patterns of interactions such as boundary shapes, connectivity, barriers, and those aspects that yield diversity but also coherence. One of the founders of the SmartCode states,

> It offers an alternative to the conventional mechanism of planning implementation (Euclidean zoning). It is essentially a new system of classification, one that organizes the elements of urbanism according to the principles of a transect-based distribution. Planners facilitate this system by learning how to allocate spatially, finding the appropriate location and juxtaposition of urban elements along a continuum of human habitats from urban to rural.[116]

Making a SmartCode an ordinance requires a willing local authority to prepare a floating zone in anticipation of the arrival of CCDPs and to carry out the necessary evaluation of their goals and aspirations in order to translate them into a regulatory form.

115 The SmartCode Version 9.2 is available from the Center for Applied Transect Studies: http://www.smartcodecentral.org/ last accessed March 8, 2011.

116 A. Duany & E. Talen, *Transect Planning* 68 J. Am. Plan. Ass'n, 245, 253–54 (2002).

4. Conclusion

There are significant barriers to the acceptance of displaced persons: no international laws protect them, there is no indication that existing laws will be extended, no domestic laws promote their acceptance, and certainly there are no relevant local laws.[117] It is only the human rights dimension[118] that adds substance to their needs and to the eventual change in law to accommodate their acceptance by a host country.

Long-term planning is necessary for their eventual acceptance because sea level rise is a conclusion that cannot be avoided. In the eventuality that there will be a need for a regulatory model for localized planning and location, New Urbanism and form-based zoning provide a bridge between the goals and aspirations of those persons displaced and the translation of those goals and aspirations into a local regulatory framework.

117 Reviewed in K.K. Moberg, *Extending Refugee Definitions to Cover Environmentally Displaced Persons Displaces Necessary Protection*, 94 Iowa L. Rev. 1107 (2009).

118 Argued persuasively by T.T.V. Duong, *When Islands Drown: The Plight of Climate Change Refugees and Recourse to Human Rights Law*, 31 U. Pa. J. Int'l L. 1239 (2010).

Part IV ESTABLISHING ACCOUNTABILITY

13 Could a Small Island Successfully Sue a Big Emitter?

Pursuing a Legal Theory and a Venue for Climate Justice

Jacob David Werksman

1. Introduction

More than twenty years after the UN General Assembly launched negotiations for a treaty on climate change, the international community has failed to agree on a legally binding instrument that comprehensively limits greenhouse gas emissions or that addresses the consequences of global warming.[1] In the context of this persistent gap in treaty law, this chapter seeks to answer two basic questions:

1. Does public international law provide a substantive basis for holding one State responsible for the impacts of its greenhouse gas emissions on the global environment and on the environment of another State? And, if so,
2. By what judicial or quasi-judicial international procedures (i.e., "litigation") could this substantive obligation be recognized and/or enforced?

1 The Intergovernmental Negotiating Committee for a Framework Convention on Climate Change (INC) was established under the auspices of the United Nations General Assembly. G.A. Res. 45/212, ¶ 1, UN Doc. A/RES/45/212 (Dec. 21, 1990).

The views expressed in this chapter are my own.

In the process of exploring answers to these questions, a third, more strategic, question is raised:

3. Would litigation brought under public international law likely promote or undermine ongoing efforts to conclude a comprehensive, legally binding agreement on climate change?

This chapter focuses on the first and second questions in an effort to suggest a substantive basis and procedural routes whereby a case or a question related to a State's responsibility for the impacts of climate change might be resolved through some form of international litigation.

As to the third question, any complainant State in a climate change action will be asking the adjudicator to clarify the existing substantive law in a progressive manner, but will also run the risk of establishing a negative precedent. The complainant would seek to confirm what many feel intuitively: if it is possible to establish a convincing causal connection between a "source State's" emissions and the consequent impacts on another State, then international law should provide the "impact State" with some form of remedy.[2] Thus it would be advisable for a climate litigant to develop a high degree of confidence that the first question will be answered affirmatively, before weighing the risks implied by the third question.

Several caveats are in order. Although there is a rich and growing academic literature in the area of state responsibility for transboundary environmental harm, the practice of States and tribunals is scant.[3]

2 This intuition is often expressed in the common law legal maxim, captured by William Blackstone, that "[i]t is a settled and invariable principle in the laws of England, that every right when with-held must have a remedy, and every injury it's [sic] proper redress." WILLIAM BLACKSTONE, VOL. 3, COMMENTARIES ON THE LAWS OF ENGLAND 109 (1775).

3 *See, generally* Alan E. Boyle, *State Responsibility and International Liability for Injurious Consequences of Acts not Prohibited by International Law: A Necessary Distinction?*, 39 INT'L & COMP. L.Q. 1 (1990); Malgosia A. Fitzmaurice, *Liability for Environmental Damage Caused to the Global Commons*, 5 REV. EUR. COMMUNITY & INT'L

International courts and tribunals are often instinctively conservative in their interpretation and application of the law, particularly when the implications of their findings could affect all members of the international community. Although science in the area of climate change is increasingly certain, State practice has been slower to respond to climate change. International law derives its authority and much of its legitimacy through the explicit or implicit consent of States.

Although the Kyoto Protocol[4] provides powerful evidence of the direction toward which the majority of the international community has aspired, the specific commitments that reflect what its Parties saw as their legal duties to reduce greenhouse gas emissions are coming to an end in 2012, without having made a significant impact on global emission trends. At present, it is uncertain whether the international community will agree to specific legally binding commitments beyond 2012. In contrast, the commitments and actions recently communicated through the Cancun Agreements and the Durban Platform for Enhanced Action under the UN Framework Convention on Climate Change (UNFCCC)[5] appear to reflect the growing political will of a much broader group of countries to act to reduce their emissions, including through the enactment of ambitious national policies and legislation and the negotiation

ENVTL. L. 305 (1996). David B. Hunter, *The Implications of Climate Change Litigation for International Environmental Law-Making* (Am. Univ. Washington Coll. of Law, Working Paper No. 2008–14, 2007), available at http://ssrn.com/abstract=1005345.

PHOEBE OKOWA, STATE RESPONSIBILITY FOR TRANSBOUNDARY AIR POLLUTION IN INTERNATIONAL LAW (2000); Richard S.J. Tol & Roda Verheyen, *State Responsibility and Compensation for Climate Change Impacts and Damages: A Legal and Economic Assessment*, 32 ENERGY POL'Y 1109 (2004).

4 Kyoto Protocol to the United Nations Framework Convention on Climate Change, Dec. 10, 1997, 37 I.L.M. 22, UN Doc. FCCC/CP/1997/7/Add.1 [hereinafter Kyoto Protocol].

5 Durban Platform for Enhanced Action, Dec. 11, 2011, UN Doc. FCCC/CP/2011/9/Add.1 *available at* http://unfccc.int/resource/docs/2011/cop17/eng/09a01.pdf; United Nations Framework Convention on Climate Change, May 9, 1992, S. Treaty Doc. No. 102–38, 1771 U.N.T.S. 107 [hereinafter UNFCCC].

of a new protocol, legal instrument, or outcome with "legal force" applicable to all parties.[6]

This analysis assumes that a convincing causal connection can be established between greenhouse gas emissions and specific environmental impacts. The nature of the relationship between greenhouse gas emissions and the predicted impacts of climate change is not like other forms of transboundary pollution, such as oil spills, where there is a directly demonstrable link between the pollutant and the impact. Nevertheless the interaction between greenhouse gas emissions, atmospheric chemistry, and global mean temperatures has likely been demonstrated with sufficient scientific confidence that it seems unlikely that an adjudicator would require a complainant, in order to obtain relief, to demonstrate what would not be possible – that a specific emission of greenhouse gases by State S directly caused the specific impact in State I.

This analysis therefore focuses on the remaining challenges of providing a convincing legal theory that can hold a "source State" responsible for the consequent damages to the "impact State" and of identifying a forum with the jurisdiction and the authority to recognize and enforce that responsibility. The issue of causation is a mixed question of fact and law, requiring both scientific evidence of cause and effect and a legal theory that recognizes that the act or omission of the source State is sufficiently proximate to the impact to hold that State responsible for the consequences. This analysis does not focus on claims that might be based on impacts to the global environment as a whole or on claims based on obligations *erga omnes*; that is, duties owed by one State to all States.

6 It must be noted, however, with regard to the Cancun Agreements that the implementation of these policies is unlikely to have a significant aggregate effect on overall temperature trends, raising questions as to whether the motivation behind these policies reflects a sense of "duty" to reduce global warming. *See* UN Env't Programme [UNEP], *The Emissions Gap Report* (Nov. 2010), *available at* http://www.unep.org/publications/ebooks/emissionsgapreport/. Significant uncertainty remains as to whether and how quickly a new legal agreement will emerge from the Durban Platform negotiations and whether it will contain specific commitments to reduce greenhouse gas emissions.

Although this analysis is of a general character, it quickly becomes apparent that it is difficult to reach even preliminary conclusions without making some specific assumptions about the source State and impact State that may be involved in the litigation. For the purposes of the discussion, the following profiles of litigant States will be assumed:

State S, the source State, is responsible for more than 30% of historic (1900–2010) carbon emissions from fossil fuel combustion and approximately 23% of cumulative global carbon emissions from fossil fuel and land use change in the same time period. In 2010 it was responsible for 22% of global carbon emissions. Between 1990, the base year for both the Convention and the Protocol, and 2010, State S's total greenhouse gas emissions have risen by 10.5%. State S's current per capita emissions of greenhouse gases amount to 6.6 tons of carbon equivalent per year, higher than any other country, and nearly twice the average of industrialized countries. State S is an Annex II[7] Party to the Convention and participated actively in the design of the Protocol. State S agreed, at Kyoto, to include a 7% reduction from 1990 levels between 2008–12 as its Protocol target. It signed the Protocol, but after a change of government in 2000, renounced it, claiming that the target is not achievable, is not fair in relation to the obligations of other countries, and would be contrary to its economic interests. State S has recently begun to recognize and directly regulate CO_2 as a pollutant. It has made a recent serious but failed attempt to enact legislation that would have reduced its greenhouse gas emissions approximately 17% from 2005 levels by 2020, and included this target in its Cancun pledge, conditional on its being able eventually to enact such a law. Its government

7 Annex II of the UNFCCC lists those countries that were members of the Organization for Economic Cooperation and Development (OECD) in 1992, when the Convention was adopted. OECD membership was used as a proxy for being highly industrialized, and therefore, under the Convention's principle of common but differentiated responsibilities and respective capabilities, Annex II countries are required to take the lead in reducing emissions and providing financial and technical support to developing countries implementing the Convention. UNFCCC, *supra* note 5, at Annex II.

continues to recognize that climate change is a serious issue that requires real action and claims to spend US$4 billion annually on climate-change-related activities and programs. State S has not consented to the general compulsory jurisdiction of the International Court of Justice either under the "optional clause" or under the UNFCCC's dispute settlement provisions.

State I, the impact State, is a small island, a least-developed country with negligible historic and present greenhouse gas emissions. It is a non-Annex I Party to the Convention and has signed and ratified the Kyoto Protocol. State I is a low-lying coral atoll and is already beginning to experience the predicted impacts associated with global warming in the form of sea level rise and with higher than average sea temperatures in the form of coastal erosion, saltwater intrusion, and coral bleaching. Like State S, State I has not consented to the general compulsory jurisdiction of the International Court of Justice either under the "optional clause" or under the UNFCCC's dispute settlement provisions.

Given the relatively inchoate state of the law and my concerns about the high political and legal risks of pursuing litigation in this context, the analysis that follows is structured in the form of propositions rather than conclusions. The propositions seek to capture, as best as possible, the elements of the substantive and procedural law relating to state responsibility for damages to the environment from greenhouse gas emissions that a climate litigant would seek to have confirmed by international litigation. Any litigation, whether contentious or nonadversarial, whether it led to binding or nonbinding consequences, enforceable or otherwise, would depend in large part on the strength of one or more of these propositions.

2. The Substantive Law on the Limitation of Greenhouse Gas Emissions

This chapter follows an analytical approach based on the International Law Commission's (ILC) *Draft Articles on State Responsibility for*

Internationally Wrongful Acts.[8] It seeks first to define States' primary international legal obligations with regard to limiting harmful greenhouse gas emissions before turning to the secondary rules of State responsibility that arise from the breach of these obligations. Under this approach, a failure to comply with a primary obligation to limit harmful levels of greenhouse gas emissions would be considered a wrongful act that would give rise to the secondary obligation to repair any damage to the environment of another State that resulted from that act.

2.1. Primary Obligations: The Duty to Prevent Harmful Levels of Greenhouse Gas Emissions

A State's primary obligation to limit greenhouse gas emissions can have its source in both treaty and customary law. The treaty instrument governing relations between States S and I is the UNFCCC. Although treaty law and customary law can coexist and support each other, it is often the case that a rule expressed in a treaty is more specific, and would prevail to the extent it conflicted with any (non-peremptory) customary law rule governing the same subject matter. Any source State that is a party to the UNFCCC is likely to argue that, by ratifying and participating in the implementation of that treaty, it has fulfilled both its treaty and customary law obligations related to climate change. In anticipation of that argument, a number of small island States, when signing or ratifying the UNFCCC, claimed that joining the Convention

> shall in no way constitute a renunciation of any rights under international law concerning state responsibility for the adverse effects of climate change, and that no provisions in the Convention can be interpreted as derogating from the principles of general international law.[9]

8 Int'l Law Comm'n [ILC], *Draft Articles on Responsibility of States for Internationally Wrongful Acts, with Commentary*, *in* Rep. of the ILC, 53d Sess., Apr. 23-June 1, June 2-Aug. 10, 2001, ¶76, U.N. Doc. A/56/10; GAOR, 56th Sess., Supp. No. 10 (2001) [hereinafter ILC, *Draft Articles*].

9 *Declarations by Parties*, UNFCCC, Fiji, Kiribati, Nauru, Papua New Guinea, http://unfccc.int/essential_background/convention/items/5410.php (last visited July 31, 2011).

There is no evidence that any other Party formally acknowledged these declarations, and their legal effect has never been authoritatively tested.

Proposition 1: *All States that are Annex II*[10] *Parties to the UNFCCC are committed to the common objective of stabilizing greenhouse gas concentrations in the atmosphere at a level that would prevent dangerous anthropogenic interference with the climate system and to taking the lead in combating climate change and the effects thereof. Such States are under a specific obligation to take measures that limit their greenhouse gas emissions and that modify longer-term trends in these emissions consistent with the objective of the Convention. Such States are also under an obligation to assist developing-country Parties that are particularly vulnerable to the adverse effects of climate change in meeting the costs of adaptation to those adverse effects.*

This proposition is based on Articles 2, 4.2(a), and 4.4 of the Convention.[11] In the context of climate litigation, this proposition could form the basis of a claim for breach of treaty or could provide a source of applicable international standards for a customary law duty of due diligence.

Proposition 2: *All States must, under rules of customary international law, exercise due diligence in limiting their emissions of greenhouse gases in such a way as to prevent damage to the global environment and to the environments of other States.*

This proposition is based on the widely recognized principle of customary international law, expressed in Principle 21 of the 1972 Stockholm Declaration,[12] as more recently restated in the 1992 Rio Declaration on Environment and Development:

10 *See* UNFCCC, *supra* note 5, at Annex II.

11 *Id.* at art. 2, 4.2(a), 4.4.

12 United Nations Conference on the Human Environment, Stockholm, Swed., June 5–16, 1972, *Declaration of the United Nations Conference on the Human Environment*, Princ. 21, UN Doc. A/CONF.48/14/Rev.1, 11 I.L.M. 1416 (June 16, 1972).

> States have, in accordance with the Charter of the United Nations and the principles of international law, the sovereign right to exploit their own resources pursuant to their own environmental and developmental policies, and the responsibility to ensure that activities within their jurisdiction or control do not cause damage to the environment of other States or of areas beyond the limits of national jurisdiction.[13]

The International Court of Justice (ICJ), in its advisory opinion on the *Legality of the Threat or Use of Nuclear Weapons*, confirmed that the "existence of the general obligation of States to ensure that activities within their jurisdiction and control respect the environment of other States or of areas beyond national control is now part of the corpus of international law relating to the environment."[14]

However, for this general obligation to take on meaning in a specific circumstance, it must be possible to describe the contours of that duty of due diligence applicable to a particular State in the context of greenhouse gas emissions and climate change impacts. Although there are many general descriptions of the concept of due diligence in treatises and academic literature, it may be useful, in this situation, to draw on the Third Restatement of Foreign Relations Law, which provides a persuasive interpretation of international law from the perspective of U.S. scholars and practitioners:

> § 601. STATE OBLIGATIONS WITH RESPECT TO ENVIRONMENT OF OTHER STATES AND THE COMMON ENVIRONMENT
>
> (1) A State is obligated to take such measures as may be necessary, to the extent practicable under the circumstances, to ensure that activities within its jurisdiction or control

13 United Nations Conference on Environment and Development, Rio de Janiero, Braz., June 3–14, 1992, *Rio Declaration on Environment and Development*, Princ. 2, U.N. Doc. A/CONF.151/26 (Vol. 1), Annex 1 (Aug. 12, 1992).

14 Legality of the Threat or Use of Nuclear Weapons, Advisory Opinion, 1996 I.C.J. 226, ¶ 29, at 241 (July 8).

a conform to generally accepted international rules and standards for the prevention, reduction, and control of injury to the environment of another State or of areas beyond the limits of national jurisdiction; and

b are conducted so as not to cause significant injury to the environment of another State or of areas beyond the limits of national jurisdiction.[15]

From this statement and the sources of international and domestic law on which it is based, it is possible to conclude the following:

Proposition 3: *The duty of a State to exercise due diligence to prevent damage to the global environment or the environment of another State extends to all activities within the source State's territory and control, including the activities of private entities. The duty extends to both acts and omissions of the State, including the failure to regulate. The duty is defined by reference to objective standards of international law applicable to the source State. The duty is not predicated on a showing of "fault" in terms of "intent to do harm."*

The UNFCCC and the Kyoto Protocol reinforce this proposition in the context of climate change, requiring each Party to account for all significant greenhouse gas emissions occurring within its territory regardless of whether the source of emissions is the private or public sector.[16] The enforcement "consequences" agreed on in preparation for the entry into force of the Kyoto Protocol take the same approach. They are automatic

15 Restatement (Third) of Foreign Relations Law of the U.S. § 601 (1987) [hereinafter Restatement].

16 Two possible exceptions to this element of the proposition are aviation and marine bunker fuels, and "emissions resulting from multilateral operations pursuant to the Charter of the United Nations." Although an Annex I Party is required to report on these emissions, they do not form a part of national totals. UNFCCC, Rep. of the Conference of the Parties on its 3d Sess., at 31, Dec. 1–11, 1997, UN Doc. FCCC/CP/1997/7/Add.1 (Mar. 25, 1998).

and calibrated to respond to the extent to which a Party has exceeded its emissions, and they do not consider "fault."[17]

Beyond these general characteristics, the duty of due diligence becomes highly contextual and must be further defined in relation to greenhouse gas emissions and their impacts. The Third Restatement points to several elements of the duty of due diligence that would have to be defined by a climate litigant with respect to an identified source State:

1. What, if any, are the sources of "generally accepted international rules and standards for the prevention, reduction, or control of injury" to the environment caused by climate change that are applicable to the source State?[18]
2. What role do the specific circumstances of the source State play in determining what kinds of measures are "practicable under the circumstances?"[19]
3. What is a sufficiently "significant injury" to trigger the duty?[20]

The question of how much due diligence states must demonstrate is the area of the law of state responsibility, as it relates to climate change, that is least developed. The international rules and standards referred to in the Restatement will have their source in State practice as reflected in both customary international law and in treaty law. What we can say is that nearly all Member States of the United Nations, including State S, have ratified the UNFCCC. At present, all States allow activities within their jurisdiction that contribute in different degrees to net global greenhouse gas emissions. Most, including State S, have policies or

17 *See* Jacob Werksman, *The Negotiations of a Kyoto Compliance System*, *in* IMPLEMENTING THE CLIMATE REGIME: INTERNATIONAL COMPLIANCE 17–37 (Olav Schram Stokke, Jon Hovi, & Geir Ulfstein eds., 2005); Geir Ulfstein & Jacob Werksman, *The Kyoto Compliance System: Towards Hard Enforcement, in* IMPLEMENTING THE CLIMATE REGIME: INTERNATIONAL COMPLIANCE 39–62 (Olav Schram Stokke, Jon Hovi, & Geir Ulfstein eds., 2005).

18 Restatement, *supra* note 15, at §601.

19 *Id.*

20 *Id.*

regulations in place that directly or indirectly affect the levels and trends in those emissions. Therefore, some level of emissions is clearly not wrongful.

State practice has thus far provided little guidance as to how the level of permissible emissions could be assessed. Article 2 of the UNFCCC indicates that atmospheric concentrations of greenhouse gases should be stabilized at safe levels,[21] and from that it would be possible to extrapolate what global emissions should amount to over the coming decades and century. In 2010, in Cancun, the Conference of the Parties (COP) to the UNFCCC recognized that

> deep cuts in global greenhouse gas emissions are required according to science, and as documented in the Fourth Assessment Report of the Intergovernmental Panel on Climate Change, with a view to reducing global greenhouse gas emissions so as to hold the increase in global average temperature below 2°C above preindustrial levels, and that Parties should take urgent action to meet this long-term goal, consistent with science and on the basis of equity.[22]

The COP also agreed to consider strengthening this long-term global goal "on the basis of the best available scientific knowledge, including in relation to a global average temperature rise of 1.5°C."[23] The scientific literature suggests that limiting global warming to 2°C would require actions to stabilize greenhouse gas concentrations in the atmosphere at 450 parts per million CO_2 equivalent or less, whereas a 1.5°C change would require the international community to stabilize at 350 parts per million. Current concentrations are estimated to be at 394 parts per million CO_2.[24]

21 UNFCCC, *supra* note 5, at art. 2.

22 UNFCCC, Rep. of the Conference of the Parties on its 16th Sess., at 3, Nov. 29-Dec.10, 2010, U.N. Doc. FCCC/CP/2010/7/Add.1 (Mar. 15, 2011).

23 *Id.*

24 National Oceanic & Atmospheric Administration, Earth System Research Laboratory, "Trends in Atmospheric Carbon Dioxide" (last visited June 7, 2012), *available at* http://esrl.noaa.gov/gmd/ccgg/trends/global.html.

Yet determining what permissible levels should be for a given State or group of States is far more challenging. The general principle that States have "common but differentiated responsibilities and respective capabilities" and that these require developed countries, in general, to undertake a greater burden than developing countries is widely accepted. What remains controversial is how this principle should be interpreted in a contemporary context where a country like China, which by many metrics remains a developing country, is in 2012 the world's second largest economy and the largest current emitter of greenhouse gases. It is therefore unclear whether the differentiation between permissible levels should be measured, for example, in terms of total emissions, per capita emissions, energy efficiency, or some other methodology.

The Kyoto Protocol's Annex B provides an indication of where the division of responsibility might lie, but this division was derived from negotiations rather than any generalizable standards or principles.[25] The more recent "commitments" and "actions" submitted by Parties and "taken note of" by the Conference of the Parties (COP) in Cancun reflect what many Parties are willing to do unilaterally (subject, in many cases, to significant qualifications and conditions) between 2010 and 2020.[26] At the subsequent Conference of the Parties in Durban in 2011, the COP agreed to launch a process to develop a protocol, another legal instrument, or an agreed outcome with legal force under the Convention applicable to all Parties and that "raises the level of ambition."[27]

It is, however, theoretically possible that a court or other adjudicator could determine in the context of specific litigation that a particular State, through its acts or omissions, failed to comply with its duty of due diligence by having exceeded a permissible level of emissions, without

25 *See, e.g.*, SEBASTIAN OBERTHÜR & HERMANN E. OTT, THE KYOTO PROTOCOL: INTERNATIONAL CLIMATE POLICY FOR THE 21ST CENTURY (1999).

26 See *The Cancun Agreements: Mitigation*, UNFCCC, http://cancun.unfccc.int/mitigation/ (last visited July 31, 2011), for a compilation of these commitments and actions.

27 UNFCCC, Rep. of the Conference of the Parties on its 17th Sess., at 3, Nov. 28-Dec.11, 2011, UN Doc. FCCC/CP/2011/9/Add.1 (15 March 2012).

that court having to determine with precision what that permissible level is. In other words, if the wrongful act was sufficiently egregious, *noncompliance* with a standard could be identified without having to describe in precise detail what *compliance* with that standard would require.

In this case, State S historically and currently emits the highest total and per capita levels of greenhouse gases. It is presumed, for the purpose of this analysis, that the adjudicator accepts that the impacts experienced by State I constitute a "serious injury" attributable to human-induced climate change, and that the adjudicator is prepared to accept a legal theory that attributes those impacts to State S's emissions.

The other elements of the standard of due diligence require a case-by-case assessment of the specific circumstances of the source State, as well as an assessment of what is practicable or possible. This standard may, for example, take into account the economic capacity of the source State. This does not, however, imply that the standard is a subjective one. It is not for the source State to determine unilaterally what is practicable or possible for it. The source State's conduct must be tested objectively against prevailing international standards applicable to that State (i.e., what could be reasonably expected of a diligent government in similar circumstances acting in accordance with its environmental obligations). State S's behavior would, therefore, be tested against the practice of its "peers;" in this case, wealthy industrialized countries that have undertaken similar international commitments in relation to climate change.

Proposition 4: *At a minimum, Articles 2, 4.2 (a), and 4.2(b) of the UNFCCC inform the relevant standard of due diligence applicable to an Annex II Party to the Convention.*[28] *The failure of a State that is an Annex II Party to the Convention, and that is historically and currently responsible for more than 20% of global emissions of greenhouse gases, to undertake actions consistent with a downward trend in those emissions constitutes a breach of that State's duty of due diligence.*

28 UNFCCC, *supra* note 5, at art. 2, 4.2(a)–(b).

There are, of course, other standards, principles, and sources of law that could inform the relevant standard of due diligence for State S. These could include the precautionary principle, the polluter pays principle, and the procedural obligation to carry out and act on environmental impact assessments. Such standards, to the extent they were found to be applicable to State S, would play a supportive role in demonstrating the basic primary obligation set out in these four propositions.

2.2. Secondary Obligations: State Responsibility for an Internationally Wrongful Act

Proposition 5: *The breach of the primary obligation described in the foregoing propositions is an internationally wrongful act, attributable to the source State and, therefore, entailing the international responsibility of that State.*

This proposition is based on rules of customary international law as reflected in the ILC Draft Articles 1 and 2, which would support a finding of an obligation of the source State to provide a remedy by way of ceasing its harmful emissions and/or by providing reparations for the injury caused by these emissions.[29]

Judicially fashioned remedies amounting to specific performance are rare in international practice, and the remedies considered here in theory would prove difficult for any court to fashion in practice. A climate litigant that was successful in establishing a breach of the source State's primary obligation may find that the court was unwilling to do more than order the disputants to return to the negotiating table.

Proposition 6: *In accordance with rules of state responsibility, the source State is under an obligation to cease those emissions of greenhouse gases that are in breach of its duty of due diligence.*

29 ILC, *Draft Articles*, *supra* note 8, at art. 1–2.

This Proposition reflects ILC Draft Article 30.[30] The duty to cease an internationally wrongful act is a duty that is independent of the duty to make reparation for the injury caused by that act. As has been indicated, it may not be necessary to establish precisely what the source State's duty of due diligence is in terms of maximum permissible emissions to show that that duty has been violated. However, to fashion a remedy based on an obligation to cease the offending behavior, an adjudicator would have to determine what would constitute a permissible level of emissions.

Although the ILC Draft Articles' rules refer to "cessation" as the appropriate remedy,[31] it may be more likely that a court or tribunal would, under the circumstances, interpret this obligation to require the source State to control and reduce its emissions over time.

Proposition 7: *The source State is under an obligation to the impact State to make full reparation for the injury caused by those levels of greenhouse gas emissions that are in excess of the source State's duty of due diligence.*

This proposition reflects ILC Draft Article 31. It sets out the general principle that a source State must "make good" the injury caused to State I by the internationally wrongful act. The application of this rule of state responsibility raises several challenges peculiar to climate change. According to ILC Draft Article 31, the accompanying commentary, and supporting literature, a State is responsible to "make good" damages only to the extent that they are caused by the internationally harmful act.[32]

This raises a number of crucial points that could provide a basis for limiting the extent of State S's responsibility for remedying the injury to State I:

30 *Id.* at art. 30.
31 *Id.*
32 *Id.* at art. 31.

1. Is the causal link between State S's emissions and the injury suffered by State I sufficiently direct and "proximate" to conclude, as a matter of law, that the emissions *caused* State I's injuries?

This is a mixed question of fact and law. A causal link could be inferred from the structure of the Convention's commitments, which call on industrialized countries to take the lead both in cutting emissions and in providing technical and financial support for adaptation. Existing international precedent and academic writings would, however, exclude reparation for injuries that are "too indirect, remote, and uncertain to be appraised."[33] Although this analysis presumes that a causal chain can successfully be established between State S's emissions and State I's impacts, previous findings of state responsibility for damage have involved comparatively much shorter causal chains between the wrongful act or omission and the complained-of harm, such as the transboundary impacts of industrial emissions from Canada on U.S. forests, which formed the basis for state responsibility in the *Trail Smelter Arbitration*.

2. Is State S responsible only for the injury caused by the level of emissions that exceed a threshold of permissible emissions?

If, as has been indicated, all States can lawfully emit a certain level of greenhouse gases, it would seem that State S could be held responsible only for any injury caused by emissions that exceeded those permissible levels. It could, however, be argued that the impacts caused by these emissions are the result of a cumulative effect and therefore inseparable. In the same way that a reckless driver who collides with a pedestrian while driving 20 mph above a 60-mph speed limit bears the full responsibility for injury caused by an 80-mph collision, it could be argued that

33 Rep. of the ILC, 53d Sess., Apr. 23-June 1, June 2-Aug. 10, 2001, at 227, UN Doc. A/56/10; GAOR, 56th Sess., Supp. No. 10 (2001) (citing Trail Smelter Arbitration (*U.S. v. Canada*), 3 R.I.A.A. 1905, 1931 (1938)).

State S must bear the responsibility for injury associated with the full impact of its cumulative emissions.

3. Can State S be held responsible for any injury that might be associated with its historical emissions, which may date from a time when the potential impacts of greenhouse gas emissions were not as well understood as they are today?

A court or tribunal might well limit State S's responsibility to those impacts considered to be reasonably foreseeable at the time the emissions were allowed to occur. Because greenhouse gases accumulate and reside in the atmosphere over long periods of time, this issue is closely related to Question 2. In other words, the application of a foreseeability requirement might place an historical cap on State S's responsibility for its cumulative emissions, holding State S responsible only for damages attributable to emissions after a date when the potential impacts of greenhouse gas emissions were widely known.

Proposition 8: *The existence of multiple source States does not preclude the impact State from recovering reparations from one source State for injury attributable to the emissions of that State.*

Any claim of state responsibility for damages caused by climate change will raise significant issues related to the involvement of multiple source states (i.e., a "plurality of responsible States"). Currently, all States allow greenhouse gas emissions from their territories. A relatively small number are responsible for emitting the largest share currently and historically. The judgments of international courts and tribunals do not have a binding precedential character for States not parties to the dispute, but a determination of state responsibility with regard to one State will have implications for the rights and obligations of all States.

It is well settled that an international court will not have jurisdiction over a potential respondent State without its consent. This gives rise to the possibility, in the context of a plurality of responsible States, that a claim would be brought in which one or more potentially responsible States were outside a court's jurisdiction. Unlike many domestic legal entities, an international court cannot force an interested respondent to join the dispute. In this circumstance, a court must weigh whether to dismiss the case on the basis that the party outside its jurisdiction is indispensable to the resolution of the case, or to proceed without it. The ICJ jurisprudence on this issue is mixed and inconclusive. It has in certain circumstances declined jurisdiction when an interested State is not joined (Monetary Gold Case)[34] and, in others, has found it could proceed (Nauru).[35] If a court does take jurisdiction over a case where one or more interested respondents are not joined, it will need a theory of responsibility for determining how to balance the interests of the claimant in securing full and adequate relief, the interests of the joined State, and the responsibilities of the State not party to the dispute.

The ICJ has never encountered the issue of a plurality of responsible States in the context of a judgment on its merits. Some commentators have called for the application of the concept of "joint and several" liability, which domestic courts have used to hold a single tortfeasor liable for injuries attributable to multiple tortfeasors.[36] The ILC has cautioned against this approach, indicating that unless it can be established that multiple responsible States are jointly responsible for the same act, "the principle of independent responsibility reflects the position under general international law."[37]

Proposition 9: *Any wrongful emissions by the impact State that are found to have contributed to the injury caused by the source State would not bar*

34 Monetary Gold Removed from Rome in 1943 (*Italy v. Fr, U.K., U.S.*), 1954 I.C.J. 19 (June 15).

35 Certain Phosphate Land in Nauru (*Nauru v. Austl.*), 1992 I.C.J. 240 (June 26).

36 *E.g.* Matthew F. Pawa, "Global Warming: The Ultimate Public Nuisance," 39 Envtl. L. Rep. 10230 (2009).

37 Certain Phosphate Land in Nauru, *supra* note 35, at 314–16.

the impact State from seeking and obtaining reparations. The source State would, however, be responsible only for the injury resulting from its own wrongful emissions.

This is the approach taken by ILC Draft Article 39 on contribution to the injury.[38] Because State I's emissions are in any case negligible, the proposition is less relevant to the case at hand.

3. Judicial and Quasi-Judicial Institutions and Procedures for Recognizing and Enforcing State Responsibility for Climate Change Impact

Because neither State S nor State I has subjected itself to compulsory and binding dispute settlement procedures, either generally before the ICJ or specifically under the UNFCCC, the greatest challenge to proceeding with climate litigation between these States is finding a venue with jurisdiction. The two routes suggested here are therefore premised on overcoming that procedural hurdle and then accepting the consequence of weaker remedies that arise from advisory or recommendatory procedures.

3.1. Case 1

Request by the UN General Assembly for an advisory opinion on whether a State that fails to take steps to limit its greenhouse gas emissions in a manner that can be shown to have contributed to an injury to the global environment or to the environment of another State is responsible to make reparation for that injury.

Venue: International Court of Justice

Cause of action: Request for an advisory opinion

38 ILC, *Draft Articles*, *supra* note 8, at art. 39.

Subject matter jurisdiction: Article 65 of the Statute of the International Court of Justice provides that "the court may give an advisory opinion on any legal question at the request of whatever body may be authorized" by the UN Charter to make such a request.[39] The question set out in Case 1, which essentially seeks to confirm Propositions 2, 3, 5, 6, and 7, is a legal question.

Standing of claimant/applicant: Article 96 of the UN Charter authorizes the General Assembly to request an advisory opinion of the Court.[40] The Court has interpreted this provision and Article 65 of the Statute liberally, taking up all UN General Assembly advisory requests made thus far.

Jurisdiction over defendant/respondent: N/A

Remedy: An advisory opinion of the ICJ is nonbinding and provides no specific remedy. It does, however, provide an authoritative interpretation that could result in the progressive development of the law of climate change and its impacts and could help bring pressure on all governments to ratify and implement the Protocol and to continue to cooperate in reducing their emissions beyond the Kyoto period.

Comment/risk analysis: Initiating a request for an advisory opinion from the General Assembly requires broad political support from both developed and developing countries. To gain support from developing countries, the question would have to be crafted in a way that would avoid the implication that it could be used to pressure them to undertake commitments before they are ready or that it could suggest that they may be held responsible for failing to do so. One large developing country is currently the highest emitter of greenhouse gases.

39 Statute of the International Court of Justice art. 65, June 26, 1945, 59 Stat. 1055, 156 U.N.T.S. 77.

40 UN Charter art. 96.

3.2. Case 2

A dispute between Parties to the UNFCCC, concerning the interpretation and application of Articles 2, 4.2(a), and 4.2(b) of the Convention in relation to State S's failure to limit its greenhouse gas emissions.

Venue: A conciliation commission established under Article 14.6 of the UNFCCC

Cause of action: Breach of treaty, Proposition 1, as applied to State S

Subject matter jurisdiction: UNFCCC conciliation procedures have subject matter jurisdiction over any dispute between two or more Parties to the UNFCCC, concerning interpretation or application of the Convention, if, twelve months after State I's notification of the dispute to State S, the dispute has not been settled by negotiation or been submitted to arbitration or adjudication.[41]

Standing of respondent/claimant/applicant: State I, as a Party to the UNFCCC, may request the establishment of a Conciliation Commission.[42]

Jurisdiction over defendant/respondent: The conciliation procedures, as set out in the UNFCCC, are compulsory and shall be submitted to the commission at the request of any of the parties to the dispute.[43]

Remedy: The procedures, as set out in the UNFCCC, can result only in a "recommendatory award, which the parties [to the dispute] shall consider in good faith."[44] This award could, however, result in a progressive clarification of the meaning of Article 2 and bring public pressure to bear on State S to demonstrate how it intends to reduce its emissions unilaterally.

41 UNFCCC, *supra* note 5, at art.14, para. 5.
42 *Id.* at art. 14, para. 6.
43 *Id.*
44 *Id.*

Comment/risk analysis: Although the UNFCCC's basic rules on conciliation are set out in Article 14, the Convention also provides that additional rules on conciliation "shall be adopted by the [COP], as soon as practicable."[45] The COP has never adopted these rules, and no Party has formally submitted a proposed draft of these rules. State S would likely argue that conciliation procedures could not move forward until these rules are adopted, and it could either seek to prevent their adoption or express its "non-acceptance" of these procedures under Article 16, Paragraph 3.[46]

45 *Id*. at art. 14, para. 7.
46 *Id*. at art. 16, para. 3.

14 Making Good the Loss

An Assessment of the Loss and Damage Mechanism under the UNFCCC Process

Ilona Millar, Catherine Gascoigne, and Elizabeth Caldwell

1. Introduction

Loss and damage caused by climate change will have profound implications on human settlements and the protection of affected persons' human rights, including their rights to health, culture, decent living standards, self-determination, nationality, and, above all, the right to life. Although each of these human rights is broadly guaranteed through pre-existing international instruments, such instruments were never formed with the specific problems posed by climate-change-induced loss and damage in mind.

Accordingly, there is no coordinated legal or administrative framework that specifically aims to manage climate-related risk, protect human lives and property from the impacts of climate change, or compensate those affected for the loss and damage caused. In particular, there is no framework that supports populations displaced by climate-change-induced loss and damage and allows them to migrate to safer territory in an orderly manner.

For many years, migration was not discussed in the context of adaptation to climate change, because many developing countries considered it a mechanism of last resort, reflecting the failure of developed countries adequately to address the mitigation challenge. However, in the face of increased and repetitive climate-induced disasters, it is becoming clear that some climate change impacts are unavoidable and that it will not be possible to protect certain communities from damage in their current locations.

Some commentators argue, in the context of responding to climate-change-induced loss and damage, that States have an obligation to prevent damage and minimize the risk of damage to other States (the no-harm principle). In this vein, such commentators note that States have a responsibility to develop further international law to address the problem of liability and compensation for the adverse effects of environmental damage caused by activities within their control.[1] Although principles such as "common but differentiated responsibilities and respective capabilities" and "the polluter pays" underpin the United Nations Framework Convention on Climate Change (UNFCCC) and the Kyoto Protocol, taking that next step to apportion liability and seek compensation for climate change impacts has not found international support.

Recognizing the absence of these necessary frameworks, a number of Parties to the UNFCCC, including members of the Alliance of Small Island States (AOSIS),[2] have called for the establishment of measures to address climate-change-induced loss and damage through the adaptation framework being discussed for the post-2012 period. As is seen in the following discussion, coming to an international consensus about detailed obligations of, and financial support for, a loss and damage mechanism under the UNFCCC's adaptation framework is plagued with logistical and political difficulties. Nevertheless, such a mechanism may

1 United Nations Conference on Environment and Development, Rio de Janeiro, Braz., June 3–14 1992, *Rio Declaration on Environment and Development*, Principle 13, 31 ILM 874 (June 14, 1992).

2 AOSIS, the Alliance of Small Island States, is a coalition of small island and low-lying coastal countries that share similar development challenges and concerns about the environment. For a complete list of members, see http://www.sidsnet.org/aosis/members.html (last visited July 25, 2011).

significantly improve the ability of particularly vulnerable countries to adapt and respond to climate change.

When referring to a loss and damage mechanism in this chapter, we are proposing the development of an institutional structure and funding mechanism that will assist vulnerable developing countries in addressing the un-avoided and unavoidable loss and damage resulting from climate change. Building on work carried out by AOSIS, such a mechanism would focus on risk management, insurance, and rehabilitation/compensation. The optimal design of such a mechanism would reduce vulnerability and enhance adaptive capacity,[3] providing adequate and predictable funding to address direct and indirect (consequential) losses.

This chapter looks at the legal basis for a loss and damage mechanism under the UNFCCC and provides an overview of the elements that may comprise a loss and damage mechanism. It considers the funding structures that may support such a mechanism and highlights how such a mechanism may help particularly vulnerable developing countries respond to the unavoidable impacts of climate change, including migration.

In our view, extending the scope of a loss and damage mechanism to cover all or part of the costs of relocating climate-change-displaced communities has the potential to provide a coherent and integrated framework that will support longer-term risk reduction strategies, as well as providing access to both public and private sources of funding.

2. Adaptation – Impacts and Vulnerability

Least Developed Countries (LDCs) and Small Island Developing States (SIDS) are recognized under the UNFCCC as being particularly

3 Assoc. of Small Island States [AOSIS], *Proposal to the AWG-LCA: Multi-Window Mechanism to Address Loss and Damage from Climate Change Impacts*, at 2, *available at* http://unfccc.int/files/kyoto_protocol/application/pdf/aosisinsurance061208.pdf (last visited Mar. 30, 2011).

vulnerable to climate change. Their particular vulnerability stems from three main conditions:

1. They have the greatest **exposure** to climate change.
2. They are most **sensitive** to climate change; and
3. They have the least developed **adaptive capacity** to climate change, in that they lack the necessary institutional, economic, and financial capacity to cope with the adverse effects of climate change (and are therefore often reliant on foreign aid in response to natural disasters).

Many SIDS have much of their territory only a few meters above sea level, with large portions of their populations living in close proximity to the coast. As sea level rises and as the intensity of extreme weather events such as hurricanes, storm surges, and floods increases, many of these countries suffer, and will continue to suffer, significant loss and damage (for a more thorough discussion of climate impacts on small island settlements, see Chapter 2). By way of example, Hurricane Ivan, which hit Grenada in 2004, damaged 90% of the housing stock and hotel rooms, devastating the tourism industry and setting economic growth back significantly.[4]

In 2008, more than 36 million people were displaced, internally and across borders, due to sudden-onset "climate-related disasters."[5] In the

4 UN Framework Convention on Climate Change [UNFCCC], *Physical and Socio-Economic Trends in Climate-Related Risks and Extreme Events, and Their Implications for Sustainable Development*, at 47, FCCC/TP/2008/3 (Nov. 20, 2008).

5 Internal Displacement Monitoring Centre [IDMC], Monitoring Disaster Displacement in the Context of Climate Change, at 9 (Sept. 22, 2009), available at http://www.unhcr.org/refworld/docid/4ab9cd4e2.html (last visited Feb. 14, 2011). The term "climate-related disasters" is used in the report to mean all events in the meteorological, hydrological, and climatological categories of the Emergency Events Database (the primary source for the study). The category "climate-related disaster" is distinct from the Emergency Events Database's definition of "climatological" disasters, which are defined as events caused by long-lived/meso- to macro-scale processes (in the spectrum from intraseasonal to multidecadal climate variability), including such events as extreme temperature, drought, and wildfire. The author notes that due to the sporadic nature of hazards such as floods and storms, this study can be considered a

same year, internal displacement due to slow-onset climate-related disasters including drought and desertification affected an additional 26.5 million people.[6] Although many of these impacts were first seen in Africa, inhabitants of SIDS are increasingly being confronted with the reality of having to leave their low-lying coastal homes. In countries such as Vanuatu, whole communities have already been relocated to higher ground.[7] Estimates of the scale of climate-change-induced human displacement vary, but Professor Norman Myers' figure of 200 million climate migrants by 2050 has become widely accepted (for more information on current displacement in small island nations, see Chapter 11).[8]

In 2008 the UNFCCC Secretariat prepared a technical paper that noted that the average annual number of climate-related disasters in the period 2000–6 was 87% higher than in the period 1987–98.[9] In addition, losses due to natural disasters in developing countries were twenty times greater as a percentage of GDP than in industrialized countries.[10] One of the reasons developing countries suffer so much more than developed countries is that many of the tools that facilitate adaptation in wealthier countries, such as access to appropriate risk sharing and risk transfer mechanisms (e.g., insurance),[11] are not readily available or affordable to

minimum estimate of displacement as a result of climate-related events and processes in 2008 only.

6 *Id.* at 12.

7 Eleanor Hall, *Vanuatu Village Relocated due to Rising Sea Levels*, THE WORLD TODAY, ABC News, Dec. 6, 2005, http://www.abc.net.au/worldtoday/content/2005/s1524755.htm.

8 International Organization for Migration, *Migration and Climate Change*, at 11, IOM MRS 31(2008) (*prepared by* Oli Brown), *available at* http://www.migrationdrc.org/publications/resource_guides/Migration_and_Climate_ Change/ MRS-31.pdf.

9 UNFCCC, *Mechanisms to Manage Financial Risks from Direct Impacts of Climate Change In Developing Countries*, ¶ 41, FCCC/TP/2008/9 (Nov. 21, 2008), *available at* http://unfccc.int/resource/docs/2008/tp/09.pdf.

10 *Id.*

11 See ECONOMICS OF CLIMATE ADAPTATION WORKING GROUP, SHAPING CLIMATE RESILIENT DEVELOPMENT: ECONOMICS OF CLIMATE ADAPTATION (2009), *available at* http://www.mckinsey.com/App_Media/Images/Page_Images/Offices/SocialSector/PDF/ECA_Shaping_Climate%20Resilent_Development.pdf.

them. As a result, these countries are reliant on financial assistance from donor countries to respond to extreme events, rather than being able proactively to manage risk and enhance resilience.

3. What is the Legal Basis for a Loss and Damage Mechanism?

The UNFCCC places obligations on developed-country Parties to provide such financial resources as are needed by developing-country Parties to facilitate their adequate adaptation to climate change (Articles 4(3)– 4(5)). Article 4(8) of the Convention, moreover, stipulates that "the Parties shall give full consideration to what actions are necessary under the Convention, including actions related to funding, insurance and the transfer of technology, to meet the specific needs and concerns of developing country Parties arising from the adverse effects of climate change and/or the impact of the implementation of response measures, especially [in the most vulnerable States]." The Kyoto Protocol reiterates the concept of adaptation and compensation for loss. For example, in Article 3(14) of the Protocol the Parties are obliged to consider measures to minimize the adverse consequences of climate change, including "the establishment of funding, insurance and transfer of technology."

However, the broadly defined nature of these obligations has meant that little progress has been made in developing specific frameworks to share risk, implement insurance mechanisms, and ultimately compensate developing countries for climate-change-induced losses and damages.

3.1. No-Harm Principle

Under international law, States have an obligation not to inflict damage or violate the rights of other States.[12] Over the last century, this principle has been progressively extended to cover transboundary

12 Richard Tol & Roda Verhayen, *State Responsibility and Compensation for Climate Change Damages – A Legal and Economic Assessment*, 32 Energy Pol'y 1109 (2004).

environmental harm, creating the widely recognized rule of customary international law,[13] the no-harm principle (for more information on the responsibilities of States for greenhouse gas emissions, see Chapters 8 and 12).

The underlying basis of the no-harm principle is that States have a responsibility not to cause damage to the environment of other States or to areas beyond their national jurisdiction. The principle, first recognized in the *Trail Smelter Arbitration* case,[14] has been applied by the International Court of Justice (ICJ) in a number of cases,[15] most recently in its 1996 advisory opinion on *The Legality of the Threat or Use of Nuclear Weapons*, where it stated,

> [T]he existence of the general obligation of States to ensure that activities within their jurisdiction and control respect the environment of other State or areas beyond national control is now part of the corpus of international law relating to the environment.[16]

In addition, the rule has been reiterated in international instruments and treaties, including Principle 21 of the 1972 Stockholm Declaration,[17] Principle 2 of the 1992 Rio Declaration,[18] and in the preamble to the UNFCCC.[19]

13 Customary international law is defined by Article 38 of the Statute of the International Court of Justice as "evidence of general practice accepted as law," Statute of the International Court of Justice art. 38, June 26, 1945, 3 Bevans 1179.

14 Trail Smelter Arbitration, which found that "no State has the right to use or permit the use of its territory in such a manner as to cause injury by fumes in or to the territory of another or of property or persons therein." Trail Smelter Arbitration, 3 R. Int'l Arb. Awards 1905, 1965 (Can.-U.S. 1941).

15 Lac Lanoux Arbitration (Fr. v. Spain), 24 ILR 101 (Fr.-Spain. 1957); Corfu Channel Case (U.K. v. Alb.), I.C.J. Rep. 4 (1949).

16 The Legality of the Threat or Use of Nuclear Weapons, Advisory Opinion, 1996 I.C.J. 226, at 242 (July 8).

17 Stockholm Declaration on the United Nations Conference on the Human Environment, Principle 22, 11 ILM 1416 (June 16, 1972).

18 United Nations Conference on Environment and Development, Rio de Janeiro, Braz., June 3–14 1992, *Rio Declaration on Environment and Development*, Principle 13, 31 ILM 874 (June 14, 1992).

19 United Nations Framework Convention on Climate Change [UNFCCC], Preamble, ¶ 9, UN Doc. A/AC.237/18 (May 9, 1992).

The no-harm principle has two parts: States have a duty to both (1) prevent transboundary environmental harm and (2) redress the damage if the transboundary harm occurs.[20] Legal obligations are therefore created before any harm has occurred, and States must right any wrong already committed. In the Chorzow Factory case,[21] the Permanent Court of International Justice (PCIJ) clearly stated that a State that breaches its international obligations has a duty of reparation that must "as far as possible, wipe out the consequences of the illegal act and re-establish the situation which would, in all probability, have existed if that act had not been committed."[22]

In 2001 the International Law Commission adopted a codified version of the law on State responsibility for breach of international law.[23] Although nonbinding, it reasserts that every breach by a State of an obligation under international law constitutes an internationally wrongful act from which specific legal consequences arise. First, the responsible State must cease to act and offer appropriate assurances and guarantees

20 The Executive Secretary, *Liability and Redress under the Convention on Biological Diversity: Review of Relevant International Legal Instruments and Issues for Consideration*, at 3, *delivered to the Workshop on Liability and Redress in the Context of the Convention on Biological Diversity*, UNEP/CBD/WS-L&R/2 (May 21, 2001).

21 The Factory at Chorzow (Germ. v. Pol.), 1928 P.C.I.J. (ser. A) No. 17, ¶ 125 (Sept. 13)

22 See WWF-UK Climate Change Programme, Beyond Adaptation: The Legal Duty to Pay Compensation for Climate Change Damage 15 (2008) *available at* http://wwf.org.uk/climate.

23 See Int'l Law Commission [ILC], *Responsibility of States for Internationally Wrongful Acts*, U.N. ILC, 53d Sess. (2001). Text adopted by the International Law Commission at its fifty-third session, in 2001, and submitted to the General Assembly in 2001. The General Assembly took note of the draft articles and commended them to the attention of Governments without prejudice to the question of their future adoption or other appropriate action, G.A. Res. 56/83, UN Doc. A/RES/56/83 (Dec. 12, 2001). In 2004, the General Assembly again commended the articles and requested the Secretary-General to invite Governments to submit their written comments on any future action regarding the articles and prepare an initial compilation of decisions of international courts, tribunals, and other bodies referring to the articles. The General Assembly also invited Governments to submit information on their practice in this regard, G.A. Res. 59/35, UN Doc. A/RES/59/35 (Dec. 2, 2004). For further detail on the development of the articles, see http://www.un.org/law/ilc/.

of nonrepetition.[24] Second, the responsible State must make full reparation for the injury caused, including restitution (that is, to reestablish the situation that existed before the wrongful act was committed); the State must compensate for the damage caused by the wrongful act; and it must provide satisfaction, either via acknowledgment of the breach or through an expression of regret or formal apology.[25]

3.2. Application to Climate Change Impacts

Implementation of the no-harm principle hinges on the fact that the principle is fault-based. Behavior must be measured against a specific standard of care. Once this standard is established, if a State fails to take proportionate measures, the no-harm principle is breached and gives rise to an obligation to pay compensation.

Verheyen and Roderick express the view that a State's failure to prevent or minimize the risk of internationally recognized climate change damage would constitute a wrongful act in international law.[26] As a consequence, States would be required to act to prevent further damage, for example through financing adaptation measures. If damage occurred that could not be avoided through adaptation measures, the only possible redress would be via compensation.[27] Although the focus of these discussions has been on the provision of monetary compensation, it is feasible to consider other forms of compensation/redress, which may include in-kind support (e.g., the provision of goods and services) or facilitating resettlement through access to land.

Climate change will in many instances result in damage to the territory and environments of States that have not significantly contributed to global greenhouse gas (GHG) emissions, and compelling arguments

24 *Id.*, Art. 30.

25 *Id.*, Art. 35–38.

26 See WWF-UK Climate Change Programme, *supra* note 22, at 18.

27 *Id.*, at 24.

can be made that States responsible for those emissions have an international obligation to prevent such harm.

3.3. Practical Limitations

In spite of the existence of a viable legal basis for liability as outlined earlier, the options available for enforcing the no-harm rule or similar actions under the tort of nuisance are limited under the UNFCCC. A claim for a breach of the no-harm principle under public international law raises several procedural and evidentiary issues.[28] Both States Parties must first accept the jurisdiction of the international tribunal,[29] which only a very limited number of the Parties to the UNFCCC have done.[30] Once initiated, claimants must then identify the damaging activity attributable to the State and link it to the resulting injury, establishing causation.[31] In 2002, Tuvalu's Prime Minister threatened to bring a claim against the United States and Australia for compensation; however, to date no international tort claims have actually been started in part because of the difficulty in proving causation.[32] Similarly, although cases have been initiated in international courts based on other legal principles, such as the Inuit case against the United States for violation

28 Chris Larson, *Racing the Rising Tide: Legal Options for the Marshall Islands*, 21 MICH. J. INT'L L. 495 (2000).

29 For example, the International Court of Justice only has jurisdiction for matters where the member states have consented, Statute of the International Court of Justice art. 36, June 26, 1945, 3 Bevans 1179.

30 Article 14(2) of the UNFCCC provides that when ratifying, accepting, approving, or acceding to the Convention a party may declare in a written instrument that it recognizes as compulsory and without special agreement, the submission of any dispute concerning the interpretation or application of the Convention to the ICJ. At the time of writing, only Cuba, the Netherlands, the Solomon Islands, and Tuvalu have made a declaration in regards to Article 14(2): UNFCCC, "Declarations by Parties," http://unfccc.int/essential_background/convention/items/5410.php, date viewed 4 April 2012. UNFCCC, *supra* note 19, Art. 14(2).

31 TOL & VERHAYEN, *supra* note 12.

32 PlanetArk.com, *Tuvalu Seeks Help in US Global Warming Lawsuit*, PLANET ARK, Aug. 30, 2002, www.planetark.com/dailynewsstory.cfm/newsid/17514/story.htm.

of human rights, these cases have not resulted in favorable orders for the complainant.[33]

In view of the difficulties associated with claims under customary international law, and more specifically under the UNFCCC, some commentators have advocated for the establishment of an international civil liability regime.[34] This framework has been used in relation to environmental issues to impose strict liability on the operator of hazardous or dangerous activities.[35] Another option, which covers impacts caused by State actors, is the establishment of a Compensation Fund or Commission,[36] similar to the United Nations Claims Commission established to handle claims against Iraq for war-related damage, including environmental damage and depletion of natural resources. However, these mechanisms have compensated only environmental damage caused by discrete events whose harm to affected individuals was indisputable. Establishing causation and quantifying loss and damage are extremely difficult in the case of climate change impacts. This difficulty is partly due to the disparate sources of emissions that collectively contribute to global warming and also to a lack of data in many developing countries (especially SIDS) of historical baselines against which to measure impacts (e.g., sea level rise, temperature variation). Notwithstanding these difficulties with causation, a loss and damage mechanism could still draw on the international compensation models as a means of coordinating resources to respond to impacts.

33 See Joanna Harrington, *Climate Change, Human Rights, and the Right to Be Cold*, 18 Fordham Envtl. L. Rev. 513 (2007).

34 Philippe Cullet, *Liability and Redress for Human-Induced Global Warming: Towards an International Regime*, 43 Stan. J. Int'l L. 1 (2007).

35 Examples include the Basel Convention on the Control of Transboundary Movements of Hazardous Wastes and their Disposal; and the 2003 Joint Liability Protocol to the 1992 UNCED Convention on Protection and Use of Transboundary Watercourses. These regimes, however, are largely concerned with damage caused by private actors.

36 Daniel Farber, *The Case for Climate Compensation: Justice for Climate Change Victims in a Complex World* 2008 Utah L. Rev. 377 (2008).

4. Background to Loss and Damage Mechanism in UNFCCC

4.1. UNFCCC

4.1.1. C (a) AOSIS 1 – Early 1990s

Calls for the introduction of mechanisms to cope with climate change-induced loss and damage arose early on in the UNFCCC negotiations.

In 1991, AOSIS introduced the concept of insurance-related actions to cope with damage related to climate change in the third session of the Intergovernmental Negotiation Committee for a Framework Convention[37] (INC). It proposed that developing countries should be compensated (i) in situations where selecting the least climate-sensitive development option involves incurring additional expense and (ii) where insurance is not available for damage resulting from climate change.[38]

The concept was fleshed out in the fourth session of the INC in which AOSIS proposed the creation of an International Insurance Pool to compensate small island and low-lying developing nations for loss and damage resulting from sea level rise.[39] Arguing that the financial burden of loss and damage suffered by such States should be distributed in an equitable manner among the industrialized developed countries, AOSIS submitted that the Pool be funded by contributions calculated according to

37 The Intergovernmental Negotiation Committee [INC] was the UN body authorized by Resolution A/RES/45/212 to undertake negotiations for a framework convention on climate change. See CLIMATE CHANGE SECRETARIAT, A GUIDE TO THE CLIMATE CHANGE CONVENTION PROCESS (2002), *available* at http://unfccc.int/resource/process/guideprocess-p.pdf [Accessed 17 February 2011].

38 Vanuatu on behalf of AOSIS, *Elements of a Framework Convention on Climate Change*, clause 19.3, *delivered to the International Negotiation Committee*, UN Doc. A/AC.237/Misc.1/Add.3 (June 18, 1991), *available at* http://unfccc.int/resource/docs/1991/a/eng/misc01a03.pdf.

39 Intergovernmental Negotiation Committee for a Framework Convention on Climate Change [INC/FCCC], Geneva, Switz., Dec. 9–20, 1991, *Consolidated Working Document*, UN Doc. A/AC.237/Misc.17/Add.9 (Dec. 19, 1991), *available at* http://unfccc.int/resource/docs/1991/a/eng/misc17a09.pdf.

a formula modeled on the 1963 Brussels Supplementary Convention on Third Party Liability in the Field of Nuclear Energy. Contributions were to be calculated based on the following:

(i) 50% – the ratio between the gross national product (GNP) of each industrialized country contributing to the fund as against the total country contributor GNPs
(ii) 50% – the ratio of each industrialized country's CO_2 emissions as against the total CO_2 emissions of the group of contributing countries[40]

Industrialized nations would contribute their calculated shares to the Pool fund if the rate of global sea level rise had reached an agreed figure ten years after the Convention came into force.

Developing countries that would be directly affected by sea level rise would be valued for insurance before the Insurance Pool was implemented. Each country would negotiate the insured value, coverage, and trigger levels with the Authority,[41] a body under the control of the Conference of Parties (COP). Each would also register all assets and interests to be covered by the scheme. For a developing country to make a claim against the Pool, the rate of global mean sea level rise and the absolute level of global mean sea level rise would need to meet agreed figures, as well as surpass agreed base levels of relative mean sea level rises for the insured area of the vulnerable country.[42]

Although the IPCC in its first assessment report identified three responses to sea level rise (retreat, accommodation, and protection),

40 *Id.* at 4.

41 The composition and role of the Authority were comparable to that of the CDM Executive Board. *See* Joanne Linnerooth-Bayer, M.J. Mace & Roda Verheyen, (2003) *Insurance-Related Actions and Risk Assessment in the Context of the UNFCCC: Background paper for UNFCCC workshops – commissioned by the UNFCCC Secretariat* (May 2003), http://unfccc.int/files/meetings/workshops/other_meetings/application/pdf/background.pdf.

42 INC/FCCC, *supra* note 39, at 4.

only the former two fell within the scope of the insurance scheme.[43] Protection was seen as a matter for adaptation and would therefore fall within the financial mechanism of the UNFCCC. AOSIS also proposed that the Authority could consider whether other measures to mitigate the loss and damage were available when assessing the payout of claims.

The AOSIS proposal was kept in the Convention's draft negotiating text until the last meeting of the INC held in New York in 1992. Although no insurance scheme ultimately made it into the text of the Convention, the Convention can be seen as requiring all its Parties to consider the requirements of developing countries (including small island States and low-lying countries) in terms of insurance-related actions necessary for the potentially adverse effects of climate change.[44] This obligation was further supported in the Kyoto Protocol in Article 3(14), where the Parties called for the implementation of Article 4(8) of the Convention and the consideration of the establishment of insurance at the first Meeting of the Parties to the Kyoto Protocol (MOP).[45]

The concept of insurance has since been considered by the Parties in a number of decisions and reports of the technical bodies to the Convention and the Protocol.[46] Specific workshops have also been held under

43 Working Group III, Intergovernmental Panel on Climate Change, Climate Change: The IPCC Response Strategies (1990).

44 M.J. Mace, *The Bali Road Map: Can It Deliver an Equitable Post-2012 Climate Agreement for Small Island States?*, 17 Rev. Eur. Community & Int'l Envtl. L. 183, 189 (2008).

45 Kyoto Protocol to the United Nations Framework Convention on Climate Change, Dec. 10, 1997, 37 ILM 22.

46 *See* UNFCCC Conference of the Parties, Kyoto, Japan, Dec. 1–11, 1997, *Action Taken by the Conference of the Parties*, Decision 3/CP.3, at 32, UN Doc. FCCC/CP/1997/7/Add.1 (Dec. 11, 1997), *available at* http://unfccc.int/resource/docs/cop3/07a01.pdf; *See also* UNFCCC Conference of the Parties, Buenos Aires, Arg., Nov. 2–14, 1998, *Action Taken by the Conference of the Parties*, Decision 5/CP.4, at 17, UN Doc. FCCC/CP/1998/16/Add.1 (Jan. 25, 1999), *available at* http://unfccc.int/resource/docs/cop4/16a01.pdf; *see also* UNFCCC Conference of the Parties, Bonn,

the UNFCCC process,[47] including the UNFCCC Expert Meeting on Adaptation for SIDS.[48] In these meetings, there were extensive discussions on the subjects of insurance, risk management, and risk reduction. Importantly, Parties continued to consider the original AOSIS Insurance Pool and highlight the need for an international insurance mechanism to help SIDS manage climate-related risk and build risk-resilient economies.

4.1.2. AOSIS 2 – Vienna Dialogue

AOSIS sought to revisit and build on its initial proposal for an insurance mechanism at the fourth session of the Ad Hoc Working Group on Further Commitments for Annex I Parties to the Kyoto Protocol and the Dialogue on Long-Term Cooperative Action to Address Climate Change by Enhancing Implementation of the Convention (Dialogue), which met in parallel.[49]

AOSIS made a formal submission in advance of the negotiations, stating that "the avoidance of climate change impacts on small island developing States must be one of the key benchmarks for assessing

Germany, Oct. 25-Nov. 5, 1999, *Action Taken by the Conference of the Parties*, Decision 12/CP.5, at 32, UN Doc. FCCC/CP/1999/6/Add.1 (Feb. 2, 2000), *available at* http://unfccc.int/resource/docs/cop5/06a01.pdf; *see also* Conference of UNFCCC, New Delhi, India, Oct. 23-Nov. 1, 2002, *Report of the Subsidiary Body for Implementation*, UN Doc. FCC/SBI/2002/17, *available at* http://unfccc.int/resource/docs/2002/sbi/17.pdf.

47 *See* Conference of the UNFCCC, New Delhi, India, Oct. 23–29, 2002, *Workshop held under* Decision 5/CP.7, UN Doc. FCCC/SBI/2002/MISC.4/Add.2., *available at* http://unfccc.int/resource/docs/2002/sbi/misc04a02.pdf.

48 See Conference of the UNFCCC, Bonn, Germany, May 7–18, 2007, *Report on the Expert Meeting for Small Island Developing States; Progress on the Implementation of Decision 1/CP.10*, UN Doc. FCCC/SBI/2007/11, *available at* http://unfccc.int/resource/docs/2007/sbi/eng/11.pdf.

49 The AWG-KP is the negotiating track under the Kyoto Protocol, and the Dialogue was the negotiating track under the Convention. Both negotiating tracks were established at COP 11/MOP 1 in Montreal in 2005. The Dialogue has since been followed by the Ad Hoc Working Group on Long-Term Cooperative Action (AWG-LCA), created at COP 13/MOP 3 in Bali in 2008.

the appropriateness of any long-term goal."[50] It also observed that adaptation funding through the Convention's funding mechanism was "extremely limited, and procedures for accessing these limited funds [were] excessively complicated." It argued that collective loss-sharing mechanisms and an international solidarity fund were needed to address high-impact extreme events.[51] This concept was picked up by the Note of the Co-Facilitators on the Dialogue process and found expression in the Bali Action Plan.[52]

4.1.3. Bali Action Plan

Under the Convention negotiations at COP 13 in Bali, Indonesia, in 2007, the Parties agreed to turn the Dialogue into a formal negotiating process – the Ad Hoc Working Group on Long-Term Cooperative Action (AWG-LCA). As part of the Bali Action Plan,[53] the Parties agreed to work toward enhanced action on adaptation through risk management and risk reduction strategies, including such risk sharing and transfer mechanisms as insurance, disaster reduction strategies, and means to address loss and damage associated with climate change impacts in developing countries.[54] This agreement ensured a future work program on the subject of risk sharing, and transfer mechanisms such as insurance were kept on the AWG-LCA adaptation agenda. Despite some resistance to specific submissions of Parties on insurance-related

50 AOSIS. *Dialogue Working Paper 14.* at 13, *submitted to the Fourth Workshop of the Dialogue on Long-Term Cooperative Action to Address Climate Change by Enhancing Implementation of the Convention* (Aug. 24, 2007), *available at* http://unfccc.int/files/meetings/dialogue/application/pdf/wp14-aosis.pdf.

51 *See also* Mace, *supra* note 44.

52 *See* UNFCCC COP 13, Bali, Indon., Dec. 3–14, 2007, *Report on the Dialogue on Long-Term Cooperative Action by Enhancing Implementation of the Convention: Note by the co-facilitators*, UN Doc. FCCC/CP/2007/4/Add.1 (Nov. 19, 2007), *available at* http://unfccc.int/resource/docs/2007/cop13/eng/04a01.pdf.

53 UNFCCC Conference of the Parties, Bali, Indon., Dec. 3–14, 2007, *Action Taken by the Conference of the Parties*, Decision 1/CP.13, at 3, UN Doc. FCCC/CP/2007/6/Add.1 (March 14, 2008), *available at* http://unfccc.int/resource/docs/2007/cop13/eng/06a01.pdf.

54 UNFCCC, *supra* note 52 at 4.

matters,[55] technical papers were drafted to facilitate work of the AWG-LCA in Poznan in December 2008.[56]

Importantly, AOSIS submitted a proposal to the AWG-LCA on a Multi-Window Mechanism to Address Loss and Damage from Climate Change Impacts, which had three interdependent components:

1. an **insurance** component – to assist particularly vulnerable developing countries in managing financial risk from increasingly frequent and severe extreme weather events
2. a **rehabilitative/compensatory** component – to address loss and damage resulting from un-avoided and unavoidable climate change impacts, such as rising sea levels, increasing land and sea temperatures, and ocean acidification
3. a **risk management** component – to support risk assessment and risk management[57]

In terms of the rehabilitative/compensatory arm of the proposed mechanism, a critical point was the need to establish baseline data for extreme weather events, historical temperatures, and sea level rise. With those baselines in place, rehabilitation or compensation payments would be triggered by changes in parameters relative to those baselines. These payments could cover economic loss, property loss and damage, loss of

55 Despite agreeing to the terms relating to insurance mechanisms in the Bali Action Plan, many Parties of the COP resisted specific submissions on the topics, failing to come to an agreement on the inputs to advance the AWG-LCA's work. A proposal for a paper on risk reduction was subsequently not supported by developing countries for fear that focus of the discussions under cl 1(c)(ii) would be altered to exclude insurance mechanisms. *See also* Mace, *supra* note 44.

56 Papers include UNFCCC, *Technical Paper: Integrating practices, tools and systems for climate risk assessment and management and strategies for disaster risk reduction into national policies and programs*, UN Doc. FCCC/TP/2008/4 (Nov. 21, 2008), *available at* http://unfccc.int/resource/docs/2008/tp/04.pdf; and UNFCCC, *Technical Paper: Mechanisms to manage financial risks from direct impacts of climate change in developing countries*, U.N. Doc. FCCC/TP/2008/9 (Nov. 21, 2008), *available at* http://unfccc.int/resource/docs/2008/tp/09.pdf.

57 AOSIS, *supra* note 3, at 3–4.

life, and environmental damage.[58] Although not raised by AOSIS, a logical extension of this scope would be to cover the costs of relocation.

The difficulty with this proposal, as noted earlier, is that many SIDS lack the baseline data on which to measure variance and its attribution to either natural or anthropogenic causes. To address these data gaps, it will be necessary to develop appropriate computer models that, over time, can be enhanced with direct observation. However, financial and technical support to put those data collection systems in place will be necessary.

AOSIS proposed that funding for the Multi-Window Mechanism be primarily derived from a "Convention Adaptation Fund" comprised of contributions from developed countries, assessed on the basis of each country's levels of GHG emissions and taking into account their respective levels of development, historical contributions to GHG emissions, and ability to pay.[59] Additional funding could be obtained through the UNFCCC's existing financial mechanisms, the Kyoto Protocol's Adaptation Fund, and bilateral or multilateral sources.[60] The level of funding available, as well as the level at which access triggers are set and how applications are prioritized, would determine whether the mechanism would be able to fully or partially compensate for impacts.

The institutional framework for the proposed mechanism involved the establishment of a Board, which would oversee the functioning of the mechanism and ensure transparent governance. The Board would be supported by a technical advisory facility and financial vehicle/facility that would provide assistance to all three adaptive components, with administrative support from the UNFCCC Secretariat.[61] Importantly, AOSIS envisaged that the financial institution would be housed outside the UNFCCC, although AOSIS provided few further details.

58 *Id.* at 7.

59 GHG emissions would be used as a proxy for historic responsibility, and GDP as a measure of capability to pay.

60 AOSIS, *supra* note 3; *Id.*at 3.

61 *Id.* at 4–5.

4.1.4. Copenhagen/Cancun/Durban

The AOSIS proposal was overshadowed by wider political events at COP 15 in Copenhagen in December 2009. The outcome of that meeting, the Copenhagen Accord, referred to the need for enhanced action on adaptation and the establishment of a comprehensive adaptation program with international support. In addition, those developed-country Parties that endorsed the Accord committed to scale up finance for adaptation and mitigation to US$100 billion per year by 2020, to be delivered through the Copenhagen Green Climate Fund. However, the Accord made no mention of the content of the adaptation program, in particular whether it would include a loss and damage mechanism.

In the absence of a final, legally binding agreement in Copenhagen, COP 15 mandated the AWG-LCA to continue its work program during 2010. In doing so, the Parties continued to discuss the scope of an enhanced adaptation framework and possible measures to address risk management, risk sharing, and rehabilitation/compensation for loss and damage.

The role of a loss and damage mechanism was again considered at COP 16 in Cancun in December 2010. The Parties revisited submissions that had been made throughout the previous year on enhanced action on adaptation and adopted Decision 1/CP.16, which invites all Parties to enhance action on adaptation under the Cancun Adaptation Framework, taking into account their common but differentiated responsibilities and respective capabilities, and specific national and regional development priorities, objectives, and circumstances by undertaking, among other things, the following:

(i) Enhancing climate-change-related disaster risk reduction strategies, taking into consideration the Hyogo Framework for Action where appropriate; early warning systems; risk assessment and management; and sharing and transfer mechanisms such as insurance, at local, national, subregional and regional levels, as appropriate

(ii) Measures to enhance understanding, coordination, and cooperation with regard to climate-change-induced displacement, migration, and planned relocation, where appropriate, at national, regional, and international levels[62]

For the first time, this decision recognized the need to strengthen international cooperation and expertise to understand and reduce loss and damage associated with the adverse effects of climate change, including impacts related to extreme weather events and slow-onset events.[63] To strengthen this cooperation, the COP decided to establish a work program, overseen by the Subsidiary Body for Implementation (SBI), to consider approaches to address loss and damage associated with climate change impacts. Parties were invited to make submissions by February 21, 2011, on the activities to be part of the work program, including the following:

(i) Possible development of a climate risk insurance facility to address impacts associated with severe weather events
(ii) Options for risk management and reduction; risk sharing and transfer mechanisms such as insurance, including options for micro-insurance; and resilience building, including through economic diversification
(iii) Approaches for addressing rehabilitation measures associated with slow-onset events
(iv) Engagement of stakeholders with relevant specialized expertise

This work program is intended to inform recommendations that will be considered by Parties at COP 18 in 2013. It was briefly discussed at

62 UNFCCC Conference of the Parties, Cancun, Mex., Nov. 29–Dec. 10, 2010 [hereafter COP16], *Action Taken by the Conference of the Parties*, Decision 1/CP.16, ¶14, UN Doc. FCCC/CP/2010/7/Add.1 (March 15, 2011), *available at* http://unfccc.int/resource/docs/2010/cop16/eng/07a01.pdf.

63 Including sea level rise, increasing temperatures, ocean acidification, glacial retreat and related impacts, salinization, land and forest degradation, loss of biodiversity, and desertification.

COP 17 in Durban. The Adaptation Committee, established by Decision 2/CP.17, may play a role in supporting the efforts of the work program.[64]

The establishment of a work program on loss and damage is a progressive and much needed step by the Parties. In response to the call, submissions were rcceived from twenty-two Parties representing ninety-eight countries, along with ten submissions from intergovernmental organizations, including inter alia the United Nations Environment Program (UNEP), United Nations International Strategy for Disaster Reduction (UNIDR), the World Bank, and the Food and Agriculture Organization.[65] In addition, as is discussed later, a number of NGOs from the academic, environment and business community have, in recent years, explored various models to provide enhanced access to insurance and to compensate particularly vulnerable developing countries for unavoidable impacts.

The content of these submissions, although primarily focused on the organization of the work program, is nevertheless very instructive in terms of their authors' early views on the scope and content of a loss and damage mechanism. There appears to be broad consensus that reducing exposure to risk by enhancing prevention and reduction measures and disaster preparedness is critical to climate risk management. There is also widespread recognition that not all risks can be prevented or reduced, meaning that risk sharing and risk transfer mechanisms also need to be considered as part of any risk management strategy.[66] It is clear that Parties are now more open to discussing the role of insurance mechanisms

64 UNFCCC Conference of the Parties, Durban, South Africa., Nov. 28- Dec. 11, 2011 [hereafter COP17], *Action Taken by the Conference of the Parties*, Decision 2/CP.16, ¶Annex V, UN Doc. FCCC/CP/2011/9/Add.1 (March 15, 2012), *available at* http://unfccc.int/resource/docs/2011/cop17/eng/09a01.pdf.

65 *See* UNFCCC, Subsidiary Body for Implementation, *Views and information on elements to be included in the work program on loss and damage*, UN Doc. FCCC/SBI/2011/MISC.1 (April 19, 2011); *and see* UNFCCC, Subsidiary Body for Implementation, *Synthesis report on views and information on elements to be included in the work program on loss and damage*, UN Doc. FCCC/SBI/2011/3 (27 April 2011).

66 *See* UNFCCC, *Synthesis report*, UN Doc. FCCC/SBI/2011/3 ¶ 12.

and engagement with the private sector. However, very few submissions raised the controversial question of compensation for unavoidable loss.

Although some Parties submitted that the work program should develop modalities and procedures for a loss and damage mechanism over the next two years, in parallel with a process of gathering information about existing tools and information gaps, others considered the establishment of specific loss and damage approaches to be premature. This reflects a common divide in the UNFCCC negotiations: on the one hand, developed-country Parties promote a process-driven solution, involving, in this instance, enhancing understanding of risks and losses facing vulnerable developing countries; identifying gaps in risk exposure data, modeling, and regulatory frameworks; synthesizing data; promoting knowledge sharing; and enhancing understanding of existing and appropriate response strategies.[67] On the other hand, developing-country Parties are seeking action-oriented programs that immediately begin to develop the rules for elements of the mechanism, in particular the risk insurance facility.[68]

Of all of the matters being considered as part of the work program on loss and damage it is the third element – rehabilitation measures associated with slow-onset events – that is perhaps the most challenging. Parties and intergovernmental organizations (IGOs) recognize that they are starting from a low knowledge base when considering response strategies, in particular those strategies that are likely to involve financial and risk transfer mechanisms.[69] Although policy responses such as coastal land use planning are likely to form part of these response strategies, relocation may also need to form part of any response.

It is interesting to note that the Multi-Window Mechanism proposed by AOSIS in 2009 addressed each of the elements raised in paragraph 28 of Decision 1/CP.16 and as such provides a useful foundation

67 *See*, *e.g.*, the submissions of Australia, Canada, Norway, Switzerland, and the United States in UNFCCC, *Views and information*, – UN Doc. FCCC/SBI/2011/MISC.1.

68 *See*, *e.g.*, the submissions of China, Grenada on behalf of AOSIS, and India in *id*.

69 *See* UNFCCC, *supra* note 65, ¶ 35–36.

for building on these concepts. The three interdependent components proposed by AOSIS – insurance, rehabilitation/compensation, and risk management[70] – combine to produce an integrated approach to risk reduction, risk transfer, risk management efforts, and, above all, adaptive capacity. The insurance component is designed to help SIDS and other particularly vulnerable developing countries manage financial risk from increasingly frequent climate-related extreme weather events that result in loss and damage. AOSIS contends that such an insurance component is necessary, because SIDS and other particularly vulnerable developing countries are least likely to be able to afford commercial insurance.[71] The rehabilitation/compensation component, in turn, is put forward by AOSIS to address the progressive negative impacts of climate change, such as sea level rise. AOSIS acknowledges, however, that the insurance component will not provide sufficient resources to reduce physical risks and exposure.[72] Finally, the risk management component is intended to support risk assessment and the development of risk management tools that could inform the insurance and rehabilitation/compensatory measures to be adopted.[73]

4.2. NGO input into the Description of a Loss and Damage Mechanism

4.2.1. Proposed Compensation Fund Model from WWF

In the lead-up to COP 14 in Poznan, Poland, World Wildlife Federation-UK engaged legal scholars Rhoda Verheyen and Peter Roderick to prepare a paper titled "Beyond Adaptation – The Legal Duty to Pay

70 UNFCCC, Ad Hoc Working Group on Long-Term Cooperative Action under the Convention, *Ideas and Proposals on the Elements Contained in Paragraph 1 of the Bali Action Plan: Submissions from Parties*, UN Doc. FCCC/AWGLCA/2008/Misc.5/Add.2 (Dec. 10, 2008), *available at* http://unfccc.int/resource/docs/2008/awglca4/eng/misc05a02p01.pdf.

71 *Id.* at 13–14.

72 *Id.*

73 *Id.*

Compensation for Climate Change Damage." It outlines the basis, under public international law, on which States may be liable to compensate other States for damage caused by climate change. The authors propose the establishment of a fund that would be financed through contributions, primarily from developed-country Parties, and would be available to those countries most vulnerable to climate change seeking redress for climate change impacts. The key elements of the model include (i) the institutional arrangements for the fund and its governance structure; (ii) the basis for constituting the fund, whether by using assessed contributions for Parties or accessing other funding sources, such as shares or proceeds linked to the Kyoto Protocol flexible mechanisms or new sources of finance; (iii) identifying the beneficiaries of the fund; (iv) the types of damage that may apply to a claim for compensation; (v) circumstances under which compensation may be payable, for example on proof of loss or where thresholds for affectation are triggered; and (vi) limits to liability.

In summary, the model endorses an approach founded on traditional heads of damage for compensation (loss of life, property, economic loss, and environmental damage). A compensation fund would be constituted with contributions from developed-country Parties and other funding sources, such as revenue from market-linked mechanisms. Compensation would be payable on proof of loss and may be linked to trigger thresholds.[74]

The model proposed by WWF-UK has many similarities to that proposed by AOSIS. However, it is primarily focused on one element of the Multi-Window Mechanism: the rehabilitative/compensatory aspect.

4.2.2. Business and Insurance Industry Contributions

Work has also been undertaken by the UNFCCC Secretariat, Parties, and Observer organizations through technical papers and workshops on the two other aspects of the AOSIS proposal: risk management

74 See See WWF-UK Climate Change Programme, *supra* note 22, at 33.

and insurance. One important contribution in this regard is the work co-authored by the United Nations International Strategy for Disaster Reduction Secretariat (UNISDR) and the Munich Climate Insurance Initiative (MCII).

In 2009, UNISDR and MCII explored the potential role of insurance in reducing disaster risk and advancing adaptation in a study titled "Adaptation to Climate Change: Linking Disaster Risk Reduction and Insurance."[75] Citing examples such as the use of micro-insurance, catastrophe bonds, and multicountry insurance pools, the authors proposed that risk transfer tools such as insurance, if properly designed, have the potential to be useful to the poor in managing the disaster risks posed by climate change. However, these initiatives need to be matched both with risk reduction measures to minimize the scale of loss and damage (e.g., improved planning and building standards) and with direct funding for both risk reduction and rehabilitation/compensation.

In their recent submissions on the work program on loss and damage, a range of organizations such as UNEP, UNDP, the World Bank, and the Inter-American Development Bank (IADB) highlighted their support for the use of financial and insurance-related risk reduction and risk transfer products. These organizations are already using these types of products: for example, the IADB's policy-based loans to enhance disaster risk preparedness in climate change adaptation strategies, or the World Bank's support for country-specific climate risk financing solutions such as the facilities implemented in Mexico and the Caribbean.

One notable point to draw from these submissions is that, although innovative initiatives are underway, they are somewhat ad hoc and show the lack of a common language relating to loss and damage between the insurance/finance sectors and the UNFCCC community. In this respect, UNEP and the UNEP Finance Initiative suggest that the UNFCCC

75 Munich Climate Insurance Initiative [MCII], *Adaptation to Climate Change: Linking Disaster Risk Reduction and Insurance, submitted to the June Session of the UNFCCC Climate Talks* (June 6, 2009), *available at* http://unfccc.int/resource/docs/2009/smsn/ngo/163.pdf.

community should learn from the risk management processes adopted by the insurance community before launching the provision of risk transfer in the form of insurance.[76]

To date, these discussions have generally settled on several insurance and non-insurance risk financing measures as offering the best opportunity to protect LDCs and SIDS. Some of these measures have already been successfully deployed in vulnerable countries. In many instances, these mechanisms leverage public sources of funding and spread risk across a wide number of potentially affected countries (in the case of regional insurance pools) or individuals (in the case of micro-insurance and micro-savings schemes) to facilitate greater access to private sector financing.

5. Insurance

Insurance for climate-change-related damage and loss is one of the most commonly contemplated financial mechanisms for dealing with this issue. The most relevant types of insurance for climate-change-related damage or loss are traditional insurance, micro-insurance, and insurance-linked securities.

5.1. Traditional Insurance

Insurance is a contractual transaction that guarantees financial protection against stipulated forms of losses in return for a premium. Where the insured experiences a loss of the pre-described type, then the insurer will pay out a pre-agreed amount. Some traditional insurance is taken out by sovereign States with private sector insurers. For example, the Turkish Catastrophe Insurance Pool (TCIP) was established after the devastating Marmara earthquake by the World Bank, Government of

76 See the submissions of UNEP in UNFCCC, *supra* note 66.

Turkey, and private sector insurers.[77] The TCIP is a unique pool in which the public and private sectors cover different risks to reduce premium costs.

Insurance pools have also been successful in providing short-term liquidity in the aftermath of disasters. One example is the Caribbean Catastrophe Risk Insurance Facility, which functions as a mutual insurance company controlled by the sixteen participating governments. It was established following the devastation wrought by the hurricane season of 2004 and is the first multicountry risk pool.[78] It provides insured countries with parametric coverage, paying out policies based on the estimated impact of an event on each country's budget. Participants pay an annual premium proportionate to their specific risk exposure and receive compensation based on the agreed levels of coverage and the occurrence of a trigger event.

5.2. Micro-Insurance

Micro-insurance, which emerged out of micro-finance, is targeted at LDCs and is characterized by both low premiums and low coverage. The response to the 1988 floods in Bangladesh offers a prominent example of micro-insurance. The NGO Proshika and the Government of Bangladesh offered a Participatory Livestock Compensation Fund (PLCF) that targeted farmers. At the time of writing, the scheme covers 10% of the population of Bangladesh for property insurance and 25% for life insurance.[79] Regrettably, however, large areas of the country and much of the population are still vulnerable to similar natural disasters.

77 UNFCCC, "Risk Management Approaches to Address Adverse Effects of Climate Change – Insurance," http://unfccc.int/adaptation/adverse_effects_and_response_measures_art_48/items/4971.php.

78 *Id.*

79 *Id.*

5.3. Insurance-Linked Securities

Insurance-linked securities are insurance policies tied to capital markets, allowing for the dispersion of risk. Insurance-linked securities are generally tied to the occurrence of a specified event, which will trigger the payment of insurance. The most common type of insurance-linked security is the catastrophe (cat) bond. Cat bonds are bonds whose payouts are pegged to an index of natural catastrophe risks or specific trigger conditions.

The Mexico Catastrophe Bond (MCB) is a high-profile example of a cat bond. In 2006 the Mexican government insured its catastrophe reserve fund, the Fondo de Desastres Naturales (FONDEN), against earthquakes, to which the country is particularly vulnerable.[80] FONDEN involves a hybrid of reinsurance and a catastrophe bond. The resulting contract was linked to a parametric trigger, expressed in terms of magnitude and depth of seismicity. Such a strategy meant that the MCB was able to provide coverage of US$450 million for a lower cost than the reinsurance itself.[81]

The insurance mechanisms listed here offer several promising approaches for risk management. However, the application of these tools in particularly vulnerable countries also faces conceptual and practical hurdles. Historically, insurance has been a mechanism used by wealthier countries and private entities. There is a dearth of understanding about insurance among vulnerable States, which lack an insurance culture, and to the extent that policies do exist, they are often very conditional and have limited coverage and expensive premiums. Challenges for developing insurance tools for particularly vulnerable developing countries

80 The World Bank, "Climate Finance in the Urban Context," November 2010, http://wbi.worldbank.org/wbi/Data/wbi/wbicms/files/drupal-acquia/wbi/578590revised0101Public10DCFIB0141A.pdf.

81 Wolfgang Karl Härdle & Brenda L'opez Cabrera, *Calibrating CAT Bonds for Mexican Earthquakes*, Center for Applied Statistics and Economics, Humboldt-Universität zu Berlin, http://www.uibk.ac.at/fakultaeten/volkswirtschaft_und_statistik/forschung/natcatrisk/natcatrisk_cabrera1.pdf.

within an adaptation framework include making products affordable, incentivizing the uptake of insurance, and reconsidering who actually covers the premiums. In some instances, products will be able to be marketed to individuals, but in many instances it may be States that explore coverage or participation in insurance pools. Both situations require a determination of who covers the costs of the premiums. In this respect, it may be feasible to use funds and guarantees that are backed by developed countries.

Insurance is widely used in the context of weather-related disasters, where loss and damage are relatively simple to identify and quantify. However, determining when claims may be made for slow-onset events is more complex. Over time there is an extremely high probability that these types of events will occur. In many ways this diminishes the utility of insurance, with insurers either unwilling to provide insurance or being willing to provide insurance only at very high premiums. Nevertheless, there is some scope to insure by reference to milestones that may or may not be met (i.e., sea level rise of 50 cm above a baseline by a particular date). By providing an element of uncertainty linked to timing, insurers are still able to manage risk associated with these events. With the exception of drought insurance, which is triggered by events like crop failure, there are few examples of insurance being used to mitigate the risk of events like sea level rise. As noted in the original AOSIS proposal, triggers could be designed relative to predetermined baselines, but challenges would be faced in setting baselines, determining the scope of loss, and assessing the extent to which the person suffering the loss could have mitigated that loss over time.

6. Non-Insurance Financing Mechanisms

In part to make up for the shortfalls of insurance mechanisms, the UNFCCC has also considered the utility of non-insurance risk-financing mechanisms as a source of compensation for climate-change-induced loss and damage.[82] Forms of non-insurance risk-financing mechanisms include informal risk sharing, micro-savings and micro-credit, and post-disaster assistance.

6.1. Informal Risk Sharing

Informal risk sharing involves financial arrangements of reciprocal exchange among community members or persons linked by kinship ties. Remittances (where foreign workers send money back to their home country) are a major source of finance for LDCs and other vulnerable States. Remittances often spike in the wake of natural disasters.

6.2. Micro-Savings and Postcalamity Credit

Micro-savings arrangements involve individuals creating their own personal savings. These types of savings are receiving increasing assistance from micro-finance institutions (MFIs), banks, and other agencies.

Micro-credit involves the provision of small loans to vulnerable communities for the purposes of some kind of development. Micro-credit is especially valuable after natural disasters, because local money lenders often increase their rates at such times. Postdisaster credit may be larger in scope, as where an entity issues disaster victims bonds and other debt instruments that transfer the burden to future periods.

Informal risk sharing and micro-savings both place the financial burden of risk management on individuals and communities rather than on the host State or third-party States that may have a responsibility for the circumstances that caused the loss and damage. Yet these arrangements are very effective at providing immediate finance to people directly suffering from the adverse effects of climate-related disasters. In particular, the efficacy of their delivery channels should be appreciated: transfer

82 UNFCCC, *supra* note 9, at 87–101.

agencies and micro-finance institutions have been especially successful in conveying third-party funds to affected areas.

6.3. Postdisaster Assistance

Postdisaster assistance comes in the form of the affected country's government relief, which often supports specific projects targeted at recovery. Some governments have established a postcalamity fund, which accumulates in years without catastrophes and is reduced in years that experience disasters. Governments of vulnerable countries, however, have heightened reliance on foreign governments' aid or foreign aid donated by individuals and nongovernment organizations.

As part of the loss and damage program being developed by SBI, there will be further exploration of the potential for using these types of tools (and others). One key principle that is likely to inform the loss and damage mechanism is that risk reduction and risk transfer strategies will be country (and in some instances regionally) driven. In other words, different countries may choose to explore the use of a variety of these tools, as they find appropriate.

In the context of the Multi-Window Proposal from AOSIS discussed earlier, AOSIS puts forward insurance as a possible mechanism of postdisaster relief. However, the AOSIS proposal does not detail who is to pay the premiums for such insurance. Some may argue that international funds should assist with contributions for premiums. As detailed in the next section, however, such funds are already stretched.

In all likelihood, vulnerable countries will be required to use a combination of insurance and non-insurance financing mechanisms to share the risks associated with climate-change-induced damage. Perhaps the greatest factor likely to influence the financial mechanisms ultimately put in place is the ability of more developed countries to underwrite the risks in the form of insurance or assist with adaptation in the form of postdisaster assistance. The precise scale of funds required to mitigate climate-related risks is still very much an open question.

7. Funding for Loss and Damage

In recent years a number of institutions have sought to quantify the amount of finance that will be required to support climate-change-impact adaptation in developing countries. Estimates range from €10–20 billion (US$14–28)[83] to US$50 billion annually.[84] Although some commentators have also attempted to quantify the costs of responding to climate-change-related disasters and slow-onset events, which may include relocating mass populations displaced by climate change, the figures remain unreliable and do not take the costs of climate change beyond the range of estimates detailed earlier.[85]

Existing climate change funds available through the UNFCCC financial mechanism[86] and the Adaptation Fund have long been recognized as being inadequate to fully compensate LDCs for climate-change-induced damages and losses. The underfunding is primarily a result of the fact that contributions to many of the existing funds by States are largely voluntary. Where contributions are mandatory or are linked to market mechanisms such as the Adaptation Fund, there is the potential to generate a predictable stream of revenue. However, as the Adaptation Fund demonstrates, the size of the fund is constrained by the limited

83 The Secretary-General's High-Level Advisory Group on Climate Change Financing, *Report* (Nov. 5, 2010), *available at* http://www.un.org/wcm/webdav/site/climatechange/shared/Documents/AGF_reports/AGF%20Report.pdf.

84 Robert Bailey, Oxfam, The Right to Survive in a Changing Climate (2009), *available at* http://www.oxfam.org.uk/resources/policy/climate_change/downloads/right_to_survive_climate.pdf.

85 See discussion on costs in Jane Mcadam & Ben Saul, An Insecure Climate for Human Security? (2008)

86 Including the Global Environment Facility (GEF) – an operating entity of the UNFCCC's financial mechanism; Least Developed Countries Fund (LDCF) – established to assist Least Developed Countries (LDCs) under the Convention; Special Climate Change Fund (SCCF) – established under the UNFCCC to finance activities complementary to those funded by the resources allocated to the GEF; Bilateral, regional, and multilateral channels – funding provided for in keeping with developed countries' obligations under Article 11.5 of the UNFCCC.

scale of the clean development mechanism of which it receives a share of proceeds.[87]

Enhancing funding for loss and damage has recently gained traction within the international negotiations of the Convention and Kyoto Protocol. At COP 15, the Parties that endorsed the Copenhagen Accord pledged to scale up finance for mitigation and adaptation to US$100 billion per year by 2020. The Parties proposed that multilateral funding for adaptation be delivered through new funding arrangements, including a new Green Climate Fund (GCF).[88] The GCF would be established as an operating entity of the Convention's financial mechanism and would support projects, programs, policies, and other activities in developing countries related to mitigation, including REDD-plus, adaptation, capacity-building, technology development, and transfer.

This pledge was anchored in Cancun at COP 16, where Decision 1/CP.16 established the GCF and fleshed out further operational requirements of the Fund.[89] The Fund was launched in Durban at COP 17, where Parties approved the governing instrument of the GCF through Decision 3/CP.17.[90] One major change to GCF's design was an amendment to the way it was applied. Instead of limiting the GCF to mitigation projects like REDD-plus, adaptation, capacity-building, technology

87 UNFCCC Conference of the Parties, Marrakech, Morocco, Oct. 29- Nov. 10, 2001, *Action Taken by the Conference of the Parties*, Decision 10/CP.7, at 52, UN Doc. FCCC/CP/2001/13/Add.1 (Jan. 21, 2002), *available at* http://unfccc.int/resource/docs/cop7/13a01.pdf.

88 UNFCCC Conference of the Parties, Copenhagen, Den., Dec. 7–19, 2009, *Action Taken by the Conference of the Parties*, Decision 1/CP.15, at 4, UN Doc. FCCC/CP/2009/11/Add.1 (Mar. 30, 2010), *available at* http://unfccc.int/resource/docs/2009/cop15/eng/11a01.pdf#page=4.

89 COP 16, *Outcome of the work of the Ad Hoc Working Group on Long-Term Cooperative Action under the Convention*, Decision 1/CP.16, ¶ 102–111, UN Doc. FCCC/CP/2010/7/Add.1 (Mar. 15, 2011), *available at* http://unfccc.int/resource/docs/2010/cop16/eng/07a01.pdf#page=2.

90 UNFCCC Conference of the Parties, Durban, South Africa., Nov. 28- Dec. 11, 2011 [hereafter COP17], *Action Taken by the Conference of the Parties*, Decision 3/CP.16, ¶p.55, UN Doc. FCCC/CP/2011/9/Add.1 (March 15, 2012), *available at* http://unfccc.int/resource/docs/2011/cop17/eng/09a01.pdf.

development, and transfer areas, the Fund would be applied using thematic windows, thus allowing it potentially to extend to adaptation as well as mitigation.[91] The windows will be designed by a Transitional Committee made up of forty member States. The Fund will be governed by the Green Climate Board, consisting of twenty-four members, represented by equal numbers of developing- and developed-country members.[92] The COP further decided that an independent secretariat shall provide administrative support to the Fund.[93] A trustee will administer the assets of the Green Climate Fund in a manner that is separate and apart from the assets of the trustee and only for the purpose of, and in accordance with, the relevant decisions of the Green Climate Board.[94] The World Bank was invited by the COP to serve as the interim trustee of the Fund, subject to a review three years after the Fund has been operationalized.[95]

Among the issues that the Transitional Committee will have to address is the question of balance between funding for mitigation and for adaptation.[96] Because existing climate funds are skewed toward supporting mitigation action,[97] there is likely to be a strong expectation that adaptation should be a predominant feature of the thematic windows. This offers the possibility for addressing loss and damage caused by the impacts of climate change. However, such a mechanism will have to compete with a number of other priority areas for developing countries.

Parties have repeatedly stated that the US$100 billion pledged by 2020 will comprise a mix of both public and private funding. To date,

91 *See* International Institute for Sustainable Development, *Summary of the Cancun Climate Change Conference*, Earth Negotiations Bull., Dec. 13, 2010, at 1, *available at* http://www.iisd.ca/download/pdf/enb12498e.pdf.

92 UNFCCC COP 16, *supra* note 62, ¶ 103.

93 *Id.* ¶ 108.

94 *Id.* ¶ 104.

95 *Id.* ¶ 107.

96 *See* Neil Bird et al., Overseas Development Institute, Design Challenges for the Green Climate Fund (2011), *available at* http://www.odi.org.uk/resources/download/5256.pdf.

97 *See* MCII, *supra* note 75.

there has been no clear statement of how this split will be determined. However, several options to scale up financing have been mooted, including extending the share of proceeds from the clean development mechanism to other market-based mechanisms, and imposing levies on sectors such as aviation. To make financial contributions obligatory rather than voluntary, groups of Parties such as AOSIS and observer groups like WWF and the Climate Action Network have proposed that clear liability needs to be attached to certain States, which will fund the majority of the compensation. This is perhaps the most controversial aspect of a loss and damage mechanism. However, other elements of a compensation model will also need to be agreed on, including the following:

- how a compensation fund might be administered
- how compensation is to be constituted
- who the beneficiaries of the compensation should be
- what the heads of compensable damage should be
- the circumstances in which compensation is payable
- whether any limits should be placed on liability

The proposals included in the Multi-Window Mechanism by AOSIS and the options set out in the WWF-UK paper provide a sound starting point for discussions about the content of the compensation aspect of a loss and damage mechanism. However, when one considers the circumstances in which compensation may be payable, it is inevitable that relocation costs will also be factored in.

Many developed-country governments are highly concerned about any mechanism that, even through a proxy, suggests liability for damage arising in either the past or the future. Thus it will be extremely difficult to include references to responsibility, liability, and compensation in UNFCCC decisions and any future climate agreement. For example, developed-country Parties supportive of a work program on loss and damage have given SBI two years to develop that program and will likely resist any efforts to implement the funding aspects of a mechanism until that program is in place. The developed-country submissions on the work program for loss and damage discussed earlier reflect this resistance, with Parties such as Australia and Canada reluctant to preempt the form of and funding for the mechanism before preparatory research and evaluation of options have been completed.

8. Need for In-Kind Support

Other forms of support for regions and persons suffering climate-change-induced loss and damage need to be considered in contexts where monetary compensation may be inappropriate. In-kind support could be relevant in such circumstances, complementing additional funding provided through international climate funds and insurance mechanisms.

Examples of the type of in-kind support include capacity-building and technology transfer to support the deployment of risk management tools and assist in developing response plans for extreme weather events as well as slow-onset events. Policy measures, both in particularly vulnerable States and also less vulnerable developed and developing countries, are also important – particularly when considering climate-induced migration.

Governments are already testing programs of assistance and initiating changes in migration and settlement policies. New Zealand, for example, has introduced an immigration category for the Pacific whereby 75 people from Tuvalu, 75 from Kiribati, and 250 from Tonga may immigrate to New Zealand each year based on their employment potential.[98] The program does not specifically focus on environmental factors, but there has been discussion of the need for resettlement of islanders

98 See Immigration New Zealand, Pacific Access Category: http://www.immigration.govt.nz/migrant/stream/live/pacificaccess (last visited Feb. 22, 2011). The participants must also be between 18 and 45 years old, have an acceptable level of spoken and written English, meet minimum income requirements, undergo a health check, and have no history of illegal entrance.

affected by rising sea levels.[99] These agreements are spelled out in more detail in Chapter 11.

In Australia, the Greens Party introduced a bill to Parliament in 2007 to establish a "climate refugee visa."[100] A criterion of the proposed visa was that the applicant be displaced as a result of a climate-change-induced environmental disaster. The Minister of Immigration was to make a determination that a disaster existed based on the following criteria:

- the geographical scope of the disaster
- adaptation options and long-term sustainability
- the capability of the country and neighboring countries to absorb displaced persons
- international efforts to provide assistance

Although the bill was defeated, the initiative remains a key platform of the Greens Party's immigration and refugee policy.[101]

Although there are no examples of legislation or policies that address migration of persons from slow-onset climate change, Sweden and Finland do accept environmental migrants within their immigration policies on asylum grounds. In those countries, persons who are unable to return to their native country may qualify as a person in need of protection because of an environmental disaster. Underlying these assistance programs are calls from small island States, such as Kiribati, for urgent discussions on migration for the people displaced by climate change.[102]

99 *See* Susan Martin, Internat'l Org. for Migration, Background Paper WMR 2010: Climate Change and International Migration (2010), *available at* http://publications.iom.int/bookstore/free/WMR2010_climate_change_migration.pdf.

100 *See* Migration (Climate Refugees) Amendment Bill, 2007 (Austl.), *available at* http://www.comlaw.gov.au/Details/C2007B00149.

101 The Greens Party, Policies: Immigration and Refugees, ¶ 12, *available at* http://greens.org.au/policies/care-for-people/immigration-and-refugees (last visited Feb. 22, 2011).

102 The Pacific Conference of Churches has called for a regional immigration policy giving citizens most affected by climate change rights to resettle in other Pacific island nations or Pacific regional countries of their choice. *See* Maryanne Loughry & Jane McAdam,

Policy support, through internal and external migration programs at national and regional levels, will play an important part in addressing climate-change-induced loss and damage. However, these policies and programs also have a cost, which is ordinarily borne by the host government. Providing funding (where required) or other incentives to participate in these migration programs could be explored through the loss and damage work program. That is, although the focus of the Decision 1/CP.16 loss and damage work programs is largely on enabling the adaptation of societies to the effects of climate change on their own environments, to the extent that such environments cannot be salvaged, migration arguably comes within the ambit of addressing loss and damage.

9. Conclusions – Optimal Design of a Loss and Damage Mechanism

Vulnerable States, particularly LDCs and SIDs, will face increasing challenges in managing future climate impacts and related risks. In high climate change scenarios, climate-related losses could increase by up to 200% as soon as 2030.[103] A substantial proportion of vulnerable States will also confront the additional impact of forced displacement as the severity of climate change increases.

Impacts on vulnerable developing States can be grouped into three categories: avoidable impacts, impacts that are not avoided because of a country's inability to respond, and unavoidable impacts. To manage the impacts of climate change effectively, an adaptive framework must address all categories, including unavoidable impacts such as migration.

Traditionally, the international community has concentrated on the first two categories of impacts, providing funding and capacity-building

Kiribati – Relocation And Adaptation, 31 Forced Migratino Rev. 51 (2008), *available at* http://www.fmreview.org/FMRpdfs/FMR31/FMR31.pdf; *see also* Pacific Council of Churches Conference on Protection and Reparations for "Climate Refugees": Imperatives and Options: Summary of Discussions, Chavannes-de-Bogis, Switz., May 4–5, 2010, *Summary of Discussions* (*prepared by* Kasia Snyder et al.), *available at* http://www.pcc.org.fj/docs/Climate%20Change%20Refugees%20Conf.pdf.

103 *See* Economics of Climate Adaptation Working Group, *supra* note 11.

support to developing countries to build community resilience and adaptive capacity to climate change. The need for an expanded and strengthened response has been widely accepted by the international community; however, to date serious gaps remain in existing adaptation protection schemes that fail to consider the unavoidable impacts of climate change. Already we are seeing instances of land loss and internal displacement in small-island developing States.

A mechanism such as the Multi-Window Mechanism proposed by AOSIS offers a staged approach to loss and damage due to climate change. It addresses three different but complementary components: insurance, rehabilitation/compensation, and risk management. It provides a range of measures to address both the drivers and responses to un-avoided and unavoidable impacts of climate change, including management of financial risk from extreme weather events, adaptation to slow-onset events, and assessment of future development options. This multifaceted approach mitigates the isolationism in the international community's current response to the impacts of climate change.

The discussion about a loss and damage mechanism within the UNFCCC has only recently found its voice, largely through the efforts of the LDCs and SIDS in bringing it to the AWG-LCA/SBI agenda. In considering the potential scope of such a mechanism, the agreed focus of SBI, which will be elaborated between 2012 and 2015, is likely to include consideration of the following:

- options for risk management and reduction; risk sharing and transfer mechanisms such as insurance; and resilience building, including through economic diversification
- potential development of a risk insurance facility to address impacts associated with severe weather events
- possible rehabilitation measures in response to slow-onset events

Each of these elements should play an important part in the development of a loss and damage mechanism. Such a mechanism will also need to be sufficiently flexible to enable vulnerable countries to access relevant support, which may come in the form of knowledge, technology, insurance, finance, compensation, and other in-kind support.

A loss and damage mechanism in any form builds on the existing obligations under the UNFCCC and Kyoto Protocol, thereby enhancing the current discussions on the availability of insurance-related products and other financial mechanisms to support adaptation. It will also crystallize existing obligations under international law to protect other States from environmental damage.

Fundamentally, a coherent and integrated framework that draws on longer-term risk reduction strategies and allows access to both public and private sources of funding is needed to address the complex and diverse impacts of climate change suffered by vulnerable developing countries. Migration brings a moral imperative to the debate over appropriate adaptive frameworks. Pertinently, climate-induced displacement and options for risk management and reduction are now also contemplated by decisions of Parties to the UNFCCC and the Kyoto Protocol. The future work program of the international climate change regimes offers a valuable platform to address a comprehensive loss and damage mechanism. This opportunity to address the gap in adaptation frameworks should not be lost.

15 Ocean Acidification

International Legal Avenues under the UN Convention on the Law of the Sea

Dean Bialek and Judah Ariel

1. Introduction

There is overwhelming evidence that unrestrained anthropogenic emissions of greenhouse gases have the potential to cause climate change with disastrous consequences. In recent years, global atmospheric concentrations of carbon dioxide (CO_2) have reached 385 parts per million (ppm), compared to preindustrial levels of around 280 ppm.[1] Half of this increase has occurred in the last thirty years.[2]

Achievement of current government targets and pledges to reduce greenhouse gas (GHG) emissions would lead the world toward a 3.5°C increase in global average temperature above preindustrial levels by the end of the century.[3] If these targets and pledges are not met, it is

1 Ocean Acidification: A Summary for Policymakers from the Second Symposium on the Ocean in a High CO2 World 3, Monaco, Oct. 6–9, 2008 (Maria Hood et al. eds., 2009) [hereinafter Symposium].

2 Climate Action Tracker, *Delay in Climate Decisions Will Cost More, as We Head to 3.5 Degrees C of Warming, Say Scientists*, Dec. 6, 2011, http://climateactiontracker.org/news/96/Delay-in-climate-decisions-will-cost-more-as-we-head-to-3.5-degrees-C-of-warming-say-scientists.html.

3 Eur-Oceans Knowledge Transfer Unit, Ocean Acidification – the Other Half of the CO2 Problem (2007).

We would like to thank David Wei, Julia Christian, and Nico Tyabji for their valuable assistance with background research and editing successive versions of this chapter.

possible that the global average temperature will increase by up to 7°C above preindustrial levels during that same time period.[4]

Although global warming is the most commonly referenced result of increased atmospheric concentrations of GHGs, ocean acidification is often referred to as the other half of the CO_2 problem. As the Earth's largest carbon sink, the oceans absorb a large proportion of atmospheric CO_2, which in turn produces carbonic acid. This increases the acidity of the oceans, producing potentially devastating effects for the oceans, marine life, and its human uses, including food security. Worse yet, threats to the oceans – such as global warming, ocean acidification, and deoxygenation – occur in feedback loops that reinforce each other.[5]

The science behind ocean acidification is easily understood and predictable, perhaps even more so than the causal linkages between GHG emissions and global warming. The last decade has seen a surge in research and understanding of the phenomenon and its potential impacts. These impacts include the degradation of coral reefs and the possible collapse of entire segments of the marine food web.

The cost of the many impacts of business-as-usual emissions on the oceans has been projected to total nearly US$2 trillion by 2100.[6] In the tropical Pacific, ocean acidification, together with ocean warming and deoxygenation, may reduce fisheries' catches by an additional 30% beyond estimated reductions from climate change.[7] Other economic

4 Matthias Hofmann & Hans-Joachim Schellnhuber, *Ocean Acidification: A Millennial Challenge* 3 Energy & Envtl. Sci. 1883, 1883 (2010).

5 Stockholm Environment Institute, Valuing the Ocean: Draft Executive Summary 8, *available at*: http://www.sei-international.org/mediamanager/documents/Publications/SEI-Preview-ValuingTheOcean-DraftExecutiveSummary.pdf.

6 Stockholm Environment Institute, Valuing the Ocean Environment: Economic Perspectives, *available at*: http://www.sei-international.org/mediamanager/documents/Publications/SEI-Preview-ValuingTheOceanEnvironment-Economic-Perspectives.pdf.

7 Stockholm Environment Institute, The Pacific Ocean: A Case for Coordinated Action, *available at*: http://www.sei-international.org/mediamanager/documents/ Publications/SEI-Preview-ThePacificOcean-ACaseForCoordinatedAction.pdf.

impacts of ocean acidification, such as to coral reefs and mollusk fisheries, are also being felt and quantified.[8]

Like climate change, ocean acidification is a "global commons" problem, but the traditional atmospheric focus of efforts to counter temperature increases has left ocean acidification without a natural policy home.

The cross-cutting causes and effects of ocean acidification lend themselves to treatment under several different international regimes, including the UN Framework Convention on Climate Change (UNFCCC), the UN Convention on Biological Diversity (CBD), and the Law of the Sea Convention (LOSC). Soft-law fora, including the UN General Assembly and various bodies under its authority, may also have a role to play.

Public awareness of the problem is weak, making it difficult for any groundswell of public interest to make ocean acidification a serious driver of policies to reduce GHG emissions. One international strategy to address this gap would be for a group of countries particularly vulnerable to the impacts of ocean acidification – such as atoll island countries and archipelagic States reliant on the health of coral systems for their livelihoods and tourism – to champion an international response, including the possible threat of legal action in the international courts.

Although international litigation is often seen as a relatively aggressive diplomatic step and might be criticized as an unhelpful distraction from efforts to negotiate a global response under the UNFCCC, it would likely generate significant public attention in the international policy sphere, particularly if initiated by a like-minded group of particularly vulnerable countries against one or more major GHG emitters.

Legal action based on the marine-pollution-related obligations under the LOSC appears to offer the most fertile ground for pursuing a case addressing CO_2 emissions. But international litigation of this nature does

8 *See, e.g.*, Elizabeth Grossman, *Northwest Oyster Die-offs Show Ocean Acidification Has Arrived*, YALE ENVT. 360, Nov. 21, 2011, http://e360.yale.edu/feature/northwest_oyster_die-offs_show_ocean_acidification_has_ arrived/2466/; Daiju Narita et al., *Economic Costs of Ocean Acidification: A Look into the Impacts on Global Shellfish Production*, CLIMATIC CHANGE, Nov. 28, 2011.

not come without its complications and difficulties – scientific, legal, and political – and might therefore be considered as a longer-term objective. Much could be done in advance, or indeed while building a legal case, to raise public awareness, educate the broader populace, and, through a targeted diplomatic campaign, bring the seriousness of ocean acidification to the attention of policymakers. Indeed, as an interim strategy, a request to an international court or tribunal for an advisory opinion may help clarify the relevant legal principles, including the scope and content of the relevant international obligations, such as those under the LOSC. This would in turn enhance the chances of successful international litigation if and when it arises in the future.

In Part 2 of this chapter, we summarize the current science on ocean acidification. Part 3 explores the rationale for an international legal action to address ocean acidification, and Part 4 addresses the problems with bringing an action under the most relevant international legal regimes. Part 5 argues that the LOSC contains both the substantive obligations and dispute-settlement procedures to underpin an international claim. In Part 6, we provide a detailed assessment of the legal issues involved in bringing a contentious action related to ocean acidification under the LOSC, and in Part 7, we consider the option of seeking an advisory opinion from the International Tribunal for the Law of the Sea (ITLOS) on the application of key LOSC obligations to the causes and impacts of ocean acidification.

2. The Science and Risks of Ocean Acidification

Oceans are the Earth's largest carbon sink, estimated to have absorbed between one-quarter and one-third of all CO_2 emissions caused by human activity over the past 200 years.[9] The Intergovernmental Panel on Climate Change (IPCC) has estimated that the oceans continue to

9 Secretariat of the Convention on Biological Diversity, *Scientific Synthesis of the Impacts of Ocean Acidification on Marine Biodiversity*, at 10, Technical Paper No. 46 (2009) [hereinafter Scientific Synthesis].

absorb approximately 30% of CO_2 released into the atmosphere each year.[10] As atmospheric concentrations of CO_2 increase, marine concentrations will continue to rise as well. The net absorption of CO_2 by the oceans will only slow and eventually cease when atmospheric concentrations stabilize and equilibrium is reached between CO_2 in the atmosphere and in the ocean surface.[11] However, at higher levels of ocean acidification, the absorptive capacity of the oceans is reduced, making it more difficult to balance CO_2 levels in the atmosphere and hydrosphere.[12]

As CO_2 levels in the oceans increase, they cause significant changes to ocean chemistry. CO_2 entering the ocean reacts with water to produce carbonic acid. Over the last 150 years, this process has caused an increase in ocean acidity of around 30%.[13] These changes are unprecedented, with the oceans reaching higher levels of acidity than at any time in the past 20 million years and with the increase in acidity occurring faster than at any time during the past 300 million years.[14] Because changes in ocean pH move in lockstep with increases in global CO_2 concentrations, they can be predicted with high degrees of accuracy using models of marine chemistry and ocean cycles.[15] By the end of the century, the oceans might

10 Kenneth L. Denman et al., Intergovernmental Panel on Climate Change [IPCC] Working Group I, *Couplings Between Changes in the Climate System and Biogeochemistry*, *in* CLIMATE CHANGE 2007: THE PHYSICAL SCIENCE BASIS (Susan Solomon et al. eds., 2007).

11 Michiel Schaeffer & Bill Hare, *Ocean Acidification: Causes and Consequences* (May 10, 2010) (unpublished briefing paper), *available at* https://sites.google.com/a/climateanalytics.org/test/variousfiles/20100510 ShortbriefingonOceanacidification.pdf.

12 Scientific Synthesis, *supra* note 9, at 60.

13 Quirin Schiermeier, *Earth's Acid Test*, NATURE, Mar. 10, 2011, at 154.

14 Symposium, *supra* note 1, at 3; Alex Morales, *Oceans Acidifying Fastest in 300 Million Years due to Emissions*, BLOOMBERG, Mar. 2, 2012, http://www.bloomberg.com/news/2012–03–01/oceans-acidifying-fastest-in-300-million-years-due-to-emissions.html.

15 James C. Orr et al., *Research Priorities for Understanding Ocean Acidification*, OCEANOGRAPHY, Dec. 2009, at 182, 182; Scientific Synthesis, *supra* note 9, at 9.

reach levels of acidity up to 150% to 200% greater than in preindustrial times.[16]

Rapid ocean acidification carries with it severe risks for ocean ecosystems, particularly those elements reliant on carbonates for growth.[17] Between 55 and 65 million years ago, when ocean acidification last reached the levels anticipated by the end of this century, geological evidence indicates a period of mass extinction of marine organisms.[18] Today, however, acidification is progressing ten times faster than during this earlier period of mass extinction.[19]

At particular risk from ocean acidification are corals and shellfish.[20] Should average atmospheric concentrations of CO_2 reach 450 ppm, it is projected that the growth of coral reefs around the world will slow considerably, and at concentrations of 550 ppm, it is anticipated that coral reefs will begin to dissolve (for further discussion on the effects of ocean acidification on coral reefs, see Chapter 2).[21] Even under current pledges for GHG emission reductions, projections indicate that we may reach 550 ppm within the next fifty years.[22]

Coral reefs provide habitat to one-quarter of the world's fish species and generate billions of dollars in export and tourism revenue for many countries.[23] Ocean acidification has already been observed to reduce rates of coral growth[24] and to increase the vulnerability of corals to high-temperature events, the impacts of local pollution, and mass coral bleaching. Small, low-lying atoll countries, such as Kiribati, the Maldives,

16 Schiermeier, *supra* note 13, at 155; Stockholm Environment Institute, Draft Executive Summary *supra* note 5, at 4.

17 Schaeffer and Hare, *supra* note 11.

18 Schiermeier, *supra* note 13, at 155; Symposium, *supra* note 1, at 3.

19 Schiermeier, *supra* note 13, at 155.

20 Interacademy Panel on International Issues, Statement on Ocean Acidification (June 2009), *available at* http://www.interacademies.net/9075.aspx.

21 Schaeffer and Hare, *supra* note 11.

22 *Id.*

23 *See generally* Royal Society, Ocean Acidification Due to Increasing Atmospheric Carbon Dioxide 33–34 (2005).

24 Schaeffer and Hare, *supra* note 11.

Tuvalu, and the Republic of the Marshall Islands, are particularly exposed to a multitude of ocean acidification's most serious impacts, including damage to foundational corals, surrounding reefs and knock-on effects for fisheries, food security, and tourism.

Ocean acidification also threatens the health of plankton, pteropods, mollusks, and other creatures with calcium carbonate shells or exoskeletons, which could have far-reaching effects on marine food webs.[25] Additionally, increased CO_2 accumulation may disturb the metabolism, reproduction, and other biological processes in marine animals, including fish.[26] The chain reactions that may be triggered by ocean acidification could reverberate throughout the marine environment, jeopardizing the food security and economic prospects for millions of the world's poorest, many of whom depend on the oceans for their sustenance and livelihoods.[27]

As researchers anticipated, the effects of ocean acidification are appearing first in the cold polar waters of the Arctic and Southern Oceans. However, interactions with other stressors and natural phenomena make it difficult to predict the actual impacts of acidification in specific geographical areas or on particular aspects of the marine environment. For example, it has been suggested that the interaction of increasingly acidic surface waters and seasonal upwelling currents may be contributing to harmful impacts in certain locations, such as the dramatic rise in oyster mortality off the Pacific Northwest of the United States since 2005.[28]

Analysis of past geological events suggests that increased ocean acidity may only be reversed over tens of thousands of years, with the recovery of ecosystems and biological diversity taking much longer.[29] A recent assessment suggests that the long-term survival of coral reefs will require

25 Symposium, *supra* note 1, at 5.

26 *Id.*

27 *Id.* at 6.

28 Schiermeier, *supra* note 13, at 155.

29 Scientific Synthesis, *supra* note 9, at 9; Symposium, *supra* note 1, at 3.

reductions to below 350 ppm by the end of the century.[30] However, prompt action might still avoid the most significant damages if levels of atmospheric CO_2 can be stabilized at below 450 ppm by the middle of the century.[31]

Despite an increased awareness of the potential damage, scientific understanding of the broader impacts of ocean acidification is at an early stage.[32] A high-level international scientific conference recently concluded that "[d]espite advances in understanding the impacts of elevated CO_2 concentrations on a wide range of marine organisms, we are still unable to make meaningful projections of impacts on marine ecosystems and fisheries as a whole, or to identify thresholds beyond which marine ecosystems may not recover."[33]

3. The Rationale for an International Legal Action Addressing Ocean Acidification

Following the disappointing outcomes from the UNFCCC Copenhagen Climate Summit in December 2009, and some tentative signs of progress the following year at the Sixteenth Conference of the Parties (COP 16) in Cancun, countries agreed at the Seventeenth Conference of the Parties (COP 17) in Durban to develop and adopt "a protocol, another legal instrument or an agreed outcome with legal force" applicable to all Parties by 2015, which will "come into effect and be implemented from 2020."[34] Although this mandate marks the beginning of a new phase in international negotiations under the UNFCCC, many negotiators and observers of the process remain pessimistic that a new legal framework

30 J.E.N. Veron et al., *The Coral Reef Crisis: The Critical Importance of <350 ppm CO2*, 58 MARINE POLLUTION BULL. 1428 (2009).

31 Schaeffer and Hare, *supra* note 11.

32 Symposium, *supra* note 1, at 7.

33 *Id.*

34 UNFCCC Conference of the Parties, Establishment of an Ad Hoc Working Group on the Durban Platform for Enhanced Action, Decision 1/CP.17, UN Doc. FCCC/CP/2011/9/Add.1 (Dec. 11, 2011).

can deliver the emission reductions necessary to achieve the internationally agreed goal of holding the increase in global average temperature below 2°C above preindustrial levels.

With the impacts of climate change increasingly apparent and accelerating, attention in some quarters has turned to possible alternatives to multilateral negotiations and to additional strategies, such as international legal action, that would advance the goal of preventing "dangerous anthropogenic interference in the climate system."[35] For example, Palau is currently working to build support at the United Nations General Assembly for a resolution to seek an advisory opinion from the International Court of Justice on the responsibilities and obligations of a State under international law to ensure that its greenhouse gas emissions do not cause damage to the environment of another State or States.[36] Of course, the main cause of ocean acidification – CO_2 emissions – is the same as that for global warming, and so a successful legal action to make one or more major emitters accountable for the impacts of one would necessarily benefit efforts to address the other.

The political realities of international relations and the historically cautious use of international courts suggest that global environmental issues as complex as climate change and ocean acidification will ultimately need to be resolved through international negotiation, rather than by judicial edict. Nevertheless, international legal strategies and actions, including those intended to address the damage caused by ocean acidification, can affect the political context for international negotiations in several ways.

First, bringing the issue of ocean acidification to international courts would likely raise its profile within the UNFCCC negotiations, a setting where the chemical impacts of CO_2 levels on ecosystems have thus far received scant attention. This would no doubt have the effect of moving

35 UNFCCC art. 2.

36 UN News Center, Palau seeks UN World Court opinion on damage caused by greenhouse gases, UN News Center, http://www.un.org/apps/news/story.asp?NewsID=39710& Cr=pacific+island&Cr1 (Sept. 22, 2011).

ocean acidification up the international agenda, in turn intensifying calls for a more urgent and effective policy response to peak CO_2 emissions and then ensure their reduction.

Second, legal action has the potential to reframe international climate diplomacy and negotiations around the concepts of actual and imminent harm, including the harm already resulting from sea level rise and ocean acidification, rather than more uncertain and distant impacts. Such a shift would sharpen attention on the plight of vulnerable small island States – those bearing the greatest burdens from current and imminent harms – and strengthen arguments in favor of financial mechanisms to compensate and insure against emission-related loss and damage.

Third, international jurisprudence resulting from a legal action may help clarify the legal obligations of States in a way that reframes international climate negotiations by defining behavioral standards from which any proposed derogation must be justified. Against the backdrop of an international system that imposes general obligations not to cause serious harm to the environment of other States, the clarification and elaboration of such standards are likely to strengthen the leverage of those States demanding more robust and effective limits on CO_2 emissions.

Fourth, a successful claim for monetary damages against one or more major emitters for loss and damage suffered or already certain to occur from ocean acidification[37] would entail the relevant polluter or polluters paying for the damage caused by their emissions. This would exemplify the costs associated with the negative environmental externalities of CO_2 emissions and highlight the need to internalize those costs, particularly

37 Recent scientific studies already confirm that concentrations of CO2 in the atmosphere and the oceans are causing observable changes to states' environments and natural resources. *See* IPCC, Climate Change 2007, *supra* note 10. Furthermore, the long time lags associated with the world's carbon cycle means that, even under the most optimistic emission reduction scenarios, a significant quantum of harm caused by ocean acidification is a *fait accompli*.

if the remedies ordered included compensation for ongoing harms or opportunities to obtain such compensation in the future. Compensation from those most responsible would potentially provide access to financial and other resources aimed at making the injured State whole in accordance with the foundational principle of legal redress for harm.[38] Thereafter, the mere threat or initiation of a claim for damages would be more likely to incentivize behavioral change by the major CO_2 emitters.

Yet the decision to bring an international action against certain major-emitting States entails political risks and could affect negatively the political and negotiating dynamic between major emitters and other States in ways that are difficult to predict. For example, the political message of the inadequacy of results from the UNFCCC implicit in the initiation of an international legal action runs the risk of damaging an already fragile international process and may have the effect of isolating claimant States in the UNFCCC and other fora. In addition, vulnerable States considering possible legal action would need to bear in mind broader bilateral and regional considerations, including interests in the spheres of trade, foreign policy, and development assistance.

Finally, a successful legal action could also serve to highlight for States the risks associated with belonging to treaty regimes that include strong or non-derogable dispute settlement mechanisms. For example, the initiation of a legal action against one or more major emitters under the LOSC dispute settlement procedures – especially a successful one – could set back, for many years, ongoing efforts to secure U.S. ratification of the LOSC.[39] It might also serve to cement the United States' apparent unwillingness, and that of other States, to participate in a strong, enforceable, and compliance-based international climate change regime.

38 Issues involved in demonstrating compensable damages and the apportionment of liability are dealt with in Section 5.5, *infra*.

39 *Cf. Rising Prospects Reported for the Law of the Sea Convention*, OCEAN L. DAILY (Rule of Law Committee for the Oceans, Washington, DC), Feb. 5, 2011, noting the prospect of U.S. ratification of LOSC).

4. Applicable International Regimes

Ocean acidification is often seen as a phenomenon within the sphere of impacts caused by climate change and the resultant warming of the oceans. This view is misleading and inaccurate. As described earlier, ocean acidification is rather a second, parallel impact resulting from higher atmospheric concentrations of CO_2. With this in mind, it is worth recalling that there are several international regimes that potentially apply to ocean acidification, depending on whether the regulatory focus is on causes, effects, or both.

Before turning to the LOSC, this section offers a very brief and non-exhaustive overview of other relevant multilateral regimes, as well as customary international law, as possible bases for an international legal action to seek redress for damage caused by ocean acidification.

4.1. The UN Framework Convention on Climate Change

The UNFCCC is the main international instrument for addressing the causes and effects of climate change. As described elsewhere in this book, the "ultimate objective" of the UNFCCC is to achieve "the stabilization of greenhouse gas concentrations in the atmosphere at a level that would prevent dangerous anthropogenic interference with the climate system,"[40] which is in turn defined as "the totality of the atmosphere, hydrosphere, biosphere and geosphere and their interactions."[41]

In seeking to achieve this ultimate objective, countries are to be guided by the principle of "common but differentiated responsibilities and respective capabilities," which is elaborated in the UNFCCC to mean that developed-country Parties should take the lead in combating climate change and its adverse effects.[42] The specific needs and special circumstances of developing-country Parties, especially those that are

40 UNFCCC, *supra* note 34, art 2.

41 *Id.*

42 UNFCCC, *supra* note 34, at art. 3.

particularly vulnerable to the adverse effects of climate change, should be given full consideration, and Parties should take a precautionary approach in preventing the causes of climate change and mitigating its adverse effects.

In December 1997, States Parties to the UNFCCC adopted the Kyoto Protocol, which commits industrialized countries and countries with economies in transition (known together as "Annex I countries") to achieve reductions in emissions of a basket of six GHGs. Under the Kyoto Protocol, Annex I countries agreed to reduce their overall emissions to 5.2% below 1990 levels during the five-year period 2008 to 2012 (the "first commitment period"), with individual targets varying country by country. Developing-country ("Non-Annex I") Parties have no such emission reduction commitments. The Kyoto Protocol entered into force on February 16, 2005, and now has 192 States Parties following the withdrawal of Canada in December 2011.[43] Despite being Party to the UNFCCC and signing the Kyoto Protocol in 1997, the United States has never ratified it. In Durban, the States Parties to the Kyoto Protocol agreed that a second commitment period under the Kyoto Protocol will begin on January 1, 2013, but both Japan and the Russian Federation indicated that they do not intend to be bound by an emission reduction commitment under the Kyoto Protocol after the first commitment period expires at the end of 2012.[44]

Although the UNFCCC or the Kyoto Protocol does not mention ocean acidification, the scope of the Convention appears to cover activities and impacts related directly to it as a phenomenon affecting the

43 *See* RTT News, Canada under Fire over Withdrawal from Kyoto, Dec. 14, 2011, http://www.un.org/apps/news/story.asp?NewsID=40714; UN News Center, Canada's withdrawal from Kyoto Protocol regrettable – UN climate official, UN News Center, Dec. 13, 2011, http://www.un.org/apps/news/story.asp?NewsID=40714.

44 UNFCCC, Conference of the Parties serving as the meeting of the Parties to the Kyoto Protocol, Decision 1/CMP.7, UN Doc. FCCC/KP/CMP/2011/10/Add.1 (Dec. 11, 2011).

"climate system," including the release of CO_2 (atmosphere), and its impacts on the oceans (hydrosphere) and marine life (biosphere).[45] The Convention also addresses climate change as the net result of "emissions from sources" less "removals by sinks,"[46] and it defines a sink as "any process, activity or mechanism which removes a greenhouse gas . . . from the atmosphere."[47] Clearly, the definition of sinks includes the natural oceanic absorption of atmospheric CO_2 that is ultimately the cause of ocean acidification.

However, recourse to the UNFCCC as the basis for an international legal action related to ocean acidification is problematic for two main reasons. First, the UNFCCC's provisions have only general and largely unquantifiable obligations to adopt "national policies and take corresponding measures" to mitigate climate change. Second, the UNFCCC's dispute settlement procedures provide for negotiation and conciliation, but not compulsory binding adjudication of disputes. Rather, disputes can only be referred to the International Court of Justice (ICJ) or an arbitral tribunal with the *ex ante* consent of the Parties to the dispute.[48]

Although the Kyoto Protocol has more quantifiable obligations in the form of economy-wide emission reduction commitments for Annex I countries and a set of compliance procedures established by the Marrakesh Accords,[49] the UNFCCC's relatively weak dispute settlement procedures apply *mutatis mutandis* to disputes arising under the Kyoto Protocol.[50] Unsurprisingly, no dispute under the UNFCCC or the Kyoto Protocol has ever been settled pursuant to these provisions.

45 UNFCCC, art. 1.3.

46 *Id.* at art 3.3.

47 UNFCCC, art 1.8.

48 *Id.* at art. 14.

49 UNFCCC Conference of the Parties Procedures and Mechanisms Relating to Compliance under the Kyoto Protocol, Decision 24/CP.7, at 64, UN Doc. FCCC/CP/2001/13/Add.3 (Nov. 10, 2001).

50 Kyoto Protocol to the United Nations Framework Convention on Climate Change, art. 19, Dec. 11, 1997, 2303 U.N.T.S. 148.

4.2. The Convention on Biological Diversity (CBD)

Like the UNFCCC, the CBD was adopted at the Rio Earth Summit in June 1992, but it entered into force earlier than the UNFCCC, in December 1993. The CBD currently has 193 States Parties, but again, as with the LOSC, the United States has not ratified it. The three main objectives of the CBD are the conservation of biological diversity, the sustainable use of its components, and the fair and equitable sharing of the benefits from the use of genetic resources.[51]

Ocean acidification has garnered increasing attention at the CBD's biennial Conferences of the Parties (COPs) due to concern over the impacts of climate change on marine biodiversity. Decisions at COP 4 (1998) and COP 5 (2000), respectively, expressed concern and urged action to address coral bleaching driven by the changing climate.[52] By 2008, Parties to the CBD began to see ocean acidification as a problem independent of climate change, and at COP 9 in 2008, Parties requested the Secretariat to prepare a synthesis report of "available scientific information on ocean acidification and its impacts on marine biodiversity and habitats, which is identified as a potentially serious threat to cold-water corals and other marine biodiversity."[53]

The resulting report concluded that if current emission trajectories were to be carried forward, then by 2100, 70% of cold-water corals – key refuges and feeding grounds for commercial fish species – would be exposed to corrosive waters, resulting in significant disruptions to large components of the marine food web.[54] In response to this and other accumulating science, the Parties at COP 10 in Nagoya, Japan, requested the CBD Executive Secretary to develop and coordinate a "series of joint expert review processes" to monitor and assess the impacts of ocean

51 Convention on Biological Diversity (CBD), art. 1, June 5, 1992, 1760 U.N.T.S. 79.

52 CBD, *Scientific Synthesis of the Impacts of Ocean Acidification of Marine Biodiversity*, at 5, *Note by the Executive Secretary*, UN Doc. UNEP/CBD/SBSTTA/14/INF/8 (Mar. 26, 2010).

53 CBD, Conference of the Parties, Decision IX/20, ¶ 4, UN Doc. UNEP/CBD/COP/9/29, (Oct. 9, 2008).

54 CBD, *Scientific Synthesis*, *supra* note 52.

acidification on marine and coastal biodiversity, and to widely disseminate its outputs, including transmission to the UNFCCC.[55] An expert meeting was held in October 2011 to elaborate these processes.[56]

Although ocean acidification has gained currency in the CBD as an important environmental concern, the Convention lacks any clear or enforceable standards of conduct applicable to CO_2 emissions. Moreover, Article 27 of the CBD, like Article 14 of the UNFCCC, only provides for referral of a dispute to the ICJ or arbitration subject to the prior written consent of the Parties involved.[57]

4.3. The IMO London Convention and MARPOL

The Convention on the Prevention of Marine Pollution by Dumping of Wastes and Other Matter 1972 and its 1996 Protocol (London Convention and Protocol) forms the legal regime developed through the International Maritime Organization (IMO) to prevent and address marine pollution caused by dumping or disposal of waste at sea.[58] Because "dumping" is defined in the Convention as "deliberate disposal into the sea,"[59] it has relevance for ocean acidification only as it relates to active CO_2 sequestration (such as sub-seabed storage of CO_2), rather than the passive absorption of CO_2 emissions, which causes the vast majority of acidification.[60]

For similar reasons, the scope of the International Convention for the Prevention of Pollution from Ships 1973 (MARPOL) – preventing

55 CBD, Conference of the Parties, Decision X/29, ¶ 66, UN Doc. UNEP/CBD/COP/10/27.

56 CBD, Subsidiary Body for Scientific, Technical, and Technological Advice, *Report of the expert meeting to develop a series of joint expert review processes to monitor and assess the impacts of ocean acidification on marine and coastal biodiversity*, UN Doc. UNEP/CBD/SBSTTA/16/INF/14 (Mar. 15, 2012).

57 CBD, *supra* note 52, art. 27.

58 1996 Protocol to the Convention on the Prevention of Marine Pollution by Dumping of Wastes and Other Matter, Nov. 29, 1972, A.T.S. 1985 No 16.

59 *Id.* at art. 1, Sec. 4.1.1.

60 Rachel Baird et al., *Ocean Acidification: A Litmus Test for International Law* 3 CARBON & CLIMATE L. REV. 459, 465 (2009).

pollution of the marine environment through the direct "discharge" of harmful substances or effluents from vessels into the ocean – is too narrow to be a strong platform for action to address ocean acidification.[61] Although MARPOL includes "emitting" within its definition of "discharge," emissions from shipping globally account for only 1.5% of global greenhouse gases from anthropogenic sources,[62] and the contribution of any one country's vessels to the atmospheric concentrations of CO_2 currently generating ocean acidification would likely be insufficient to warrant a regime response.

4.4. Customary International Law

In addition to the treaty-based regimes considered earlier, there is a large body of legal scholarship dedicated to the potential bases in customary international law for interstate legal action relating to the causes and effects of climate change.

A recent review of this material concluded that, although challenges exist in applying customary international law – such as the prohibition on causing significant transboundary damage (or the "no-harm" rule) – to a problem as complex and cross-cutting as climate change, a credible case can be made that States bearing the greatest historic responsibility for climate change, due to accumulated historic emissions, have and are continuing to commit an international legal wrong.[63] Yet the review concluded that only "in a limited number of possible scenarios, there are procedural means to pursue an inter-state litigation before an international judicial forum, in particular the International Court of Justice (ICJ)"[64] (for further analysis of the responsibilities of states for greenhouse gas emissions, see Chapters 8 and 13).

61 International Convention for the Prevention of Pollution from Ships as amended by the Protocol of 1978, Feb. 17, 1978, 1340 U.N.T.S. 61.

62 Pew Center on Global Climate Change, Marine Shipping (2010).

63 Christophe Schwarte with Ruth Byrne, *International Climate Change Litigation and the Negotiation Process* (FIELD, Working Paper, 2010), *available at* http://www.field.org.uk/files/FIELD_cclit_long_Oct.pdf.

64 *Id.*, at 23.

The jurisdiction of the ICJ over such disputes is based on the consent of the States concerned, and only those States that have made a prior declaration consenting to the court's compulsory jurisdiction can be brought before it against its will. Neither of the world's two largest emitters – the United States and China – have such a declaration in force. Moreover, it is doubtful that customary international legal obligations alone are sufficiently precise to form the basis for injunctive relief or the award of damages to a State for specific losses attributable to climate change.

Similar considerations may apply to the problem of ocean acidification, but there are some important differences as well. In particular, ocean acidification has a more direct and scientifically verifiable causal relationship with higher atmospheric concentrations of CO_2, as well as more localized and defined impacts than many of the detrimental impacts of climate change. Ocean acidification also directly engages specific obligations under the LOSC, which brings with it near-universal participation, the clarity of substantive treaty obligations not to pollute the marine environment (as well as the incorporation of standards of customary international law), and a compulsory and nonderogable dispute settlement regime with binding consequences.

Bearing in mind the political imperatives and objectives outlined in Part 3, as well as the legal and jurisdictional realities discussed here, the following analysis focuses exclusively on the potential pursuit of an adversarial claim or a request for an advisory opinion under the LOSC's compulsory dispute settlement system.

5. The LOSC and the Duty to Prevent Ocean Acidification

Described on its adoption in 1982 as the "constitution of the oceans,"[65] the LOSC defines the rights and responsibilities of the international

65 Tommy B. Koh, President, Third U.N. Conference on the Law of the Sea, Remarks at the Final Session of the Conference at Montego Bay (Dec. 6 and 11, 1982), *available at* http://www.un.org/Depts/los/convention_agreements/texts/koh_english.pdf.

community in its interactions with the world's oceans, including the conservation and management of the marine environment and its natural bounty.[66] The LOSC entered into force in 1994 and currently has 160 States Parties. The most notable exception is the United States, which signed the LOSC but has not ratified it. Nevertheless, many of the LOSC's provisions, including those relating to the pollution of the marine environment, are generally regarded as having achieved the status of customary international law and are therefore binding on all States.[67]

In contrast to the international regimes discussed in Section 4, the LOSC imposes broad and enforceable duties that address directly the harm of ocean acidification and its cause – anthropogenic CO_2 emissions. In general, the LOSC obliges States to take measures to prevent, reduce, and control pollution of the marine environment from all sources, including through the atmosphere and from land-based sources.[68] It also requires that special protection be provided for rare or fragile ecosystems, such as coral reefs and other vulnerable marine ecosystems known to be particularly susceptible to the impacts of ocean acidification.[69]

Importantly, the LOSC defines pollution broadly to include any substance that "results or is likely to result in such deleterious effects as harm to living resources and marine life."[70] But what makes these duties so potentially potent is their coexistence with the Convention's provisions for "responsibility and liability,"[71] as well as the Convention's compulsory dispute settlement system, which may offer the most promising opportunity for pursuing international legal action against major CO_2

66 As a comprehensive regime, LOSC also addresses, among other subjects, issues of sovereign and economic rights at sea, navigation, exploitation of natural resources, scientific cooperation, and governance of the seabed beyond national jurisdiction.

67 Patricia Birnie et al., International Law & the Environment 387 (2009).

68 United Nations Convention on the Law of the Sea (LOSC) art. 194.3, Dec. 10, 1982.

69 *Id.*at art. 194(5).

70 *Id.* at art. 1(4).

71 *See id.* at art. 235.

polluters for damage caused to the environment by anthropogenic CO_2 emissions and resultant ocean acidification.

As explained in the following analysis, and subject to detailed procedural requirements, the LOSC dispute settlement system also provides for the option of seeking an advisory opinion from the ITLOS to clarify the rights and obligations contained in the LOSC and their application in certain situations, such as a possible international legal action relating to ocean acidification.

5.1. Relevant Obligations under the LOSC

The LOSC provisions related to environmental protection contain both general and specific obligations of conduct. States are responsible for the fulfillment of their obligations and "shall be held liable in accordance with international law."[72]

The provisions of the LOSC most relevant to the causes and impacts of ocean acidification are found in Part XII, "Protection and Preservation of the Marine Environment," as outlined as follows.

5.1.1. General Obligations

The general obligations of marine environmental protection under the LOSC are found in Articles 192 and 194:[73]

72 LOSC, *supra* note 68, art. 235. The English text generally preserves the distinction between responsibility, the discharge of obligations under international law, and liability, the regime of reparations for violating an international obligation. Myron Nordquist et al., eds., Univ. of Va. Ctr. for Oceans L. & Pol'y, 4 United Nations Convention on the Law of the Sea, 1982 : A Commentary § 235.10(a) (1991) [hereinafter 4 Virginia Commentary] For a more in-depth linguistic analysis of these terms, *see* Responsibilities and Obligations of States Sponsoring Persons and Entities with Respect to Activities in the Area, ¶¶ 66–67 (Advisory Opinion of Feb. 1) (Int'l Trib. L. of the Sea 2011) [hereinafter Obligations of States], *available at* http://www.itlos.org/case_documents/2011/document_en_379.pdf.

73 Unless otherwise clear from the context, all references to an "article" or "annex" are references to the LOSC.

- Article 192 declares that States have an obligation to protect the marine environment.
- Article 194(1) provides that States shall take all measures necessary to "prevent, reduce and control pollution of the marine environment from any source, using ... the best practicable means at their disposal and in accordance with their capabilities."[74] Such measures are to be taken individually or jointly, as appropriate.
- Article 194(2) provides that States shall take all measures necessary "to ensure that activities under their jurisdiction or control" are carried out so as not to cause damage by polluting other States and their environments and to ensure that pollution from such activities does not spread beyond the areas where the State exercises sovereign rights.
- Article 194(3) clarifies that these obligations extend to all sources of pollution of the marine environment and the measures taken must include those designed to minimize the release of harmful substances from land-based sources or through the atmosphere "to the fullest possible extent."
- Article 194(5) further clarifies that the measures taken shall include those required to "protect or preserve rare or fragile ecosystems as well as the habitat of threatened or endangered species and other forms of marine life."

Article 192 is both phrased and understood to be declaratory of generally applicable international law[75] including the no-harm

74 Pollution of the marine environment is defined in LOSC Article 1(4) as "the introduction by [humans], directly or indirectly, of substances or energy into the marine environment ... which results or is likely to result in such deleterious effects as harm to living resources and marine life ..., hindrance to marine activities, including fishing and other legitimate uses of the sea."

75 *See* The Secretary-General, *Law of the Sea: Protection and Preservation of the Marine Environment*, ¶ 29, *Report to the General Assembly*, UN Doc. A/44/461 (Sept. 18, 1989); 4 VIRGINIA COMMENTARY § 192.8. Structurally, Article 192 is paired with Article 193, which affirms States' sovereign rights to exploit their natural resources "in accordance with their duty to protect and preserve the marine environment." This provision

principle, whereas Article 194 elaborates on the content and scope of those obligations.[76]

5.1.2. Obligations of Specific Conduct

In addition to the general obligations established in Articles 192 and 194, the subsequent sections of Part XII outline the more specific obligations necessary to fulfill the obligation to prevent or mitigate damage caused by pollution to the marine environment, including particular obligations for which States can be held responsible:[77]

- Article 204 requires States, to the extent practicable, to monitor the risks or effects of pollution on the marine environment. More specifically, it obliges States to keep under surveillance those activities they permit in order to determine the risk of pollution.[78]
- Article 206 establishes that, when a State has a reasonable basis for believing a planned activity within its jurisdiction or control may cause "substantial pollution" or "significant and harmful changes to the marine environment," it must assess the activity's potential effects and communicate the results to other States through competent international organizations.[79]

appears primarily to provide context for the remainder of Part XII, rather than establishing its own independent rights. This pairing reflects the balance that has historically been struck in international environmental law. *See, e.g.*, United Nations Conference on Environment and Development, Rio de Janiero, Braz., June 3–14, 1992, *Rio Declaration on Environment and Development*, Principle 2, UN Doc. A/CONF.151/26/Rev.1 (Vol. I), Annex I (Aug. 12, 1992).

76 Secretary-General, *supra* note 75, ¶ 30; *see also* 4 Virginia Commentary § 194.1, .10(h).

77 *Cf.* Obligations of States ¶¶ 121–23 (discussing a similar distinction in Part XI of LOSC).

78 A provision to limit the geographic scope of this article to areas under a state's sovereignty or control was rejected during the drafting. *See* 4 Virginia Commentary § 204.4.

79 The primary difference between Articles 204 and 206 is that the former is an ongoing process, whereas the latter is an assessment, akin to an Environmental Impact Statement under the U.S. National Environmental Policy Act, of a particular project before it has begun. *See id.* § 206.1. The method of communication is provided for in Article 205.

- Article 207 requires States to adopt laws and regulations (as well as take all other measures necessary) to "prevent, reduce, and control" pollution of the marine environment from land-based sources.[80]
- Article 212 elaborates the obligations under Article 207 to apply them to pollution of the marine environment *from or through the atmosphere*. In addition to aircraft registered in a State, this obligation is "applicable to air space under [the State's] sovereignty."[81]
- Article 235(2) requires a State to ensure that recourse is available in accordance with the State's legal system for damage done by persons under the State's jurisdiction.[82]

5.1.3. LOSC Obligations Are Relevant to Ocean Acidification

A plain reading of both the general and specific LOSC obligations relating to the prevention, reduction, and control of marine pollution suggests that they are relevant to the causes and effects of ocean acidification.

The vast majority of historical atmospheric emissions of CO_2 – known to be the chief cause of ocean acidification – has come from land-based activities undertaken within national jurisdictions and would appear to constitute "pollution of the marine environment from or through the atmosphere" as set out in Article 212.

What remains is to assess which countries might be well placed to bring an action based on the failure of others to fulfill these

80 Articles 207 to 212 are designed to establish the relationship between international and national regulatory regimes in a manner that fleshes out the various provisions of Article 194. *Id.* § 207.1. However, the relationship set out preserves the obligations to adopt the required national measures, merely "taking [international rules] into account." *Id.* § 207.7(a).

81 *See also* LOSC, *supra* note 68, arts. 213–14, 222 (duty to enforce laws adopted in accordance with Articles 207–08 and 212).

82 This provision is understood to apply to recourse by private actors against private actors. However, because of the exhaustion of local remedies principle, it could affect a state's ability to espouse a claim on behalf of a national. It does not appear relevant to a state's ability to bring a claim against another state for the latter's violation of an international obligation. *See* 4 Virginia Commentary § 235.10(b).

obligations, which countries the action should be brought against, and in which forum. It is also necessary to consider carefully what thresholds of harm and causation need to be established to prove a violation of international obligation so that an appropriate remedy might be sought.

6. Bringing a Contentious Action under the LOSC

The LOSC dispute settlement procedures, set out in Part XV of the Convention, provide for the nonconsensual adjudication between Parties of "any dispute concerning the interpretation or application of [the] Convention." After seeking to settle the dispute by peaceful means, it may be submitted for adjudication to any international court or tribunal with jurisdiction.

Most notably, every State Party to the LOSC is subject to the mandatory jurisdiction of at least one such court or tribunal simply by virtue of having ratified the Convention. But the LOSC does not specify any one forum for all disputes arising out of the interpretation and application of the Convention. There are a number of possibilities under Part XV, always determined by reference to what the Parties to the dispute in question have chosen.

6.1. The Potential Parties

All Parties to the LOSC have access to the Convention's dispute settlement mechanisms.[83] As noted earlier, the States with the greatest potential interest in bringing a claim based on harm suffered due to ocean acidification are likely to include small atoll island countries and archipelagic States reliant on the health of coral reefs for seafood-based livelihoods

83 LOSC, *supra* note 68, pt. XV. Although there are certain exceptions, *see id*. arts. 297–98, they are not relevant to this discussion.

and tourism. Most States fitting these criteria – including Kiribati, the Maldives, the Marshall Islands, and Tuvalu – are States Parties to the LOSC.[84]

In considering potential respondents, both historical and current levels of CO_2 emissions are relevant: historical emissions go to responsibility for harm suffered, whereas current and future emissions are relevant to the prevention of future harm.

The leading historical emitters, defined here as those States responsible for more than 5% of cumulative CO_2 emissions between 1850 and 2008,[85] are the United States, the European Union,[86] China, and Russia.[87] The United States and EU have each been responsible for approximately one-quarter of cumulative historical emissions, with no other State emitting more than 10% of the total.

84 *See* UN Division of Ocean Affairs and the Law of the Sea, Status of the United Nations Convention on the Law of the Sea, http://www.un.org/Depts/los/reference_files/status2010.pdf (consolidated listing of LOSC parties as of November 30, 2010) [hereinafter Status Report].

85 All data regarding CO_2 emissions are taken from the World Resource Institute's Climate Analysis Indicators Tool, *available at* http://cait.wri.org/cait.php. The 150-year time frame reflects the availability of data. However, the choice of historical time frames does not materially affect this analysis.

86 Although the European Union is not a State Party itself, it is Party to the LOSC as an international organization. References to "states" should therefore be understood to include the E.U. However, competence with regard to issues governed by the LOSC is allocated between the EU and Member States pursuant to European law. *See* European Community [EC], Competence of the European Communities with Regard to Matters Governed by the Convention on the Law of the Sea, Declaration Made Pursuant to Article 2 of Annex IX to the Convention (Dec. 7, 1984), *available at* http://www.un.org/Depts/los/convention_agreements/convention_declarations.htm#European Community. Upon signature (describing division of competence between European institutions and member states); EC, Declaration Made Pursuant to Article 5(1) of Annex IX to the Convention and to Article 4(4) of the Agreement (Apr. 1, 1998), *available at* http://www.un.org/Depts/los/convention_agreements/convention_declarations.htm#European Community Declaration made upon formal confirmation (same).

87 Germany meets this threshold independently as well but is included within the European Union.

Considering only emissions since the entry into force of the LOSC – that is, the period from 1983 to 2007 – the only emitters responsible for more than 5% cent of cumulative emissions are the United States, EU (as a group), China, and Russia. These same four entities are still the leading annual emitters according to 2008 data, although China has now surpassed the United States and EU in total CO_2 emissions. If land use changes are taken into account, Brazil and Indonesia also exceed 5% of annual global emissions, according to 2005 data. All of these major emitting States are Parties to the LOSC, with the exception of the United States.[88]

6.2. LOSC Dispute Settlement Procedures

Prior to bringing a claim, States first must seek to settle their disputes through any peaceful means listed in Article 33 of the UN Charter; that is, negotiation, inquiry, mediation, conciliation, arbitration, judicial settlement, resort to regional agencies or arrangements, or other peaceful means of their own choice.[89]

If these efforts at settlement fail, a Party may seek compulsory settlement of the dispute, subject to the choice of procedure mechanism contained in Article 287.[90] Under Article 287, a Party may declare, on accession to the LOSC or thereafter, that it accepts the jurisdiction of any of the adjudicatory bodies or mechanisms listed in Article 287: the ITLOS, the ICJ, or one of two forms of ad hoc arbitral tribunal constituted in accordance with Annexes VII and VIII to the LOSC. A Party may bring a claim against one or more Parties in any forum to which both or all have consented, without requiring any dispute-specific agreement. If there is no overlap in the fora consented to by the parties to the dispute, arbitration under Annex VII serves as a default

88 Status Report, *supra* note 84.
89 LOSC, *supra* note 68, art. 280.
90 *Id.*

mechanism for which no specific or additional declaration of consent is required.[91]

Only a few States have made declarations under Article 287 indicating their preferred choice of procedure. For example, from among the major emitters listed earlier, only the Russian Federation has made any declaration regarding choice of dispute settlement procedures.[92] Therefore, the likelihood of an overlapping choice of a preferred forum between likely claimants and major emitters is very low, and any ocean-acidification-related claim under the LOSC would almost certainly default to an Annex VII tribunal, unless the parties specifically agreed otherwise.

6.2.1. ITLOS and the Annex VII Arbitral Tribunal

As the permanent tribunal established under the Convention, the ITLOS has default jurisdiction over all requests for "provisional measures" under Article 290, pending the establishment of an Annex VII tribunal.[93] This procedure is available to States in situations where a quick decision is necessary to "preserve the respective rights of the parties" or "to prevent serious harm to the marine environment, pending the final decision."[94] But the tribunal may prescribe provisional measures only "if it considers that *prima facie* the tribunal which is to be constituted would have jurisdiction and that the urgency of the situation so requires."[95]

91 LOSC, *supra* note 68 *Id.* at art. 287.

92 Under Article 287, the Russian Federation chose an Annex VII arbitral tribunal "as the basic means for the settlement of disputes concerning the interpretation or application of the Convention," and an Annex VIII special arbitral tribunal for "matters relating to fisheries, the protection and preservation of the marine environment, marine scientific research, and navigation, including pollution from vessels and dumping." Declaration by the Russian Federation on signature of UNCLOS, United Nations Treaty Collection Database, http://treaties.un.org/Pages/ViewDetailsIII.aspx?&src=TREATY&mtdsg_no=XXI~6&chapter=21&Temp=mtdsg3&lang=en#EndDec .

93 LOSC, Art 209.

94 LOSC, art 290.1.

95 LOSC, art. 290.5.

Of the eighteen contentious claims that the ITLOS has heard since the LOSC entered into force in 1994, nine have been summary procedures regarding the prompt release of vessels and three have been requests for provisional measures. The Tribunal has addressed environmental issues in respect of only two requests for provisional measures: the *Southern Bluefin Tuna* and *MOX Plant* cases, both of which are discussed later. Thus, even including the recent Seabed Disputes Chamber advisory opinion in *Obligations of States*,[96] the ITLOS has had very limited opportunities to address the environmental obligations in the LOSC.

Of the six cases considered by an Annex VII arbitral tribunal, five were hosted at the Permanent Court of Arbitration[97] and one, *Southern Bluefin Tuna*, at the International Center for the Settlement of Investment Disputes.[98] The arbitrators appointed in these cases have tended to be prominent figures in public international law arbitration, some with a specialization in oceans issues or the law of the sea.

Under LOSC Annex VII, a proceeding is begun by means of written notification, including a statement of the claim and the grounds on which it is based. A five-member panel is then constituted – one member appointed by each side, and three agreed on jointly or, if the parties cannot agree, appointed by the President of the ITLOS. Other procedures applicable under Annex VII are left primarily to the discretion of the arbitral tribunal itself, unless otherwise agreed by the parties.[99]

96 *See* Seabed Disputes Chamber of the International Tribunal for the Law of the Sea, Responsibilities and Obligations of States Sponsoring Persons and Entities with Respect to Activities in the Area, February 1, 2011.

97 Permanent Court of Arbitration, *Ad Hoc Arbitration under Annex VII of the United Nations Convention on the Law of the Sea*, http://www.pca-cpa.org/showpage.asp?pag_id=1288.

98 *See* S. Bluefin Tuna – Aus. and N.Z. v. Jap., 39 I.L.M. 1359 (Award on Jurisdiction and Admissibility of Aug. 4) (U.N. Convention L. of the Sea Annex VII Arb. Trib. 2000) [hereinafter *Southern Bluefish Tuna* Decision].

99 *See* LOSC, *supra* note 68. *Id.* at Annex VII, art 5.

6.2.2. Joinder of Parties

Questions have been raised regarding the ability of one or more claimant(s) to bring a single case against multiple respondents, particularly in the absence of a special agreement among the parties or the authority of a tribunal to join the proceedings against multiple respondents. Such procedural questions are highly relevant to a claimant's ability to prove causation when there are multiple contributors to harm, as is the case with ocean acidification.

Although the institution of one action against multiple parties has been exceedingly rare in noncriminal proceedings before international courts, nothing in the applicable governing texts appears to preclude such an action, and both the ITLOS constituting documents and Annex VII to the Convention clearly contemplate multiparty proceedings.

For example, Article 47 of the ITLOS Rules of Procedure authorizes the Tribunal to join proceedings in different cases without requiring the parties' consent,[100] although this authority (as is the case with the ICJ's joinder authority) has been used to consolidate identical claims by different States only against a single respondent.[101] Annex VII, however, provides more explicit authority for the filing of a claim against multiple respondents, stating in Article 1 that submission of a dispute requires the provision of notice to "the other party *or parties*" to the dispute.[102]

Although separate bilateral claims could potentially lead to the establishment of different arbitral panels depending on the preferences of the various parties concerned, Annex VII does not appear to include a mechanism for one panel to require the joinder of a respondent in an action before a second panel, absent the consent of the parties involved. Thus, in pursuing an action in an Annex VII tribunal, institution of a

100 Rules of the Tribunal, Doc. No. ITLOS/8, art. 47 (Mar. 19, 2009) [hereinafter ITLOS Rules].

101 *See Southern Bluefin Tuna* Decision, ¶ 4.

102 LOSC, *supra* note 68, Annex VII, art. 1 (emphasis added).

single claim against multiple states would seem a better method for the claimant to achieve a consolidated proceeding.

6.2.3. The Potential Effects of Other Treaty Dispute Settlement Provisions on Jurisdiction under the LOSC

In general, LOSC Article 288 makes clear that a tribunal established under the Convention's procedures is competent to decide challenges to its own jurisdiction.

Parties to the LOSC maintain the freedom to agree on dispute settlement procedures other than those specifically provided by the LOSC.[103] Under Article 281, when the Parties have agreed to settle their dispute by alternative means, the regular dispute settlement procedures under LOSC Part XV can apply only "when no settlement has been reached by recourse to such means and the agreement between the parties does not exclude any further procedure." Simply put, if the parties to a dispute have agreed separately (for example, in another treaty related to ocean acidification) to resolve their dispute by certain means, the LOSC procedures will only apply when those other means have failed to resolve the dispute. Specifically, one could foresee a respondent to an action arguing that, in ratifying the UNFCCC and its dispute settlement procedures, the disputant States intended to exclude the application of LOSC procedures to disputes arising out of activities that are regulated under the UNFCCC regime.

The meaning and application of Article 281 as a jurisdictional bar were first tested in *Southern Bluefin Tuna*, in which Australia and New Zealand initiated proceedings against Japan for a purported violation of LOSC obligations relating to the conservation and management of straddling fish stocks. Japan argued that the case related to annual catch limits agreed under the 1993 *Convention for the Conservation of Southern Bluefin Tuna* (CCSBT),[104] a regional fisheries agreement that included

103 *Id.* at art. 280.

104 Convention for the Conservation of Southern Bluefin Tuna, May 10, 1993, http://www.ccsbt.org/docs/pdf/about_the_commission/convention.pdf.

its own dispute settlement provisions and was therefore intended to exclude the application of the LOSC procedures within the meaning of Article 281.

In addressing the request for provisional measures, the ITLOS found initially that Australia and New Zealand had made a prima facie showing of jurisdiction; it noted that "the fact that the 1993 CCSBT applies between the parties does not exclude their right to invoke the provisions of the [LOSC] in regard to the conservation and management of southern bluefin tuna."[105]

At the merits phase, however, the majority of the Annex VII tribunal found that it lacked jurisdiction to hear the case.[106] The tribunal considered the 1993 CCSBT and its dispute settlement procedure to constitute an "alternative means of settling disputes" within the meaning of Article 281. In reaching this finding, the tribunal confirmed that a dispute existed under both the LOSC and the CCSBT, but decided to read the provision in the CCSBT authorizing consensual arbitration as implicitly excluding recourse to the LOSC compulsory dispute settlement procedures. Thus, the tribunal found that the requirements of Article 281 had not been met and that it lacked jurisdiction to hear the case.[107]

More than a year later, in *MOX Plant*, the United Kingdom argued at the ITLOS that Ireland's claims were governed by dispute settlement procedures agreed to by the parties under the Convention for the Protection of the Marine Environment of the North-East Atlantic[108] and the EURATOM Treaty.[109] As in the *Southern Bluefin Tuna* case, ITLOS

105 S. Bluefin Tuna (Aus. v. Jap., N.Z. v. Jap), 117 I.LR. 148, ¶ 51 (Order on Provisional Measures of Aug. 27.) (Int'l Trib. L. of the Sea 1999).

106 S. Bluefin Tuna – Aus. and N.Z. v. Jap., 39 I.L.M. 1359 (Award on Jurisdiction and Admissibility of Aug. 4) (UN Convention L. of the Sea Annex VII Arb. Trib. 2000).

107 *Id.* ¶¶ 56–59.

108 Convention for the Protection of the Marine Environment of the North-East Atlantic, Sept. 22, 1992, 2354 U.N.T.S. 67; 32 I.L.M. 1069 (1993).

109 Treaty Establishing the European Atomic Energy Community, Mar. 25, 1957, 2010 O.J. (C 337) 1 (consolidated version through Mar. 30, 2010).

again issued an order for provisional measures and, in doing so, found that the jurisdictional standard in Article 281 had been met:

> [T]he dispute settlement procedures under the [other treaties] deal with disputes concerning the interpretation or application of those agreements, and not with disputes arising under [the LOSC];
>
> [E]ven if the [other treaties] contain rights or obligations similar to or identical with the rights or obligations set out in the Convention, the rights and obligations under those agreements have a separate existence from those under [the LOSC];
>
> [T]he Tribunal is of the opinion that, since the dispute before the Annex VII arbitral tribunal concerns the interpretation or application of [the LOSC] and no other agreement, only the dispute settlement procedures under [the LOSC] are relevant to that dispute[.][110]

Viewed in the light of more recent ITLOS proceedings, the decision of the arbitral tribunal in *Southern Bluefin Tuna* is perhaps best understood as an outlier that is unlikely to be followed in future proceedings of a similar nature.

Turning to the potential application of Article 281 in an action based on ocean acidification, it is plausible and perhaps even likely that a respondent or group of respondents would argue that the UNFCCC constitutes the overarching legal regime governing the impacts of CO_2 emissions on the oceans and that it already provides its own dispute settlement procedures intended to cover "any dispute concerning the interpretation or application of the [UNFCCC]."[111] The UNFCCC, however, lacks any specific references to marine pollution or harm to the marine environment. As explained earlier, ocean acidification is not strictly a climate-related problem, and the application of *lex specialis derogat legi generali*[112] tends to support the contention that the rights and obligations

110 MOX Plant (Ire. v. U.K.), 126 I.L.R. 260, ¶¶ 49–50, 52 (Order on Provisional Measures of Dec. 3) (Int'l Trib. L. of the Sea 2001) [hereinafter *Mox Plant* Provisional Measures]. The case was eventually withdrawn by Ireland in response to a judgment by the European Court of Justice.

111 *Id*. art. 14(2).

112 "Specific laws prevail over general laws."

established by the LOSC are more closely related to what is strictly an oceanic phenomenon. As in *MOX Plant*, the fact that the UNFCCC addresses sources and sinks of CO_2 does not render an ocean-acidification-related claim inadmissible under the LOSC.

6.3. The Nature and Scope of General LOSC Obligations and Thresholds of Harm

6.3.1. The Nature of the Duty Established by Articles 192 and 194: Due Diligence or More?

The wording of Articles 192 and 194, outlined earlier, is suggestive of obligations entailing strict liability.[113] However, multiple reputable sources, including the authoritative Virginia Commentary on the Law of the Sea, suggest that they instead establish obligations of due diligence.[114] This understanding appears to be consistent with a reading of the general LOSC obligations as reflecting existing customary rules on transboundary environmental harm.[115] Consistent with the general practice of States to apply strict liability only to activities that are inherently, particularly, and imminently hazardous,[116] the practical effect of such a reading may be to absolve States of responsibility for environmental damage when they have taken the "necessary and practicable measures."[117]

In its recent advisory opinion in *Obligations of States*, the ITLOS Seabed Disputes Chamber discussed the scope and content of the

113 *See, e.g.*, LOSC, *supra* note 68, art. 194(1) ("States shall take ... all measures ... necessary to prevent, reduce, and control pollution of the marine environment.").

114 *See, e.g.*, 4 Virginia Commentary. § 235.5–235.7 (discussing the rejection of negotiating drafts providing for strict liability).

115 *See id.*, § 194.10(e).

116 *See* Int'l L. Comm'n [ILC], Survey of Liability Regimes Relevant to the Topic of International Liability for Injurious Consequences Arising out of Acts Not Prohibited by International Law, ¶¶ 19–22, 26, 59, UN Doc. A/CN.4/543 (June 24, 2004) (*prepared by* the Secretariat) [hereinafter Liability Regimes].

117 Restatement (Third) of Foreign Relations Law § 601 comment d (1987); *see also* LOSC, *supra* note 68, art. 194(1) ("using ... the best practicable means for this purpose and in accordance with their capabilities").

obligation in international instruments "to ensure"[118] a particular outcome and prevent a different one from occurring, primarily by requiring States to exercise their power over entities, including private enterprises, under their control.[119] According to the Tribunal, the State's "obligation 'to ensure' is not an obligation to achieve, in each and every case, [compliance with treaty obligations by private entities]. Rather, it is an obligation to deploy adequate means, to exercise best possible efforts, to do the utmost, to obtain this result."[120] The opinion goes on to characterize the obligation as one "of conduct," rather than "of result" (i.e., an obligation of due diligence)[121] and cites LOSC Article 194 as an example of such an obligation.[122]

The content of the obligation of due diligence depends on the particular circumstances in question. In the past, the content of the obligation has been determined by a balancing of interests.[123] As confirmed by the Seabed Disputes Chamber, due diligence is a "variable concept."[124] It may change over time in light of new scientific or technical knowledge and can vary in relation to the risks of the activity at issue.[125] It is clear, however, that the standard has become more demanding over time,[126] particularly as environmental awareness has grown. For example, the ICJ held in *Gabčíkovo-Nagymaros Project* that, as the nature of environmental awareness evolves and leads to new environmental norms,

118 LOSC, *supra* note 68, art. 194(2) ("States shall take all measures necessary to ensure that activities under their jurisdiction or control are so conducted as not to cause damage by pollution to other States and their environment.").

119 *Obligations of States* ¶ 108. This opinion is particularly important, as it is the first explication of these provisions by an authoritative LOSC body.

120 *Id.* ¶ 110.

121 Id. ¶¶ 110–12 (citing Pulp Mills on the River Uruguay).

122 *Id.* ¶ 113. The opinion also finds further support in the meaning of the French, Spanish, Arabic, Chinese, and Russian words used where the English text uses "ensure." *Id.* ¶ 114.

123 Liability Regimes, *supra* note 118, ¶ 215.

124 Obligations of States ¶ 117.

125 *Id.*

126 *See id.* ¶¶ 226–33 (discussing the meaning of the *Trail Smelter*, *Corfu Channel*, and *Lac Lanoux* precedents).

"such new norms have to be taken into consideration [and] given proper weight, not only when States contemplate new activities but also when continuing with activities begun in the past."[127]

The content of the "due diligence" obligation to prevent environmental damage was recently discussed by the ICJ in *Pulp Mills on the River Uruguay* and is worth quoting here at length:

> [T]he principle of prevention . . . has its origins in the due diligence that is required of a State in its territory. It is every State's obligation not to allow knowingly its territory to be used for acts contrary to the rights of other States. A State is thus obliged to use all the means at its disposal in order to avoid activities which take place in its territory, or in any area under its jurisdiction, causing significant damage to the environment of another State. . . .
>
> [T]he obligation . . . is an obligation to act with due diligence in respect of all activities which take place under the jurisdiction and control of each party. It is an obligation which entails not only the adoption of appropriate rules and measures, but also a certain level of vigilance in their enforcement and the exercise of administrative control applicable to public and private operators, such as the monitoring of activities undertaken by such operators, to safeguard the rights of the other party. The responsibility of a party to the [treaty] would therefore be engaged if it was shown that it had failed to act diligently and thus take all appropriate measures to enforce its relevant regulations on a public or private operator under its jurisdiction.[128]

The court thus identifies four elements to the due diligence requirement: (i) the use of all means at a State's disposal, (ii) the adoption of appropriate rules, (iii) vigilance in enforcement, and (iv) ongoing monitoring. Additionally, as confirmed in *Obligations of States*, the substance of the general obligations of due diligence under the LOSC is given

127 Gabčíkovo-Nagymaros Project (*Hung. v. Slov.)*, 1997 I.C.J. 7, at 74 (Sept. 25).

128 Pulp Mills on the River Uruguay (*Arg. v. Uru.*), 2010 I.C.J. 1, ¶¶ 101, 197 (Apr. 20) (internal citations omitted).

further meaning by reference to the specific obligations provided under the Convention.[129]

Finally, the advisory opinion in *Obligations of States*, in perhaps its most novel and unprecedented finding,[130] pointed out that a "precautionary approach" is an "integral part" of the general obligation of due diligence.[131] Although never explicitly stating that the precautionary principle has achieved the status of law, the Chamber recognized a "trend" in this direction and, nonetheless, found the "precautionary approach" to be relevant to the interpretation of the LOSC more broadly.[132]

6.3.2. The Threshold of Harm

International law does not require evidence of damage to one State as a precondition to indicating the breach of another State's obligation – any

129 *See Obligations of States* ¶ 119.

130 Whereas the Chamber cites to Pulp Mills On the River Uruguay ¶ 164, the ICJ merely found the precautionary approach to be a relevant interpretive principle because such an approach was reflected in the agreement to be interpreted. Here, although the Chamber also finds a precautionary approach implicit in decisions and regulations issued by LOSC instruments, to find such an approach more broadly relevant as context under Article 31 of the VCLT is a progressive step. Obligations of States ¶ 135. *But cf.* Appellate Body Report, *European Communities – Measures Concerning Meat and Meat Products (Hormones)*, ¶ 123 n.93, WT/DS26/AB/R, WT/DS48/AB/R (Jan. 16, 1998) (citing the ICJ's failure in Gabĉíkovo-Nagymaros to include the precautionary principle among the new norms that it recognized as reason not to find it to have achieved the status of a principle of international law); Panel Report, *European Communities – Measures Affecting the Approval and Marketing of Biotech Products*, § 7.88, WT/DS291/R, WT/DS292/R, WT/DS292/R (Sept. 29, 2006) (same, citing *EC-Hormones*).

131 *See* Obligations of States ¶¶ 131–35. This is a new development since the ITLOS order on provisional measures in Southern Bluefin Tuna whereby Judge Treves, in his separate opinion, expressed disappointment that it did not recognize the necessity of a precautionary approach but understood its reluctance to recognize a precautionary principle in law. Southern Bluefin Tuna, 117 I.LR. at 179 (separate opinion of Treves, J.). *But cf.* Obligation of States ¶132 (citing the Southern Bluefin Tuna provisional measures order as implicitly recognizing the precautionary approach).

132 Obligations of States ¶ 135.

such condition must derive from the obligation itself.[133] However, international courts and tribunals have traditionally held that duties to prevent environmental damage only apply above a threshold level of harm, often referred to as "significant" harm.[134] According to the ILC, "significant" is defined as "something more than 'detectable' but need not be at the level of 'serious' or 'substantial.'"[135] As such, "significant" has often been characterized as a *de minimis* requirement.[136]

The LOSC appears to depart from this "significance" limitation, however, at least with regard to pollution-specific obligations, such as those under Article 194.[137] The LOSC defines "pollution" as "the introduction by [humans], directly or indirectly, of substances or energy into the marine environment ... which results or is *likely to result* in such deleterious effects as harm to living resources and marine life."[138] Thus, the duty in Article 194(1) to prevent, reduce, and mitigate pollution is applicable in circumstances involving a likelihood of harm, rather than any past or currently appreciable harm. Similarly, the reference in Article 194(2) to "damage by pollution" can best be understood as addressing any covered polluting activities that result in "deleterious effects," not only those above a certain threshold of harm or "significance."

The Virginia Commentary does not indicate any intent by the drafters to incorporate extrinsic limitations on the object of Article 194 (unlike the standard of conduct required).[139] In contrast, the test for

133 *See* Obligations of States ¶ 178; ILC, Draft Articles on Responsibility of States for Internationally Wrongful Acts with Commentaries art. 2, comment 9, U.N. Doc. A/56/10, ¶ 77 [hereinafter DASR].

134 *See, e.g.*, Pulp Mills on the River Uruguay ¶ 101.

135 ILC, Draft Articles on Prevention of Transboundary Harm from Hazardous Activities with Commentaries art. 2 comment 4, 53d Session, Report to the General Assembly, ¶ 98, U.N. Doc. A/56/10 (2001).

136 Christina Voigt, *State Responsibility for Climate Change Damages*, 77 Nordic J. Int'l L. 1, 9 (2008).

137 There is less evidence that Article 192, which merely states a customary norm, departs from the traditional "significance" threshold.

138 LOSC, *supra* note 68, art. 1(4) (emphasis added).

139 *See* 4 Virginia Commentary § 194.10(f).

granting provisional measures under the LOSC is the "prevention of *serious* harm to the marine environment,"[140] demonstrating that had the drafters intended to limit the scope of covered harms, they could have done so expressly.

Moreover, judicial precedent confirms that the LOSC can apply to future or potential harms, as well as past ones. In *MOX Plant*, Ireland claimed that the UK had failed to take measures required by both the general and specific obligations of the Convention vis-à-vis a radioactive site that had yet to become operational and, therefore, could not yet have caused injury.[141]

The forward-looking nature of certain duties is also found in the preventive nature of a due diligence requirement and the principle of sustainable development, as expressed by the ICJ in *Pulp Mills on the River Uruguay* and *Gabĉíkovo-Nagymaros*. Similarly, in *Obligations of States*, the ITLOS Seabed Disputes Chamber found that the existence of past damages was necessary for State liability regarding activities in the International Seabed "Area" – a requirement it termed an exception to the customary rule – only because of an explicit reference to "damage" in the text.[142]

6.4. Establishing the Breach of International Obligation

A State only incurs international responsibility for an internationally wrongful act. In the case of an obligation to prevent environmental damage, "what is internationally wrongful then is allowing... the harm to occur."[143] In other words, there must be a legally recognized causal

140 LOSC art. 290 (emphasis added).

141 *See* MOX Plant (Ire. v. U.K.), 126 I.L.R. 314, 327 (Order on Suspension of Proceedings and Request for Further Provisional Measures of June 24) (Perm. Ct. Arb. 2003) (stating that future harms could be considered at the merits stage).

142 Obligations of States ¶ 178.

143 ROSALYN HIGGINS, PROBLEMS AND PROCESS: INTERNATIONAL LAW AND HOW WE USE IT 165 (1994).

connection between the failure to exercise due diligence and the resulting or likely harm.

The extent to which international law includes an accepted standard for what constitutes a legally recognized cause is unclear. However, the notion of an unbroken chain of effects where the injury suffered is a normal or predictable result of the allegedly wrongful act appears to have the greatest support. The idea that proximity, or the length of the causal chain, is a limit on finding cause-in-law has had some support in international judicial decisions,[144] but this is increasingly seen as an outdated view.[145] What seems to have emerged in practice is an analysis of whether the act or omission was a cause-in-fact of the harm, regardless of proximity, though limited by whether the consequences were a logical or foreseeable result.

One formulation of the legal cause standard was expressed by the U.S.-German Mixed Claims Commission after World War I. Describing the legal standard as a rule of general application, the Commission stated that "[i]t matters not whether the loss be directly or indirectly sustained[,]" nor "how many links there may be in the chain of causation so long as there is a clear, unbroken connection between [the causal] act and the loss complained of."[146]

The ILC Articles on State Responsibility appear to have adopted a similar approach, in particular in its rejection of the distinction between direct and indirect loss. In the Second Report on State Responsibility (the "Second Report"), the then-ILC Special Rapporteur Gaetano Arangio-Ruiz wrote that "it is necessary for [the harm] to be linked to an unlawful act by a relationship of cause and effect[,] and an injury is so linked to an unlawful act whenever the normal and natural course of events would indicate that the injury is a logical consequence of the act or

144 *Id.* ¶ 40 (quoting IAN BROWNLIE, SYSTEM OF THE LAW OF NATIONS: STATE RESPONSIBILITY 224 (1983)).

145 *See id.* ¶¶ 41–43.

146 Administrative Decision No. II (U.S. v. Ger.), 7 R.I.A.A. 23, 29 (Mixed Cl. Comm'n 1923).

whenever the author of the unlawful act could have foreseen the damage his act would cause."[147]

The recognition by the Seabed Disputes Chamber that the precautionary approach is a relevant element of the obligation of due diligence may also be useful in lowering the degree of foreseeability required. Because the precautionary approach means that uncertainty over cause and effect should not prevent the regulation of potentially hazardous activities, a logical corollary is that less *ex ante* certainty is necessary for the harmful effects to be considered normal or predictable.

6.4.1. A State Is Responsible for Its Breach Even if the Injury Would Not Have Occurred but for Other Contributory Causes

In addressing whether the existence of multiple causes affects a State's responsibility for an injury, one must be careful to distinguish among a number of similar concepts: the responsibility of multiple States for the same internationally wrongful act, the responsibility of multiple States for an internationally wrongful act that gives rise to a single injury, the responsibility of one State for an internationally wrongful act that would not have occurred but for other factors, multiple factors not constituting internationally wrongful acts causing a single injury, and the allocation of liability for an injury with causes including one or more internationally wrongful acts.

Claiming that multiple States have violated Articles 192 and 194 in their respective contributions to damage caused by ocean acidification would require characterizing the situation as either a single internationally wrongful act by multiple States or as multiple internationally wrongful acts (among other factors) contributing to a single injury.

The ILC's Articles on State Responsibility provide that, when several States are responsible for the same internationally wrongful act, the responsibility of each State may be invoked independently.[148]

147 Gaetano Arangio-Ruiz, Second Report on State Responsibility by the Special Rapporteur, ¶ 37, UN Doc. A/CN.4/425 (June 9, 1989) [hereinafter Second Report].

148 DASR, *supra* note 133, at art. 47(1).

The official commentaries counsel further caution with making analogies to municipal law concepts such as "joint and several,"[149] because the general principle of international law is that each State is responsible for its international wrongs, with such responsibility not diminished by the responsibility of other States.[150] However, in illustrating the principle that multiple States can be responsible for a single wrongful act, the examples used in the relevant commentary refer primarily to actions taken by States jointly or in coordination, such as the establishment of a joint commission to manage a boundary river.[151]

More on point is official Comment 8 to the Articles on State Responsibility, which clarifies that "situations can also arise where several States by separate internationally wrongful conduct have contributed to causing the same damage. For example, several States might contribute to polluting a river by the separate discharge of pollutants."[152] The official commentary goes on to cite *Corfu Channel*[153] as supporting the proposition that a State can be held internationally responsible for failure to exercise due diligence, even though such failure would not have caused any damages but for the actions of a third State.[154] By way of example, in *Corfu Channel*, "it was not suggested that Albania's responsibility for failure to warn [the U.K. that Yugoslavia had mined Albanian waters] was reduced, let alone precluded, by reason of the concurrent responsibility of a third State. In such cases, the responsibility of each

149 *Id.* art. 47 comment. 3. However, the commentaries are explicit that joint and several responsibility is not foreign to international law, but application of the concept rather depends on the circumstances and specific obligation.

150 A procedural corollary of this principle would be the absence of any joinder requirement.

151 *See id.* at art. 47 comments. 2, 4.

152 Noyes and Smith refer, as well, to a situation whereby "several states may independently pollute the environment in a way that, in the aggregate, seriously damages another state" as an illustration of multiple wrongful acts contributing to a single injury. John E. Noyes & Brian D. Smith, *State Responsibility and the Principle of Joint and Several Liability*, 13 YALE J. INT'L L. 225, 228 (1998).

153 Corfu Channel (U.K. v. Alb.), 1949 I.C.J. 4 (Apr. 9).

154 DASR, *supra* note 133, at art. 47 cmt. 8.

participating State is determined individually, on the basis of its own conduct and by reference to its own international obligations."[155]

Elsewhere, the official commentary suggests that "[i]n certain circumstances the wrongfulness of a State's conduct may depend on the independent action of another State. . . . [T]he conduct of the other State may be relevant or even decisive in assessing whether the first State has breached its own international obligations."[156]

This reasoning is not convincing for three reasons. First, as discussed, *Corfu Channel* supports the proposition that a State can be held responsible for violating an international obligation even if the violation would not have occurred but for the acts of other States. Second, the legal obligations established under the LOSC provide a context for assessing whether a State's actions are a cause-in-law of an injury. These obligations are analogous to the "duty to prevent" at issue in *Corfu Channel.* In other words, had the State fulfilled its obligations, it could not have been a contributory cause to the injury at issue.

Finally, as elucidated in the Second Report, to constitute a cause of the harm, an act must meet a certain contributory threshold, described as being a "decisive, though not exclusive" cause. Although "decisive" could be read to mean "necessary," the Salvioli and Eagleton passages cited in the Report indicate that "decisive" instead refers to a cause sufficient to have contributed a tangible share of the injury.[157]

Bearing in mind the ambiguity in the legal standard governing multiple causes of harm, and the current scientific uncertainties[158] regarding thresholds for defining a contributory cause-in-fact, the most

155 *Id.*

156 *Id.* ch. 4, cmt. 4.

157 Second Report, *supra* note 147, ¶ 44 & nn. 87–88 (*citing* Gabriele Salvioli, *La Responsabilité des Etats et la Fixation des Dommages et Intérêts par les Tribunaux Internationaux*, *in* 28 Hague Academy of International Law, Recueil des Cours 231, 245–45 (1929); Clyde Eagleton, The Responsibility of States in International Law 203 (1928)).

158 *See infra*. Part 5.4.3.

prudent course would be for a State or States harmed by ocean acidification to bring a consolidated action against all or most of the historically responsible major emitters. This would at least allow a tribunal to assess the comparative contribution of each respondent State and evaluate the resultant damage in a holistic manner.

6.4.2. The Scientific Evidence Necessary to Show a Causal Contribution to Ocean Acidification

Defining the threshold above which a State's emissions can be considered sufficient to have contributed a share of the harm from ocean acidification is largely a question of fact. Scientifically, it depends on the precision of existing models that correlate the relationship between (i) atmospheric emissions and marine concentrations of carbon, (ii) marine carbon concentrations and changes in ocean pH levels, and (iii) the various physical effects of ocean pH levels on the marine environment and marine natural resources subject to the sovereign rights of a coastal State. In other words, the question is: when are the cumulative historical and current CO_2 emissions of the respondent State(s) sufficient to show a likely contribution to ocean acidification *and its current or projected demonstrable impacts*? Robust models of the first two relationships exist, but scientific understanding of the relationship between ocean pH and the specific harms it causes, although improving rapidly, is still at an early stage. Any claim brought by an affected State(s) would presumably need to be supported by credible expert scientific evidence demonstrating particular instances of harm suffered as a result of ocean acidification.

6.4.3. Claims Limited to Article 194(1) and the Specific Obligations of Conduct May Simplify the Causation Inquiry

In contrast to the outcome of "damage by pollution" prohibited in Article 194(2), paragraph 1 of the same Article targets generally the pollution of the marine environment, which, as discussed, relates merely to substances *likely* to cause harm. Because CO_2 emissions appear to fall within this definition, any State could be said to have breached Article 194(1) by failing to exercise due diligence in taking measures necessary to prevent, reduce, and control CO_2 emissions, which are known to be likely to result in deleterious effects to the marine environment as a result of ocean acidification. Under this approach, each State commits its own international wrong, with the contributions of other States serving merely to establish the conditions necessary for another State's emissions to constitute harmful pollution.

Finally, the text of Article 194(1) provides a further useful means to distinguish between those States whose emissions constitute a treaty violation and those whose emissions do not by requiring States to use only "the best practicable means at their disposal and in accordance with their capabilities."[159] The related notion of due diligence would similarly serve this function. Thus, the causal question would deal not with the relationship between a State's CO_2 emissions and the resulting harm, but only between a State's failure to use the means at its disposal to prevent the introduction of CO_2 into the oceans via the atmosphere.

6.5. Principles of Liability and Remedies

6.5.1. Available Remedies Depend on the Nature of the Wrongful Act

As discussed earlier, an injury's multiple causes do not vitiate the principle of individual state responsibility. However, the multiplicity of causes may affect the available remedies and the liability of a responsible State. Article 47(2) of the ILC Articles on State Responsibility provides only that an injured State cannot collect more in compensation than it suffered in damages, and it implicitly recognizes that there are multiple ways in which the obligation of reparation can be allocated among responsible States.[160] Because the LOSC does not establish a special regime of

159 LOSC, art. 194(1).

160 The reference in Paragraph 2(b) to "any right of recourse" reflects that different liability regimes can establish different substantive standards and procedures for the payment of compensation.

liability,[161] reference to foundational principles of international law may be helpful.

As an initial consideration, the responsible State incurs an obligation to cease the wrongful conduct and to provide assurances of nonrepetition.[162] In this case, the obligation of cessation would be an obligation to cease the failure to exercise due diligence in preventing pollution of the marine environment and to comply with all specific obligations in that regard. The practical ramifications of a judicial declaration to that effect, if followed, might mean the implementation of domestic measures in the respondent State(s) to reduce CO_2 emissions.[163]

The basic principle of remedy is that the victim of an internationally wrongful act is entitled to full reparation.[164] To achieve full reparation, however, three types of remedies are available: restitution, compensation, and satisfaction. Restitution means such measures as can restore the wronged State to the position it would have been in if not for the wrongful act.[165] Second, to the extent full restitution is not possible, the wronged State is entitled to monetary compensation.[166] Finally, the injured State may be entitled to "satisfaction," which could "consist [of] an acknowledgement of the breach, an expression of regret, a formal apology or another appropriate modality."[167] In general, satisfaction is the form of reparation designed to address moral injuries and other noncompensable harms.

161 See LOSC, *supra* note 68, art. 235 (providing for liability in accordance with international law).

162 *Id.* art. 30. Technically, under the Articles, cessation and nonrepetition are not remedies, but separate legal duties brought about by the wrongful act.

163 Once liability is established, the unprecedented practical scope of the duty of cessation would provide a very strong incentive for a negotiated settlement on remedies – that is, an international agreement to reduce carbon emissions.

164 DASR, *supra* note 133, at art. 31.

165 *Id.* at art. 35. It is unclear whether there are forms of restitution that could be made, other than by compensation.

166 *Id.* at art. 36.

167 *Id.* at art. 37.

If a claimant State successfully demonstrates harm resulting from a violation of its rights and the breach of another State's obligations, it is entitled to restitution and compensation. However, because the bulk of ocean acidification's harmful effects have yet to occur, proving a compensable injury based on harms already suffered may be difficult at this time. This difficulty arises due both to the nature of the harm and the uncertainty inherent in attributing harms to a certain level of cumulative emissions within a complex climate system. For example, although the Interacademy Panel on International Issues has found that the effects of ocean acidification are already observable (including that coral calcification rates have declined in recent decades), it also stated that "attributing causes for these impacts among multiple drivers (acidification, warming, pollution, etc.) is a challenge."[168]

This is not to say that there is no possible claim for compensation based on the current impacts of ocean acidification, particularly given the "deleterious effects" standard discussed earlier. Changes in calcification rates or even the measurable changes in ocean pH could certainly be considered deleterious effects. If a claimant State were able to connect the changes that have already been observed to some present material harm, a claim for restitution or compensation could be made.[169] Such harms could include impacts on coral reefs and on the tourism and fishing industries that rely on them. An additional claim could be for the resources reasonably expended to mitigate and adapt to the impacts of ocean acidification. But even if compensation could be awarded, determining the quantum of damages due for environmental damage remains challenging because of the difficulty in assessing "non-use" values, such as for biodiversity.[170]

Regardless of whether compensation is currently available, a claimant State will likely have a greater interest in prospective remedies.

168 Interacademy Panel on International Issues, *supra* note 20.

169 Although it does not appear that compensation is clearly precluded for moral injury, monetary payments are not the usual remedy for non-monetizeable damages.

170 See DASR, *supra* note 133, at art. 36 comment 15.

While necessarily guided by the primacy of restitution, the most attractive remedies may be those that complement the claimants' diplomatic, political, and environmental objectives; for example, the creation of an ocean acidification monitoring regime or the establishment of an insurance mechanism to cover the costs of adaptation and compensation for future slow-onset damages.

Independent of the degree of harm suffered, establishing the breach of an international legal obligation brings into play the duties of cessation, non-repetition, and satisfaction. In the case of ocean acidification, cessation would mean taking the "due diligence" steps necessary within the responsible State's capabilities to ensure that its emissions are no longer contributing to ocean acidification.

Whereas proving a breach would only require evidence that a State's cumulative emissions are sufficient to have contributed to ocean acidification with "deleterious effects," defining the duty of cessation would require the much more precise determination of the level and time frame of emission reductions necessary for a State to no longer contribute to an increased likelihood of harm. Some cutting-edge studies are beginning to suggest the levels of aggregate emission reductions required (i.e., sufficient to stabilize atmospheric CO_2 concentrations below 450 ppm and eventually return to 350 ppm),[171] but these estimates are unlikely to be sufficiently robust or certain such that a tribunal would rely on them to require massive emission reductions. However, they may be sufficient for a tribunal to decide on remedies that reduce a State's emissions in a manner consistent with "safe" emissions trajectories emerging from the best available and evolving science.

Most notably in *Trail Smelter*, *Gabĉíkovo-Nagymaros*, and *MOX Plant*, tribunals have been willing to impose remedies designed to prevent the occurrence or recurrence of harm, including environmental monitoring, an obligation to negotiate a compromise regime in good faith, restrictions on the emission of air pollutants, and a regime of

171 *See* Veron, *supra* note 30.

indemnification for future injuries.[172] In an ocean acidification claim, similar remedies could be sought in lieu of, or in addition to, monetary compensation for damages already suffered. Additionally, the principle of individual state responsibility allows prospective remedies to be imposed on each responsible State, limited only by the need to avoid overcompensation.

6.5.2 When the Proportion of Harm Attributable to Each State Can Be Determined, Payment of Proportional Damages Is Appropriate

The primary rule, as recounted in the Second Report on State Responsibility, favors "the payment of partial damages, in proportion to the amount of injury presumably to be attributed to the wrongful act and its effects."[173] The rule finds further support in the commentary to the ILC Articles on State Responsibility, which suggests that when an injury is divisible, proportional compensation is the appropriate remedy.[174] In the case of ocean acidification, cumulative historical CO_2 emissions would be the metric most suitable for such apportionment.

However, when proportional attribution of the injury is not possible or advisable, there is also some support for allowing full damages to be recovered against any responsible State.[175] Citing *Corfu Channel*

172 *See* Trail Smelter (U.S. v. Can.), 3 R.I.A.A. 1905, 1974–78 (Awards of Apr. 16, 1938 and Mar. 11, 1941) (1949); Gabčíkovo-Nagymaros Project ¶ 140; MOX Plant, 126 I.L.R. at 278.

173 Second Report, *supra* note 147, ¶ 44 & notes 87–89. It has also been suggested that proportionality is an appropriate principle when the causal connection between the wrongful act and the harm suffered is uncertain. Michael G. Faure & André Nollkaemper, *International Liability as an Instrument to Prevent and Compensate for Climate Change*, 43A Stan. J. Int'l L. 123, 163–65 (2007). In the case of ocean acidification, the direct relationship between carbon emission and ocean pH limits the impact of this sort of factual uncertainty. However, the reduction of damages proportional to background risk or natural factors may be appropriate. *See id.* at 161–62.

174 DASR, *supra* note 133, at art. 31 comment 13.

175 *Id.* at comment 12. This rule, despite sounding like joint and several responsibility – *see id.* n. 471 ("This approach is consistent with the way in which these issues are generally dealt with in national law.").

and *United States Diplomatic and Consular Staff in Tehran*,[176] the commentary states that even when "the injury in question was effectively caused by a combination of factors, only one of which is to be ascribed to the responsible State, international practice and the decisions of international tribunals do not support the reduction or attenuation of reparation for concurrent causes."[177]

In fleshing out this principle, Noyes and Smith argue that joint and several allocation of liability for compensation has effectively emerged as an international norm for certain circumstances, including nondivisible injuries and situations when proportional compensation would leave injured States undercompensated (for example, when claims cannot effectively be brought against all States whose wrongful acts contributed to the injury).[178] Both of these situations might justify a claim for full damages against any State responsible for ocean acidification; the former if historical emission records preclude a proportional attribution of damage, and the latter because, with the United States beyond the reach of an LOSC mechanism, any claim would likely leave an injured State undercompensated.

6.5.3 Compensation Due Depends on the Character of the Breach

It is a basic rule of international law that an act is wrongful only if it violated an international obligation in force at the time of the act.[179] One possible application of this principle would be to hold a State liable for the impact of its CO_2 emissions only after it became a party to the LOSC. This would in turn require an analysis of whether emissions from a respondent State subsequent to the entry into force of the LOSC have

176 United States Diplomatic and Consular Staff in Tehran (U.S. v. Iran), 1980 I.C.J. 3 (May 24).

177 DASR, *supra* note 133, at art. 31 comment 12.

178 Noyes & Smith, *supra* note 152, at 242–52 ("If a rule of joint and several liability does not constitute a 'general proposition' rising to the level of a general principle, it nevertheless may constitute a component or subsidiary rule of the accepted general principle of law recognizing full reparation for injury.")

179 DASR, *supra* note 133, at art. 13.

caused or are likely to cause sufficient ocean acidification to result in observable and measurable damage to another State.

However, an obligation of due diligence may already have applied prior to the entry into force of the LOSC if one assumes that the LOSC merely codified a preexisting obligation under customary international law. What that obligation would have required at any given time would depend, in addition to the general considerations, on when a State knew that its activities would produce harmful effects as a result of the increased acidity of the oceans.

Additionally, when the primary obligation requires a State to prevent an event – such as the general obligations under the LOSC to prevent pollution of the marine environment – the breach of that obligation is considered to occur when the event occurs.[180] It may be possible to show that CO_2 emitted decades ago continues to pollute the marine environment, and therefore, even though such emissions occurred prior to the entry into force of the LOSC, the obligations under the LOSC had come into force by the time the atmospheric CO_2 concentrations had reached levels likely to cause harm (i.e., when the breach occurred).[181]

Bearing in mind the centrality of the "due diligence" obligation, it may be a stronger argument to allege a "breach consisting of a composite act"[182] (i.e., a series of actions or failures to act that together constituted the failure of due diligence). For a composite act, a breach is considered to occur if enough acts, taken together, are sufficient to constitute a breach. If so, the breach is then considered to have begun with the first act or omission in the series.[183] Thus, if the State has breached an LOSC

180 *See id.* at art. 14(3).

181 A similar approach was taken by the WTO Panel in *European Communities and Certain Member States – Measures Affecting Trade in Large Civil Aircraft*, ¶ 7.64, WT/DS316/R (June 30, 2010), which found that an obligation not to cause "adverse effects" under the Agreement on Subsidies and Countervailing Measures applied to any situation that existed when the agreement came into force, even if that situation arose as a result of acts that occurred before that time.

182 *See* DASR, *supra* note 133, art. 15.

183 *Id.*

obligation in force at the time, it can be held responsible for the entire series of acts leading to the unlawful marine pollution.

6.6. Conclusions on the Likelihood of Success

The foregoing discussion reveals that any claim for breach of environmental duties under the LOSC implicates a series of novel procedural and legal issues. We recognize that few of these questions have clear-cut answers and that there are ample legal arguments that might lead to different conclusions not contemplated here.

Certainly, narrower claims or forms or relief make for a greater likelihood of success and may build the legal groundwork for future decisions. For example, a succession of cases would limit the number of independent legal issues on which the claimant would have to prevail in any single case.

However, given the current state of the science and the lack of certainty regarding specific and localized damages likely to result from ocean acidification, establishing the necessary causal linkages may pose the most difficult obstacle. Although our scientific understanding of ocean acidification is improving rapidly, a more prudent or cautious approach might involve addressing the legal issues separately from the burden of factual proof or the assessment of damages. This may be possible by means of pursuing an advisory opinion in the ITLOS.

7. The Advisory Opinion Option

When compared to adversarial proceedings, an advisory opinion tends to avoid many of the procedural and jurisdictional challenges and could be introduced in a way that avoids the need for difficult judicial findings on cutting-edge scientific facts. Depending on the questions asked, it might also avoid a range of legal questions relating to causation, liability, remedies, and their application to the facts at a time when many of the issues are likely to present difficult choices for an international tribunal.

The various procedural requirements and political considerations related to the seeking of an ITLOS advisory opinion are considered briefly in this section.

7.1. Availability and Legality

Although the LOSC authorizes the ITLOS Seabed Disputes Chamber to issue advisory opinions on issues related to the International Seabed "Area,"[184] the full Tribunal's authority to issue advisory opinions on other LOSC-related legal questions is limited to the procedure provided in ITLOS Rule 138.[185] Notably, this authority was added to the Rules by the Tribunal's judges, and not the States Parties in the Working Group responsible for the original draft of the Rules.[186]

The strongest legal basis for this broad advisory authority is Article 21 of the ITLOS Statute,[187] which defines the Tribunal's jurisdiction as including "all matters specifically provided for in any other agreement which confers jurisdiction on the Tribunal." For this reason, advisory opinions under Rule 138 are limited to when "an international agreement related to the purposes of the Convention specifically provides for the submission to the Tribunal of a request for such an opinion."

The jurisdiction of ITLOS to issue advisory opinions has been raised on several occasions at meetings of LOSC States Parties and during

184 LOSC, *supra* note 68, art. 191.

185 ITLOS Rules, *supra* note 100, art. 138.

186 Doo-young Kim, Advisory Proceedings before the International Tribunal for the Law of the Sea as an Alternative Procedure to Supplement the Dispute-Settlement Mechanism under Part XV of the United Nations Convention on the Law of the Sea, Issues in Legal Scholarship 2010, at 2, http://www.bepress.com/ils/iss11/art6.

187 *Id.* at 3–5; *but see* Tafsir Malick Ndiaye, *The Advisory Function of the International Tribunal for the Law of the Sea*, 9 Chinese J. Int'l L. 565, 581–82 (2010) (arguing that the agreement of states is a stronger legal basis for the tribunal's advisory jurisdiction than any legal provision); Ki-Jun You, *Advisory Opinions of the International Tribunal for the Law of the Sea: Article 138 of the Rules of the Tribunal, Revisited*, 39 Ocean Dev. & Int'l L. 360, 362 (2008) (concluding that Article 21 does not provide a sufficient legal basis).

debates in the UN General Assembly. However, no strong objection appears to have been raised, and a number of States have expressed support for Rule 138.[188] Authoritative commentators, including several judges on the Tribunal, have also affirmed the existence of a sound legal basis for Rule 138 in the LOSC.[189]

7.2. Requirements and Procedure

Rule 138 establishes four requirements governing requests for advisory opinions: (i) there must be an international agreement; (ii) the agreement must be related to the purposes of the Convention; (iii) it must provide specifically for the submission of a request for an advisory opinion to the ITLOS; and (iv) the opinion sought must be on a question of a legal nature.[190]

Because no request for an advisory opinion under Rule 138 has yet been made, little guidance exists as to how the ITLOS would deal with several important procedural and jurisdictional issues. However, it is worth noting that Rule 138(3) provides for the general application of the procedures established for the Seabed Advisory Chamber's advisory jurisdiction, which includes the right of States to offer written and oral testimony.

7.2.1. An "International Agreement" and the Purposes of the LOSC

It has been suggested that the first requirement – an international agreement – although certainly encompassing a bilateral or multilateral treaty, may also include an agreement between a State and an international organization[191] or, less likely, a non-governmental entity.[192] It has also

188 Ndiaye, *supra* note 187, at 582–83; You, *supra* note 187, at 363.

189 *See* Kim, *supra* note 186, at 3–4 (citing Judge Jose Luis Jesus & Shabtai Rosenne); Ndiaye, *supra* note 187, at 581–83.

190 All language quoted in this section is from ITLOS Rules, *supra* note 103, art. 138.

191 You, *supra* note 187, at 367.

192 Kim, *supra* note 186, at 7.

been suggested that the Meeting of States Parties to the LOSC could make such a request.[193] Whatever the interpretation, this is a significantly lower threshold than would be required, for example, to request an advisory opinion from the ICJ, which can come only from an authorized organ of the UN or by a resolution passed by a simple majority of the UN General Assembly.[194]

As for the second requirement – that the agreement be related to the purposes of the Convention – it is unclear whether this refers to an agreement on the same subject matter as the Convention or specifically for the purpose of implementing it.[195] Additionally, it is not clear whether an ad hoc agreement between countries for the sole purpose of seeking an advisory opinion from the ITLOS would meet this standard or whether an agreement with a more substantive purpose is required.

For a small group of particularly affected States, such as low-lying atoll nations, one possible approach would be to conclude an agreement – treaty-level or otherwise – on addressing the causes and impacts of ocean acidification. The agreement could provide for information sharing and cooperation among these States, particularly in the area of scientific monitoring and ongoing assessment of environmental impacts, while also authorizing a State or a body established under the agreement to seek an advisory opinion from the ITLOS on legal issues related to ocean acidification.

7.3. The Legal Question(s)

A request for an advisory opinion must be made in the form of a precise legal question or set of questions.[196] The entity making the request

193 *See* Ndiaye, *supra* note 187, at 583–84 & n.72.

194 *See* U.N. Charter, art. 96.

195 *Cf.* Ndiaye, *supra* note 187, at 584; Kim, *supra* note 186, at 6–7.

196 *See* ITLOS Rules, *supra* note 100, art. 138(3) (applying the provisions of Articles 130–37, governing advisory opinions of the Seabed Dispute Chamber, to advisory opinions under Rule 138, *mutatis mutandis*); *id.* art. 131(1) ("A request for an advisory opinion on a legal question . . . shall contain a precise statement of the question."); *see also* Kim, *supra* note 186, at 5–6.

has the flexibility to frame the questions and the underlying factual assumptions, although it appears that the ITLOS has some discretion to either accept or reject the request.[197] As the ICJ explained recently, if a question is susceptible to a legal answer, it constitutes a legal question, notwithstanding its political context or ramifications.[198]

The issue of ocean acidification presents a number of possible questions on which an advisory opinion could be sought, from the fact-specific to more abstract questions of international environmental law. For example, questions might relate to the precise scope and nature of the obligations in LOSC Articles 192 and 194 as they relate to the causes and impacts of ocean acidification, the extent of a State's responsibility for emissions by private entities operating in its territory or under its jurisdiction, or the legal consequences of a failure to comply with such obligations.

7.4. An ITLOS Advisory Opinion: Disadvantages and Advantages

Of course, an advisory opinion could not provide actual relief to the States suffering or at risk of harm from ocean acidification, nor would it have the same deterrent or legally binding effect as a contentious claim. But as noted earlier, an ITLOS advisory opinion could provide some of the benefits of a declaratory ruling, such as clarification of the relevant legal obligations, the confirmation of relevant legal norms, and additional diplomatic leverage in international negotiations, all without foreclosing the possibility of a later adversarial claim.

Moreover, the very nature of a request for an advisory opinion circumvents likely political difficulties associated with the bringing of a contentious action targeted at one or more major emitters, including impacts

197 *See* ITLOS Rules, *supra* note 100, art. 138(1) ("The Tribunal *may* give an advisory opinion") (emphasis added); *cf.* Obligations of States ¶¶ 47–49 (declining to decide whether the use of "shall" in the relevant authorizing clause eliminates the Seabed Dispute Chamber's discretion regarding acceptance of requests).

198 *See* Accordance with International Law of the Unilateral Declaration of Independence in Respect of Kosovo, Advisory Opinion, 2010 I.C.J. ¶ 25 (July 22) (cited approvingly in Obligations of States ¶ 39).

on broader strategic relations and foreign policy priorities. It would provide the requesters with the opportunity to carefully frame their question(s), including the assumption of certain facts or causal linkages that may prove more difficult to demonstrate in an adversarial proceeding. Finally, the fact that an advisory opinion does not entail immediate consequences for States also gives the tribunal more flexibility to "push the envelope" in defining the scope and content of LOSC obligations.

8. Conclusion

Ocean acidification represents the "second big impact" of escalating concentrations of atmospheric CO_2, and is relevant to any discussion of possible international legal action to address the causes of climate change.

Given the fragile state of negotiations under the UNFCCC, a key objective of any claimant State(s) will be to use the initiation of legal action to build political pressure and a sense of urgency to deliver the environmentally necessary reductions of CO_2 emissions. As discussed earlier, even the threat of a well-considered international legal action may offer some scope to refocus negotiations on the impacts of unbridled CO_2 emissions and to reallocate resources toward a better understanding of the long-term risks and associated costs.

Although the interface of ocean acidification, the precise treaty obligations of the LOSC and the compulsory nature of its dispute settlement system offers a plausible and thus far largely unexplored legal window of opportunity, this chapter has illustrated the complex scientific, procedural, technical, and legal issues that would need to be navigated in preparing a contentious case against major emitters for impacts suffered as a result of anthropogenic CO_2 emissions and the resulting ocean acidification. Potential claimants would be well advised to ensure that their science is credible and their specific harms demonstrable. No doubt this will take time, and so it is worth considering various interim strategies, legal and otherwise.

This chapter argues that a legally cautious but diplomatically deft first step may be for States particularly affected by ocean acidification to seek an advisory opinion from the ITLOS. Although it lacks the binding effect of a judgment, an advisory opinion that clarifies relevant international legal obligations could still be expected to affect the political dynamic and focus of negotiations under the UNFCCC and to turn greater attention to the impacts of growing CO_2 emissions on the plight of the most vulnerable countries in the world. It would avoid the need to deal with the many uncertain legal issues extraneous to the actual merits of an alleged international breach, particularly those related to liability, and it may avoid the damage to bilateral relations that is sometimes the byproduct of a targeted action in international courts.

As noted in Part 7, one possible source for a request for an advisory opinion would be for a group of particularly vulnerable, low-lying atoll island States to conclude an international agreement relating to ocean acidification that makes express reference to the possibility of seeking judicial advice from the ITLOS. Such an agreement need not be complicated and could likely be concluded in a matter of months. This may be the best way for those most threatened by ocean acidification to quickly take matters into their own hands and to reap the benefits of a cooperative approach to tackling the risks of ocean acidification before they become insurmountable.

16 Securing Planetary Life Sources for Future Generations

Legal Actions Deriving from the Ancient Sovereign Trust Obligation

Mary Christina Wood, Stephen Leonard, Daniel Bartz, and Nicola Peart

1. Introduction

> [T]he day would not be too far when all else would be lost not only for the present generation, but also for those to come – generations which stand to inherit nothing but parched Earth incapable of sustaining life.
>
> *Oposa v. Factoran*, Philippines Supreme Court, 1993[1]

Climate change threatens the continued existence of sovereign States located on drowning islands across the planet. Current estimates have found that global sea level will rise by at least one meter above 1990 levels by 2100 (for an in-depth discussion of these projections, see Chapter 2). Put another way, greenhouse gas pollution, most of it from the industrialized world, will cause massive waste of Earth's property. As the planet approaches irrevocable climate thresholds,[2] the failure of polluting countries worldwide to slash carbon emissions has profound

1 Oposa v. Factoran, G.R. No. 101083 (S.C., July 30, 1993) (Phil.), *reprinted in* Jan G. Laitos et al., Natural Resources Law (2006).

2 *See generally* James Hansen et al., *Target Atmospheric CO2: Where Should Humanity Aim?* 2 Open Atmospheric. Sci. J. 217 (2008), *available at* http://www.benthamscience.com/open/toascj/openaccess2.htm.

We wish to thank Professor Melissa Powers for providing insights and materials on the matter of energy regulation and Antonio A. Oposa, Simon Baczkiewicz, Paul Wexler Sande, Jennifer Stallkamp, Andrew Welle, and Christine V. Urgello for valuable research assistance.

implications for intergenerational justice and international equity.[3] As Kiribati President Anote Tong observed, "The fate of our people, our culture, our memories is at stake. I don't think anyone wants to drown a people, yet that is what will happen."[4]

Existing domestic and international legal mechanisms have failed to force abatement of greenhouse gas pollution. This chapter draws on the public trust doctrine as a legal framework from which sovereign obligations flow, both to domestic citizens, but also to citizens of other countries and future generations worldwide. The chapter highlights emerging international and national legal regimes that recognize a public trusteeship over a widening range of environmental resources.[5] Within this framework, nations are shown to be *co-tenant trustees* of the global commons and, hence, owe fiduciary duties toward beneficiaries – present and future citizens of the planet – to protect Earth's life sources and the natural inheritance of generations to come. Climate change, as a consequence of the failure of co-trustee nations to take precautionary action, constitutes a breach of this duty to safeguard and manage the trust.

Section 2 begins by explaining the public trust doctrine and its recognition in domestic legal systems worldwide, as well as on the international level. Section 3 presents the atmosphere as a public trust asset and briefly explains a cutting-edge litigation approach called Atmospheric Trust Litigation (ATL), now being used on the domestic level to force countries to curb carbon emissions. Although the ATL approach

3 Maxine Burkett, *Just Solutions to Climate Change: A Climate Justice Proposal for a Domestic Clean Development Mechanism*, 56 Buff. L. Rev. 169, 192–93 (2008); *see generally* Edith Brown Weiss, *Climate Change, Intergenerational Equity, and International Law*, 9 Vt. J. Envtl. L. 615 (2008).

4 Rhett A. Butler, *As a Tiny Island Nation Makes a Big Sacrifice, Will the Rest of the World Follow Suit?*, Mongabay.com (Sept. 15, 2010), *available at* http://news.mongabay.com/2010/0915-kiribati_interview_anote_tong.html.

5 For discussion of international trust concepts, *see generally* Peter H. Sand, *Sovereignty Bounded: Public Trusteeship for Common Pool Resources?*, 4 Global Envtl. Pol., Feb. 2004, at 47.

gives rise to myriad potential claims, this chapter explores potential jurisdiction available to drowning nations under the United States' Alien Tort Claims Act (ATCA).[6] Section 4 characterizes the World Heritage Convention (WHC)[7] as a global trusteeship over designated sites of world heritage and looks at enforcement possibilities of fiduciary obligations through that structure.

This chapter does not begin to exhaust all possible claims arising within the trust framework but rather highlights the global obligation of fiduciary care for the planet's irreplaceable atmosphere and suggests potential avenues of enforcement. With both the ATL claims and WHC claims, there remain many questions – particularly those involving procedural barriers – that necessarily fall outside the scope of this summary treatment. Nevertheless, it is worth noting that, with a disastrous climate tipping point looming, the law must respond in some fashion or be rendered irrelevant as the planet hurtles toward a point of no return into a state of runaway heating and climate chaos. Courts must see beyond narrow procedural defenses and interpretations to find laws whose purposes address the crisis at hand. The ancient trust obligation offers a forceful mandate of fiduciary care that drowning island nations may draw on in their epochal struggle to protect their future.

2. The Public Trust as a Global Frame for Climate Responsibilities

2.1. The Trust Principle

A trust is essentially a bifurcation of legal and beneficial ownership of property. A trustee is the legal owner of the trust property and has management powers over the trust assets, but is subject to the duty to

6 28 U.S.C. § 1350 (2011).

7 *See* Convention Concerning the Protection of the World Cultural and Natural Heritage, Nov. 16, 1972, 1037 U.N.T.S. 151 [hereinafter "World Heritage Convention" or "WHC"], *available at* http://whc.unesco.org/archive/convention-en.pdf.

exercise those powers for the exclusive benefit of the beneficiary, who is the equitable or beneficial owner of the trust property.[8] Under private trusts, trustees are subject to strict fiduciary obligations, the most fundamental of which is to "protect the trust property against damage or destruction" and "to do all acts necessary for the preservation of the trust *res* which would be performed by a reasonably prudent man."[9] Trustees must not sit idle in the face of threats but must vigilantly guard against waste and damage to the trust assets.[10] Trustees who fail to protect the property against waste, or indeed fail in any of their obligations as guardians of the trust assets, are liable to the beneficiaries, who can in turn seek legal remedies to restore or modify the trust or to receive compensation for lost assets.[11]

The public trust doctrine applies this principle of fiduciary obligation to crucial natural resources, characterizing them as public assets held by the governments in trust for present and future citizens. Governments must adhere to strict fiduciary standards in managing the assets because of the unique characteristics and central importance of such resources to society. The public trust doctrine does not merely encompass government's standard general obligation to act for the public good; rather, it calls forth a "special and more demanding obligation which it may have

8 *See In re Orr*, 180 F.3d 656, 660 (5th Cir. 1999), 76 Am. Jur. 2d *Trusts* § 331 (2005) ("[A] trustee's first or primary duty is to act wholly for the benefit of the trust, and to administer the trust solely in the interest . . . of the beneficiaries.")

9 George G. Bogert & George T. Bogert, The Law of Trusts & Trustees § 582 (rev. 2d ed. 1980).

10 George T. Bogert, Trusts § 99 (6th ed. 1987) ("The trustee has a duty to take whatever steps are necessary . . . to protect and preserve the trust property from loss or damage." (citation omitted)); *id.* § 107 ("The trustee . . . is liable for damages if he should have known of danger to the trust, could have protected the trust, but did not do so."); 76 Am. Jur. 2d *Trusts* § 606 (2005) (noting it is within the "power, and a duty of the trustee, to initiate actions . . . for the protection of the trust estate").

11 *See* 76 Am. Jur. 2d *Trusts* § 404 (2005) ("trustee . . . must not suffer the estate to waste or diminish, or fall out of repair"), *Hogg v. Walker*, 622 A.2d 648, 653 (Del. 1993), ("A trustee's liability for a breach of trust is personal in character with all the consequences and incidents of personal liability.")

as a trustee of certain public resources."[12] As a fundamental concept of property law, the public trust doctrine presents principles of fiduciary obligations that parallel private trust law, albeit on a sovereign level.[13] However, the public trust doctrine itself is not a creature of private trust law but is rather an expression of sovereign understanding between citizens and their government as to the classification of certain natural property as public goods. Thus, public trust law can prove forceful even in countries (such as those with civil law systems) that do not recognize private trusts.[14] The public trust doctrine is found in varied forms including common law decisions, statutory iterations, and constitutional provisions.[15]

The origins of the public trust doctrine are ancient; they are considered as "old as humanity" and existing "from the inception of humankind."[16] They speak to public ownership of natural resources held

12 Joseph L. Sax, *The Public Trust Doctrine in Natural Resource Law: Effective Judicial Intervention*, 68 MICH. L. REV. 471, 478 (1970).

13 *Idaho Forest Indus. v. Hayden Lake Watershed Improvement Dist.*, 733 P.2d 733, 738 (1987) (noting the administration of public trust "is governed by the same principles applicable to the administration of trusts in general").

14 *See, e.g.*, Hanno Kube, *Private Property in Natural Resources and the Public Weal in German Law – Latent Similarities to the Public Trust Doctrine?* 37 NAT. RESOURCES J. 857, 860–62 (1997) (discussing civil law in countries that lack private trust mechanisms).

15 *See* Mary Christina Wood, *Atmospheric Trust Litigation around the World*, in FIDUCIARY DUTY AND THE ATMOSPHERIC TRUST (Ken Coghill, Charles Sampford, & Tim Smith eds., 2012), earlier draft *available at* http://law.uoregon.edu/assets/facultydocs/mwood/atmospherictrustlitigation.pdf.

16 Justice Christopher Weeramantry, *Justice Can Be Shortsighted*, 15 OUR PLANET, no. 3, 2004, at 7, 8, *available at* http://www.ourplanet.com/imgversn/153/images/Our_Planet_15.3_english.pdf; *Excerpts from Landmark Judgments in the Field of Environment*, 15 OUR PLANET, no. 3, 2004, at 17, *available at* http://www.ourplanet.com/imgversn/153/images/Our_Planet_15.3_english.pdf; *see also generally* Gabcikovo-Nagymaros Project (*Hung. v. Slovk.*), 1997 I.C.J. 88 (September 25) (separate opinion of Vice-President Weeramantry) (discussing how ancient civilizations employed sustainable methods of natural resource extraction for the benefit of future generations); *Oposa v. Factoran*, G.R. No. 101083 (Phil. S.C., 1993), *excerpted in* LAITOS ET AL., *supra* note 1, at 441–44.

in common. An early legal source, the Justinian code of sixth-century Rome, iterated the doctrine of *res communes* that held certain property as common to humankind: "the air, running water, the sea, and consequently the shores of the sea... [and] the right of fishing in a port, or in rivers, is common to all men."[17] *Res communes* were excluded from private control so as to be available for public purposes.[18] In England, the concept appeared in common law and was inscribed in the Magna Carta, prohibiting the King from conveying absolute private title to lands under navigable waterways. The Crown reserved a *jus publicum* interest, "by which the King held the land in his sovereign capacity as a representative of all the people."[19]

By 1896, the trust doctrine was firmly entrenched in American jurisprudence. That year the U.S. Supreme Court in *Geer v. Connecticut*, stated,

> [T]he power or control lodged in the state, resulting from this common ownership, is to be exercised, like all other powers of government, as a trust for the benefit of the people, and not as a prerogative for the advantage of the government as distinct from the people, or for the benefit of private individuals as distinguished from the public good.... [T]he ownership is that of the people in their united sovereignty.[20]

Another case, *Illinois Central Railroad Co. v. Illinois*, deemed the lodestar of American trust jurisprudence, declared in 1892 that submersible lands along navigable waterways were held in public trust for

17 Code Just. 2.2.1, -.2 (Justinian I 529), *available at* http://www.fordham.edu/halsall/basis/535institutes.html#I.%20Divisions%20of%20Things.

18 Patricia Kameri-Mbote, *The Use of the Public Trust Doctrine in Environmental Law*, 3/2 L. Env't & Dev. J. 195, 197 (2007) *available at* http://www.lead-journal.org/content/07195.pdf.

19 *United States v. 1.58 Acres of Land*, 523 F. Supp. 120, 123 (D. Mass. 1981) (describing origins of trust); *see also Gann v. Free Fishers of Whitstable* [1865], 11 Eng. Rep. 1305 (H.L.) 192 (U.K.).

20 *Geer v. Connecticut*, 161 U.S. 519, 529 (1896).

the benefit of citizens: "It is a title held in trust for the people of the State, that they may enjoy the navigation of the waters, carry on commerce over them, and have liberty of fishing therein, freed from the obstruction or interference of private parties."[21]

The North Carolina Supreme Court in *Shepard's Point Land Co. v. Atlantic Hotel* noted in 1903 that "the decisions are numerous which declare that ... property is held by the state, by virtue of its sovereignty, in trust for the public."[22] The public trust remains a vibrant doctrine in modern U.S. jurisprudence, reflected in a plethora of cases at the state level.[23] In a 1998 case, the New York Supreme Court expressed the essence of the trust when it declared,

> [I]n enacting environmental mandates ... we are merely discharging our obligation under the societal contract between "Those who are dead, those who are living and those who are yet to be born." ... This generation's duty has been discharged merely by setting aside this land for their use under the doctrine of the public trust.[24]

In 1970, the legal scholar, Joseph Sax, published an influential article that suggested that citizens invoke the public trust to fight exploitation

21 *Ill. Cent. R.R. Co. v. Illinois*, 146 U.S. 387, 452 (1892). *See also Phillips Petroleum Co. v. Mississippi*, 484 U.S. 469, 476 (1988) (holding that Mississippi took title to submerged lands in the Gulf of Mexico on entering the Union).

22 *Shepard's Point Land Co. v. Atlantic Hotel*, 44 S.E. 39, 42 (N.C. 1903).

23 For a broad survey of cases, see Robin Kundis Craig, *A Comparative Guide to the Western States' Public Trust Doctrines: Public Values, Private Rights, and the Evolution toward an Ecological Public Trust*, 37 Ecology L.Q. 53 (2010); Robin Kundis Craig, *A Comparative Guide to the Eastern Public Trust Doctrine: Classifications of States, Property Rights, and State Summaries*, 16 Penn St. Envtl. L. Rev. 1, 21–24 (2007); Sax, *supra* note 12; Mary Christina Wood, *Advancing the Sovereign Trust of Government to Safeguard the Environment for Present and Future Generations (Part I): Ecological Realism and the Need for a Paradigm Shift*, 39 Envtl. L. 43 (2009). For cases and materials on the public trust, see Laitos et al., *supra* note 1, at 622–54.

24 *W.J.F. Realty Corp. v. State*, 672 N.Y.S.2d 1007, 1012 (Sup. Ct. 1998) (citation omitted), *aff'd*, 710 N.Y.S.2d 249 (App. Div. 1999).

of resources that should rightfully be protected as common property.[25] Sax identified three basic justifications of the public trust doctrine:

(1) "[C]ertain interests are so intrinsically important to every citizen that their free availability tends to mark the society as one of citizens rather than of serfs;"[26]
(2) "[C]ertain interests are so particularly the gifts of nature's bounty that they ought to be reserved for the whole of the populace;"[27] and
(3) "[C]ertain uses have a peculiarly public nature that makes their adaptation to private use inappropriate."[28]

Today, in the face of collapsing ecosystems and a heating climate, scholars worldwide are delving into foundational trust principles to hold government to basic obligations of safeguarding resources necessary to the survival of present and future citizens.[29] The public trust has broad applicability, appearing in various forms in countries throughout the world. Courts in several different countries have expressed the trust concept with respect to tidelands, beaches, waters, groundwater, wildlife, parks, recreation areas, forests, and indeed, in some jurisdictions, all natural resources.[30] Although air has not explicitly been included in the

25 Sax, *supra* note 12.

26 *Id.* at 484.

27 *Id.*

28 *Id.* at 485.

29 Catherine Redgwell, Intergenerational Trusts and Environmental Protection (1999); David Takacs, *The Public Trust Doctrine, Environmental Human Rights, and the Future of Private Property*, 16 NYU E.L.J. 711, 762–64 (2008); Edith Brown Weiss, *The Planetary Trust: Conservation and Intergenerational Equity*, 11 Ecology L.Q. 495, 558–59 (1984).

30 *See Oposa v. Factoran*, G.R. No. 101083 (S.C., July 30, 1993) (Phil.), *as excerpted in* Laitos *et al.*, *supra* note 2, at 443–44 (trust extends to old growth forests); *Mehta v. Kamal Nath*, 1 SCC 388 (1997) (India) (public trust extends to forests and ecologically fragile lands); William Araiza, *Democracy, Distrust, and the Public Trust: Process-Based Constitutional Theory, the Public Trust Doctrine, and the Search for a Substantive Environmental Value*, 45 UCLA L. Rev. 385, 404 (1997); Richard J. Lazarus, *Changing Conceptions of Property and Sovereignty in Natural Resources: Questioning the Public Trust Doctrine*, 71 Iowa L. Rev. 631, 648 (1986); *supra* note 23 and sources cited

public trust *res*, the U.S. Supreme Court made clear in *Georgia v. Tennessee Copper Co.* that "the state has an interest independent of and behind the titles of its citizens, in all the earth and air within its domain."[31] The widening scope of protected resources is hardly surprising given judicial recognition that the doctrine must be flexible enough to accommodate ever-changing needs of society.[32]

The trust obligation presents a striking legal angle in a world that is increasingly dominated by neoliberal, pro-market thinking.[33] By invoking a property interest held in common by citizens and future generations, the public trust provides a check against the singular privatization of crucial resources that is so often the tool of market ideology.[34] As one commentator observes, trust law "harness[es] the strength of private property rights in the communal interest for environmental protection purposes."[35] Globally, the trust concept holds inspiring populist vigor to

therein. In the United States, the Supreme Court has made clear that states have latitude in defining the trust. *See Phillips Petroleum Co. v. Mississippi*, 484 U.S. 469, 494 (1988).

31 *Georgia v. Tenn. Copper Co.*, 206 U.S. 230, 237 (1907) (holding Tennessee copper companies liable for noxious fumes that drifted across state lines). The passage was cited in *Massachusetts v. E.P.A.*, 549 U.S. 497, 518–19 (2007).

32 *See e.g. Matthews v. Bay Head Improvement Ass'n.*, 471 A.2d 355, 365 (N.J. 1984) (the public trust doctrine "is not to be fixed or static but one to be molded and extended to meet changing conditions and needs of the public it was created to benefit." [internal quotation marks omitted]).

33 *See generally* M. Shamsul Haque, *The Fate of Sustainable Development under Neo-Liberal Regimes in Developing Countries*, 20 Int'l Pol. Sci. Rev. 197 (1999), *available at* http://profile.nus.edu.sg/fass/polhaque/ipsr-sd.pdf.

34 *See Lake Mich. Fed'n v. U.S. Army Corps of Eng'rs*, 742 F. Supp. 441, 445 (N.D. Ill. 1990) ("[T]he public trust is violated when the primary purpose of a legislative grant is to benefit a private interest."). *See also* Kevin Gray, *Equitable Property*, 47 Curr. Legal Probs 157, 199–200 (1994) ("For the first time it becomes meaningful to claim on behalf of the citizenry a 'property' interest comprising enforceable access to such inherent public goods as clean air, unpolluted rivers and seaways, ozone regeneration, recreational enjoyment of wild country, and the sustainable development of land and marine areas."); *see also* David Takacs, *supra* note 29, at 715–17 (describing breadth of doctrine).

35 Redgwell, *supra* note 29, at 1.

counter powerful market-state forces that would exert climate hegemony over drowning island nations.

This is not to claim that the public trust doctrine is an unassailable legal tool certain to provide adequate redress for irreparable harms like climate change. The public trust largely depends on courts for enforcement, and judicial disposition often rides on a number of factors quite apart from legal doctrine. Certainly, the public trust gives rise to some vexing questions regarding whether particular resources ought to be privatized or held public, the duties of the trustee (particularly in the case of conflicting public purposes), and the role courts should take in trust enforcement. But though the trust is not immune to legal dilemmas, its basic premise nevertheless holds impressive force worldwide, as discussed in the following sections. Absent a property-based, duty-infused framework, nations typically approach global environmental problems as a matter of political discretion, with the result that the more powerful (and polluting) nations abdicate what is logically their fair share of responsibility to protect the common resource. A legal framework of sovereign obligation represents a marked departure from the present policy framework that legitimizes the political discretion to do nothing in response to a mounting threat of runaway climate heating.

2.2. Modern Application of the Public Trust Doctrine

As a concept of sovereignty, the public trust doctrine operates on two levels. First, it provides a positive legal obligation on the part of government to protect resources within the particular domestic jurisdiction for the benefit of that sovereign's present and future generations. In this regard, courts and commentators have aptly described the trust as an "attribute of sovereign authority" with inalienable characteristics.[36] Professor Coplan observes, "Public trust principles have been described

36 *See, e.g.*, *In re* Water Use Permit Applications, 9 P.3d 409, 442–43 (Haw. 2000) ("[H]istory and precedent have established the public trust as an inherent attribute of sovereignty."); *see also* Ill. Cent. R.R. Co. v. Illinois, 146 U.S. 387, 453, 460 (1892) ("The state can no more abdicate its trust over property in which the whole people

as an essential attribute of sovereignty across cultures and across millennia."[37] Some have suggested this iteration of the trust as an expectation of democratic governance.[38] On this level, trust enforcement generally takes the form of actions brought by citizen beneficiaries against their own governments.[39]

The second level involves transboundary natural assets and therefore invokes an intersovereign construct. By defining natural assets in property terms (rather than as abstract environmental rights), the trust creates a framework within which various sovereigns have legally based rights vis-à-vis one another with respect to shared resources such as the oceans and atmosphere. Sovereigns are aptly described as co-trustees and co-tenants of commonly held global trust property.[40] In landmark

are interested . . . than it can abdicate its police powers in the administration of government and the preservation of the peace. . . . Every legislature must, at the time of its existence, exercise the power of the state in the execution of the trust devolved upon it."); *United States v. 1.58 Acres of Land*, 523 F. Supp. 120, 124 (D. Mass. 1981) (applying public trust to both federal and state governments and holding that the trust, as a limitation on sovereignty, "can only be destroyed by the destruction of the sovereign"); *See generally* Douglas L. Grant, *Underpinnings of the Public Trust Doctrine: Lessons from Illinois Central Railroad*, 33 Ariz. St. L.J. 849 (2001) (explaining the public trust doctrine as part of the constitutional reserved powers doctrine, which prevents any one legislature from taking acts that would compromise a future legislature's ability to exercise sovereignty on behalf of the people); Jan S. Stevens, *The Public Trust: A Sovereign's Ancient Prerogative Becomes the People's Environmental Right*, 14 U.C. Davis L. Rev. 195, 196 (1980) (noting jurisprudence "in the form of declarations that the public trust is inalienable as an attribute of sovereignty no more capable of conveyance than the police power itself"); Wood, *supra* note 23, pt. V.B (describing trust as an "[i]nalienable [a]ttribute of [s]overeignty" that is "[d]erived from the [p]eople").

37 Karl S. Coplan, *Public Trust Limits on Greenhouse Gas Trading Schemes: A Sustainable Middle Ground?* 35 Colum. J. Envtl. L. 287, 311 (2010).

38 *See id. See generally* Wood, *supra* note 15.

39 *See, e.g., National Audubon Society v. Superior Court of Alpine County*, 33 Cal.3d 419, 447 (1983) (holding that the National Audubon Society can enforce the public right in water in California's Mono Lake by compelling reconsideration of water diversions in light of the sovereign public trust obligation).

40 The construct of co-tenancy applied to the atmosphere originated in Mary Christina Wood, *Atmospheric Trust Litigation*, *in* Adjudicating Climate change: Sub-National, National and Supranational Approaches (William C.G. Burns, & Hari M. Osofsky eds. 2009). Professor Fox-Decent adopts the approach in the context of a human rights analysis of climate obligations. *See generally* Evan Fox-Decent, *From*

U.S. Indian treaty fishing cases, the Ninth Circuit Court of Appeals characterized tribal and state sovereigns as analogous to "co-tenants" in a shared salmon fishery that migrates between State and tribal borders.[41] Another case, *United States v. 1.58 Acres of Land*, described the federal government and a State as "co-trustees" of public trust property (submerged lands in Boston Harbor).[42] The characterization of *co-tenant trustee* is particularly helpful with respect to global assets such as atmosphere, because it invokes property-based duties from common law that accompany the trust and co-tenancy relationships. Both impose a duty against waste, or derogation, of the common asset. This duty was firmly recognized by the Ninth Circuit:

> Co-tenants stand in a fiduciary relationship one to the other. Each has the right to full enjoyment of the property, but must use it as a reasonable property owner. A co-tenant is liable for waste if he destroys the property or abuses it so as to permanently impair its value. A court will enjoin the commission of waste.
>
> By analogy, neither the treaty Indians nor the State on behalf of its citizens may permit the subject matter of these treaties to be destroyed.[43]

Fiduciary States to Joint Trusteeship of the Atmosphere 19 (McGill University Working Paper, 2010), *available at* http://papers.ssrn.com/sol3/papers.cfm?abstract_id=1698755# (applying a fiduciary theory of international governance to climate, noting "the idea that states are essentially joint trustees of the earth's atmosphere").

41 *Puget Sound Gillnetters Ass'n. v. U.S. Dist. Court for W. Dist of Wash.*, 573 F. 2d 1123, 1126 (9th Cir. 1978) (holding that the treaty established "something analogous to a cotenancy, with the tribes as one co-tenant and all citizens of the Territory (and later of the state) as the other."); *United States v. Washington*, 520 F.2d 685, 686, 690 (9th Cir. 1975) (applying co-tenancy construct, by analogy, to Indian fishing rights).

42 This formulation [set forth in *Illinois Central Railroad*] recognizes the division of sovereignty between the state and federal governments of those aspects of the public interest in the tideland and the land below the low-water mark that relate to the commerce and other powers delegated to the federal government and are administered by Congress in its capacity as trustee of the *jus publicum*, whereas those aspects of the public interest in this property that relate to nonpreempted subjects reserved to local regulation by the states are administered by state legislatures in their capacity as co-trustee of the *jus publicum. United States v. 1.58 Acres of Land*, 523 F. Supp. 120, 123 (D. Mass. 1981).

43 *See Washington*, 520 F.2d at 685.

Apart from these reciprocal obligations arising from the co-tenancy relationship, co-trustees in private trust law also have a mutual, correlative duty not to waste the common asset. As one leading treatise describes classic trust law in the private context, the powers of co-trustees are "held jointly by the trustees and all must unite in their exercise."[44] Co-trustees in private trust law have duties to take action against other co-trustees to protect the estates for the beneficiaries.[45] The duty operates to the benefit of all beneficial owners of the trust.

Characterizing nations as *co-tenant trustees* of one integral global asset (such as the atmosphere) suggests that all nations owe duties toward all citizen beneficiaries of that asset. In this way, the trust construct elevates obligations out of confined domestic boundaries and suggests a strong foundation for international obligations toward planetary citizens – present and future members of humankind having an equal stake in the integrity of the global asset. After a description of the public trust worldwide on both the domestic sovereign level and the international trans-sovereign level, this chapter applies the co-tenant trustee construct to atmospheric trust litigation and to World Heritage Sites, suggesting avenues by which citizens or leaders of drowning islands might enforce the basic duty against waste.

2.2.1. Domestic Assertions of the Trust in Nations Worldwide

The public trust doctrine is apparent in legal systems of countries worldwide.[46] Although an extensive review is beyond the scope of this chapter,[47] some national examples outside of the United States (which

44 Bogert, *supra* note 10, § 91.

45 *Id.* § 92 ("It is the duty of one trustee to protect the trust estate from any misfeasance by his cotrustee, upon being made aware of the intended act, by obtaining an injunction against him.... It is, of course, disagreeable to take a co-trustee by the throat...[but may be necessary] in order to keep the estate intact") (citation omitted).

46 *See generally* Michael C. Blumm & R.D. Guthrie, *Internationalizing the Public Trust Doctrine: Natural Law and Constitutional and Statutory Approaches to Fulfilling the Saxion Vision*, 44 U.C. Davis L. Rev. (forthcoming 2012), *available at* http://ssrn.com/abstract=1816628; Wood, *supra* note 15.

47 For more extensive treatment, from which this discussion is adapted, see Wood, *supra* note 15.

have been discussed earlier) demonstrate the basic applicability of the trust concept. These examples show that the trust principle is reflected in case law, statutes, and constitutions. Although some legal sources do not explicitly use the words "public trust," they evoke a firm sovereign duty to present and future generations by excising a realm of property from privatization in order to provide lasting benefits to the public.

In 1993, the Philippine Supreme Court invoked the public trust doctrine to halt rainforest logging.[48] Though the Philippine government claimed complete discretion to allow private companies to cut the last 2.8% of the country's remaining forests, the Court implicitly recognized the rainforest as a public trust asset:

> Every generation has a responsibility to the next to preserve that rhythm and harmony for the full enjoyment of a balanced and healthful ecology....
>
> [T]he right to a balanced and healthful ecology...belongs to a different category of rights [than civil and political rights] altogether for it concerns nothing less than self-preservation and self-perpetuation...the advancement of which may even be said to predate all governments and constitutions. As a matter of fact, these basic rights need not even be written in the Constitution for they are assumed to exist from the inception of humankind. If they are now explicitly mentioned...it is because of the well-founded fear of its framers that unless the right to a balanced and healthful ecology and to health are mandated as state policies by the Constitution itself...the day would not be too far when all else would be lost not only for the present generation, but also for those to come – generations which stand to inherit nothing but parched earth incapable of sustaining life.[49]

In 1997 the Supreme Court of India rendered a strong iteration of the public trust doctrine in a landmark case, *M.C. Mehta v. Kamal Nath*:

48 *Oposa v. Factoran*, G.R. No. 101083 (S.C., July 30, 1993) (Phil.), *as excerpted in* Laitos et. al. *supra* note 1, at 443–44.

49 *Id.*

> The State is the trustee of all natural resources which are by nature meant for public use and enjoyment. [The] public at large is the beneficiary of the sea-shore, running waters, airs, forests and ecologically fragile lands. The State as a trustee is under a legal duty to protect the natural resources. These resources meant for public use cannot be converted into private ownership.[50]

South Africa has enshrined environmental rights for future generations in its constitution, stating in Article 24 that everyone has the right "to have the environment protected, for the benefit of present and future generations."[51] And in 2006 Kenya's High Court compared the right to a healthy environment with the right to life and declared a public trust duty in natural resource management.[52]

Although the Canadian High Court has not directly applied the public trust doctrine, in *British Columbia v. Canadian Forest Products Limited* (2004), it recognized that the doctrine forms an appropriate common law basis for recovering natural resource damages to a public resource.[53] Finding declarations of public rights and ownership in "running water, air, the sea and the shores of the sea" in both the Institutes of Justinian and H. de Bracton's influential treatise on English law, the Court stated, "By legal convention, ownership of such public right was vested in the Crown, as [was] authority to enforce public rights of use."[54]

Outside of the common law tradition, the public trust doctrine is also manifest in civil code laws recognizing public ownership of resources

50 *M.C. Mehta v. Kamal Nath* (1997), 1 SCC 388, ¶ 34 (India), *cited in* Takacs, *supra* note 12, nn.135–37.

51 Takacs, *supra* note 29 at 740.

52 *Waweru v. Republic of Kenya*, (2006) 1 L.R.K. 677, ¶¶ 31, 32, 34, 38, 40 (C.C.K.) (Kenya), *available at* http://www.chr.up.ac.za/index.php/browse-by-subject/339-kenya-waweru-v-republic-2007-ahrlr-149-kehc-2006-.html.

53 *British Columbia v. Canadian Forest Products Ltd.*, [2004] 2 S.C.R. 74, ¶¶ 64–83 (Can.), *available at* http://scc.lexum.org/en/2004/2004scc38/2004scc38.html.

54 *Id.* ¶¶ 74–75.

(such ownership providing the basis of sovereign obligation).[55] France, for instance, upholds common ownership of certain public assets such as navigable waters, streams, sea coasts, and parks.[56] Commentators have suggested parallels between the German concept of public weal and the public trust.[57] Ukraine has a constitution that evokes the public trust by underscoring public rights to natural resources and a concomitant state duty of protection.[58] Latin American countries, too, have an analog to the public trust doctrine.[59] Both Ecuador and Bolivia have embedded "rights of Nature" in their constitutions, resembling trust iterations.[60] Several nations recognize the rights of future generations in their constitutions or statutes.[61]

Two commentators summarizing this thematic prevalence note, "The basic legal concept of public trust exists in many nations under a variety of labels.... The principles of public trust are such that they can be understood and embraced by most countries of the world."[62] Indeed, some commentators have ascribed the pervasiveness of the trust principle to

55 Mary Turnipseed et al., *Reinvigorating the Public Trust Doctrine: Expert Opinion on the Potential of a Public Trust Mandate in U.S. and International Environmental Law*, ENV'T, Sep.–Oct. 2010, at 6, 12.

56 Ved P. Nanda & William K. Ris, Jr., *The Public Trust Doctrine: A Viable Approach to International Environmental Protection*, 5 Ecology L.Q. 291, 305 (1976) (also noting a duty of conservation in the public interest).

57 Kube, *supra* note 14, at 858.

58 Конституція України [CONSTITUTION] art. 13 (Ukr.) ("The land, its mineral wealth, atmosphere, water and other natural resources within the territory of Ukraine, the natural resources of its continental shelf, and the exclusive (maritime) economic zone, are objects of the right of property of the Ukrainian people.... Property shall not be used to the detriment of the person and society. The State ensures the protection of the rights of all subjects of the right of property.").

59 Thomas T. Ankersen, *Shared Knowledge, Shared Jurisprudence: Learning to Speak Environmental Law Creole (Criollo)*, 16 TUL. ENVTL. L.J. 807, 812–13 (2003).

60 John Vidal, *Bolivia Enshrines Natural World's Rights with Equal Status for Mother Earth*, THE GUARDIAN, April 10, 2011, http://www.guardian.co.uk/environment/2011/apr/10/bolivia-enshrines-natural-worlds-rights; Jennifer Koons, *Following Pa. Mining Town's Example, Ecuador OKs Constitution Giving Rights to Nature*, GREENWIRE (September 30, 2008), *available at* http://celdf.org/article.php?id=185.

61 *See* REDGWELL, *supra* note 29.

62 *See* Nanda & Ris, *supra* note 56, at 306.

likely origins in natural law,[63] which is the foundation for legal systems worldwide. The striking commonality of the trust concept in so many legal systems makes possible an international, orchestrated strategy of domestic litigation brought by citizens against their own governments to enforce protection of global natural assets such as the atmosphere.[64] But equally important to drowning islands is the second level of trust application – possible international legal recourse against polluting nations of the world based on nations' obligations as co-tenant trustees to protect the public trust.

2.2.2. International Manifestation of Trust Interests

Trust principles have forceful application on the international level in organizing correlative duties among nations toward global assets. Peter Sand observes, "A transfer of the public trust concept from the national to the global level is conceivable, feasible, and tolerable.... The essence of transnational environmental trusteeship... is the democratic *accountability* of States for their management of trust resources in the interest of the beneficiaries – the world's 'peoples.'"[65] In a comprehensive article, Nanda and Ris suggest that the public trust doctrine, appropriately applied, "offers a viable approach to international environmental protection."[66] Indeed, there have been significant scholarly attempts to suggest international environmental trusts and identify their features.[67] The ability of international tribunals and domestic courts applying international law to effectuate trust principles is a question bound to be at

63 *See* Wood, *supra* note 23 (Advancing the Sovereign Trust Part I), at 70.

64 *See generally* Wood, *supra* note 15 (setting forth atmospheric trust litigation strategy for use in nations worldwide).

65 *See* Sand, *supra* note 5, at 57–58 (emphasis in original). *See also* Daniel Bodansky et al., *Invoking State Responsibility in the Twenty-First Century*, 96 AM. J. INT'L L. 798 (2002) (discussing the emerging trend of recognizing States' obligations to individuals under international law).

66 Nanda and Ris, *supra* note 56, at 292.

67 *See, e.g.*, SUSAN J. BUCK, THE GLOBAL COMMONS (1998); REDGWELL, *supra* note 29; PETER SAND, TRANSITIONAL ENVIRONMENTAL LAW: LESSONS IN CHANGE (1999); Weiss, *supra* note 29.

the cutting edge of climate litigation as threatened nations seek effective and binding legal redress for climate harm inflicted by overconsuming and polluting nations.

International law has continuously raised the general question of whether a common law equitable doctrine (such as the trust principle) can be applied in an international tribunal. To this end, Article 38 of the statute governing the International Court of Justice (ICJ) mandates the specific sources of law that the judges can use to reach a decision and provides for "the general principles of law recognized by civilized nations."[68] Judge Arnold McNair observed in his separate opinion in the landmark *South West Africa* case before the ICJ that "international law has recruited, and continues to recruit, many of its rules and institutions from private systems of law."[69] An enormous body of literature addresses the nuances of this question but generally asserts that Article 38 refers to both principles of international law and various domestic systems of law,[70] in particular to the "fundamental law of justice and injustice" and "all the principles of common law."[71] In the *South West Africa* case, Judge McNair described the trust concept as among "general principles of law recognized by civilized nations" and identified three defining trust principles:

(1) The trustee is precluded from administering the property for his own benefit.
(2) The trustee is under a legal obligation to carry out the trust or mission confided to him for the benefit of some other persons or for some public purpose.

68 I.C.J. Statute art. 38.

69 International Status of South West Africa, Advisory Opinion, 1950 I.C.J. 128, 148 (July 11) (separate opinion of Judge McNair).

70 *See* Manley Ottmer Hudson, The permanent court of International Justice: a Treatise 528 (1934).

71 *See* Permanent Court of International Justice Advisory Committee of Jurists, The Hague, Neth., June 16–July 4, 1920, *Proces-Verbaux of the Proceedings of the Committee* 310, 316 (1920), *available at* http://ia600308.us.archive.org/28/items/procsverbauxof00leaguoft/procsverbauxof00leaguoft.pdf.

(3) Any attempt by the trustee to absorb the property entrusted to him into his own patrimony would be illegal.[72]

In a famous advisory opinion issued by the ICJ in 1996 on *The Legality on the Threat of Use of Nuclear Weapons*, Judge Christopher Weeramantry also announced a global trust concept, stating, "Juristic opinion is now abundant, with several major treaties appearing upon the subject and with such concepts as intergenerational equity and the common heritage of mankind being academically well established.... [Another] principle[] of environmental law... [is] the principle of trusteeship of earth resources."[73] Weeramantry recognized that trust principles, along with the rights of future generations, have "woven themselves into international law through major treaties, through juristic opinion and through general principles of law recognized by civilized nations. Among treaties may be mentioned [the World Heritage Convention]."[74] Two years later, in the 1997 ICJ case of *Gabčíkovo-Nagymaros Project (Hungary/Slovenia)*, Judge Weeramantry classified the trust concept as the first principle of modern environmental law where he stated in a separate opinion,

> [L]and is never the subject of human ownership, but is only held in trust, with all the connotations that follow of due care, wise management, and custody for future generations. The first principle of modern environmental law – the principle of trusteeship of earth resources – is thus categorically formulated in this system.[75]

72 *International Status of South West Africa*, 1950 I.C.J. at 149 (separate opinion of Judge McNair).

73 Legality of the Threat or Use of Nuclear Weapons, Advisory Opinion, 1996 I.C.J. 225, 455, 502 (July 8) (dissenting opinion of Judge Weeramantry). Judge Weeramantry relies on the mandatum in civil law, African customary law, Hindu law, and Islamic Law as evidence of the universality of the trust concept. *But see International Status of South-West Africa*, 1950 I.C.J. at 132 (refusing to apply civil law concept).

74 *Legality of the Threat or Use of Nuclear Weapons*, 1996 I.C.J. at 455 (dissenting opinion of Judge Weeramantry).

75 Gabčíkovo-Nagymaros Project (*Hung. v. Slovk.*) 1997 I.C.J. 7, 108 (September 25) (separate opinion of Judge Weeramantry).

Although the international judiciary can be expected to continue to consider and define the existence of a global trust, it is possible to locate the principles of trusteeship in a range of international law instruments that provide expressions of shared sovereign property rights limited by obligations of custodianship. A strong echo of the trust is found in the United Nations Framework Convention on Climate Change (UNFCCC), which declares a duty of nations to "protect the climate system for the benefit of present and future generations of humankind."[76] The 1992 Convention on Biological Diversity (CBD) affirms that "the conservation of biological diversity is a common concern of humankind" and that "States are responsible for conserving their biological diversity and for using their biological resources in a sustainable manner"[77] (for additional discussion of the CBD, see Chapter 15). The World Heritage Convention described later protects sites of "outstanding universal value for present and future generations of humanity."[78] Ramon Reyes examines the emergence of a fiduciary principle in international law with respect to the duties that administering nations owe to territories under the trustee system.[79] He draws on the League of Nations Covenant, the United Nations Charter, and other sources to demonstrate "the existence and extent of the fiduciary duty at customary international law."[80] Professor Michael Blumm

76 United Nations Framework Convention on Climate Change art. 3, May 9, 1992, S. Treaty Doc. No. 102–38, 1771 U.N.T.S. 107, *available at* http://unfccc.int/resource/docs/convkp/conveng.pdf.

77 Convention on Biological Diversity pmbl., June 5, 1992, 1760 U.N.T.S. 79, *available at* http://www.cbd.int/convention/articles/?a=cbd-00.

78 UN Educ., Scientific and Cultural Org. [UNESCO], Intergovernmental Comm. for the Protection of World Cultural and Natural Heritage, Operational Guidelines for the Implementation of the World Heritage Convention, ¶ 49, UN Doc. WHC. 08/01 (2008), *available at* http://whc.unesco.org/archive/opguide08-en.pdf.

79 Ramon E. Reyes, Jr., *Nauru v. Australia: The International Fiduciary Duty and the Settlement of Nauru's Claims for Rehabilitation of its Phosphate Lands*, 16 N.Y.L. Sch. J. Int'l & Comp. L. 1 (1996).

80 *Id.*, at, 41 (1996). *See also generally* Evan J. Criddle & Evan Fox-Decent, *The Fiduciary Duty of Jus Cogens*, 34 Yale J. Int'l L. 331 (2009) (advancing the theory that the peremptory power of *jus cogens* international law to supersede domestic law stems from the fiduciary duty that States owe to their citizens).

explores the trust doctrine in depth in a number of countries and concludes that it is "close to becoming considered a general principle of international law."[81] Birnie, Boyle, and Redgwell note that the multilateral environmental agreement regime of governance "enables States to exercise a fiduciary or trusteeship role in the protection of the environment, other species, and future generations."[82] Other commentators develop the trust concept in the international oceans context.[83]

The trust is distinct from many other international norms (such as piracy, violation of safe conduct, and infringement on the rights of ambassadors) that merely mediate relationships between different citizen nationals in an international setting. As an attribute of sovereignty applicable to nation-states and their subdivisions, the trust seemingly presents an intrinsic international law norm,[84] forming a central beam in the very architecture that supports international environmental law. As some commentators observe,

> [T]he most convincing characterization [of international environmental law] is no longer that of neighborly relations, but of environmental trusteeship, with certain institutional similarities to the protection of social and economic rights, and a comparable concern for community interests at a global level, not merely those of states inter se.[85]

81 Blumm & Guthrie, *supra* note 46, draft at 8. *See also* Turnipseed et al., *supra* note 55, at 12 (functional equivalents of public trusteeship are evident in many civil law systems).

82 Patricia W. Birnie et al., International Law and the Environment 84 (2009).

83 *See generally* Gail Osherenko, *New Discourses on Ocean Governance: Understanding Property Rights and the Public Trust*, 21 J. Envtl. L. & Litig. 317 (2006); Mary Turnipseed et al., *The Silver Anniversary of the United States' Exclusive Economic Zone: Twenty-Five Years of Ocean Use and Abuse, and the Possibility of a Blue Water Public Trust Doctrine*, 36 Ecology L.Q. 1 (2009).

84 As Professor Charles Wilkinson observed in a broad survey of the trust, "[t]he real headwaters of the public trust doctrine . . . arise in rivulets from all reaches of the basin that holds the societies of the world." Charles F. Wilkinson, *The Headwaters of the Public Trust: Some Thoughts on the Source and Scope of the Traditional Doctrine*, 19 Envtl. Law 425, 431 (1989).

85 Birnie et al., *supra* note 82, at 39.

The next two sections apply these international principles of trusteeship in two contexts of global policy. Section 3 applies the public trust doctrine to the planet's atmosphere. Section 4 applies the trust to World Heritage Sites that will suffer harm from planetary heating. Both sections suggest a co-tenancy in global assets (the atmosphere and World Heritage sites) that gives rise to co-tenant trustee remedies on the part of drowning island nations. Although certainly novel, both types of actions invoke equitable concepts deriving from timeless principles of property law, applying them to global assets that stand to be destroyed permanently as a result of pollution originating in overconsuming nations.

3. Atmospheric Trust Litigation for Drowning Islands

Atmospheric Trust Litigation (ATL) draws on public trust principles to compel greenhouse gas pollution abatement in countries worldwide.[86] The ATL approach is described in detail in other writings and is only summarized here.[87] ATL presents the atmosphere as a global public trust asset, the legal beneficiaries of which are present and future citizens of the world.[88] Invoking the public trust, ATL asserts a sovereign fiduciary obligation to prevent waste or degradation of the atmosphere from greenhouse gas pollution that now threatens catastrophic climate change. It seeks to accomplish what has thus far eluded the centralized international diplomatic treaty-making process in a decentralized fashion through the courts: judicial injunctive and declaratory relief as a last resort to confront a global pollution emergency that has been all but orphaned by politically driven international treaty negotiations.

86 An international focus of ATL was formulated in Wood, *supra* note 15. For earlier writings developing ATL, see Mary Christina Wood, *Atmospheric Trust Litigation*, in CLIMATE CHANGE READER (W.H. Rodgers, Jr. and M. Robinson-Dorn eds., 2011); Wood, *supra* note 40.

87 *See supra* note 86.

88 *Id.*

Other writings more fully characterize the atmosphere as a public trust asset.[89] In brief, courts have always looked to the needs of the public as the primary guiding factor in extending public trust protection to natural resources.[90] In the landmark case of *Illinois Central Railroad Co. v. Illinois*, the U.S. Supreme Court classified the shore of Lake Michigan as a public trust asset, describing it as property in which "the whole people are interested."[91] The Court reasoned, "The trust with which they are held, therefore, is governmental ... follow[ing] necessarily from the *public character of the property*."[92] It need hardly be stated that the atmosphere is essential to all civilizations and to human survival across the globe. As one climate analyst put it, carbon reduction in the atmosphere is necessary to avert "the end of life as we know it on this planet."[93] Leading court opinions express ecological balance as a public trust purpose – an approach that necessarily includes protection of the atmosphere[94] to ensure human survival.

89 *Id.*; *see also* Gerald Torres, *Who Owns the Sky?*, 19 Pace Envtl. L. Rev. 515 (2002); Fox-Decent, *supra* note 40.

90 Charles F. Wilkinson, *The Public Trust Doctrine in Public Land Law*, 14 U.C. Davis L. Rev. 269, 315 (1980) ("The public trust doctrine is rooted in the precept that some resources are so central to the well-being of the community that they must be protected by distinctive, judge-made principles.").

91 *Ill. Cent. R.R. Co. v. Illinois*, 146 U.S. 387, 453–56 (1892).

92 *Id.* at 455–56 (emphasis added).

93 *See* Joe Romm, *Is 450 ppm (Or Less) Politically Possible? Part 0: The Alternative is Humanity's Self-Destruction*, Climate Progress (April 26, 2008), *available at* http://thinkprogress.org/romm/2008/04/26/202588/is-450-ppm-or-less-politically-possible-part-0-the-alternative-is-humanitys-self-des-truction/.

94 *See, e.g.*, *Nat'l. Audubon Soc'y. v. Superior Court*, 658 P.2d 709, 712 (Cal. 1983) [hereinafter Mono Lake Opinion] (preservation of ecological functions is a public trust right.); *In re* Water Use Permit Applications, 9 P.3d 409, 458 (Haw. 2001) (The trust demands "maintenance of ecological balance."). For analysis, see Z.J.B. Plater, R.H. Abrams, & W. Goldfarb, Environmental Law and Policy: A Coursebook on Nature, Law and Society (1992) ("Mono Lake may stand for the proposition that the Public Trust Doctrine is capable of reaching out and encompassing the ecological values of an entire functioning ecosystem.").

Recognition of air as a public asset dates back at least to Roman law. The Institutes of Justinian compiled in 529 CE declared, "By the law of nature these things are common to mankind – the air, running water, the sea, and consequently the shores of the sea."[95] Given the essential nature of air, it is unsurprising that numerous judicial decisions, state and national constitutions, and codes across the world recognize air as part of the *res* of the public trust.[96] Long ago, the U.S. Supreme Court held in *Georgia v. Tennessee Copper Co.* that "the State has an interest independent of and behind the titles of its citizens, in all the earth and air within its domain."[97] The U.S. Supreme Court cited this decision in the *Massachusetts v. EPA* climate case in 2007.[98] Several commentators have urged application of trust principles to the atmosphere.[99] As Gerald Torres wrote, "[T]he traditional rationale for the Public Trust doctrine provides a necessary legal cornerstone ... to protect the public interest in the sky."[100]

On the international level, the UNFCCC, negotiated in 1992 and signed and ratified by the United States and most nations of the world, provides a clear expression of the atmosphere as a public trust asset. In

95 Code Just. 2.1.1 (Justinian I 529) (*cited in Nat'l. Audubon Soc'y*, 658 P.2d at 718).

96 *See* Wood, *supra* note 40, § 1.4. *See also* Haw. Const. art. XI, § 1 ("[T]he State and its political subdivisions shall conserve and protect Hawaii's natural ... resources, including land, water, air, minerals and energy sources.... All public natural resources are held in trust by the State for the benefit of the people."). Federal statutory law includes air as a trust asset for which the federal government, states, and tribes can gain recovery of natural resource damages. Comprehensive Environmental Response, Compensation, and Liability Act (CERCLA), 42 U.S.C. § 9601 (2011) (defining air as among the natural resources subject to trust claims for damages).

97 *State of Georgia v. Tennessee Copper Co.*, 206 U.S. 230, 237 (1907).

98 *Massachusetts v. EPA*, 549 U.S. 497, 518–519 (2007).

99 *See, e.g.*, Coplan, *supra* note 37, at 320 ("[A]s governments seek to privatize rights to atmospheric assets through tradable emissions rights, the public trust doctrine should naturally extend to protect previously unpossessable interests in the atmospheric commons."); Sax, *supra* note 12, at 556–57 (urging application of doctrine to "controversies involving air pollution"); Torres, *supra* note 89; Wood, *supra* note 41.

100 Wood, *supra* note 40, at 108–9.

archetypical trust language calling on nations to "protect the climate system for the benefit of present and future generations of humankind,"[101] the UNFCCC sets forth a standard for fiduciary care by obliging nations to prevent "*dangerous anthropogenic interference with the climate system.*"[102] It acknowledges duties of co-tenant trustee nations to act according to their "common but differentiated responsibilities."[103] Although subsequent international treaty negotiation processes have failed to produce adequate carbon reduction, the UNFCCC nevertheless serves as a compelling iteration of global atmospheric trust responsibilities among co-tenant nations and can buttress ATL cases brought in domestic courts.

ATL defines the fiduciary obligation to protect the atmospheric trust by reference to a scientifically defined global prescription to restore atmospheric equilibrium at 350 parts per million (ppm) concentration of carbon dioxide in the atmosphere.[104] This figure represents the maximum concentration that leading scientists believe will limit planetary heating to 1.5 degrees Celsius (above preindustrial temperatures) – thought by these scientists to be the highest safe target.[105] In May 2011, Dr. James Hansen, along with other leading climate scientists,

101 UNFCCC, *supra* note 76, art. 3, ¶ 1.

102 *Id.* art. 2 (emphasis added).

103 The term is explored as an iteration of the waste principle. *Id.* art. 3, ¶ 1 (*cited in* Wood, *supra* note 15, draft at 30).

104 James Hansen et al., *The Case for Young People and Nature: A Path to a Healthy, Natural, Prosperous Future* (submitted for publication, 2012) posted at http://ourchildrenstrust.org/sites/default/files/20110505_CaseForYoungPeople.pdf *See* Wood, *supra* note 15, at 27.

105 For discussion of temperature increase associated with 350 ppm as contrasted with 450 ppm, see Wood, *supra* note 15, at n. 144–47, 156–59, and accompanying text. *See also generally* Hansen et al., *supra* note 2. Despite potentially obliterating consequences to island nations from a 450-ppm concentration, some scientists and policy makers still continue to calibrate reduction trajectories to that numeric goal. *See e.g.*, Nat'l Research Council, Limiting the Magnitude of Future Climate Change: America's Climate Choices Panel on Limiting the Magnitude of Future Climate Change 48, *available at* http://www.nap.edu/catalog/12785.html.

issued a path-breaking paper that set forth a trajectory of global carbon emission reduction to serve as the marker for ATL remedies. They presented projections showing that a global decline of 6% in fossil fuel emissions, beginning in year 2012, would lower the atmospheric concentration of carbon dioxide (CO_2) to 350 ppm by the end of the century, if coupled with a corresponding effort to extract roughly 100 gigatons of CO_2 (GtC) from the atmosphere through reforestation and improved forestry/agricultural practices (measures that augment "carbon sinks" so as to naturally draw carbon out of the atmosphere).[106] Unlike the political targets that dominate international treaty negotiations and domestic legislative processes, this two-part scientific prescription (carbon reduction and sink enhancement) calibrates to the planet's *actual requirements* to restore the homeostatic function of the atmosphere.

Defining the fiduciary duty by reference to this scientific prescription,[107] ATL actions against sovereign governments seek remedies that take the form of carbon accountings and enforceable plans for carbon reduction (along with sink enhancement).[108] The plaintiffs in such actions do not ask courts to prescribe how the reduction will be accomplished. The court's role, rather, is to set a numerical fiduciary pathway of reduction and then defer to the sovereign defendant as to how the reduction will be accomplished – subject to the continuing jurisdiction and supervision of the court to ensure that it actually happens.[109] By invoking domestic jurisdiction in courts worldwide, ATL aims to

106 James Hansen et al., *supra* note 104. *See also* Paul Baer, Tom Athanasiou, & Sivan Kartha, *A 350 ppm Emergency Pathway*, Greenhouse Dev. Rts. 1–4 (2009), *available at* http://www.undp-adaptation.org/undpcc/files/docs/publications/a-350-ppm-emergency-pathway.pdf (carbon emission reduction trajectory calibrated to achieve 350 ppm without sink drawdown assumptions).

107 The fiduciary duty of carbon reduction necessarily varies, however, in terms of the time frame for accomplishing near-zero emissions. *See* Wood, *supra* note 15, at 28.

108 *See* Wood, *supra* note 15; Wood, *supra* note 40; *see also infra* note 112.

109 In that manner, ATL strikes a balance between meaningful judicial intervention and appropriate deference to the political branches.

jumpstart global carbon reduction before the world crosses the calamitous climate tipping point.[110]

3.1. Atmospheric Trust Litigation as a Global Strategy for Climate Redress

ATL provides two causes of action to support this remedy. The first – a national-level approach developed more extensively in other writings – allows citizens to sue their own governments to achieve carbon reduction.[111] In the first week of May 2011, this cause of action materialized in an unprecedented "hatch" of lawsuits and petitions filed on behalf of youth plaintiffs simultaneously across the United States. The actions, coordinated by a nonprofit organization, Our Children's Trust, consisted of a federal lawsuit, nine state law suits, thirty-nine petitions for state rule-making, and one notice of intent to sue – a legal campaign covering all states in the United States.[112] Outside of the United States, an ATL lawsuit was filed at the same time in Ukraine.[113] The suits and petitions declared a sovereign atmospheric trust duty and called on government to produce carbon accountings and plans for annual emission reductions of at least 6% (as well as to embark on reforestation and improved soil

110 *See generally* Wood, *supra* note 15.

111 *See* Wood, *supra* note 15, at 12–14.

112 The complaints and petitions are compiled on the website of Our Children's Trust. *See Legal Action*, Our Children's Trust, http://ourchildrenstrust.org/page/31/legal-action (last visited Sep. 3, 2012). *See also Media Coverage*, Our Children's Trust, http://ourchildrenstrust.org/page/36/media (last visited Sep. 3, 2012). The initial "hatch" of ATL actions corresponded with 125 youth-led marches (called the "iMatter Marches") in cities across the world, organized by the youth climate organization, Kids vs. Global Warming. iMatter March (2011), http://imattermarch.org/. Our Children's Trust maintains updated information on the ATL litigation, and ATL petitions, on its website.

113 Complaint, *Environment People Law v. Cabinet of Ministers of Ukraine* (complaint filed May 2011), *available at* http://ourchildrenstrust.org/sites/default/files/ATL%20Ukraine.pdf.

practices to achieve carbon drawdown from the atmosphere) – a remedy incorporating the scientific prescription described earlier.[114]

On May 31, 2012, the United States District Court for the District of Columbia dismissed the federal lawsuit in *Alec L. v. Lisa Jackson*.[115] Judge Wilkins granted a motion to dismiss filed by the federal defendants and intervenors – the National Association of Manufacturers and several other California companies and trade associations – based on the lack of subject matter jurisdiction. In so doing, Judge Wilkins reasoned that the public trust doctrine is founded on state law, rather than federal law, and therefore, the lawsuit did not raise a necessary federal question needed to invoke the court's federal question jurisdiction.

The court went on to state that even if there were federal jurisdiction over the case, under the Supreme Court's decision in *American Electric Power v. Connecticut*,[116] the Clean Air Act has displaced federal common law claims for greenhouse gases. Judge Wilkins found that the same reasoning applies here – where Congress has given EPA the job of setting limits for GHGs, it would not be appropriate for the federal courts to do so on their own.

Judge Wilkins concluded by saying,

> While the issues presented in this case are not ones that this Court can resolve by way of this lawsuit, that circumstance does not mean that the parties involved in this litigation ... have to [end the discussion at the close of the case]. All of the parties seem to agree that protecting and preserving the environment is a more than laudable goal, and the Court urges everyone involved to seek (and perhaps even seize) as much common ground as courage, goodwill and wisdom might allow to be discovered.[117]

The *Alec L.* case will probably be appealed, and several of the other similar lawsuits are moving forward.

114 *See* Hansen, *supra* note 104, at 10–11 (explaining prescription).
115 *Alec L. v. Jackson*, 11-cv-02235, 2012 U.S. Dist. LEXIS 75791 (D.D.C. May 31, 2012).
116 *Amer. Elec. Power Co. v. Connecticut*, 131 S. Ct. 2527, 2537, 180 L. Ed. 2d 435 (2011).
117 *Alec L. v. Jackson*, 11-cv-02235, 2012 U.S. Dist. LEXIS 75791 at *20.

Although drowning island nations are not the huge carbon emitters of the world, domestic ATL lawsuits in those countries are crucial for a number of reasons. First, courts of drowning island nations are perhaps best positioned to issue forceful declarations of an atmospheric trust – and such a step can prove instrumental in creating the framework for global reduction. Second, if the planet is to achieve near-zero carbon emissions, every country must ultimately participate in the effort, despite differing time frames for doing so. Courts of drowning island nations may require their nations' governments to take the initial steps of planning and identifying funding sources for the transition to a near-zero carbon emissions economy. Moreover, they can require governments to develop adaptation plans and set a cost on such plans, a step that may also advance global understanding of the devastating and unmanageable impacts to island nations. Third, ATL suits may pave the way for natural resource damage recovery actions prosecuted by island governments against atmospheric polluters using their own domestic laws and legal systems.[118] Funds gained from polluters could substantially help finance the transition to a new low-carbon economy and finance adaptation. ATL suits may also be crucial for addressing the reforestation and agricultural practices that make up the "drawdown" component of the scientific prescription for returning the atmosphere to 350 ppm equilibrium.[119]

Although it is true that not all developing countries have common law iterations of the public trust, ATL lawyers in those countries may use existing environmental statutes and constitutional provisions that harness sovereign fiduciary concepts as hooks for an

118 It is a settled principle of trust law that trustees have the affirmative duty to recoup monetary damages against third parties that destroy trust assets. RESTATEMENT (SECOND) OF TORTS § 177 (1959). *See also Commonwealth of Puerto Rico v. S.S. Zoe Colocotroni*, 628 F.2d 652, 672 (1st Cir. 1980); *Canadian Forest Prod. v. British Columbia*, [2004] S.C.R. 74 (Can.). The discussion of natural resource damages is beyond the scope of this chapter.

119 Such practices must be implemented to achieve the 100 GtC necessary for a return to 350 ppm. See Hansen, *supra* note 104, at 10–11.

atmospheric trust suit.[120] A legal observer in Bangladesh, for example, suggests a public trust duty implied by (1) the Constitution's fundamental protection of the Right to Life; (2) the Constitution's expression of "absolute trust and faith" as core principles of government policy; (3) the land tenure system, which places the government as "common owner of land"; and (4) several statutes such as the 1927 Forest Act and the 1973 Bangladesh Wild Life (Preservation) statute that provide further evidence of a responsibility to protect Bangladeshi natural resources in trust.[121] Lawyers in Ukraine, a civil law jurisdiction, brought an ATL suit against the Cabinet of Ministers of Ukraine and other governmental defendants based on the Constitution's declaration of air and other natural resources as property of the Ukrainian people, as well as on the UNFCCC.[122] By tapping into the varied iterations of trustlike principles in their own countries, lawyers in drowning nations can play a crucial role in advancing an urgent global litigation approach to the climate crisis using domestic processes.

The second ATL cause of action, subject to more analysis later, arises from the co-tenant trustee duty to protect a shared global public trust. As noted earlier, under common law, co-tenant trustees have a reciprocal and correlative duty not to waste their common asset.[123] Within the United States, for example, this duty seemingly gives rise to a cause of action against recalcitrant co-tenant sovereigns.[124] Scaling up to the global level, drowning island nations, as co-tenant trustee sovereigns, should justifiably have legal recourse against grossly polluting industrialized nations for their wholesale abdication of fiduciary responsibility toward the shared atmospheric trust.

120 Wood, *supra* note 15, nn.126–28 and accompanying text (noting, "[s]uch statutory or constitutional provisions can fasten an atmospheric trust lawsuit to the existing legal structure of the nation").

121 Sharaban Tahura Zaman, *Atmospheric Trust Litigation*, Campaign for Sustainable Rural Livelihoods, *available at* http://csrlbd.org/resource/column/atmospherictrustlitigation (last visited Aug. 18, 2011).

122 *See* Complaint, *supra* note 113.

123 *See supra* notes 10–11, 44–45, and accompanying text.

124 Wood, *supra* note 40, sec. 2.

The obvious question, however, is one of forum. International tribunals such as the ICJ may render important declaratory judgments, but the enforcement mechanisms for actual carbon reduction and sink enhancement are tenuous.[125] Another possibility is to locate a jurisdictional tool that provides for redress against a co-tenant trustee nation (or its major polluters) using that nation's own domestic legal system. Although there may be multiple possibilities, the following discussion briefly outlines one avenue of potential jurisdiction within the United States that invokes jurisdiction under the Alien Tort Claims Act (ATCA). Leaders or citizens of drowning island nations might be positioned to bring an ATL claim against major electric utility polluters as third-party tortfeasors to the atmospheric trust.[126] Although the following discussion is hardly exhaustive, it illustrates the realm of possibility that emerges from the application of a public trust framework to the climate crisis. Indeed, ATCA jurisprudence remains in flux, as indicated by the U.S. Supreme Court's grant of certiorari in *Kiobel v. Royal Dutch Shell* on the issue of whether corporations may be held liable under the statute.[127] After hearing oral arguments, the Court ordered a rehearing on the issue of whether the ATCA provides jurisdiction for torts occurring outside the United States.[128]

3.2. The United States Alien Tort Claims Act as a Jurisdictional Statute for Drowning Islands

The ATCA, enacted in 1789 as part of the Judiciary Act, reads, "[t]he district courts shall have original jurisdiction of any civil action by an alien for a tort only, committed in violation of the law of nations or a

125 The ICJ is explored further *infra* with respect to a World Heritage Claim arising out of thc public trust construct.

126 *See infra* Section III.B.1.

127 *Kiobel v. Royal Dutch Petroleum Co.*, 621 F.3d. 111, 145 (2d Cir. 2010), *cert. granted*, 132 S. Ct. 472, 181 L. Ed. 2d 292 (2011).

128 *Kiobel v. Royal Dutch Petroleum Co.*, 2012 U.S. LEXIS 1998 (U.S. Mar. 5, 2012) (order directing parties to file supplemental briefs).

treaty of the United States."[129] The statute appears to be the only plausible jurisdictional tool allowing citizens or leaders of drowning island nations to seek climate redress in U.S. courts.

Interestingly, the ATCA was passed with an overriding policy objective to prevent international conflict by incorporating customary international law and international legal norms into the domestic legal system.[130] Prior to the ATCA, the federal courts lacked jurisdiction over suits by aliens, and the Continental Congress was "hamstrung by its inability to 'cause infractions of treaties, or of the law of nations to be punished.'"[131] The need to enforce international law (or "the law of the nations") was first brought into focus with an incident in 1784, in which a French citizen, De Longchamps, violently accosted the Secretary of the French Legion in Philadelphia.[132] The Secretary had no judicial remedy in the United States for the assault, because at the time there was no jurisdiction over suits brought by aliens.[133] The matter erupted into a crisis of international diplomatic affairs.[134] Unwilling to find itself in a similar situation in the future, the First Congress enacted the ATCA.[135] The framers of the ATCA realized that federal courts, by internalizing international norms in the domestic legal system, could act as an important safety valve for diffusing international conflicts before they boil over.[136]

129 28 U.S.C. § 1350 (1948). For background on the statute, see *Sosa v. Alvarez-Machain*, 542 U.S. 692, 713–24 (2004).

130 Professor Harold Hongju Koh describes the process of international norm internalization as a reciprocal exchange between the international community and domestic courts. Harold Hongju Koh, *Is International Law Really State Law?*, 111 Harv. L. Rev. 1824, 1857 (1998). International norm internalization, however, is not without controversy. *See generally* Curtis A. Bradley & Jack L. Goldsmith, *Customary International Law as Federal Common Law: A Critique of the Modern Position*, 110 Harv. L. Rev. 815 (1997).

131 *Sosa*, 542 U.S. at 716 (quoting J. Madison, Journal of the Constitutional Convention 60 (E. Scott ed. 1893).

132 *Id.*

133 *Id.* at 717.

134 *Id.* n.11.

135 *Id.*

136 For background, see Koh, *supra* note 130, n.87 and accompanying text; *see also infra* note 137.

As the modern Supreme Court recognizes, the ATCA was born out of a desire to regulate behavior "threatening serious consequences in international affairs."[137]

If De Longchamps' assault on a French diplomat compromised U.S. foreign policy two centuries ago, the pollution from individual coal-fired plants within the United States creates an exponentially greater threat to global security. Human-caused global climate change threatens catastrophe on a scale that would destabilize international relations – the same type of harm (in far greater magnitude) that the ATCA was intended to prevent. Leading reports describe geopolitical upheaval wrought by climate change as posing "a serious threat to America's national security."[138] A 2009 report co-authored by the United Nations University Institute for Environment and Human Security drew attention to the impending crisis of environmental refugees displaced by climate change.[139] It warned that "[i]n coming decades, climate change will motivate or force millions of people to leave their homes in search of viable livelihoods and safety. Although the precise number of migrants and displaced people may elude experts for some time, the mass of people on the move will likely be staggering and surpass any historical antecedent."[140] Although the Supreme Court has in other cases counseled great caution in expanding the ATCA's reach, many policy concerns at the forefront of those cases pale in comparison to the consequences resulting from the utter lack of judicial relief to drowning islands and other co-tenant trustee

137 *Sosa*, 542 U.S. at 715 ("It was this narrow set of violations of the law of nations, admitting of a judicial remedy and at the same time threatening serious consequences in international affairs, that was probably on minds of the men who drafted the [ATCA] with its reference to tort.").

138 National Security and the Threat of Climate Change, CNA Corporation 44 (2007), *available at* http://securityandclimate.cna.org/report/National%20Security%20and%20the%20Threat%20of%20Climate%20Change.pdf.

139 Koko Warner et al., In Search of Shelter: Mapping the Effects of Climate Change on Human Migration and Displacement 2 (2009), *available at* http://www.ciesin.columbia.edu/documents/clim-migr-report-june09_media.pdf.

140 *Id.*

nations for harm to one of the planet's most crucial life sources – the atmosphere.[141]

3.2.1. ATCA Jurisdiction for an ATL Drowning Islands Claim

The ATCA is, on one hand, very generous in its jurisdictional grant to aliens aggrieved by torts committed in violation of treaties or the law of nations; it opens the doors of federal district courts to aliens suffering torts anywhere in the world.[142] The statute is at the same time stingy with respect to who can be sued. Neither the U.S. government nor its agencies are proper defendants under the ATCA.[143] Moreover, foreign sovereigns cannot be sued under the ATCA.[144] Cases such as the Second Circuit's recent decision in *Kiobel v. Royal Dutch Petroleum Co.* have cast considerable doubt over the viability of a suit against corporate defendants.[145] Courts in other circuits have reached the opposite conclusion, however, holding that corporations are within the ATCA's jurisdiction.[146] The U.S. Supreme Court stepped in to resolve the circuit split when it granted certiorari in *Kiobel*.[147] After hearing oral

141 Any ATCA claim will undoubtedly encounter the political question doctrine, an issue beyond the scope of this chapter. It should be noted that the justifications behind the ATCA (averting international conflict) establish a unique context for evaluating the political question defense.

142 *See, e.g.*, *Kadic v. Karadzic*, 70 F.3d 232, 238 (2d Cir. 1995).

143 *See Sosa*, 542 U.S. at 698 (establishing a bifurcated scheme whereby torts against the U.S. government and its agencies are prosecuted under the Federal Tort Claims Act, and torts against private parties are brought under the ATCA).

144 The Foreign Sovereign Immunity Act of 1976 (FSIA), not the ATCA, provides "the sole basis for obtaining jurisdiction over foreign states in [U.S.] courts." *See Arg. Rep. v. Amerada Hess Shipping Corp.*, 488 U.S. 429, 434 (1989).

145 *See Kiobel v. Royal Dutch Petrol. Co.*, 621 F.3d. 111, 145 (2d Cir. 2010) (ATCA does not extend jurisdiction to corporate defendants.).

146 *See Sinaltrainal v. Coca-Cola Co.*, 578 F.3d 1252, 1263 (11th Cir. 2009); *Flamano v. Firestone Natural Rubber Co.*, No. 10–3675, slip op. at 15 (7th Cir. 2011); *Doe v. Exxon Mobile Co.*, No. 09–7125, slip op. at 85 (D.C. Cir. 2011); *Sarei v. Rio Tinto*, No. 09–56381, slip op. at 6 (9th Cir. 2011) (holding that the ATCA does not impose a general bar to corporate liability).

147 *Kiobel v. Royal Dutch Petroleum Co.*, 621 F.3d. 111, 145 (2d Cir. 2010), cert. granted, 132 S. Ct. 472, 181 L. Ed. 2d 292 (2011).

arguments, the Court ordered a rehearing on the issue of whether the ATCA grants jurisdiction for torts committed in other countries.[148] Nevertheless, in order not to be a dead letter, the ATCA must allow suits against *some defendants*. As the following discussion suggests, the ATCA still seemingly affords jurisdiction against private parties or municipal entities behaving as state actors for some torts.[149] Although the jurisdictional needle is a narrow one to thread, it may prove open to ATL suits against private or municipal utilities operating coal-fired plants. As explained later, these utilities cause irrevocable damage to the atmospheric trust.

In *Sosa v. Alvarez-Machain*, the Supreme Court created a framework to assess the viability of claims brought under the ATCA.[150] By the statute's own terms, the claim must allege a tort committed "in violation of the law of nations or a treaty of the United States."[151] The *Sosa* court indicated that, to be actionable, any modern tort must be as universally accepted and well defined as were the paradigmatic nineteenth-century torts recognized under customary international law.[152] These ancient torts were violation of safe conduct, infringement on the rights of ambassadors, and piracy.[153] The *Sosa* Court made clear, however, that the ATCA was not shackled by the three traditional international tort categories and could accommodate novel cases – though expansion beyond these three torts is subject to "vigilant doorkeeping" (with particular concern for judicial restraint).[154] The Court held, however "that a

148 *Kiobel v. Royal Dutch Petroleum Co.*, 2012 U.S. LEXIS 1998 (U.S. Mar. 5, 2012) (order directing parties to file supplemental briefs).

149 *Kadic*, 70 F.3d at 245 (incorporating the state action jurisprudence of 42 U.S.C. § 1983 into the ATCA context). Some torts, including piracy, are actionable under the ATCA absent state action. *Id.* at 239.

150 *Sosa*, 542 U.S. 732–33.

151 28 U.S.C. § 1350 (1948).

152 *Sosa*, 542 U.S. at 732.

153 *Id.* at 715.

154 *See id.*, at 724–29. The Court listed several policy concerns associated with the judicial role, two of which are particularly applicable to an atmospheric trust claim.

single illegal detention of less than a day" is not a tort cognizable under the ATCA.[155]

A universally accepted international law norm derives from various sources of international law, including customary international law, treaties, and the objective works of legal scholars.[156] Although the Supreme Court has justifiably expressed reluctance in expanding the ATCA, nevertheless, the ability of the courts to recognize new causes of action under the statute ensures that international law continues to be internalized in domestic U.S. law and does not ossify.

ATL attorneys seeking ATCA jurisdiction on behalf of drowning island alien plaintiffs must establish the public trust obligation as a universal international law norm. This should be possible to achieve, and in so doing they may draw on its manifestation in natural law, in domestic legal systems around the world, and in international organizing principles and treaties such as the UNFCCC, the CBD, and the WHC. To be actionable under the ATCA, the public trust claim must also have "definite content."[157] As discussed earlier, the doctrine draws from a substantial body of private trust law that supplies basic standards of fiduciary behavior, together with a well-developed body of public trust law that applies these standards to sovereign trustees. Such standards have provided definite content to guide fiduciary behavior since the beginning

First, the Court expressed concern for collateral consequences that may have unusually significant implications in the realm of foreign affairs, which is typically the domain of the representative branches. But in the case of climate crisis, continued inaction by the political branches creates the gravest potential for collateral consequences. Second, the Court expressed hesitancy to recognize international law norms without the benefit of congressional guidance. In climate cases, however, the U.S. Congress has already ratified the UNFCCC, which sets forth atmospheric trust principles.

155 *Id.* at 738.

156 *See The Paquete Habana*, 175 U.S. 677, 700 (1900); *United States v. Smith*, 18 U.S. 153, 160–61 (1820) (Customary international law "may be ascertained by consulting the works of jurists, writing professedly on public law; or by the general usage and practice of nations; or by judicial decisions recognising [sic] and enforcing that law.").

157 *Sosa*, 542 U.S. at 732.

of the U.S. legal system.[158] At its core, the doctrine requires protection from "substantial impairment" to the trust.[159] As discussed earlier, this standard finds elucidation in the UNFCCC, which obliges nations to prevent "*dangerous anthropogenic interference with the climate system*,"[160] necessarily amounting to "substantial impairment" of the trust. The ATL approach of defining the atmospheric fiduciary obligation according to a scientifically prescribed standard of carbon reduction is seemingly geared toward the ATCA's test of "definite content."[161]

An ATL claim brought under the ATCA necessarily assumes a different posture from other ATL actions, because, due to the limitations of the ATCA, it must be brought against a non-governmental party. A standard public trust claim is brought against the government trustee for failing to protect the asset.[162] But classic trust principles do allow recourse against private tortfeasors causing damage to the trust, particularly where the trustee itself is not performing its fiduciary responsibility of preventing the damage. Under settled principles of private trust law, third parties may be held liable for damage to trust property where the trustee takes no action to abate the harm.[163] As one treatise notes,

> Every property owner, whether his interest is legal or equitable, has a right to have third persons refrain from injuring or appropriating the subject of his property right. It is therefore a truism that a trust beneficiary, being the owner of an equitable property right, has the support of the courts in his claim that strangers shall not cause damage to the trust res. . . . If the trustee refuses to bring an action against a third

158 *See supra* note 23 and cases cited therein. Commentators draw on such standards in confronting modern, complex environmental problems. *See* Nanda & Ris, *supra* note 56, at 301–2 ("Developments in the application of the [American] doctrine have broadened its potential scope and have made it a strong and flexible mechanism for confronting complex environmental problems.").

159 *See In re* Water Use Permit Applications, 9 P.3d 409, 451 (Haw. 2000).

160 *See* UNFCCC, *supra* note 76, art 2.

161 *See supra* note 157.

162 *See, e.g.*, In re *Water Use Permit Applications*, 9 P.3d at 409.

163 "The action should be brought by the trustee but the beneficiary may sue in equity if the trustee refuses or neglects to bring a suit." Bogert, *supra* note 10, § 166.

> person who has injured or misappropriated the trust property... the beneficiary may maintain a suit in equity against the third person in which the trustee is made a party defendant. To wait until... the present trustee saw fit to act... would endanger the cause of action. The necessities of the case entitle the beneficiary to proceed directly, in place of the trustee, to enforce the cause of action which the trustee has.[164]

ATL attorneys may use this line of doctrinal reasoning to build a case against private utility tortfeasors damaging the atmospheric trust. The climate crisis presents a matter of egregious trustee malfeasance, as the United States has failed to comprehensively regulate greenhouse gas emissions even in face of a well-established and known threat of catastrophic planetary harm and the annihilation of entire island nations. Although there are many categories of major private tortfeasors causing pollution damage to the atmospheric trust, it is most crucial to bring the utilities operating coal-fired plants, as a class of polluters, under immediate judicial supervision. The sheer magnitude of their carbon pollution overshadows emissions by many other possible defendants.[165] As a leading climate scientist, Dr. James Hansen, stated in testimony before Congress in 2007, the most critical action for saving the planet at this time is preventing construction of more coal-fired power plants.[166]

164 *Id.* at 609–11. One recent California case cast some doubt over the ability of citizen beneficiaries to sue third parties directly for damage to the trust. *Ctr. for Biological Diversity v. FPL Grp.*, 83 Cal. Rptr. 3d 588, 602 (Cal. Ct. App. 2008). In that case, however, the plaintiffs had not first asked the state, as trustee, to take appropriate action, a deficiency the court found significant. *Id.*

165 In 2010, coal-fired electricity generation accounted for 1,827.3 Tg CO_2 Eq of emissions, out of a total of 5,406.8 Tg CO_2 Eq for all emissions from fossil fuel combustion in the United States. The second highest source of CO_2 emissions was petroleum consumption for transportation, accounting for 1,700.4 Tg CO_2 Eq. U.S. Envtl. Prot. Agency, Draft Inventory of U.S. Greenhouse Gas Emissions and Sinks: 1990 – 2010 § 3-4, ¶ 12 (2012), *available at* http://www.epa.gov/climatechange/emissions/downloads12/Complete%20Report%20(Main%20Text).pdf.

166 James Hansen, Dangerous Human-Made Interference with Climate, Testimony Before Select Committee on Energy Independence and Global Warming, U.S. House of Representatives 18 (April 26, 2007) (*available at* http://www.columbia.edu/~jeh1/testimony_26april2007.pdf).

3.2.2. Private Utilities as State Actors for the Purposes of ATCA Jurisdiction

ATL attorneys arguing for ATCA jurisdiction over electric utility companies must argue that such companies effectively act as instrumentalities of the State. Courts have borrowed standards from 46 U.S.C. §1983 jurisprudence to determine whether an entity is considered a state actor under the ATCA.[167] Municipally owned power plants have generally been deemed state actors, because they are government owned and operated.[168] Thus, the ATCA would seemingly give federal courts jurisdiction over them without the need to apply the state action test.[169] As to private utilities, several tests fashioned with reference to §1983's state action standard could be applied persuasively to a climate action brought under the ATCA.[170] Though courts have severely limited ATCA jurisdiction over private corporations,[171] private utilities fall within a class of actors that is nearly *sui generis*. The State is entangled on many levels with their operations due to the fact that electricity provision is an essential public service. Private utilities in some states operate

167 *See Kadic v. Karadzic*, 70 F.3d 232, 245 (2d Cir. 1995).

168 *See Craft v. Memphis Light, Gas and Water Div.*, 534 F.2d 684, 687 (6th Cir. 1976), *aff'd*, 436 U.S. 1 (1978); *Gunter v. Long Island Power Authority/Keyspan*, No. 08 CV 498(RRM)(LB), 2011 WL 1225791, at *7–*8 (E.D.N.Y. Feb. 15, 2011), *adopting report and recommendation*, 2011 WL 1154382 (E.D.N.Y. Mar. 29, 2011). *See also Davis v. Weir*, 497 F.2d 139, 143–144 (5th Cir. 1974) (finding that a municipal water utility is obviously a state actor).

169 *See, e.g., Newton v. Se. Alabama Gas Dist.*, 708 F. Supp. 1254, 1257 (M.D. Ala. 1989) ("[I]f the [utility company] is found to be a public entity, then it is unnecessary for the court to address whether its activities are 'private action' or 'state action.'") (citing *Rendell-Baker v. Kohn*, 457 U.S. at 838, 102 S.Ct. at 2769 (1982)); *see also City of Canton v. Harris*, 489 U.S. 378, 389 (1989) (municipality may be found liable under 42 U.S.C § 1983 when the violation is a result of official custom or policy of the government). *Cf. Lopez v. Richardson*, 647 F. Supp. 2d 1356, 1363–64 (N.D. Ga. 2009) (finding that there is nothing in the ATCA that says a municipality may not be found liable under the ATCA).

170 For the Court's most recent listing of state action tests used for purposes of § 1983, see *Brentwood Acad. v. Tennessee Secondary Sch. Athletic Ass'n.*, 531 U.S. 288, 295–96 (2001).

171 *See supra* note 145.

in heavily regulated markets.[172] Moreover, some utilities remain publicly owned.[173]

Of the several tests federal courts use to determine state action under §1983, four are instructive as to atmospheric trust claims brought against private power companies under the ATCA. The *joint participation* or *in concert* test finds state action when it can be sufficiently shown that the defendant "operated as a willful participant in joint activity with the State or its agents"[174] or "act[ed] together with state officials, or with significant state aid"[175] in an effort to deprive the plaintiff of his or her rights. Acquiescence or approval by state officials will not satisfy the joint action test.[176] Rather, the presence of officials must have influenced or have been a necessary part of the conduct. The *public function* test recognizes that "state action can exist where a private entity performs a function [that is] traditionally the exclusive prerogative of the State."[177] The function must be shown to have been a legislative mandate on the State, which the State then assigned to a private entity. Courts have not been persuaded by arguments that, because a utility provided a "public service" and was forced to provide that service "on a reasonably

172 Congress began to deregulate the energy markets in the 1990s. *See, e.g.* OpenSecrets, *Electricity Deregulation*, available at http://www.opensecrets.org/news/issues/electricity/index.php (last visited March 18, 2012). However, twenty-seven states still have regulated energy markets. *See* Quantum Gas Power Services, LTD., *Energy Deregulated States in the United States, available at* http://www.quantumgas.com/list_of_energy_deregulated_states_in_united_states.html (last visited on March 18, 2012); Chrislynn Energy Services, *Energy Deregulated States in the United States, available at* http://www.chrislynnenergy.com/services/deregulation.html (last visited on March 18, 2012).

173 In 2005, coal-fired public utilities in the United States produced more than 2,000 terawatts/hour of generation. International Energy Agency, *Energy Efficient Indicators for Public Electricity Production from Fossil Fuels* (2008), available at www.iea.org/papers/2008/En_Efficiency_Indicators.pdf.

174 *Gorman-Bakos*, 252 F.3d at 551–52.

175 *Kadic v. Karadzic*, 70 F.3d 232, 245 (2d Cir. 1995).

176 *Beanal v. Freeport-McMoRan*, Inc., 969 F. Supp. 362, 377 (E.D. La. 1997), *aff'd.*, *Beanal v. Freeport-McMoran, Inc.*, 197 F.3d 161 (5th Cir. 1999).

177 *Id.* at 379.

continuous basis," it served a public function.[178] The *symbiotic relationship* test will pertain when it can be shown that the government "'has so far insinuated itself into a position of interdependence' with a private entity that 'it must be recognized as a joint participant in the challenged activity.'"[179] The two must be financially and physically interdependent. This relationship must be exceedingly close, however. The U.S. Supreme Court has held that state licensing requirements and regulations are not sufficient to transform a private entity into a state actor for the purpose of invoking the strictures of the Fourteenth Amendment.[180] The Court has also suggested, however, that regulation may be a *factor* in determining state action.[181] The last test is the *pervasive entwinement* test, which finds that a private entity is so "overborne by the pervasive entwinement of public institutions and public officials in its composition and workings [that] there is no substantial reason to claim unfairness in applying constitutional standards to it."[182] This test requires showing a great deal of personal involvement by public officials and some financial connection with the government (aside from taxation). Regulation, regardless of how extensive it is, will not be sufficient to demonstrate pervasive entwinement.[183]

178 *Jackson v. Metro. Edison Co.*, 419 U.S. 345, 352–354, 95 S. Ct. 449, 454–456 (1974).

179 Beanal, *supra* note 17664, at 377–78 (quoting *Burton v. Wilmington Parking Authority*, 365 U.S. 715, 725 (1961) citation omitted).

180 *See, e.g., Moose Lodge No. 107, v. Irvis*, 407 U.S. 163, 175–77 (1972) (private club was not a state actor by virtue of it receiving a liquor license from the state and being required to comply with extensive regulation).

181 *See Jackson v. Metro. Edison Co.*, 419 U.S. 345, 350–51) (1974) ("Nor does the fact that the regulation is extensive and detailed, as in the case of most public utilities,. . . . [create state action]. It may well be that acts of a heavily regulated utility with at least something of a governmentally protected monopoly will more readily be found to be 'state' acts than will the acts of an entity lacking these characteristics.") (citation omitted). The *Jackson* Court held, however, that regulation of a private utility, without more, was insufficient to make it a state actor. *Id.*, at 358–59.

182 *Brentwood Acad. v. Tenn. Secondary Sch. Athletic Ass'n.*, 531 U.S. 288, 298 (2001).

183 *Id.* (private statewide association governing high school athletics of public and private school found to be state actor).

As noted earlier, these tests have been developed in the context of §1983 actions for violations of individual Constitutional rights. Within this context, courts have taken a generally restrictive view of holding private power plants accountable as state actors.[184] The context of international climate harm is markedly different, however, because it carries exponentially magnified consequences to society globally. Although more elaborate discussion is well beyond the scope of this chapter, it will be up to ATL lawyers to sift through the case law and make convincing arguments as to why private power companies meet the state action tests based on the intricate legal and regulatory relationships between the utility industry and state governments, particularly those shaping the choice of fossil fuel versus renewable energy sources. As a general matter, government's role in allowing private utilities to pollute the atmospheric trust resource falls into a category of heightened state involvement distinct from the realm of social, individual, and financial affairs that characterize the §1983 context. The difference lies in the trustees' duty, implicated in natural resources management but not in other realms of government function. A trustee, sovereign or private, can never abdicate responsibility over the *res*, though it may allow some use of the *res*. As a basic fiduciary obligation, the trustee must exercise continuing, close supervision over any use of the *res*. The duty of supervision, by its very nature, significantly involves the trustee in the action affecting the *res*.[185] Anything less than significant involvement would violate fiduciary

184 *See Jackson*, *supra* note 180, at 358–59 (1974); *Spickler v. Lee*, 63 F. App'x. 2, 2 (1st Cir. 2003); *Mays v. Buckeye Rural Elec. Co-op., Inc.*, 277 F.3d 873, 880 (6th Cir. 2002); *Taylor v. Consol. Edison Co. of New York, Inc.*, 552 F.2d 39, 46 (2d Cir. 1977). *But see Barnett v. Carberry*, 420 Fed. App'x 67, 69 n.1 (2d Cir. 2011) ("[T]his Court presumes, without deciding, that the private utility company defendants engaged in state action.").

185 *See, e.g.*, *Nat'l Audubon Soc'y v. Superior Ct.*, 658 P.2d 709, 712 (1983) (*en banc*) (public trust requires state to have continuous supervision and control over navigable water and land under it); *In re* Water Use Permit Applications, 9 P.3d 409, 453 (Haw. 2000) ("The continuing authority of the state over its water resources precludes any grant or assertion of vested rights to use water to the detriment of public trust purposes."); *see generally* Coplan, *supra* note 37, at pp. 307–12.

standards, because a trustee may not simply walk away from a trust he or she is charged with managing.[186]

3.2.3. *The Viability of an ATCA Claim after* American Electric Corp. (AEP) v. Connecticut

On its face, the relief sought in an ATCA claim against major carbon-emitting electric utilities roughly parallels that sought in *American Electric Corp., Inc. v. Connecticut* (*AEP v. Connecticut*), a climate nuisance case decided by the U.S. Supreme Court after the time of this writing but before publication.[187] What follows is a necessarily brief discussion of that ruling's effect on the viability of an ATCA claim. In *AEP v. Connecticut*, States and other plaintiffs brought a federal common law public nuisance claim against major carbon dioxide emitters (four private power companies and the federal Tennessee Valley Authority – together the five largest carbon-dioxide emitters in the United States). The Court held that the federal nuisance action was "displaced by the federal legislation [the Clean Air Act] authorizing [the Environmental Protection Agency (EPA)] to regulate carbon-dioxide emissions."[188] Although, as explained later, the decision should have no direct bearing on ATL suits brought against sovereign governments for failure to regulate greenhouse gas emissions, many might hasten to conclude that the *AEP* ruling eliminates the possibility of a trust suit brought against coal-fired plants under the jurisdiction of the ATCA due to the parallelism in the relief sought between the two types of claims. Though both seek injunctive relief in the form of emission reduction, there are crucial distinctions between the third-party nuisance claims and the third-party

186 See 76 Am. Jur. 2d *Trusts* § 404 (2005) (A trustee representing beneficiaries in succession is under a duty to successive beneficiaries to act with due regard to their respective interests and to preserve trust property for remainderpersons.")

187 *Conn. v. Am. Elec. Power Co., Inc.*, 406 F. Supp. 2d 265 (S.D.N.Y. 2005), *rev'd.*, 582 F.3d 309 (2nd Cir. 2009), *rev'd.*, *Am. Elec. Power Co., Inc. v. Conn.*, 131 S. Ct. 2527 (2011).

188 *Am. Elec. Power Co., Inc.*, 131 S. Ct. at 2537.

trust tortfeasor claims, suggesting that the *AEP* case still leaves space for an ATCA case.

One basic difference between the ATCA third-party trust claim and the nuisance claim is that the former is based on an international norm (duties arising from co-tenant trusteeship); indeed, the statute limits jurisdiction to violations of the "law of nations or a treaty of the United States."[189] The nuisance claim in *AEP v. Connecticut*, by contrast, is decidedly domestic in nature, based on federal common law. As a technical matter, a federal statute such as the Clean Air Act (CAA) seemingly does not displace an international set of obligations deriving, as this chapter has suggested, from global co-trusteeship of a planetary asset. The clear policy objectives of the ATCA to thwart international torts that cause injury to aliens and thereby threaten international conflict is far removed from the domestic nuisance context, which deals with rights among citizens, States, and corporations subject to U.S. laws like the CAA.

Even more importantly, the sovereign trust obligation on which third-party tortfeasor liability is based is fundamentally different from the nuisance principle. It is necessary to explore these differences as a predicate to looking at derivative ATL claims against utility tortfeasors under the ATCA. Importantly, the *AEP* Court recognized the viability of federal common law in matters of national concern, such as interstate pollution.[190] The trust framework is based on the duty of *a sovereign* to protect and secure the integrity of the public asset as a whole, not only for

189 *See* 28 U.S.C. § 1350 (1948). In *Sosa v. Alvarez-Machain*, the Court recognized the teaching of *Erie Railroad Co. v. Tompkins* that "[t]here is no federal common law," *Erie R.R. Co. v. Tompkins*, 304 U.S. 64, 78 (1938), but held that new causes of action within the ATCA may be recognized under international law. *Sosa*, 542 U.S. 692, 729 (2004). In *AEP*, the Court signaled a retraction from the *Erie* statement that there is no federal common law and recognized that there is, indeed, federal common law in areas of national concern, *Am. Elec. Power Co., Inc.*, 131 S. Ct. at 2535 (explaining, "In the wake of *Erie*, however, a keener understanding developed."); however, the shift away from *Erie* does not change the scope of the ATCA, which is limited to claims recognized under international law.

190 *Id.*

present generations but for future ones as well. A public nuisance claim, by contrast, focuses exclusively on *individual parties* causing "substantial and unreasonable interference with public rights" enjoyed by the present generation alone.[191] The differences between the two make a nuisance claim susceptible to displacement in a way that a trust claim against a sovereign cannot be. As the *AEP* Court noted, "The test for whether congressional legislation excludes the declaration of federal common law is simply whether the statute 'speak[s] directly to [the] question' at issue."[192] Whereas the nuisance claim is directed toward third parties and is therefore displaced by a regulatory scheme directed toward the same kind of harm caused by the same class of party, a trust claim focuses on the more fundamental question of whether the *sovereign trustee* is meeting its duty toward the citizens by adequately protecting the asset held in public trust. A trust claim lodged against a sovereign does not invade the actual sphere of regulation – deciding what activities to curtail and to what extent – but rather demands an accounting of the bottom-line effectiveness of the regulation (combined with other tools) to achieve asset protection. An ATL claim against government simply does not put courts in the position of making policy determinations as to *who should be regulated and to what extent* – an unavoidable inquiry embedded in the "unreasonable" standard of a nuisance claim.

It is also important to recognize that regulation alone does not answer the question of whether the sovereign has fully met its fiduciary duty to the public beneficiaries. Even when a government enacts regulations to prevent harm to the trust, the basic trust question remains whether such regulation is adequate to protect the asset for present and future generations. It is nearly inconceivable that pollution regulation under the CAA would suffice to meet the atmospheric trust obligation. A sovereign would have to implement other types of policy (such as tax incentives, land use reform, infrastructure changes, reforestation, soil

191 *See Am. Elec. Power Co., Inc.*, 131 S. Ct. at 2534.

192 *See id.* at 2537 (citation omitted).

improvements, and a host of other measures) to carry out the fiduciary obligation of achieving a 6% annual carbon emission reduction and carbon drawdown from the atmosphere. In this sense, no matter how extensive a regulatory scheme under the CAA may be, the scheme itself does not "speak directly to the question" of whether the regulation alone is adequate to meet fundamental fiduciary duties of asset protection.[193] As the Supreme Court of Arizona once said in underscoring the judicial role in enforcing the trust, "The check and balance of judicial review provides a level of protection against improvident dissipation of an irreplaceable res."[194]

As a related matter, the very distinct legal standards flowing from the nuisance and trust frameworks put judges in markedly different positions when determining appropriate relief. The nuisance prohibition against "unreasonable" interference with a public right asks judges to determine the "reasonable" level of emissions for the particular defendants before the court, an inquiry that invites review of many economic and social concerns, seemingly without any reference to the overall needs of the atmosphere. In the *AEP* case, for example, the trial court was asked to decide what level of emission reduction is "practical, feasible, and economically viable."[195] The Supreme Court deemed this question much more appropriate for regulation than judicial resolution, noting that the CAA's scheme vested the EPA with the power to engage in the determinations of who should be regulated and to what extent.[196] Based on the presence of a statutory scheme designed to make tradeoffs and determine appropriate objects of regulation, and of action on that statutory scheme by the EPA, the Court was inclined to find that the nuisance claim was displaced by the statutory scheme. By contrast, a

193 *Id.*

194 *Ariz. Ctr. for Law in the Pub. Interest v. Hassell*, 837 P.2d 158, 169 (Ariz. 1991). *See also generally Lake Mich. Fed'n. v. U.S. Army Corps of Eng'rs.*, 742 F. Supp. 441 (N.D. Ill. 1990).

195 *Am. Elec. Power Co., Inc.*, 131 S. Ct. at 2540 (citation omitted).

196 *Id.* at 2539–40.

public trust claim asks whether the sovereign is protecting the functionality of the asset in a manner that will provide benefits not only to present generations but also to future generations of citizens, who are also the rightful beneficiaries of the trust.[197] The sovereign fiduciary obligation to prevent *substantial impairment* to the *res*[198] speaks to the essential purpose of the trust to continue providing ecosystem services and support life into the future. Such a standard demands a biophysical definition of asset function and protection. To this end, ATL calibrates the fiduciary obligation to the scientific prescription established by leading climate scientists calling for a 6% reduction of carbon emissions globally, beginning in 2012, in order to achieve 350 ppm by the end of the century.[199]

Although these distinctions seemingly put a large wedge between the *AEP* case and ATL cases against sovereigns, the question of whether *AEP* applies to an ATL claim brought against trust tortfeasors is still very difficult to answer, because such an ATL claim aims to reduce emissions of third-party polluters – exactly the same kind of polluters that were defendants in *AEP*. However, the basic differences between trust law and nuisance law also distinguish the posture of the private defendants in the two types of cases. The nuisance claim necessarily focuses on tangible *harm* to the public's "rights"; it logically requires some linkage between the defendants' action and the plaintiffs' enjoyment of rights.[200] The defendants in *AEP* contended that because carbon pollution is mixed in the atmosphere, "emissions in New Jersey may

197 *See Geer v. Connecticut*, 161 U.S. 519, 534 (1896) ("[I]t is the duty of the legislature to enact such laws as will best preserve the subject of the trust, and secure its beneficial use in the future to the people of the state."); *Ariz. Ctr. for Law in the Pub. Interest*, 837 P.2d at 169 ("The beneficiaries of the public trust are not just present generations but those to come.").

198 *See supra* note 159 and accompanying text (discussing substantial impairment).

199 *See supra* note 106 and accompanying text (discussing atmospheric fiduciary obligation).

200 In this respect, the modern climate nuisance cases are vastly more difficult than the nineteenth-century cases involving pollution between neighboring states.

contribute no more to flooding in New York than emissions in China."[201] The nuisance frame treats the atmosphere as merely a conduit for harm originating from the defendant and ending up with the plaintiff; the presence of multiple polluters thus makes causation a tricky matter. An ATL claim, by contrast, characterizes the atmosphere as an integral asset that belongs to sovereigns in co-tenancy. Under the framework developed in this chapter, any harm to the joint property is *directly a tort*, without regard to where or how the climate consequences may land. Pollution of the atmospheric trust impairs the asset's ability to function and provide life-support to the planet as a whole. Although the *AEP* Court did not dwell on the causation dilemmas inherent in the domestic nuisance claim, that concern undoubtedly contributed to the Court's finding of displacement, because that holding was based in large part on the complexities of choosing among emission sources to regulate and the Court's preference to leave such policy decisions to an agency.[202]

As a further matter, an ATL claim against third-party tortfeasors is fundamentally different from a nuisance claim, because it arises only when the trustee fails to protect the trust against third-party harm. It asks the court to step in under emergency circumstances to enjoin third-party harm where the trustee has been recalcitrant or has been sitting idle in the face of a clear threat – both of which are situations that aptly describe the present circumstances surrounding climate crisis. If and when the government acts *in adequate fashion to fully meet its trust responsibility*, there would be no cause of action. In this regard, the ATL claims look at regulatory delay in a much different light from nuisance claims. Whereas the Supreme Court was content with EPA's promise (at the time of writing) of a forthcoming regulation in May 2012, an ATL claim would likely characterize that delay, as well as foreseeable delays thereafter, as posing an imminent threat to the atmospheric trust in light of

201 *Am. Elec. Power Co., Inc.*, 131 S. Ct. at 2536.

202 *See id.* at 2538 ("The critical point is that Congress delegated to EPA the decision whether and how to regulate carbon-dioxide emissions from power plants; the delegation is what displaces federal common law.").

substantial ongoing pollution. There is, moreover, no guarantee that the regulation, even if delivered when promised, will be adequate on its face to meet the fiduciary obligation of a 6% annual emission reduction, that implementation of the regulation will not be derailed through political pressure or court challenges, or that there will be effective enforcement of its prohibitions. The regulatory history under U.S. environmental laws has been plagued with dysfunction on all scores.[203] Thus, ATL appears to be a more promising tool for recompensing climate change than nuisance actions.

Although regulatory delay might not always justify a trust claim against major third-party tortfeasors, it does so for two reasons in the case of climate change. First, the danger from delay is extraordinary. Scientists have repeatedly warned that further delay in carbon emission reduction may push the planet past the "tipping point," already recognized by courts as having catastrophic potential for the planet.[204] Because the planet's CO_2 levels already threaten "dangerous climate change," as described by scientists,[205] courts are justified in intervening to force an emission reduction from major emission sources until the federal government adequately regulates them. Second, because it is now well established that some portion of emissions remain in the

203 *See generally* Craig Collins, Toxic Loopholes: Failures and Future Prospects for Environmental Law (2010).

204 *See Ctr. for Biological Diversity v. Nat'l Highway Traffic Safety Admin.*, 508 F.3d 508, 523 (9th Cir. 2007) ("Several studies also show that climate change may be non-linear, meaning that there are positive feedback mechanisms that may push global warming past a dangerous threshold (the 'tipping point')." (citation omitted)). *See also Green Mountain Chrysler v. Crombie*, 508 F. Supp. 2d 295, 313–17 (D. Vt. 2007) (finding Dr. Jim Hansen's testimony as to the tipping point reliable under the *Daubert* test, stating, "[the] 'tipping point' theory posits that at a certain point the changes associated with global warming will become dramatically more rapid and out of control.... [D]rastic consequences, including rapid sea level rise, extinctions, and other regional effects, would be inevitable with a two to three degrees Celsius warming expected if no limits are imposed and emissions continue at their current rate. Such changes could happen quickly once a tipping point is passed.").

205 Hansen, *supra* note 166.

atmosphere for hundreds or thousands of years,[206] and the pollution cannot simply be "cleaned up" as is the case in more discrete contexts (hazardous waste disposal on land, for example), the ongoing emissions from major sources represent irrevocable harm to the atmospheric trust property that might never be capable of remediation.

In sum, these concerns differentiate the posture of an ATL claim from a nuisance claim against third parties. Although the Supreme Court in *AEP* expressed reservations about opening the courthouse doors to a potential flood of cases, that concern is mitigated in an ATCA trust context because of the absence of jurisdiction over most conventional corporate defendants (ones that do not meet the test of state actors as explained earlier). Those courts that fully understand the perils of continued carbon pollution may be inclined to distinguish ATL claims from the *AEP* claims; they arguably have a sound basis for doing so.

As noted earlier, the U.S. District Court for the District of Columbia, in *Alec L. v. Jackson*, took a much different view from that just expressed, and found that *Connecticut v. AEP* meant that the Clean Air Act displaces ATL claims. That decision is likely to be appealed to the U.S. Court of Appeals for the District of Columbia.

4. World Heritage Sites as Global Trust Assets

A second possible avenue of legal recourse for drowning nations derives from the global co-tenancy framework evident in the World Heritage Convention (WHC). The WHC arose from an international effort aimed to protect ancient Egyptian temples and monuments from impending flooding as a result of construction of the Aswan high dam on the Nile River.[207] Adopted in 1972, the WHC recognizes that certain cultural and natural heritage locations around the world hold outstanding universal value and warrant preservation as part of the world

206 *See* Hansen et. al., *supra* note 2, at 220.

207 *About the World Heritage Convention*, UNESCO World Heritage Convention, http://whc.unesco.org/en/convention/#Brief-History; (last visited March. 24, 2012).

heritage of humankind.[208] In view of the magnitude and gravity of the dangers threatening world heritage sites, the WHC obliges the international community as a whole to participate in their protection.[209] In so doing, the WHC establishes, on the international level, what is best characterized as a global trusteeship over designated sites with duties incumbent on nations as co-trustees of the heritage. As the following discussion explains, the WHC presents an international obligation and process that drowning nations might invoke to curb greenhouse pollution causing climate change.

4.1. World Heritage Convention

There are currently more than 900 designated cultural and natural world heritage sites across 153 countries.[210] Many of these environments are vulnerable to catastrophic damage from predicted climate change. In the Queensland tropical rainforests in Australia, for example, scientists warn, "[W]e may be facing an unprecedented loss in biodiversity in any montane biota, an environmental catastrophe of global significance."[211] On the Great Barrier Reef, ocean acidification and warming sea temperatures are causing massive coral bleaching, with grave losses to biodiversity.[212] In Waterton-Glacier International Peace

208 WHC, *supra* note 7, pmbl.

209 *Id.*

210 *World Heritage List*, UNESCO World Heritage Convention, http://whc.unesco.org/en/list/ (last visited Aug. 19, 2011).

211 Stephen E. Williams, Elizabeth E. Bolitho, & Samantha Fox, *Climate Change in Australian Tropical Rainforests: An Impending Environmental Catastrophe*, 270 Proc. R. Soc. Lond. B 1887, 1889 (2003), *available at* http://www.ncbi.nlm.nih.gov/pmc/articles/PMC1691452/pdf/14561301.pdf.

212 *Climate Change*, Australian Government: Great Barrier Reef Marine Park Authority (2006), *available at* http://www.gbrmpa.gov.au/corp_site/key_issues/climate_change/climate_change_and_the_great_barrier_reef (last visited Aug. 19, 2011); *Climate Change Impacts on Coral Reefs*, Australian Government: Great Barrier Reef Marine Park Authority, *available at* http://www.gbrmpa.gov.au/outlook-for-the-reef/climate-change/what-does-this-mean-for-habitats/coral-reefs (last visited Aug. 19, 2011).

Park in Canada, glaciers are melting so fast that scientists predict their complete disappearance by 2030.[213] In India and Bangladesh, a 45-cm rise in global sea levels will eradicate 75% of the Sundarbans, the world's largest estuarine mangrove forest.[214] There are currently at least 125 cultural and natural world heritage sites identified as threatened by climate change.[215]

World Heritage sites are places designated for their "outstanding universal value," defined as "cultural and/or natural significance which is so exceptional as to transcend national boundaries and to be of common importance for present and future generations of all humanity ... [and] the international community as a whole."[216] Such sites are best understood and characterized as global trust assets invoking shared sovereign public trust duties. In U.S. common law interpreting the public trust doctrine, for example, courts have conferred trust protection to sites and resources that are deemed part of the "people's common heritage."[217] On the international level, the positive iterations found in the WHC dovetail with the more general concepts recognizing the common heritage of humankind.[218]

213 Myrna H.P. Hall & Daniel B. Fagre, *Modeled Climate-Induced Glacier Change in Glacier National Park: 1850–2100*, *in* 53 Bioscience 131, 137–38 (2003).

214 United Nations Environment Programme [UNEP], World Conservation Monitoring Centre, The Sundarbans: Bangladesh, *available at* http://www.unep-wcmc.org/medialibrary/2011/06/24/ac3e5a3a/Sundarbans%20Bangladesh.pdf (last updated May 2011).

215 UNESCO World Heritage Convention, *Issues Related to the State of Conservation of Properties Inscribed on the World Heritage List: The Impacts of Climate Change on World Heritage Properties*, ann. 4, § 42 U.N. Doc. WHC-06/30.COM/7.1 (2006), *available at* whc.unesco.org/uploads/activities/documents/activity-393–1.doc.

216 UNESCO, *supra* note 78, ¶ 49.

217 *Nat'l Audubon Soc'y v. Superior Court*, 658 P. 2d 709, 724 (Cal. 1983) (noting "the duty of the state to protect the people's common heritage ... surrendering that right of protection only in rare cases when the abandonment of that right is consistent with the purposes of the trust.").

218 WHC, *supra* note 7, art. 6(1). For commentary on the "common heritage of mankind," see Alexandre Kiss & Dinah Shelton, International Environmental Law 273–85, 324 (1986).

In Australia, courts have expressly conferred public trust status on national parks, some of which are designated as World Heritage Sites.[219] Some courts have approached World Heritage site management controversies with basically the same restrictions on private benefit that are reflected in public trust law limiting private control and benefit from public trust resources.[220] In an Australian case, *Booth v. Bosworth*, the court held that "it would be a rare case in which a Court could be satisfied that the financial interests of private individuals, or even the interests of a local community, should prevail over interests recognized by the international community ... as being of international importance."[221]

4.2. A Co-Tenant Trustee Framework of Protection for Global Heritage

The WHC presents what could be defined as a co-tenant trustee structure over many of the world's most outstanding natural and cultural features. The heritage designation is decided by the World Heritage Committee and is not subject to veto by the State in which the site is located.[222] This absence of state control over the listing process reflects a global trusteeship over such sites, one elevated beyond national jurisdictional boundaries. Once a site is listed, the burden falls on the entire global community to participate in its protection.[223] World Heritage sites are under the guardianship of all parties to the WHC – not merely the nation in which

219 *Willoughby City Council v. Minister for Admin'g the Nat'l Parks and Wildlife Act*, (1992) 78 LGERA 19, 34 (Austl.).

220 *See* Sax, *supra* note 12, at 490 ("When a state holds a resource which is available for the free use of the general public, a court will look with considerable skepticism upon *any* governmental conduct which is calculated *either* to reallocate that resource to more restricted uses *or* to subject public uses to the self-interest of private parties."); *Lake Mich. Fed'n. v. U.S. Army Corps of Eng'rs.*, 742 F. Supp. 441, 445 (N.D. Ill. 1990).

221 *Booth v. Bosworth*, (2001) 114 FCR 39, 68 (Austl.); *see also Commonwealth v. Tasmania* (1983) 158 CLR 1 (Austl.) (restricting government actions that would affect world heritage areas in Tasmania).

222 WHC, *supra* note 7, art. 6.

223 *Id.*

the site is located. The sites are managed for the benefit of "present and future generations of all humanity" – who, from a trust perspective, are the beneficiaries of the global heritage *res*.[224]

Article 4 of the WHC defines protective obligations toward designated heritage sites, establishing a recognized duty "of ensuring the identification, protection, conservation, presentation and transmission to future generations of the cultural and natural heritage . . . situated on [the] territory. . . . [The party] will do all it can to this end, to the utmost of its own resources."[225] This protective obligation mirrors the traditional public trust doctrine, explained by one leading U.S. state supreme court case as imposing (in the context of water resources) the "duty to ensure the continued availability and existence of its water resources for present and future generations."[226] Construing the WHC as a public trust framework would require a strict judicial approach that is customary in trust cases – as distinguished from the deferential judicial review typical of administrative law cases.[227] To this end, the High Court of Australia found legally binding obligations from Articles 4 and 5 of the Convention:

> [T]he obligation under Art. 4 leaves no discretion in a party as to whether it will abstain from taking steps in discharge of the "duty" referred to. . . . [A]nd the question whether it is unable to take a particular step within the limits of its resources is a justiciable question. No doubt the allocation of resources is a matter for each party to decide and the allocation of resources for the discharge of the obligation may thus be said to be discretionary, but the discretion is not at large.[228]

224 *See supra* note 216 and accompanying text.

225 WHC, *supra* note 7, art. 4.

226 *In re* Water Use Permit Applications, 9 P.3d 409, 451 (Haw. 2000).

227 For discussion on the differences in judicial approach, see Sandy K. Fairfax and Andrea Issod, *Trust Principles as a Tool for Grazing Reform: Learning from Four State Cases*, 33 Envtl. L. 341, 391 (2003) ("Trust law puts an entirely different spin on issues of discretion, which it discusses in terms of prudence.").

228 *Commonwealth v. Tasmania*, (1983) 158 CLR 1, ¶ 41 (Austl.); *see also* Erica J. Thorson, *On Thin Ice: The Failure of the United States and the World Heritage Committee to*

The WHC states that "each State Party ... undertakes not to take any deliberate measures which might damage directly or indirectly the cultural and natural heritage ... situated on the territory of other States Parties to this Convention,"[229] an obligation considered to be legally binding.[230] The agreement effectively sets up a global *res* of world heritage with obligations transcending domestic boundaries; it positions nations as co-tenant trustees over world heritage sites and invokes duties akin to the prohibition against waste to common assets. In so doing, the WHC creates a construct in positive law that is remarkably parallel to the suggested common law construct underlying global atmospheric trust litigation, described earlier.

4.3. Enforcing Co-Tenant Obligations Not to Waste the Common Assets through Atmospheric Pollution

Joining World Heritage duties with public trust obligations may forge new causes of action to address climate change. Continued atmospheric pollution is clearly a deliberate measure being permitted by State laws and in many circumstances undertaken directly by States themselves through state-owned utilities – causing direct and indirect harm to the cultural and natural heritage of the world. In terms of domestic fora, citizens may combine a World Heritage claim with ATL actions designed to force carbon reduction. On the horizontal jurisdictional level, drowning island claimants seeking relief in U.S. courts under the ATCA might draw on the World Heritage Convention (to which the United States is a party) to shore up a claim that greenhouse gas pollution is in violation of "a treaty of the United States" for which ATCA jurisdiction lies.

On the international level, there is underused potential to invoke the processes, expertise, and structure of the WHC in defense of heritage

Take Climate Change Mitigation Pursuant to the World Heritage Convention Seriously, 38 ENVTL. L. 139, 161 (2008).

229 WHC, *supra* note 7, art. 6(3).

230 Thorson, *supra* note 228, at 161.

sites threatened by climate change. One of the duties of the World Heritage Committee is to keep a record of "World Heritage in Danger."[231] Although the World Heritage Committee has recognized climate change as a significant emerging threat to world heritage,[232] to date it has refused to list any heritage sites as endangered by climate change, instead deferring the issue of climate mitigation to the UNFCCC.[233] By not acting under its own authorities, the Committee has effectively ignored the comparative advantage of the World Heritage Convention in addressing this global problem.[234] Commentators have challenged the back-seat stance of the World Heritage Committee as a failure to abide by international principles that would give simultaneous effect to both the UNFCCC and the WHC.[235] The World Heritage Committee's passivity in the face of a global climate crisis is also, seemingly, an abdication of its unique duties as a global trust administrative body charged with protecting a *res* of world heritage. As noted earlier, trustees must take affirmative action to protect the trust they are charged with administering.

Commentators have urged a more direct role for the World Heritage Committee and its agencies in combating the pollution that imperils heritage sites.[236] Parties to the Convention should call on those nations directly responsible for the historic and projected future emissions to achieve deep cuts in pollution to prevent a catastrophic loss of world

231 WHC, *supra* note 7, at art. 11(4). *See also* UNESCO, *supra* note 78, ¶ 180.

232 Rep. of the World Heritage Committee, 31st Sess., June 23–July 2, 2007, pt. 1, U.N. Doc. WHC-07/31.COM/7.1, *available at* http://whc.unesco.org/archive/2007/whc07–31com-71e.pdf.

233 UNESCO World Heritage Committee decision WHC-06/30.COM/19 at paragraph 18;

234 Dr. William C.G. Burns, *Belt and Suspenders? The World Heritage Convention's Role in Confronting Climate Change*, 17 Southeastern Envtl. L.J. 359, 387 (2009) ("In the context of the climate change issues, then, the presumption should be that the World Heritage Committee has the authority to craft measures to protect World Heritage properties imperiled by climate change, including mandating greenhouse gas emissions reductions by major emitting States under Article 7 because these emissions constitute deliberate measures that threaten the heritage sites of other Parties under Article 6(3).").

235 *Id.*

236 *Id.*; *see also generally* Thorson, *supra* note 226.

heritage. This alone would extend much needed diplomatic pressure on co-trustee States Parties that are committing waste to common heritage assets.

Beyond diplomatic strategies, there may be available recourse on the international level in the International Court of Justice (ICJ). As Section 2 described, some jurists from that court have recognized an international principle of trusteeship among sovereign nations. Because the ICJ hears suits brought by nations against nations, it provides a forum for articulating reciprocal duties of co-tenant sovereign trustees toward the atmosphere and world heritage sites.[237] ICJ jurisdiction is significantly restrained by the fact that many nations have accepted only limited jurisdiction of the Court.[238] Nevertheless, a State Party could seek an advisory opinion from the ICJ.[239] Such an advisory opinion may draw on global traditions of equity that international law has not yet fully tapped,[240] such as the equitable use of planetary resources and a trusteeship for the benefit of future generations. In so doing, an advisory opinion could inform and inspire domestic enforcement actions worldwide aimed at forcing a reduction in carbon emissions.

5. Conclusion

Although global climate treaty processes have been exhaustive, unfortunately none has succeeded in forcing the pollution abatement necessary to protect life on Earth. In light of this dangerous vacuum of global responsibility, the judicial branches of countries worldwide must step in. The global character of climate crisis has been viewed as a detriment to legal accountability, but it can also be a great strength if the law views the

237 All state members of the United Nations are *ipso facto* parties to the Statute of the International Court of Justice. I.C.J. Statute art. 35(1).

238 John R. Crook, *The International Court of Justice and Human Rights*, 1 Nw. U. J. Int'l Hum. Rts. 2, 6 (2003).

239 *See* U.N. Charter art. 96(a) (process for advisory opinions).

240 *See generally* The International Court of Justice: Its Future Role after Fifty Years (A.S. Muller, David Raič, & J. M. Thuránszky eds., 1996).

atmosphere as held by nations in a co-tenant trusteeship. Such a framework gives rise to nondiscretionary obligations to protect the shared assets for future generations. Trust principles that have guided societies throughout our history now carry unprecedented force in climate crisis. This chapter has suggested domestic and international trust claims aimed at curbing global greenhouse gas pollution that may be brought on behalf of drowning islands.

Oliver Wendell Holmes Jr. once famously said, "A court decides the case first and determines the principle afterwards."[241] The trust provides a principled basis to force carbon emission reduction in time to thwart a planetary climate tipping point. Whether judges will invoke their authority depends on their ability to comprehend what is truly a mind-blowing ecological emergency, as well as their appreciation of the judicial capacity to prevent (in the words of one commentator) humanity's own "self-destruction."[242] Those who complain of jurisdictional hurdles and procedural defenses must ultimately come to terms with reality. If the law can provide no relief at this perilous point in human history, present and future generations will confront increasingly horrific conditions, helpless as their sources of life decline. The law's orientation has always aimed at preventing unjust harm. Although climate change is daunting, it arises from pollution, and pollution has been subject to some form of legal restraint for a very long time. As Holmes once declared, the law must reflect the "felt necessities of the time."[243]

241 O.W. Holmes, Jr., *Codes and the Arrangements of Law*, 44 Harv. L. Rev. 717, 725 (1931).

242 *See* Romm, *supra* note 93.

243 Oliver Wendell Holmes Jr., The Common Law 1 (Little, Brown & Co. 1923) (1881).

17 Transboundary Climate Challenge to Coal

One Small Step against Dirty Energy, One Giant Leap for Climate Justice

Maketo Robert, Leonito (Jun) Bacalando Jr., Jasper Teulings, Kristin Casper, Jan Šrytr, and Kristína Šabová

1. Introduction

The continuing debate on the response to climate change leaves grave uncertainty for island nations exposed to the peril of vanishing. Already island nations such as the Federated States of Micronesia (FSM) are facing environmental threats that are likely to be exacerbated by climate change. For example, in January 2011, unusual high tides destroyed food crops in FSM's Chuuk State, causing a statewide emergency. Two islands on the barrier reef of the Chuuk lagoon have completely vanished. The situation in late 2008 was even more disastrous when the tides surged across the FSM, devastating homes, public infrastructure, and food crops.[1] These events indicate the type of climate change impacts that will threaten the FSM and other small island nations in the near and long term. Coal is the largest single source of CO_2 emissions from fuel combustion in the world and is driving the climate change problem.[2]

1 These incidents are based on community observations. *See infra* note 170. For more information, please contact Director of the Office of Environment & Emergency Management Mr. Andrew Yatilman, FSM National Government, Palikir, Pohnpei State, http://www.fsmgov.org/ngovt.html.

2 International Energy Agency, CO_2 Emissions From Fuel Combustion (2011) *available at* http://www.iea.org/co2highlights/co2highlights.pdf.

Brown coal, also known as lignite, produces significantly more CO_2 per unit of energy than normal coal, oil, and natural gas.[3] The burning of coal is a significant threat to island nations.

Climate change threatens the enjoyment of a broad array of human rights.[4] Human rights obligations provide important protections to individuals who are affected by climate change. Both major greenhouse gas (GHG) emitting States and climate-vulnerable States have legal obligations to those whose rights are affected. Importantly, it can also be argued that those obligations extend extraterritorially.[5] Can international transboundary law help States ensure the fulfillment of human rights and save small island nations from disaster?

This chapter analyzes a legal procedure – transboundary environmental impact assessment (TEIA)[6] – that gives climate-vulnerable countries the ability to exert crucial influence on dirty energy projects. In a landmark intervention, in 2009 the FSM requested a TEIA of the

3 Volker Quaschning, Specific Carbon Dioxide Emissions of Various Fuels, http://www.volker-quaschning.de/datserv/CO2-spez/index_e.php (measuring CO_2 emissions from lignite at 0.36 kg/kWh, from normal coal at 0.34 kg/kWh, from fuel oil at 0.28 kg/kWh, and from natural gas at 0.20 kg/kWh) (last visited Aug. 8, 2011).

4 Office of the High Commission on Human Rights, *Report of the Office of the United Nations High Commissioner for Human Rights on the Relationship between Climate Change and Human Rights*, U.N. Doc. A/HRC/10/61 (Jan. 15, 2009), *available at* http://www2.ohchr.org/english/issues/climatechange/index.htm.

5 Submission from EarthJustice et al. to Office of High Commissioner on Human Rights, Universal Periodic Review of the Federated States of Micronesia (Apr. 12, 2010), *available at* http://earthjustice.org/sites/default/files/feature_expanded/2010/micronesia_upr_pdf_41486.pdf. Climate change results in direct threats to many human rights, such as the rights to be free from hunger and to health, subsistence livelihoods, maintain an adequate standard of living, life, property, housing, self-determination, security of person, access to freshwater, sanitation, culture and traditional knowledge, and a healthy environment. *Id.*

6 Transboundary environmental impact assessment (TEIA) is used to describe an environmental assessment procedure involving two or more states. Environmental impact assessment (EIA) is used to describe an environmental assessment procedure for a particular project or activity generally undertaken within the borders of one state based on domestic laws, and it may or may not include TEIA procedures. Strategic environmental assessment (SEA) generally refers to an assessment procedure for public plans or programs.

Prunéřov II lignite-fired power plant in the Czech Republic.[7] The FSM asserted its right to be heard as a sovereign stakeholder because the plant's emissions would contribute to potential climate impacts that would severely affect it.[8]

This procedure can be replicated by other climate-vulnerable States as a way of improving climate policy and decisions. Section 2 of this chapter examines the law of TEIA. Section 3 is a case study of FSM's request for a TEIA. Section 4 discusses the technical and legal obstacles to conducting TEIAs based on climate change concerns and suggests how to address those concerns. Finally, Section 5 evaluates the practical use of TEIA by sovereign island nations based on climate change concerns.

2. Evolution of the International Law Concerning Transboundary Pollution and Environmental Harm

This section traces the development of the no-harm principle and explores the evolution and contemporary understanding of one of its offshoots, TEIA.

2.1. No-Harm Principle

The "no-harm" principle is "one of the fundaments" of TEIA.[9] It is customary law governing State responsibility for transboundary

7 Letter from Andrew Yatilman, Director, Office of Environment & Emergency Management, to Ministry of the Environment of the Czech Republic, Request for a Transboundary Environmental Impact Assessment (EIA) proceeding from the plan for the modernisation of the Prunerov II power plant (Dec. 3, 2009), *available at* http://www.climatelaw.org/cases/country/case-documents/cz/FSM.request.TEIA.pdf [hereinafter FSM TEIA Request Letter].

8 *Id.*

9 Kees Bastmeijer & Timo Koivurova, *Transboundary Environmental Impact Assessment: An Introduction, in* THEORY AND PRACTICE OF TRANSBOUNDARY ENVIRONMENTAL IMPACT ASSESSMENT, 2 (Kees Bastmeijer & Timo Koivurova eds., 2008) [hereinafter THEORY AND PRACTICE OF TEIA].

pollution.[10] This principle was first enunciated in the 1941 *Trail Smelter Arbitration* and subsequently codified in the 1972 Stockholm Declaration and the 1992 Rio Declaration.[11]

The *Trail Smelter Arbitration* involved a Canadian smelter that emitted sulfur dioxide,[12] which allegedly caused harm to landowners down wind in the United States. The Tribunal stated in its damages award that

> [U]nder the principles of international law, as well as the law of the United States, no State has the right to use or permit the use of its territory in such a manner as to cause injury by fumes in or to the territory of another or the properties or persons therein, when the case is of serious consequence and the injury is established by clear and convincing evidence.[13]

The Tribunal put in place a regime to control and monitor future sulfur dioxide emissions from the facility[14] in the hope that it would "remove the causes"[15] of the transboundary controversy.

The adoption of Principle 21 in the 1972 Stockholm Declaration demonstrated an international consensus on a more nuanced version of the no-harm principle.[16] This authoritative formulation upholds the

10 Roda Verheyen, Climate Change Damage and International Law: Prevention Duties and State Responsibility, 146 (2005) [hereinafter Climate Change Damage and International Law].

11 Trail Smelter Arbitration (U.S. v. Can.), 3 R.I.A.A. 1911 (1941) [hereinafter Trail Smelter Arbitration], http://untreaty.un.org/cod/riaa/cases/vol_III/1905–1982.pdf; Declaration of the United Nations Conference on the Human Environment princ., 21, UN Doc. A/Conf.48/14/Rev. 1. (June 16, 1972) [hereinafter Stockholm Declaration]; Rio Declaration on Environment and Development princ. 2. UN Doc. A/CONF.151/26 (June 14, 1992) [hereinafter Rio Declaration]. The Arbitral Tribunal was constituted under the Convention between the United States of America and the Dominion of Canada signed at Ottawa, April 15, 1935, ratified by the two parties, and ratifications exchanged at Ottawa, August 3, 1935. Trail Smelter Arbitration, 3 R.I.A.A. at 1911.

12 *Id.* at 1915.

13 *Id.* at at 1965. *See also* Corfu Channel Case (U.K. v. Ir.) 1949 I.C.J. 4, 22 (April 9).

14 Trail Smelter Arbitration, 3 R.I.A.A. at 1966.

15 *Id.* at 1980.

16 Stockholm Declaration, *supra* note 11, at princ. 21; Theory and Practice of TEIA, *supra* note 9, at 3.

rights of States to exploit their own resources pursuant to their own environmental policies. But it also enforces the obligation to ensure that activities within a State's jurisdiction and control do not cause damage to the environment of other States or to areas beyond national jurisdiction.[17]

Principle 2 of the 1992 Rio Declaration is almost identical to Principle 21, except that it contains the word "developmental,"[18] reflecting the primary objective of the Rio Declaration: sustainable development.[19] It is considered to be a much more "significant statement of the law on transboundary environmental harm than the Stockholm Declaration."[20]

2.2. Transboundary Environmental Impact Assessment[21]

A patchwork of various laws – international agreements, International Court of Justice (ICJ) jurisprudence, and domestic laws – creates the obligation on States to conduct a TEIA when a proposed activity poses a risk of significant environmental harm across borders. This section provides a synopsis of the following sources of transboundary law: the Rio Declaration, UN Economic Commission for Europe (UNECE),

17 Stockholm Declaration, *supra* note 11, at princ. 21. The principle of preventive action is "closely related" to Principle 21 of the Stockholm Declaration. States are obliged to prevent damage to the environment and to reduce, limit, or control activities that might cause or risk damage. Where Principle 21 arises out of respect for the principle of sovereignty, the principle of preventive action seeks to minimize environmental damage generally and is supported by domestic and international environmental protection laws. PHILIPPE SANDS, PRINCIPLES OF INTERNATIONAL ENVIRONMENTAL LAW 246 (Cambridge Univ. Press 2003) (1995).

18 Rio Declaration, *supra* note 11, at princ. 2; PATRICIA BIRNIE, ALAN BOYLE, & CATHERINE REDGWELL, INTERNATIONAL LAW & THE ENVIRONMENT 137 (Oxford University Press, 2009) [hereinafter INTERNATIONAL LAW & THE ENVIRONMENT].

19 INTERNATIONAL LAW & THE ENVIRONMENT, *supra* note 18, at princ. 1, 4, 5, 7, 8, 9, 12, 20, 21, 24, 27.

20 *Id.* at 138.

21 This section refers to both EIA and TEIA provisions. If the provision discussed is referred to as an EIA, there is also an explanation as to whether it applies in a transboundary context.

Convention on Environmental Assessment in a Transboundary Context (Espoo Convention),[22] UN Protocol on Strategic Environmental Assessment (Protocol on SEA),[23] EU laws,[24] UN Framework Convention on Climate Change (UNFCCC),[25] Convention on Biological Diversity (CBD),[26] and the Law of the Sea Convention (LOSC), agreed to at the third meeting of the UN Convention on the Law of the Sea.[27] This synopsis is followed by a discussion of a recent ICJ case relevant to TEIA.

2.2.1. International Law

Principle 17 of the Rio Declaration is the "strongest evidence" of international acceptance of TEIA, although it does not explicitly mention transboundary assessments.[28] It states that EIAs "shall be undertaken for proposed activities that are likely to have a significant adverse impact[29] on the environment and are subject to a decision of a

22 Convention on Environmental Impact Assessment in a Transboundary Context, Feb. 26, 1991, 1989 U.N.T.S. 309 [hereinafter Espoo Convention].

23 Protocol on Strategic Environmental Assessment to the Convention on Environmental Impact Assessment in a Transboundary Context, UN Doc. ECE/MP.EIA/2003/2 (May 21, 2003) *available at* http://live.unece.org/fileadmin/DAM/env/eia/documents/legaltexts/protocolenglish.pdf [hereinafter SEA Protocol].

24 Council Directive 85/337/EEC, 1985 O.J. (L 175) (EC) [hereinafter EIA Directive], *available at* http://eurlex.europa.eu/LexUriServ/LexUriServ.do?uri=CONSLEG:1985L0337:20030625:EN:PDF; European Commission, Strategic Environmental Assessment – SEA, http://ec.europa.eu/environment/eia/sea-legalcontext.htm [hereinafter SEA Directive].

25 United Nations Framework Convention on Climate Change, May 9, 1992, S. Treaty Doc. No. 102–38, 1771 U.N.T.S. 107 [hereinafter UNFCCC].

26 Convention on Biological Diversity, June 5, 1992, 1760 U.N.T.S. 79 [hereinafter CBD].

27 United Nations Convention on the Law of the Sea, Dec. 10, 1982, 1833 U.N.T.S. 397 [hereinafter UNCLOS].

28 Rio Declaration, *supra* note 11, at princ. 17; INTERNATIONAL LAW & THE ENVIRONMENT, *supra* note 18, at 166.

29 Please note that the word "impact" and "effect" are used interchangeably in this chapter. As discussed later, the Espoo Convention is concerned with the "likely impact of a proposed activity on the environment," and impacts are defined as "any effect caused by a proposed activity on the environment." Espoo Convention, *supra* note 22, at art. 1(vi), 1(vii).

competent national authority."[30] Because sustainable development provides a backdrop for EIAs,[31] TEIA does not strictly prohibit activities that may cause harm but instead seeks to identify ways in which potential harm can be avoided and minimized. Although most national EIA laws focus on domestic impacts, transboundary and global impacts[32] are considered in the following discussion concerning the Espoo Convention, the Protocol on SEA, EU laws, the UNFCCC, the CBD, and the LOSC.

The Espoo Convention is a cross-cutting multilateral agreement that specifies the procedural rights and duties of states parties with regard to transboundary impacts of proposed activities.[33] It entered into force in September 1997 and currently has forty-four Parties, including the European Union.[34] EIA is defined as a "national procedure for evaluating the

30 Rio Declaration, *supra* note 11, at princ. 17.

31 The no-harm principle in Principle 2 of the Rio Declaration referred to "developmental policies," reflecting a new focus on sustainable development. *Id.* at princ. 2. In addition, Agenda 21, another outcome of the 1992 UN Conference of Environment and Development, makes several mentions of EIA. INTERNATIONAL LAW & THE ENVIRONMENT, *supra* note 18, at 166 (citing *Report on the UN Conference on Environment and Development*, ch. 8.4(d), UN Doc. A/CONF/ 151/26/Rev 1 (1992)).

32 The World Bank's Environmental Assessment policy was first established in Operational Directive (OD) 4.01 of 1989. It was expanded on in Operational Policy (OP) 4.01 and Bank Procedures (BP) 4.01 in 1999 and updated in March 2007. World Bank, Environment: Environmental Assessment Policy, http://go.worldbank.org/3LBMXIFF20 (lasted visited May 13, 2011). According to the Operational Policy, "EA takes into account the natural environment (air, water, and land); human health and safety; social aspects (involuntary resettlement, indigenous peoples, and physical cultural resources); and transboundary and global environmental aspects." *Id.*

33 *See* Wiek Schrage, *The Convention on Environmental Impact Assessment in a Transboundary Context*, in THEORY AND PRACTICE OF TRANSBOUNDARY EIA, *supra* note 9, at 29–51.

34 UN TREATY COLLECTION, Convention on Environmental Impact Assessment in a Transboundary Context, http://treaties.un.org/Pages/ViewDetails.aspx?src=TREATY&mtdsg_no=XXVII-4&chapter=27&lang=en (last visited June 27, 2011). The United States and Russian Federation are signatories but not parties bound to the Convention. *Id.* Participation in Espoo is currently limited to members of the UNECE. Article 17 of Espoo was amended in 2001 to allow any UN member state

likely impact of a proposed activity on the environment."[35] Appendix I contains a "list of activities," including thermal power stations, that are subject to assessment.[36] "Impacts" are defined as any effect caused by a proposed activity on the environment, including "human health and safety, flora, fauna, soil, air, water, climate, landscape and historical monuments" and other structures.[37] It also considers effects on cultural heritage and socioeconomic conditions.[38] Although the Convention does not cover transboundary impacts that are "exclusively of a global nature,"[39] according to Guidance on the Practical Application of the Espoo Convention, the Convention applies to long-range transboundary impacts, including "activities with linkages to climate change."[40] The Espoo Convention is much narrower than its subsequent companion agreement, the Protocol on SEA.

The Protocol on SEA seeks to ensure that environmental and health considerations are thoroughly taken into account in the development of plans and programs.[41] The definition of environmental and health effects includes climate-related considerations.[42] It embraces the assessment of transboundary climate change effects and requires the "development of measures to prevent, reduce or mitigate adverse effects."[43]

to accede to Espoo on approval by the Espoo parties, but that provision has not yet entered into force.

35 Espoo Convention, *supra* note 22, at art. 1(vi).

36 *Id.* at app. I, sec. 2.

37 *Id.* at art. 1(vii).

38 *Id.*

39 *Id.* at art. 1(viii).

40 UNECE, *Guidance on the Practical Application of the Espoo Convention* § 2.4.1, ¶ 26 (2006), http://www.unece.org/env/documents/2006/eia/ece.mp.eia.8.pdf [hereinafter *Espoo Guidance*].

41 SEA Protocol, *supra* note 23, at 1(a). To be considered a plan or program, it must be required by legislative, regulatory, or administrative provisions and subject to preparation and/or adoption by an authority or prepared by an authority for adoption, through a formal procedure, by a parliament or a government. *Id.* at art. 2(5)(a)–(b).

42 *Id.* at art. 2(7); *see also* Espoo Convention, *supra* note 22, at art. 1(vii).

43 UNECE, *Climate Change and the Protocol on SEA*, http://www.unece.org/env/eia/about/climate.html (last visited May 13, 2011).

Even before the formulation of the Espoo Convention, the EU required its Member States to conduct impact assessments. EU laws distinguish between EIAs for projects and activities, and strategic environmental assessments (SEAs) for plans and programs. Directive 85/337/EEC, as amended in Directive 97/11/EC and 2003/35/EC (EIA Directive), requires all Member States to institute EIA procedures for certain categories of public and private projects, and it has been adopted in each State's domestic legal system.[44] The EIA Directive sets out an obligation to initiate transboundary assessment for projects likely to affect another EU Member State.[45] In accordance with the Protocol on SEA, Directive 2001/42/EC (SEA Directive) sets out the relevant provisions for public plans and programs.[46]

International agreements concerning particular environmental issues also contain impact assessment provisions. The UNFCCC refers to EIAs as a mitigation tool.[47] It calls on States Parties to formulate domestic methods for conducting impact assessments, which should be conducted to "minimiz[e] adverse effects on the economy, on public health and on the quality of the environment, of projects or measures undertaken by them to mitigate or adapt to climate change."[48]

The year the UNFCCC was open for signature, the CBD Parties agreed to require the introduction of appropriate national procedures requiring EIA of proposed projects that are likely to have significant adverse effects on biological diversity, with a view toward avoiding or minimizing such effects and allowing public participation in such procedures.[49] The CBD issued EIA guidelines that recognize the

44 EIA Directive, *supra* note 24; *see also* Council Directive 97/11/EC, 1997 O.J. (L 073) (EC), *available at* http://ec.europa.eu/environment/eia/full-legal-text/9711.htm; Council Directive 2003/35/EC., 2003 O.J. (L 156), *available at* http://eur-lex.europa.eu/LexUriServ/LexUriServ.do?uri=OJ:L:2003:156:0017:0024:EN:PDF.

45 *Id.* at art. 7.

46 SEA Directive, *supra* note 24.

47 UNFCCC, *supra* note 25, at art. 4(1)(f).

48 *Id.*

49 CBD, *supra* note 26, at art. 14.

existence of adverse and wide-ranging effects that extend beyond the boundaries of individual habitats, ecosystems, or nations.[50]

As early as 1982, an assessment provision was included in the LOSC that requires States to assess the effects of planned activities under their jurisdiction and control when they have reasonable grounds for believing that these activities "may cause substantial pollution of or significant and harmful changes to the marine environment."[51] In 2011 an International Tribunal for the Law of the Sea (ITLOS) Seabed Dispute Chamber Advisory Opinion explained that nations may have an obligation to conduct an EIA when a proposed industrial activity may have a significant adverse impact on a shared resource in a transboundary context. The ITLOS concluded that the duty under international law to conduct an EIA "may also apply to activities with an impact on the environment in an area beyond the limits of national jurisdiction."[52] Again, this recent opinion reinforces international acceptance of TEIA in broad terms.

Although not binding on States Parties as are international agreements, the International Law Commission Draft Articles on Prevention of Transboundary Harm from Hazardous Activities (Draft Articles) draw on customary international law to address the issue of transboundary harm.[53] The Draft Articles concern activities that are permitted under the laws of the origin State but – as with GHG emissions – produce emissions whose cumulative effects may lead to "legally relevant damage" or which, in excess, are potentially unlawful.[54] The Draft

50 The guidelines recommend that environmental management plans and strategies contained in the environmental impact statement consider regional and transboundary impacts, taking into account the ecosystem approach. Convention on Biological Diversity, Guidelines on Biodiversity-Inclusive Environmental Impact Assessment (EIA) § 1.2(d)(c) (2005), http://www.cbd.int/doc/reviews/impact/EIA-guidelines.pdf.

51 UNCLOS, *supra* note 27, at art. 206.

52 Responsibilities and Obligations of States Sponsoring Persons and Entities with Respect to Activities in the Area, Case No. 17, Advisory Opinion of Feb. 1, 2011 ¶ 148, *available at* http://www.itlos.org/fileadmin/itlos/documents/cases/case_no_17/adv_op_010211.pdf.

53 Climate Change Damage and International Law, *supra* note 10, at 154.

54 *Id.* at 155.

Articles are not concerned with liability, but rather with preventing and minimizing significant transboundary harm.[55] They recommend that States seek solutions based on an equitable balance of interests of the origin State and the affected State by conducting a risk assessment through an EIA procedure.[56] The ICJ recognized the importance of risk assessment in the form of a TEIA in the *Pulp Mills* judgment.

2.2.2. ICJ Pulp Mills *Judgment*

Of the several notable ICJ judgments,[57] a recent one confirms the growing acceptance that TEIA qualifies as customary international law. In 2010, the ICJ held in *Pulp Mills on the River Uruguay* (*Pulp Mills*)[58] that general international law requires a State to conduct a TEIA where a proposed activity poses a risk of significant environmental harm in another State.[59] This holding, along with the Stockholm and Rio Declarations and other evidence of state practice, indicates that TEIA has attained customary international law status.[60] In addition, the court held

55 *Draft Articles on Prevention of Transboundary Harm from Hazardous Activities with Commentaries*, 2001 Y.B. Int'. L. Comm'n 58, art. 3 [hereinafter Draft Articles].

56 *Id.* arts. 7–9.

57 *See, e.g.* Gabcikovo-Nagymaros Project (Hung. v. Slovk.) 1997 I.C.J. 7 (Sept. 25); Legality of the Threat or Use of Nuclear Weapons, Advisory Opinion, 1996 I.C.J. 226 (July 8); Request for an Examination of the Situation in Accordance with Paragraph 63 of the Court's Judgment of 20 December 1974 in the *Nuclear Tests* (N.Z. v. Fr.), 1995 I.C.J. 288 (Sept. 22).

58 Pulp Mills on the River Uruguay (Arg. v. Uru.), Judgment (Apr. 20, 2010), *available at* http://www.icj-cij.org/docket/files/135/15877.pdf [hereinafter Pulp Mills];

59 The ICJ held that TEIAs in certain circumstances might be considered a "requirement under general international law" as demonstrated by the acceptance among states. Pulp Mills, ¶ 204.

60 *See also* International Law & the Environment, *supra* note 18, at 169–70 (stating "given the independent status of Rio Principle 17 and the evidence of state practice, the better view is simply that customary international law requires states to ensure that a transboundary EIA is carried out in appropriate circumstances. At least five ICJ or ITLOS cases have involved alleged failures to undertake a transboundary EIA. In some of these cases there are explicit treaty articles, while in others customary law or 'evolutionary interpretation' are relied on."). *See also* Panos Merkouris, *Case Concerning Pulp Mills on the River Uruguay (Argentina v. Uruguay): Of Environmental Impact Assessments and "Phantom Experts,"* Hague Justice Portal (July 15, 2010), http://www.haguejusticeportal.net/eCache/DEF/11/878.html (stating,

that the obligation to prevent pollution and preserve the environment demands the exercise of due diligence, which entails careful consideration of the technology used in the relevant sector for the proposed activity.[61]

The dispute revolved around Uruguay's construction of two pulp mills on the River Uruguay, a border and shared natural resource with Argentina.[62] Under the 1975 Statute of the River Uruguay,[63] Argentina and Uruguay established a commission to resolve disputes.[64] Because the States were not able to reach an agreement, Argentina invoked the compulsory jurisdiction of the ICJ and initiated dispute proceedings.[65] Argentina's application alleged that Uruguay committed multiple breaches of its obligations under the Statute[66] by authorizing, constructing, and further commissioning two pulp mills on the River Uruguay.[67] Argentina sought a court order requiring Uruguay to cease these breaches and pay compensation for the damage caused by the alleged internationally wrongful acts.[68] Whereas Argentina based its application on breaches of the 1975 Statute, the ICJ issued a judgment that emphasized the customary international law status of several norms – in particular TEIA and best available techniques (BAT).

"[w]hat this amounts to is an *expressis verbis* acknowledgement, within the text of an ICJ judgment, that the duty to undertake an EIA when there is a risk of pollution with transboundary elements has achieved customary international law status.").

61 Pulp Mills,¶ 223.

62 *Id.* ¶ 25; *see also* Donald K. Anton, *Case Concerning Pulp Mills on the River Uruguay (Argentina v. Uruguay)* pt. 1 (Aus. Nat'l U. Coll. of Law Research Paper No. 10–84, 2010), *available at* http://papers.ssrn.com/sol3/papers.cfm?abstract_id=1705810.

63 Statute of the River Uruguay, Arg.-Uru., Feb. 26, 1975, 1295 U.N.T.S. 340.

64 *Id.* ch. XIII, art. 49.

65 Pulp Mills,¶ 1.

66 Statute of the River Uruguay, *supra* note 63.

67 *Pulp Mills on the River Uruguay*, ¶ 22(1)(a)–(e); *see also* Press Release, Summary of the Judgment of 20 April 2010, Pulp Mills on the River Uruguay (Argentina v. Uruguay) (April 20, 2010), *available at* http://www.icj-cij.org/docket/files/135/15895.pdf.

68 Pulp Mills, ¶ 22(4), 24(2)(iv).

2.2.2.1. Obligation to Conduct Transboundary Environmental Impact Assessments. The ICJ considered whether the Parties properly complied with their obligations to prevent pollution and preserve the aquatic environment with respect to activities that may cause transboundary harm. The Parties agreed that it was necessary to conduct a TEIA, but they disagreed on its scope and content.[69] The court held that Uruguay was required to conduct a TEIA not only under the 1975 Statute but also under general international law.[70] However, the court did not resolve the disagreement regarding the scope and content of the TEIA.

There are three components to the ICJ's holding. First, the court considered that the 1975 Statute of the River Uruguay ought to be interpreted by taking into account "developments in international law."[71] Second, it held that "where there is a risk that the proposed industrial activity may have a significant adverse [environmental] impact in a transboundary context," an EIA is required under general international law.[72] This is an important statement of the conditions under which an EIA must be conducted. Further, the court noted that a State would not be acting with due diligence if it did not undertake an EIA on the potential effects when there is a risk that the proposed industrial activity may have a significant adverse impact.[73] The judgment did not define what would constitute a "significant adverse impact" (see Section 4.3 for a detailed discussion on the meaning of "significant"). Finally, the court did not take the opportunity to comment on the international requirements for the scope and content of the TEIAs because it found that this is

69 *Id.* at ¶ 203.

70 *Id.* at ¶ 204.

71 *Id.* (citing Dispute Regarding Navigational and Related Rights (Costa Rica v. Nicar.), Judgment, ¶ 64 (July 13, 2009), *available at* http://www.icj-cij.org/docket/files/133/15321.pdf).

72 *Id.*

73 *Id.* (stating, "Moreover, due diligence, and the duty of vigilance and prevention which it implies, would not be considered to have been exercised, if a party planning works liable to affect the régime of the river or the quality of its waters did not undertake an environmental impact assessment on the potential effects of such works.")

a decision that should be left to a State's discretion.[74] Note that the court did not apply the Espoo Convention because Uruguay and Argentina were not parties to it. In addition, the court noted that the UN Environment Programme (UNEP) EIA goals and principles are mere soft laws and provide little guidance, except for Principle 5, which recommends that "environmental effects in an EIA should be assessed with a degree of detail commensurate with their likely environmental significance."[75] The court did, however, clarify that a TEIA should be conducted before implementation of a project.[76]

2.2.2.2. Obligation to Employ Best Available Techniques. The ICJ also considered whether Uruguay must require the pulp mills to use BAT in order to comply with its 1975 Statute obligation to prevent pollution and protect and preserve the aquatic environment. The court noted that Article 41(a) provided that the Parties had to adopt a regulatory framework in keeping with the guidelines and recommendations of international bodies on BAT.[77] The court appeared to agree with Argentina's claim that, to comply with its obligations, Uruguay's domestic law must require that the pulp mills use BAT.[78] This may be interpreted as supporting the notion that States have an obligation to use BAT to fulfill their obligation of due diligence in protecting other States from transboundary environmental harm.

This recent judgment signals that TEIA of proposed industrial projects has matured into an international customary law requirement.

74 *Id.* ¶ 205.

75 UN Env't Programme, Goals and Principles of Environmental Impact Assessment, pt. 2, princ. 5 (1987), *available at* http://www.unep.org/regionalseas/publications/reports/RSRS/pdfs/rsrs122.pdf; *see also* Rep. on the UN Env't Program, 14th Sess., June 8–19,1987, at 77–78, UN Doc. A/42/25, Supp. No. 25 (1987) (adopting the Goals and Principles of Environmental Impact Assessment), *available at* http://www.unep.org/resources/gov/prev_docs/87_06_GC14_report_N8723250.pdf.

76 *Pulp Mills*, ¶ 205.

77 *Id.* ¶ 223.

78 *See id.*

In the near future the ICJ may analyze the legal responsibilities emanating from the transboundary impacts of climate change. For example, the President of the Republic of Palau announced in 2011 that, together with the Republic of the Marshall Islands, his country will call on the 193-member General Assembly to urgently seek an ICJ advisory opinion "on whether countries have a legal responsibility to ensure that any activities on their territory that emit greenhouse gases do not harm other States."[79] An advisory opinion in this regard furnishes ICJ an opportunity to elaborate and clarify the *Pulp Mills* judgment, among other issues, and to consider extending its application in view of the transboundary nature of climate impacts. Further, in the opinion of the authors, this request for an advisory opinion indicates a progressive development of international transboundary law to include climate change considerations.

2.3. Domestic Environmental Assessment Laws

Environmental assessment is also embraced at the national level. The first EIA provisions became law in 1970 as a part of the U.S. National Environmental Policy Act.[80] Two years later, EIA was discussed at the international level at the Stockholm Conference.[81] It is estimated that more than 100 countries have EIA legislation;[82] this includes both developing and developed countries.[83]

79 United Nations News Centre, *Palau Seeks UN World Court Opinion on Damage Caused by Greenhouse Gases*, United Nations (Sept. 22, 2011), http://www.un.org/apps/news/story.asp?NewsID=39710&Cr=pacific+island&Cr1=#.

80 National Environmental Policy Act of 1969, Pub.L. 91–190, 83 Stat. 852 (1970).

81 Neil Craik, The International Law of Environmental Impact Assessment: Process, Substance, and Integration 23 (2008).

82 *Id.*

83 B. Ganeshkumar Gobinath, *Modified EIA for Small and Medium Projects: An Effective Method to Make the Projects Eco-Friendly*, 1 Int'l J. Envtl Sci. 55, 57 (2010), *available at* http://ipublishing.co.in/jesvol1no12010/EIJES1005.pdf.

Impact assessments differ at the domestic and international levels. The international provisions stem from the no-harm principle and other international norms; they therefore focus on transboundary notification, disclosure, and consultation. National provisions, by contrast, primarily assess impacts within state borders. However, there are examples of national regimes that promote transboundary cooperation opportunities. The next section of this chapter discusses one example, from the Czech Republic, that was used by the FSM with some degree of success.

3. The FSM's Request for a TEIA of the Prunéřov II Power Plant

In December 2009, the FSM requested a TEIA concerning the Prunéřov II coal-fired power plant in the Czech Republic,[84] asserting its right to be heard as a sovereign stakeholder due to potential climate impacts of the plant's emissions.[85] It was the first "transregional"[86] use of transboundary EIA. This section presents an overview of the legal basis for the FSM's participation and the development of the case.

3.1. Participation of the FSM in the Transboundary EIA under Czech Law

The Czech EIA law uses a broader definition than the EU EIA Directive of parties eligible to initiate and participate in a transboundary EIA procedure. Whereas the EU EIA Directive applies where there is a potential of causing significant environmental effects "in another Member State,"[87] the Czech EIA law applies to any State whose "territory can be affected by significant environmental impacts."[88] Thus the Czech

84 *See* FSM TEIA Request Letter, *supra* note 7.

85 *Id.*

86 For the purposes of this chapter, "transregional" means across geographically distinct areas or regions of the world.

87 EIA Directive, *supra* note 24, at art. 7(1).

88 Zákon č.100/2001 Sb. art. 11(1)(b) (Czech) [on Environmental Impact Assessment and Amending Some Related Acts], *translated in* http://www.mzp.cz/ris/vislegcz.nsf/

TEIA provisions allow States outside the EU to initiate and participate in a TEIA.

The Czech EIA law imposes a duty on the Czech Ministry of Environment (Ministry)[89] that when it determines a TEIA is needed, or if an affected State requests one, it must send notification together with information on the course of the assessment and on related decisions to the affected State for its review.[90] The affected State can express its intention to participate in the TEIA and provide its viewpoint on the impacts of the plan.[91] The Ministry is then obliged to incorporate the participating State's viewpoint into the final EIA statement or provide reasons why it did not do so.[92] It is also obliged to send the final statement to the affected State within fifteen days of its issuance.[93]

According to Article 3 of the EU EIA Directive, the EIA "shall identify, describe and assess . . . the direct and indirect effects of a project on . . . soil, water, air, climate and the landscape."[94] Czech law incorporates this legal requirement in Article 2 of its EIA law. In addition, Annex No. 4 to the Czech EIA law requires a project's developer or investor to include, as an obligatory part of a project's documentation, an assessment of the project's climate impact.[95]

3.2. Development of the Case

In 2008 the Czech energy company, CEZ Group (CEZ), submitted to the Ministry notification of its plan for the comprehensive refurbishment

1C2B61C81B7D0323C1256DE10060892E/$file/2001100Sb_kv.pdf [hereinafter Czech EIA law].

89 *Id.* at art. 11(2).

90 *Id.* at art. 13(1).

91 *Id.* at art. 13(3).

92 *Id.* at art. 13(5).

93 *Id.* at art. 13(6).

94 EIA Directive, *supra* note 24, at art 3.

95 Czech EIA Act, *supra* note 87, at art. 2, ann. 4, pt. D I.2 [Impacts on the air and climate].

of its largest coal-fired power plant complex, Prunéřov.[96] The complex is the largest single industrial source of CO_2, SO_2, and NO_x emissions in the Czech Republic, emitting more than nine million tons of CO_2 annually.[97] CEZ proposed to replace three existing blocks in the larger part of the complex – Prunéřov II – and prolong its life for another twenty-five years.[98] According to the EIA documentation presented by CEZ, the new blocks were designed to reach a minimum net energy efficiency of about 38%.[99]

The Ministry began the EIA proceedings and required CEZ to consider alternatives with higher net efficiencies, between 42% and 45%, in accordance with BAT requirements.[100] CEZ, however, did not propose any alternatives and insisted on developing the project without substantial changes.[101]

96 Jan Horák, Oznámení záměru Komplexní obnova elektrárny Prunéřov II 3 × 250 MW_e dle zákona č. 100/2001 Sb., v platném znění, [Notification of a plan for the comprehensive refurbishment Prunéřov II 3 x 250MWe], 2008 *available at* Informační systém, MZP221___dokumentace.zip, http://tomcat.cenia.cz/eia/detail.jsp?view=eia_cr&id=MZP221.

97 Ministerstvo Životního Prostředí, Integrovaný Registr Znečisťování Životního Prostředí: Souhrnná Zpráva Za Rok 2009 [Annual Report 2009 of The Integrated Register of Pollution] 44, 45, 81 (2011), *available at* http://www.irz.cz/sites/default/files/Souhrnna_zprava_IRZ_2009.pdf.

98 Jan Horák, Dokumentace záměru Komplexní obnova elektrárny Prunéřov II 3 × 250 MWe dle zákona č. 100/2001Sb., v platném znění [EIA documentation on the project Complete renewal of Prunéřov II 3 × 250 MWe power plant]15 (2008) *available at* Informační systém EIA, MZP2211___dokumentace.zip http://tomcat.cenia.cz/eia/detail.jsp?view=eia_cr&id=MZP221 [hereinafter EIA Documentation].

99 EIA Documentation, *supra* note 97.

100 *Id.* According to the European Commission reference document on BAT for Large Combustion Plants, any new power plant should be equipped with technology that allows it to be operated between 42% and 45% net energy efficiency. European Commission, *Intergrated Pollution Prevention and Control: Reference Document on Best Available Techniques for Large Combustion Plants*, LCP (2006), 11, 268–269, *available at* http://ftp.jrc.es/eippcb/doc/lcp_bref_0706.pdf.

101 Ministerstvo Životní prostředí, Stanovisko k posouzení vlivů provedení záměru na Životní prostředí, [Ministry of the Environment, Statement on the Environmental Impact Assessment of Implementing the Plan, Comprehensive refurbishment of the Prunéřov II thermal power plant] 20, (Apr. 29, 2010). *available*

On December 3, 2009, the FSM Office of Environment and Emergency Management (OEEM) wrote to the Ministry to request a TEIA of the plan for the refurbishment of the Pruněřov II power plant.[102] The FSM OEEM made the request because its territory could be affected by the potential climate impacts caused by the plant's CO_2 emissions.[103] On December 21, 2009, the FSM OEEM received a letter from the Ministry that stated that comments on the project are "welcomed" and that its suggestions would be "employed and taken into consideration in issuing the statement or in further stages of the approval process."[104]

On January 4, 2010, the FSM OEEM submitted its viewpoint on the plan to the Ministry, expressing its concerns about the project and providing information on the state of the environment in the affected territory of the FSM.[105] The FSM OEEM also pointed out that the proposed plan failed to reach the minimum 42% net energy efficiency for a new power plant and that this would result in higher emissions of CO_2.[106] In addition, the climate impacts of the plan were not assessed in the EIA documentation or in the expert report to the EIA documentation.[107] The FSM OEEM posited that the aforementioned failures made the plan in substantial conflict with EU and Czech law

at Informační systém MZP2211___zaveryStan.zip, http://tomcat.cenia.cz/eia/detail.jsp?view=eia_cr&id=MZP221 [hereinafter EIA Statement].

102 FSM TEIA Request Letter, *supra* note 7.

103 *Id.*

104 Letter from Karel Blaha, Deputy Minister, Director General of the Directorate of Technical Protection of Environment the Ministry of the Environment of the Czech Republic, to Andrew Yatilman, Director, the Office of Environment & Emergency Management, (Dec. 21, 2009).

105 Letter from Andrew Yatilman, Director, the Office of Environment & Emergency Management, to Karel Blaha, Deputy Minister, the Ministry of the Environment of the Czech Republic, Viewpoint of the Federated States of Micronesia on the Complex Renovation of Pruněřov II Power Plant 3x250 MWe plan (Jan. 4, 2010), *available at* http://www.pohodacez.cz/_files/file/Viewpoint%20of%20FSM%20on%20renovation%20of%20Prunerov%20II%2 Plant.pdf.

106 *Id.* at 2.

107 *Id.* at 1.

and therefore asked the Ministry to issue a negative final statement on the EIA.[108]

In February 2010, the Ministry asked the Norwegian consulting company Det Norske Veritas (DNV) for an international third-party assessment of the project.[109] DNV concluded that the proposed project failed to comply with the Reference Documents for Best Available Techniques (BREF) minimal requirement of 42% net energy efficiency and that the corresponding local technical grounds to support this deviation had not been sufficiently explored. The difference was calculated to have an impact of 205,082 tons of CO_2 per annum (p.a.).[110]

On April 29, 2010, the Ministry[111] issued an affirmative EIA statement on the CEZ project.[112] The statement mentioned that the FSM had submitted its viewpoint and summarized the fact that it expressed disagreement with the climate protection and BAT compliance elements of the plan. The Ministry responded that the plan cannot seriously affect the environment and populations outside the Czech Republic because its realization will reduce the current emissions of major air pollutants[113]

108 *Id.* at 5.

109 DNV Climate Change Svcs., Third Party Assessment of the Comprehensive Refurbishment of the Prunéřov II Power Plant: Draft Report for the Ministry of the Environment of the Czech Republic, 3 (Mar. 17, 2010) [hereinafter DNV], *available at* http://www.mzp.cz/C1257458002F0DC7/cz/news_TZ_100318_DNV/$FILE/100317%20Final%20Report%20-%203rd%20Party%20Assessment%20DNV.pdf.

110 *Id.* at 3, 36, 112, 114.

111 The Czech government intervened in the process and called for approval of the project, and the Czech Prime Minister insisted on a quick decision. Under immense pressure, Czech Environment Minister Jan Dusik resigned on March 18, 2010, rather than approve CEZ Group's EIA. Rob Cameron, *Minister Resigns after Pressure from PM over Controversial Power Plant*, Radio Praha (Mar. 19, 2010), http://www.radio.cz/en/section/curraffrs/minister-resigns-after-pressure-from-pm-over-controversial-power-plant. Rut Bízková took his position. She previously worked as a spokeswoman for the CEZ Group in the 1990s. Tom Clifford & Gabriella Hold, *Compromised Minister Set to Approve Coal Plant Expansion*, The Prague Post (Apr. 28, 2010), http://www.praguepost.com/news/4272-compromised-minister-set-to-approve-coal-plant-expansion.html.

112 EIA Statement, *supra* note 100.

113 Such as particulate pollutants, sulfur dioxide, and nitrogen oxides.

and because the CO_2 emissions of the plant are marginal compared to global emissions. However, the Ministry did agree with the FSM that the plan should be considered as a new installation governed by the BAT minimum net energy efficiency requirement of 42%. Because this requirement would not be met by CEZ's plan, the Ministry requested that CEZ propose compensation measures to offset the additional emissions of 205,082 tons of CO2 p.a. at its other projects.[114]

3.3. Assessment of the Climate Impacts of the Plan

A properly conducted EIA should consider climate impacts. The documentation issued by CEZ did not follow this requirement and did not provide an assessment of the plan's climate impacts. At the request of the Ministry,[115] CEZ supplemented the documentation with a separate expert's report prepared by the Czech climatologist Jan Pretel.[116] The report states that GHG emissions of the plant are marginal and unable to cause serious environmental impacts such as rising sea level or melting glaciers.[117] This Ministry accepted this conclusion.[118]

Dr. Pretel's conclusion is based on a comparison of the isolated GHG emissions of the single plant to global GHG emissions. In our opinion this is misleading. Neither Czech nor EU laws provide clear guidance on either the GHG emission threshold that should trigger an analysis

114 EIA statement, *supra* note 100, at 21.

115 Letter from Jaroslava Honová, the Ministry of the Environment of the Czech Republic, to ČEZ a.s., (Mar. 9, 2009) *available at* Informační systém EIA, MZP2211___vraceni.pdf, http://tomcat.cenia.cz/eia/detail.jsp?view=eia_cr&id=xMZP221.

116 JAN PRETEL, ZNALECKÝ POSUDEK K ČÁSTI VYJÁDŘENÍ GREENPEACE ZE DNE: 11. 2. 2009 K DOKUMENTACI VLIVŮ NA ŽIVOTNÍ PROSTŘEDÍ ZÁMĚRU KOMPLEXNÍ OBNOVA ELEKTRÁRNY PRUNÉŘOV II 3 x 250 MWE 2 (Feb. 19, 2009), at 4. *available at* Informační systém EIA, MZP2211___posudek.zip, open Příloha 3, http://tomcat.cenia.cz/eia/detail.jsp?view=eia_cr&id=MZP221 [hereinafter PRETEL].

117 *Id.* at 3. Prunéřov II's share of global CO_2 emissions is only 0.021%, which is 0.0161% of global emissions of greenhouse gases.

118 EIA Statement, *supra* note 100, at 13.

of an individual project's impact on the climate or on the requirements for mitigation measures. Such a threshold should, however, be based on a comparison of the size of the GHG source relative to other sources, rather than to global GHG emissions. To do otherwise would lead to the absurd conclusion that there is no reason to reduce the GHG emissions of *any* single source of GHG emissions because of its "marginality." The same argument could consequently be used to reject emission reductions for entire States: there are 112 UN Member States with total annual CO_2 emissions lower than the Prunéřov complex.[119] It is highly plausible that the projected GHG emissions from CEZ's plan to refurbish the Prunéřov complex would meet any reasonable threshold triggering a consideration of climate impacts and mitigation measures, especially considering that it is the single largest contributor of GHG emissions in the country.

In addition, because of the complex nature of climate change, it is impossible to establish direct causation between a particular source of emissions and its impact on specific places. Therefore, it is the *cumulative* impact of CO_2 emissions that should be considered when determining the "significance" of environmental impacts (see Section 4.3 for an in-depth discussion on this point).[120]

The Czech Ministry was progressive and innovative in its interpretation of "affected state" and "significant environmental impacts" when accepting FSM's request for a transboundary EIA. Implicitly, the Ministry recognized that the territory of the FSM may be affected by the coal-fired power plant and thus recognized its status as an "affected state" as defined under Czech law. However, in the final stages of the proceeding,

119 Data comparison from 2007. United Nations, *UN Millenium Development Goals Indicators*, *available at* http://mdgs.un.org/unsd/mdg/SeriesDetail.aspx?srid=749&crid= (last updated June 23, 2010).

120 *See generally* Memorandum from Nancy H. Sutley, Chair, Council on Environmental Quality, to Heads of Federal Departments and Agencies on Draft NEPA Guidance on Consideration of the Effects of Climate Change and Greenhouse Gas Emissions (Feb. 18, 2010), *available at* http://ceq.hss.doe.gov/nepa/regs/Consideration_of_Effects_of_GHG_Draft_NEPA_Guidance_FINAL_02182010.pdf [hereinafter NEPA Guidance].

the Czech authorities ignored their legal duty to formally communicate the results of the EIA to the FSM.

In summary, the Ministry reasonably recognized and acted on FSM's concerns expressed during the TEIA. An expert assessment of impacts of the plan on climate was provided within the frame of the EIA procedure and the Ministry evaluated FSM's arguments within the final EIA statement. Although the Ministry concluded (erroneously, in our opinion) that CO_2 emissions from the plant could not cause any significant transboundary effect due to the "marginality" of the CEZ plan in terms of global GHG emissions,[121] it did confirm that the energy efficiency requirement was not met and therefore required CEZ to offset the additional emissions.

The success of FSM's intervention lies in creating a unique example of state TEIA practice between non-neighboring countries, specifically, between a small island nation and an EU Member State. It is also a pioneering example of extending the traditional use of TEIAs (e.g., particulate air pollution or water pollution) to the assessment of climate change impacts. Though very progressive, such state practice is in accordance with the EU law and the Protocol on SEA; furthermore, such practice lends support to the applicability of the Espoo Convention to activities that contribute to climate change.[122]

4. Transboundary EIAs and Greenhouse Gas Emissions

Despite the customary international law requirement to conduct TEIAs in appropriate circumstances,[123] a TEIA is not typically conducted for proposed activities, projects, plans, or programs that will emit GHG emissions potentially affecting a country located thousands of miles away

121 Pretel, *supra* note 115, at 4. For a critique of this classification, see discussion *supra* Section III(B).

122 *See* EIA Directive, *supra* note 24; SEA Directive, *supra* note 24; SEA Protocol, *supra* note 23; Espoo Convention, *supra* note 22.

123 *See* section II(B)(2).

from the source of emissions. However, such a process is grounded in international law. This section discusses three obstacles to transregional TEIAs concerning climate change impacts and presents legal bases for overcoming these obstacles, drawing on the sources of law discussed in Section 2 and the example of state practice in Section 3 on how the law can be interpreted and applied. The conclusions drawn in this section remain unsettled and are open for debate.

4.1. Can Non-Neighboring Countries Trigger a TEIA Process?

The determining factor for triggering a TEIA procedure is the potential transboundary impact and not the location of the parties.[124] Wide-ranging impacts that span multiple countries, or are even of a global nature, may trigger the notification, information sharing, and consultation obligations under the Espoo Convention and the Protocol on SEA.[125] The transboundary provisions in the EU EIA and SEA Directives apply to all EU Member States.[126] The ILC Draft Articles specifically state that a common border is not necessary.[127]

As the ICJ held in *Pulp Mills*, TEIA applies to shared resources.[128] Although the legal status of the climate system is undefined,[129] the UNFCCC website states that the "atmosphere is a shared resource, part of the 'global commons.'"[130] Increasingly, impact assessments consider the climate system as a shared resource. One reason to view the climate

124 *Id.* 1(vi). *But see* Theory and Practice of TEIA, *supra* note 9 at 34 (explaining that a central consideration as to the applicability of Espoo Convention is the "likely area of impact relative to the border").

125 *See* sec. II(B)(1).

126 EIA Directive, *supra* note 22, at art. 7(1); SEA Directive, *supra* note 23, at art. 7(1).

127 Draft Articles, *supra* note 55, art. 2(c).

128 Pulp Mills, ¶ 204.

129 International Law & the Environment, *supra* note 18, at 336.

130 *Understanding Climate Change: A Beginner's Guide to the UN Framework Convention*, UNFCCC (Dec. 1994) http://unfccc.int/essential_background/background_publications_htmlpdf/items/1661.php.

system as a shared resource is its limited capacity to bear increasing GHG emissions.[131]

Thus, if the climate system is considered a shared resource, a non-neighboring State can engage in TEIA procedures based on climate change impacts. Even if the climate system is not considered a shared resource, the TEIA remains a viable option, although the non-neighboring State may have a greater burden in justifying its transboundary participation.

4.2. Do GHG Emissions Have a Transboundary Effect?

To trigger the application of TEIA provisions, there must be some type of transboundary effect. This requirement raises two potential obstacles in terms of GHG emissions.

First, there is the issue of physical flow – the movement of a specific pollutant from one State to another. Specifically, the issue is whether GHGs have transboundary effects even though the physical flow of greenhouse gas emissions from one State to another is not directly identifiable. Some argue that the no-harm principle originates from the concept of transboundary pollution and "assumes that a cause-and-effect chain between two States is clearly discernible."[132] Unlike demonstrating causation in a domestic tort case[133] or a claim for damages under international law,[134] TEIA requires the assessment and identification of

131 Climate Change Damage and International Law, *supra* note 10, at 165.

132 *Id.* at 166.

133 Negligence generally consists of four elements: duty, breach of duty, causation, and damages. Black's Law Dictionary 1063 (8th ed. 2004).

134 "A claim for damages under international [law] must involve the following steps: (i) Identifying the damaging activity attributable to a state, (ii) establishing a causal link between the activity and the damage, (iii) determining either a violation of international law or a violation of a duty of care (due diligence), which is (iv) owed to the damaged state. Step (v) in a court of law would be to quantify the damage caused and relate it back to the activity. While this may be doable, we focus on the first four items, as they are sufficient to establish general state responsibility." S. J. Tol et al., *State Responsibility and Compensation for Climate*

likely transboundary harm. However, international agreements concerning TEIA indicate that physical flow is not a requirement, and thus it can be argued that transboundary effect includes globally and locally felt impacts of climate change.[135] Although it is difficult to determine the cause-and-effect relationship between GHG emissions and harm caused to a particular country or a "specified area of likely impact,"[136] scientific understanding is improving and increasingly able to demonstrate that "climate change is acting here and now to load the dice towards more extreme weather."[137]

Second, there is the issue of attribution. Climate change is part of a complex process.[138] The accumulation of GHGs in the atmosphere causes climate change and may lead to impacts such as air temperature increase and sea level rise.[139] Therefore, these impacts have a "transboundary effect." As long as state contributions are distinguishable, TEIAs are relevant.[140] It should be noted that the complex environmental change caused by GHG emissions is comparable to changes caused by long-distance air pollution. The Convention on Long-Range Transboundary Air Pollution (LTRAP) defines long-range transboundary air

Change Damages: A Legal and Economic Assessment, 32 ENERGY POL'Y 1109, 1111 (2004).

135 CLIMATE CHANGE DAMAGE AND INTERNATIONAL LAW, *supra* note 10, at 166.

136 THEORY AND PRACTICE OF TEIA, *supra* note 9, at 34.

137 *Oxford University Researchers Say Climate Change and Floods Linked*, GREEN ECONOMY (Mar. 1, 2011), http://uk.ibtimes.com/articles/20110301/oxford-university-researchers-say-climate-change-and-floods-linked.htm (quoting Myles Allen). *See also* Pardeep Pall et al., *Anthropogenic Greenhouse Gas Contribution to Flood Risk in England and Wales in Autumn 2000*, 470 NATURE 382 (2011), *available at* http://www.nature.com/nature/journal/v470/n7334/full/nature09762.html.

138 DAVE OWEN, *Climate Change and Environmental Assessment Law*, 33 COLUM. J. ENVTL. L. 57, 60 (2008).

139 Tol, et al, *supra* note 133, at 1112.

140 CLIMATE CHANGE DAMAGE AND INTERNATIONAL LAW, *supra* note 10, at 167 (citing ILA, 64th Conference, Queensland, Aus., 1990 at 282 f). Such cumulative emissions had been excluded from the application before because the contributions on individual countries could not be ascertained in fact. ILA, 60th Conference, Montreal, Can., 1982, at 159).

pollution as that which originates from one State "at such a distance that it is not generally possible to distinguish the contribution of individual emission sources or groups of sources."[141] As with climate change, attribution has been an issue with long-range air pollutants where all States contribute to the problem to varying degrees.[142] Yet, as with long-distance air pollution, the cumulative contribution of States to the problem of climate change should not prevent the application of TEIA laws.

4.3. Are GHG Emissions Likely to Have a Significant Transboundary Impact?

To determine whether GHG emissions are likely to have a significant transboundary impact, a fact-specific determination of the terms *likelihood* and *significance* must be conducted.

4.3.1. Likelihood

The threshold to trigger the assessment of the transboundary effects of GHG emissions should be very low for three reasons: the general notification requirements, the guidance provided by the *Pulp Mills* judgment concerning the dispute between Uruguay and Argentina, and the application of the precautionary principle.

The general understanding is that notification is required when there is a *possibility* of significant adverse transboundary impacts, no matter how uncertain.[143] The practice of the parties in *Pulp Mills* suggests that when a large-scale industrial activity presents a known risk of potentially significant pollution, a TEIA should be conducted "even if the likely risk is a small one."[144]

141 Convention on Long-Range Transboundary Air Pollution art. 1(b), Nov. 13, 1979, 1302 U.N.T.S. 217 [hereinafter LTRAP], *available at* http://www.unece.org/env/lrtap/full%20text/1979.CLRTAP.e.pdf [hereinafter LTRAP].

142 *See* International Law & the Environment, *supra* note 18, at 342.

143 Theory and Practice of TEIA, *supra* note 9, at 35.

144 International Law & the Environment, *supra* note 18, at 171.

The precautionary principle, which is now widely recognized as part of customary international law,[145] aids in the determination of likelihood. Related to the no-harm principle, the precautionary principle requires that, where there are reasonable grounds for concern that serious or irreversible damage might occur, effective and proportional measures should be taken to prevent environmental degradation even in the absence of scientific certainty.[146] The Rio Declaration put the global spotlight on the principle.[147] The UNFCCC requires parties to "take precautionary measures to anticipate, prevent or minimize the causes of climate change and mitigate its adverse effects."[148]

There is little doubt that ever-increasing CO_2 levels in the atmosphere and the associated changes in climate are producing adverse impacts.[149] Uncertainty lies in determining the specific impacts on an affected country or locality. Doing so requires sophisticated modeling, and we must recognize our limited ability to accurately predict climate change impacts. Nevertheless, we should assume that conducting TEIAs of large-scale coal projects, plans, and programs is necessary in light of

145 Arie Trouwborst, *The Precautionary Principle in General International Law: Combating the Babylonian Confusion*, 16 Rev. of Eur. Community & Int'l Envtl L. 185, 187–88 (2007).

146 *Id.* at 188.

147 Rio Declaration, princ. 15 ("In order to protect the environment, the precautionary approach shall be widely applied by States according to their capabilities. Where there are threats of serious or irreversible damage, lack of full scientific certainty shall not be used as a reason for postponing cost-effective measures to prevent environmental degradation.").

148 UNFCCC, *supra* note 25, at art. 3(3).

149 Richard B. Alley et al., *Summary for Policymakers*, *in* Climate Change 2007: The Physical Science Basis: Contribution of Working Group I to the Fourth Assessment Report of the Intergovernmental Panel on Climate Change 5 (Susan Solomon et al. eds., 2007) ("Warming of the climate system is unequivocal, as is now evident from observations of increases in global average air and ocean temperatures, widespread melting of snow and ice, and rising global average sea level.") [hereinafter *Summary for Policymakers*].

the particular threats that small island nations face,[150] such as sea level rise.[151]

4.3.2. Significant Impact

To the extent that TEIAs are required, this obligation does not apply to minor or transitory impacts[152] but rather to those that are significant, serious, substantial, disastrous, or irreversible. There are several different approaches to determining significance.

First, the Draft Articles require the potential harm to be significant.[153] To be "significant," the impact must be "more than 'detectable' but need not be at the level of 'serious' or 'substantial.'"[154] Two types of risk fall within this definition: risks with a high probability of causing significant transboundary harm (e.g., industrial air pollution) and risks with a low probability of causing disastrous transboundary harm (e.g., nuclear power plant meltdown).[155] Activities that increase GHG emissions should be covered under this definition because the accumulation of GHGs, which causes climate change, already has, and will increasingly have, devastating consequences worldwide and in particular for small island nations.

150 *See infra* Section 5 for further discussion on the use of TEIA procedures by small island nations.

151 The Arctic Monitoring and Assessment Programme (AMAP), a working group of the Arctic Council, released an assessment report on the impacts of climate change on Snow, Water, Ice, and Permafrost in the Arctic [hereinafter AMAP]. One of the primary findings is that global sea level is projected to rise by 0.9 to 1.6 m (2.95 to 5.25 ft) by 2100. Arctic Monitoring and Assessment Programme, Snow, Water, Ice, and Permafrost in the Arctic: Executive Summary 11 (2011), *available at* http://amap.no/swipa/.

152 International Law & the Environment, *supra* note 18, at 171.

153 Draft Articles, *supra* note 55, at art. 1, 2(a),

154 *Id.* at art. 2, cmt. (4).

155 International Law & the Environment, *supra* note 18, 141–42 (citing Draft Articles, *supra* note 55, at art. 2, and commentary, ILC Report(2001) GAOR A/56/10, 387 paras (3) & (4)).

A second method of determining significance is to categorize types of activities. The Espoo Convention facilitates the process by listing the activities, projects, and plans subject to TEIA requirements.[156]

A third option is for laws to explicitly state that climate change has a significant impact worthy of assessment. Domestic EIA laws and regulations are helpful in determining whether GHG emissions rise to a level of being significant. The Czech EIA law requires investors to include a climate change assessment in the project documentation.[157] The U.S. Council on Environmental Quality (CEQ) regulations state that a finding of significance "requires considerations of both context and intensity."[158] "Context" analyzes the impact of the activity in various settings, including its short- and long-term effects.[159] "Intensity" refers to the severity of the impact, including whether the action is individually insignificant but has cumulatively significant impacts.[160] Interpreting the laws and regulations, U.S. courts have required the consideration of GHG emissions.[161] Recently, the CEQ produced draft guidance recommending that climate change be considered throughout the environmental assessment process.[162] A presumptive threshold of 25,000 metric tons of CO_2 equivalent emissions triggers a quantitative analysis.[163]

156 Espoo Convention, *supra* note 22, at app. I.; SEA Protocol, *supra* note 23, at app. Ann. I.

157 Czech EIA Law, supra note 87, at art. 2, ann. 4, pt. D I.2 [Impacts on the air and climate].

158 40 C.F.R. § 1508.27.

159 *Id.* at § 1508.27(a).

160 *Id.* at § 1508.27(b)(2)-(9).

161 *Massachusetts v. EPA*, 549 U.S. 497, 499 (2007) (stating "EPA's argument rests on the erroneous assumption that a small incremental step, because it is incremental, can never be attacked in a federal judicial forum. Yet accepting that premise would doom most challenges to regulatory action."); *Ctr. for Biological Diversity v. Nat'l Highway Traffic Safety Admin.*, 538 F.3d 1172, 1217 (9th Cir. 2008) (stating "[t]he impact of [GHG] emissions on climate change is precisely the kind of cumulative impacts analysis that NEPA requires agencies to conduct.")

162 NEPA Guidance, *supra* note 119.

163 *Id.* at 1.

To summarize, the existing laws governing TEIA do not limit such assessments to countries sharing a border and can be interpreted as supporting transregional assessments because local GHG emissions can produce potential effects anywhere in the world. The necessary causal relationship does not depend on demonstrating physical flow but rather is a probability function determined by climate science modeling. Attributing the effects to a particular source of GHG emissions is best understood through a cumulative contribution regime. Further, the *likelihood* that GHG emissions from large-scale undertakings, including projects, plans, and programs, will produce impacts can be demonstrated through climate modeling. These impacts, which may be irreversible for small island nations, will most likely be characterized as *significant* under any reasonable definition.

5. Practical Challenges and Opportunities for Island Nations in Conducting TEIA

Already, island nations such as the FSM are threatened by tidal surges, destruction of food crops, coastal erosion, submersion of islands, coral bleaching, harm to migratory marine species due to rising sea temperature, and reduced water supply to rivers.[164] These environmental threats are likely to be exacerbated by climate change. One islet in Pohnpei State of the FSM has reportedly vanished, another has split due to coastal erosion,[165] and two more islands in the Chuuk State lagoon have now been completely submerged.[166] The FSM and other island nations cannot

164 *Summary for Policymakers*, *supra* note 148; AMAP, *supra* note 150.

165 Gary Kubota, *Micronesia Vanishing as the Climate Warms Up*, Star Bulletin, Mar. 4, 2001, http://archives.starbulletin.com/2007/03/04/news/story05.html .

166 Disappearance of small islands in the FSM is not always documented. The submerging of the islands of Piianu and Piiseemeeu in Chuuk State is based on visual observations by local communities that had in the past utilized the islands as fishing grounds. For more information, please contact Director Mr. Andrew Yatilman, Palikir, Pohnpei State, FM 96941, http://www.fsmgov.org/ngovt.html. Reports of erosion of island

afford to wait. To ensure survival, these States must actively combat climate change using existing international law principles and must explore emerging legal principles that could be more effective in addressing the unprecedented impacts of climate change.

A TEIA furnishes an important tool in making critical predictions about the costs, risks, and benefits involved in a project. This tool can raise public awareness of the effects of climate change. The TEIA can also be a test in determining whether a State has fulfilled its duty of due diligence to avoid causing harm to another State. The manner by which TEIA is undertaken, and the opportunity it affords to those affected States who can participate in the process, becomes very important. This section discusses the practicality of TEIAs especially for sovereign island nations.

5.1. No Clear and Defined Standards

Standards for conducting environmental assessment are often loosely defined or even undefined, and different planners and analysts conduct EIAs and TEIAs differently. It is nearly impossible to establish one comprehensive procedure for impact assessment "[d]ue to variation in political regimes, natural systems, and cultural values."[167] Interestingly, in the *Pulp Mills* judgment, the ICJ considered the 1987 UNEP Goals and Principles of Environmental Impact Assessment to be nonbinding EIA guidelines.[168] Nevertheless, if uniform and consistent standards

atolls can be found on the following sites: http://water.usgs.gov/wrri/06grants/progress/2006GU81B.pdf and http://map.sopac.org/data/virlib/TR/TR0268.pdf.

167 Chinthaka Mendis, Sovereignty vs. Trans-boundary Environmental Harm: The Evolving International Law Obligations and the Sethusamuduram Ship Channel Project 43 (2006) (citing, Yusuf J. Ahmad & George K. Sammy, Guidelines to Environmental Impact Assessment in Developing Countries 9 (1999)), http://www.un.org/Depts/los/nippon/unnff_programme_home/fellows_pages/fellows_papers/mendis_0607_sri_lanka.pdf.

168 Pulp Mills, ¶ 203, 205 (citing UNEP, Governing Council, 14th Sess., Dec. 14–25, 1987, *Goals and Principles of Environmental Impact Assessment*, UN Doc. UNEP/WG. 152/4, Annex (1987)).

of TEIA are to be achieved among States, multilateral cooperation is needed.

5.2. Circularity of the Obligation to Conduct a TEIA

Another difficult question that relates to the duty to conduct a TEIA is the apparent circularity of the obligation. The general understanding is that notification of a potentially affected State is required when there is a possibility, no matter how uncertain, of significant adverse transboundary impacts. The initiating State must determine, prior to conducting a TEIA, the nature or magnitude of the risk posed by an activity, plan, or program. This presents the following dilemma:

> The reason for doing an EIA is to find out whether harm is likely; yet, in most cases the obligation to do one and to notify other states only arises once it is known that it is likely, or even that it will occur. This is both circular and potentially self-defeating if taken literally.[169]

The problem becomes all the more complex when it comes to the cumulative impacts of GHG emissions on affected States that are geographically far removed from the originating State. In the case of transregional affected States, such as the FSM in the Prunéřov II TEIA proceeding, some of the challenges relate to the application of legal standing, duty of notification, information sharing, and determining likely impact.

Because of the lack of customary international law principles addressing these specifics, the conduct of TEIAs for small island nations remains amorphous and uncertain. Possible solutions include the development of TEIA treaties that explicitly cover climate change impacts and, as an initial step, for the States to consider becoming parties to the Espoo Convention[170] and the Protocol on SEA. Moreover, existing

169 International Law & the Environment, *supra* note 18, at 171.

170 Participation in ESPOO is currently limited to members of the UNECE. Article 17 of Espoo Convention was amended in 2001 to allow any UN Member State to accede to

treaties such as the UNFCCC could be expanded to explicitly cover or provide further details on the practical implementation of TEIA. The fact that the ICJ now recognizes the TEIA requirement as part of general international law because of its growing acceptance among States strongly indicates the need for clarification as to how this legal obligation can be implemented using a more predictable and possibly uniform standard. The UNFCCC as a framework convention, or its protocol, can further develop and put in place an agreed mechanism that gives effect to a recognized legal obligation.

5.3. Vague Notification and Consultation Procedures

Once it is established that a TEIA is required for a proposed activity, the next step is to determine the scope of the EIA and to what extent "affected States" should be involved in the process. There is no universal rule regarding the procedures governing the notification process and the extent to which an affected State is allowed to participate in the TEIA, if at all. Again, as the ICJ clarified,

> it is for each State to determine in its domestic legislation or in the authorization process for the project, the specific content of the environmental impact assessment required in each case, having regard to the nature and magnitude of the proposed development and its likely adverse impact on the environment as well as to the need to exercise due diligence in conducting such an assessment.[171]

Thus, the notification and consultation procedures remain vague for States that are not party to the transboundary agreements.

The Espoo Convention, the Protocol on SEA, and the Draft Articles are examples of attempts to define predictable procedures and

ESPOO on approval by the Espoo Convention parties, but that provision has not yet entered into force. Espoo Convention, *supra* note 22, at art. 17.

171 Pulp Mills on the River Uruguay (Arg. v. Uru.), Judgment, ¶ 205 (Apr. 20, 2010), *available at* http://www.icj-cij.org/docket/files/135/15877.pdf.

guidelines in conducting TEIAs.[172] Moreover, LRTAP[173] has set forth certain principles in regard to air pollution,[174] and the parties to the Convention have further agreed to exchanges of information.[175] These conventions, however, are legally binding only on the States Parties; for non-Parties, these instruments will be persuasive at best, if accorded any weight at all. Hence, an alternative for non-States Parties is to consider either negotiating a multilateral agreement of the type exemplified by the Espoo Convention and the Protocol on SEA or joining existing agreements, if permitted.

5.4. Implications on State Sovereignty

Until international standards and procedures are crafted and agreed, affected States that wish to participate in a TEIA face a legal dilemma in protecting their interests in a practical context. There is no doubt that the no-harm principle, as customary international law, is an obligation that every State should respect. Yet, how it is given effect in the context of another customary duty of conducting a TEIA, where there is significant adverse impact on the environment of another State, remains a difficult question in view of the disparity among individual state practices pursuant to their domestic laws.

Returning to the Prunéřov II case, the domestic laws of the Czech Republic provide the procedures[176] by which an affected State may participate in the TEIA that the Czech Republic was under duty to perform. Procedurally, the action taken by the FSM in participating in the process was consistent with the 2010 *Pulp Mills* judgment that TEIA is a requirement under the general international law.[177]

172 *See* SEA Protocol, *supra* note 23; Espoo Convention, *supra* note 22; Draft Articles, *supra* note 55.

173 LRTAP, *supra* note 140.

174 *Id.* at art. 2, 6.

175 *Id.* at art. 8.

176 Czech EIA law, *supra* note 87, at art. 13.

177 *Pulp Mills*, ¶ 204.

On the one hand, by participating in the process, the FSM could be seen as having used and hence subjected itself to Czech Republic law in order to enforce an international law duty of no-harm and the duty to undertake a TEIA. On the other hand, the FSM could be seen as having availed itself of an opportunity to participate in the assessment process during its critical stage, as allowed by Czech Republic law. The Czech Republic nevertheless retains the authority to decide whether the proposed project should proceed, whereas the FSM exercises a legal right, pursuant to international law principles, to present its views on a matter that it has determined would have likely significant impacts on its environment and national interests. A difficult balancing act may arise consisting of asserting a State's right pursuant to international law versus the practical necessity of enforcing such a right through the aid of another State's legal mechanisms.

A possible approach to preserve both corollary rights and duties is for the States involved to negotiate and agree on a mutually acceptable procedure or process that is consistent with their respective domestic laws, as exemplified in the case of the Baltic Sea Gas Pipeline (BSGP) TEIA. BSGP was a natural gas pipeline transmission system that passed through the exclusive economic zone (EEZ) of five countries: Denmark, Finland, Sweden, Germany, and the Russian Federation. All except Russia were parties to the Espoo Convention. How the Convention should apply to Russia in view of its lack of ratification became an issue. It was resolved through a series of negotiations, resulting in Russia's agreement that the Espoo Convention was applicable to the extent permitted by Russian domestic law.[178]

Ultimately, the 2010 *Pulp Mills* judgment, the Stockholm and Rio Declarations, and evidence of state practice now afford island nations the opportunity to meaningfully contribute to the decision-making process around large-scale GHG-polluting activities, plans, and programs in

178 1 Timo Koivurova &Ismo Pölönen, *Transboundary Environmental Impact Assessment in the Case of the Baltic Sea Gas Pipeline*, 25 Int'l. on Me & Coastal L.151, 158 (2010), available at http://agc-wopac.agc.gov.my/e-docs/Journal/0000017402.pdf.

at least some countries (although these principles remain to be tested in much of the world). The FSM's intervention in the EIA process of the Prunéřov II lignite-fired power plant in the Czech Republic is just one example of where participation was made possible by progressive application of international law principles.[179] The ICJ *Pulp Mills* decision also exemplifies progress in international environmental law and leaves open the possibility of future development that will result in a further refinement of state TEIA obligations. Beyond those examples, generally accepted principles governing state obligations, particularly the no-harm principle enunciated in the *Trail Smelter Arbitration*,[180] and the duty of diligence imposed on sovereign States, now seem clear. The applicability of these principles to particular sets of facts will continue to be tested on a case-by-case basis.

6. Conclusion

Climate change is one of the biggest threats facing our planet. Although many of its impacts are predicted to occur in the future, small island nations like the FSM are experiencing some of those harms now. If the predictions become reality, loss of sovereign territory is at stake. Energy choices made in major GHG-emitting countries will either mitigate or exacerbate the climate threats to these threatened nations. That fact alone makes low-lying countries and island nations stakeholders in the decisions made by governments across the globe.

The TEIA provides a valuable procedure through which affected countries can provide meaningful input into energy decision-making processes. TEIA was initially developed to evaluate impacts of the physical flow of pollutants into nearby countries. The TEIA requested by the

179 Note that this is not always reciprocal. For example, although U.S. laws on impact assessment apply to projects constructed in the FSM, there are no explicit provisions for the FSM participate in TEIA concerning large-scale projects on U.S. territory. Compact of Free Association Amendments, F.S.M.-U.S., tit. I, art. VI, § 161, Dec. 17, 2003, 117 Stat. 2765.

180 Trail Smelter Arbitration, 3 R.I.A.A. at 1915.

FSM in the case of the Prunéřov II coal-fired power plant demonstrates that this mechanism can be successfully deployed to highlight the potential climate impacts of the GHG emissions of dirty energy projects on the other side of the earth and to encourage decision makers to reduce emissions.

This chapter demonstrates that the FSM found solid footing in international law for this ground-breaking legal action. Of course, as with any novel application of legal principles, there remain practical challenges to be resolved. However, other climate-affected States may consider replicating the FSM's effort to improve climate policies of and decisions taken by major GHG-emitting countries.

Index